VOLUME SEVENTY EIGHT

ADVANCES IN PHARMACOLOGY

VOLUME SEVENTY EIGHT

ADVANCES IN PHARMACOLOGY

Vascular Pharmacology: Smooth Muscle

Edited by

RAOUF A. KHALIL

Brigham and Women's Hospital, Harvard Medical School, Boston, MA, United States

Serial Editor

S.J. ENNA

University of Kansas Medical Center School of Medicine, Kansas City, Kansas, United States

Managing Editor

LYNN LECOUNT

University of Kansas Medical Center School of Medicine, Kansas City, Kansas, United States

ELSEVIER

ACADEMIC PRESS

An imprint of Elsevier

elsevier.com

Academic Press is an imprint of Elsevier
50 Hampshire Street, 5th Floor, Cambridge, MA 02139, United States
525 B Street, Suite 1800, San Diego, CA 92101-4495, United States
The Boulevard, Langford Lane, Kidlington, Oxford OX5 1GB, United Kingdom
125 London Wall, London, EC2Y 5AS, United Kingdom

First edition 2017

ISBN: 978-0-12-811485-8
ISSN: 1054-3589

Publisher: Zoe Kruze
Acquisition Editor: Zoe Kruze
Editorial Project Manager: Shellie Bryant
Production Project Manager: Vignesh Tamil
Cover Designer: Greg Harris

Typeset by SPi Global, India

CONTENTS

CONTRIBUTORS

J.T. Baeten
The Center for Cardiovascular Research and The Heart Center at Nationwide Children's Hospital, The Ohio State University, Columbus, OH, United States

R.R. Chandran
Yale Cardiovascular Research Center, Section of Cardiovascular Medicine, Yale University School of Medicine, New Haven, CT, United States

J.M. Dave
Yale Cardiovascular Research Center, Section of Cardiovascular Medicine, Yale University School of Medicine, New Haven, CT, United States

A.M. Evans
Centre for Integrative Physiology, College of Medicine and Veterinary Medicine, University of Edinburgh, Edinburgh, United Kingdom

S.A. Fisher
University of Maryland School of Medicine, Baltimore, MD, United States

D. Ghosh
University of California, Davis, CA, United States

R. Ginnan
Department of Molecular and Cellular Physiology, Albany Medical College, Albany, NY, United States

D.M. Greif
Yale Cardiovascular Research Center, Section of Cardiovascular Medicine, Yale University School of Medicine, New Haven, CT, United States

A.K. Grover
McMaster University, Hamilton, ON, Canada

W.F. Jackson
Michigan State University, East Lansing, MI, United States

R.A. Khalil
Brigham and Women's Hospital, Harvard Medical School, Boston, MA, United States

B. Lilly
The Center for Cardiovascular Research and The Heart Center at Nationwide Children's Hospital, The Ohio State University, Columbus, OH, United States

R. Mazurek
Yale Cardiovascular Research Center, Section of Cardiovascular Medicine, Yale University School of Medicine, New Haven, CT, United States

A. Misra
Yale Cardiovascular Research Center, Section of Cardiovascular Medicine, Yale University School of Medicine, New Haven, CT, United States

M.F. Navedo
University of California, Davis, CA, United States

M. Nieves-Cintrón
University of California, Davis, CA, United States

M.A. Nystoriak
Diabetes and Obesity Center, University of Louisville, Louisville, KY, United States

M.P. Prada
University of California, Davis, CA, United States

H.C. Ringvold
Brigham and Women's Hospital, Harvard Medical School, Boston, MA, United States

F.Z. Saddouk
Department of Molecular and Cellular Physiology, Albany Medical College, Albany, NY, United States

L.F. Santana
University of California, Davis, CA, United States

A.Q. Sheikh
Yale Cardiovascular Research Center, Section of Cardiovascular Medicine, Yale University School of Medicine, New Haven, CT, United States

H.A. Singer
Department of Molecular and Cellular Physiology, Albany Medical College, Albany, NY, United States

A.U. Syed
University of California, Davis, CA, United States

T. Szasz
Augusta University, Augusta, GA, United States

R.C. Webb
Augusta University, Augusta, GA, United States

PREFACE

Vascular smooth muscle constitutes a major portion of the tunica media in the vascular wall. Large vessels such as the aorta have several layers of vascular smooth muscle, thus lending strength to the vessel wall against hemodynamic forces and changes in blood pressure. Small resistance vessels have few layers of vascular smooth muscle, but play a critical role in the control of the myogenic response, microvessel diameter, vascular resistance, and blood pressure. The vessel diameter is also controlled by various neurotransmitters and circulating vasoactive hormones that act on specific vascular receptors and activate different postreceptor signaling pathways. Ca^{2+}-dependent myosin light chain phosphorylation is a major determinant of vascular smooth muscle contraction. In addition, Ca^{2+}-sensitization pathways could enhance the myofilament force sensitivity to Ca^{2+} and vascular contraction. These signaling pathways are tightly controlled by various intracellular ion channels, pumps, and regulatory proteins, as well as extracellular vascular modulators and hormones in order to maintain normal vascular tone. Disturbances in these control mechanisms could cause vascular hyperreactivity and aberrant vascular smooth muscle growth and proliferation, and in turn lead to vascular dysfunction and disorders such as hypertension and coronary artery disease.

This volume of *Advances in Pharmacology* focuses on vascular smooth muscle, its role in the regulation of vascular function, and its dysregulation in vascular disorders. The first part of the volume will cover important topics on the basic function of vascular smooth muscle and the regulation of intracellular Ca^{2+} by the sarcoplasmic reticulum, Ca^{2+} channels, membrane potential, potassium channels, Ca^{2+} pumps, and ion exchangers. The second part of the volume will focus on Ca^{2+}-dependent and Ca^{2+}-sensitization pathways of vascular smooth muscle contraction including Ca^{2+}–calmodulin-dependent protein kinase, protein kinase C, and Rho kinase. The third part of the volume includes reviews on vascular cell development and notch signaling in vascular health and disease, and the diversity in vascular smooth muscle phenotype and function in different blood vessels. These important and up-to-date reviews were written by outstanding researchers and clinician–scientists from highly recognized academic centers and teaching hospitals from all over the world, thus providing different viewpoints and perspectives on vascular biology, function, and disease. Thanks to the excellent work of our contributing authors, and the careful review of our superb reviewers and Editors, we were able to assemble these important and timely topics in a very clear, concise, and informative fashion. I encourage every researcher, clinician, student, and trainee with interest in the vascular field to read this state-of-the-art synopsis on vascular smooth muscle.

I would like to take this opportunity to express my deepest gratitude to Dr. S.J. Enna, the Editor of the *Advances in Pharmacology* series, who gave me the opportunity to be the Editor of this special and important volume. I also would like to thank our terrific Managing Editor, Ms. Lynn LeCount, and our first-rate Editorial Staff who spared no effort to ensure the highest quality and clearest presentation of all the articles. I also would like to acknowledge our contributing authors not only for their excellent contributions but also for their help in taking some of the reviewers' responsibilities and providing helpful comments and constructive criticism to their fellow contributors and colleagues. I particularly wish to thank our readers for their sincere interest in vascular pharmacology and vascular smooth muscle. I encourage all of you to contact me directly if you would like to provide feedback or have any questions, comments, suggestions, ideas, or criticism that could further enhance our knowledge in the vascular field and help us achieve our ultimate goal to meet the highest expectations of our readers.

Raouf A. Khalil
Brigham and Women's Hospital, Harvard Medical School,
Boston, MA, United States

CHAPTER ONE

Nanojunctions of the Sarcoplasmic Reticulum Deliver Site- and Function-Specific Calcium Signaling in Vascular Smooth Muscles

A.M. Evans*
Centre for Integrative Physiology, College of Medicine and Veterinary Medicine, University of Edinburgh, Edinburgh, United Kingdom
*Corresponding author: e-mail address: mark.evans@ed.ac.uk

Contents

Abstract

Vasoactive agents may induce myocyte contraction, dilation, and the switch from a contractile to a migratory–proliferative phenotype(s), which requires changes in gene expression. These processes are directed, in part, by Ca^{2+} signals, but how different Ca^{2+} signals are generated to select each function is enigmatic. We have previously proposed that the strategic positioning of Ca^{2+} pumps and release channels at membrane–membrane junctions of the sarcoplasmic reticulum (SR) demarcates cytoplasmic

Advances in Pharmacology, Volume 78
ISSN 1054-3589
http://dx.doi.org/10.1016/bs.apha.2016.10.001

nanodomains, within which site- and function-specific Ca^{2+} signals arise. This chapter will describe how nanojunctions of the SR may: (1) define cytoplasmic nanospaces about the plasma membrane, mitochondria, contractile myofilaments, lysosomes, and the nucleus; (2) provide for functional segregation by restricting passive diffusion and by coordinating active ion transfer within a given nanospace via resident Ca^{2+} pumps and release channels; (3) select for contraction, relaxation, and/or changes in gene expression; and (4) facilitate the switch in myocyte phenotype through junctional reorganization. This should serve to highlight the need for further exploration of cellular nanojunctions and the mechanisms by which they operate, that will undoubtedly open up new therapeutic horizons.

1. INTRODUCTION

Vasoactive agents induce, in a manner opposed by their physiological antagonists, contraction or dilation of vascular smooth muscles and the switch from a contractile to a migratory–proliferative phenotype(s), which requires changes in gene expression. All of these processes are regulated, in part, by Ca^{2+} signals. Therefore, myocytes must provide for the generation of different Ca^{2+} signals that select for one or a combination of functions. The aim of this chapter is to elaborate on the "panjunctional sarcoplasmic reticulum" hypothesis (van Breemen, Fameli, & Evans, 2013). In essence this hypothesis is built around the proposal that cellular membrane–membrane nanojunctions are formed by the sarcoplasmic reticulum (SR) at defined "target sites" to deliver highly localized and functionally segregated Ca^{2+} signals, with the functional specification of a given signal determined both by the constraints on diffusion imposed by the nanojunction itself and by the Ca^{2+} release channels and transporters incorporated within a given junctional complex. Nanojunctions have been defined by their distance of separation, and the S/ER is now known to form such nanojunctions not only with the plasma membrane (PM) but also with a variety of organelles including lysosomes, mitochondria, and the nucleus (van Breemen et al., 2013). The underlying mechanisms of signal generation are likely more elaborate in nature than we presently envisage, but evidently rely on the strategic targeting to their designate nanojunction of macromolecular complexes that incorporate different types of Ca^{2+} transporters and release channels, each of which will be characterized by different kinetics and affinities for Ca^{2+} that can be modulated by respective binding partners, second messengers, and enzymes (Clark et al., 2010). The capacity for this is clear when one considers, for example, the fact that pharmacoresponse coupling in a single type of arterial or arteriolar smooth muscle provides for the gating of at least two of the three known S/ER resident IP_3 receptors (IP_3R1–3) by inositol 1,4,5-trisphosphate (IP_3) (Berridge, 2008a, 2008b; Grayson, Haddock, Murray, Wojcikiewicz, & Hill, 2004; Hirose, Kadowaki, & Iino, 1998; Westcott, Goodwin, Segal, & Jackson, 2012; Zhao et al., 2010), up to three S/ER resident ryanodine receptor subtypes (RyR1–3) (Herrmann-Frank, Darling, & Meissner, 1991; Kinnear et al., 2008; Westcott et al., 2012; Yang et al., 2005) by Ca^{2+} and/or cyclic adenosine diphosphate-ribose (cADPR) (Lee, 2004; Ogunbayo et al., 2011; Wilson et al., 2001), and the gating of the endolysosome targeted two pore channel (TPC) subtypes 1 and 2 by either NAADP (Boittin, Galione, & Evans, 2002; Calcraft et al., 2009; Kinnear, Boittin, Thomas, Galione, & Evans, 2004; Ogunbayo, Ma, Zhu, & Evans, 2015), phosphatidylinositol-3,5-bisphosphate ($PI3,5P_2$) (Wang et al., 2012), or by the dissociation of mTOR from TPC2 (Cang et al., 2013). It is important to note, however, that the expression pattern of Ca^{2+} release channels and pumps expressed by smooth muscle cells may vary between regions in the same vascular bed (Westcott et al., 2012) and from one vascular bed to another (Clark et al., 2010; Dabertrand, Nelson, & Brayden, 2012; Lifshitz et al., 2011; Westcott et al., 2012; Zhao et al., 2010). Likewise, variations in the prevalence of nanojunctions, their ultrastructure and their molecular machinery could explain, in part, both smooth muscle heterogeneity (Boittin, Dipp, Kinnear, Galione, & Evans, 2003; Dabertrand et al., 2012; Kur, Bankhead, Scholfield, Curtis, & McGeown, 2013; Pucovsky & Bolton, 2006; Westcott et al., 2012; Zhao et al., 2010) and plasticity (Gilbert, Ducret, Marthan, Savineau, & Quignard, 2014; van Breemen et al., 2013).

2. WHAT DEFINES A NANOJUNCTION?

The first nanojunction described in terms of function was an intercellular junction, namely the neuromuscular junction. In this respect, it is important to note that the pre- and postjunctional membranes are approximately 20 nm apart and extend more or less parallel to each other over several hundred nanometers, because it is the organization of these junctional membranes that proved to be so critical to the coordination of neuromuscular transmission by the release, receptor interactions, and reuptake of acetylcholine (Del Castillo & Katz, 1954, 1956; Fatt & Katz, 1950, 1951). Given this fact, it is surprising to note that so little attention has been paid to the role in intracellular signaling of the plethora of junctions formed between intracellular membranes, beyond acknowledging that there are "contact sites," either through their ability to direct and regulate Ca^{2+} fluxes or the action of other messengers. The one notable exception is of course in the process of excitation–contraction coupling in skeletal and cardiac muscles, where the junctional complexes formed between T-tubules of the sarcolemma and terminal cisternae of the SR are well documented and are equally critical to the coordination of excitation–contraction coupling as is the neuromuscular junction. In each instance, the junctional membrane pairs are separated by 20 nm or less (Franzini-Armstrong, 1964; Ramesh, Sharma, Sheu, & Franzini-Armstrong, 1998; Rosenbluth, 1962), as are the pre- and postjunctional membranes of the neuromuscular junction. There is, however, one important distinction between skeletal and cardiac muscles that must be mentioned in terms of the mechanism by which Ca^{2+} is mobilized during excitation–contraction coupling, and this will undoubtedly have resonance with respect to our future understanding of the versatility of signaling across all intracellular nanojunctions. Against earlier predictions, it is now evident that the sarcolemma–SR nanojunctions of skeletal muscle allow for the transfer, through noncovalent association, of electrostatic charge between sarcolemma resident dihydropyridine receptors and SR resident RyR1s (see, for example, Feldmeyer, Melzer, Pohl, & Zollner, 1990; Ma, Gonzalez, & Chen, 1996; Nakai et al., 1996), which ultimately gates Ca^{2+} release from the SR via RyR1. By contrast the sarcolemma–SR nanojunctions of cardiac muscle support what is generally regarded as the classical form of junctional coupling, namely agonist–receptor interactions, by targeting Ca^{2+} influx to clusters of RyR2 located on the terminal cisternae of the SR, which, in turn, triggers a propagating Ca^{2+} wave and contraction by further Ca^{2+}-induced Ca^{2+} release from the SR via arrays of RyR2 clusters (Cheng, Lederer, & Cannell, 1993; Soeller, Crossman, Gilbert, & Cannell, 2007). Whatever the mechanism of transduction, electrostatic, or agonist–receptor gating, these sarcolemma–SR junctions represent the archetypal intracellular nanojunctions, with each one being separated by ~20 nm and clearly designed to accurately deliver Ca^{2+} to one defined target above all else. Already there appears some consistency of form and function, so why 20 nm?

Following 10 years of intense, exciting, and enjoyable discussions on experimental data and modeling outcomes with Casey van Breemen and Nicola Fameli (which often occurred weekly by Skype and all other means possible), it is now my firm assertion that all active nanojunctions constitute two biological membranes that are separated by 50 nm or less and thus demarcate a well-defined cytoplasmic nanospace, and on these matters the three amigos concur. To maintain junctional integrity, it is now clear to us and we hope it will soon be to all that the optimal operating limit for intracellular nanospaces is 10–50 nm in width at most, typically a few 100 nm in extension, and that on either side the membrane pair will contain complementary ion transporters and channels that serve to deliver and/or receive specified Ca^{2+} signals. As described previously, it is evident that both the ultrastructure, the electrostatic properties of each membrane of the junctional pair and the macromolecular composition of transport molecules embedded in their limiting membranes will ensure that cytoplasmic cation concentrations, Ca^{2+} in particular, are locally determined. Ca^{2+} may thus target "receptive sites" of different affinities with great accuracy and modulate function appropriately, whether we consider myocyte contraction or gene expression (van Breemen et al., 2013). In this respect, it is important to highlight one further property of nanojunctions that our modeling has revealed. Briefly, outcomes suggest that a 30–50 nm operating limit for junctional separation, above which there will be loss of junctional integrity and inadequate control of ion movements within the junctional space, i.e., free diffusion of Ca^{2+} away from the junction will be permitted. Put another way, a junctional separation of 50 nm or less confers the capacity to restrict Ca^{2+} diffusion and thus "hold" Ca^{2+} within the junction and direct its transfer across the cytoplasmic nanospace demarcated by the junctional membrane pair (Fameli, Ogunbayo, van Breemen, & Evans, 2014).

2.1 The PM–SR Nanojunctions of Vascular Smooth Muscles

It is now 40 years since the first narrow cytoplasmic spaces (~20 nm across) were identified between the PM and the superficial SR of smooth muscles (Devine, Somlyo, & Somlyo, 1972; Gabella, 1971). This led to the "superficial buffer barrier" hypothesis (van Breemen, Chen, & Laher, 1995; van Breemen & Saida, 1989), which posited that restricted diffusion from such PM–SR "nanospaces" allows the superficial SR to limit direct Ca^{2+} flux from the PM to the myofilaments (Van Breemen, 1977). This hypothesis has now received support from studies on a variety of smooth muscle types. Moreover it is now apparent that PM–SR nanojunctions have the capacity to coordinate the delivery of Ca^{2+} to and the removal of Ca^{2+} from the SR (Poburko, Kuo, Dai, Lee, & van Breemen, 2004; Rembold & Chen, 1998; Shmigol, Eisner, & Wray, 1999; White & McGeown, 2000; Yoshikawa, van Breemen, & Isenberg, 1996; Young, Schumann, &

Zhang, 2001), and in doing so not only serve to regulate SR luminal [Ca^{2+}] but also hyperpolarization and relaxation, depolarization and vasomotion, and may perhaps in this way also influence gene expression.

However, our understanding of how smooth muscles accomplish site- and function-specific Ca^{2+} signaling remains in its infancy, because recent efforts have primarily focused on the temporal characteristics of unitary and macroscopic Ca^{2+} signals (Hill-Eubanks, Werner, Heppner, & Nelson, 2011), and relatively has little attention given to the possibility that Ca^{2+} may be targeted to cytoplasmic nanospaces created by membrane–membrane nanojunctions of the SR (van Breemen et al., 2013). Nevertheless, it is evident that PM–SR nanojunctions must be polymodal (Nazer & Van Breemen, 1998b; van Breemen et al., 1995) in that they may support either vasoconstriction or vasodilation (Fig. 1). This is clear from the fact that the superficial SR retains the capacity not only to empty when overloaded with Ca^{2+} (Nazer & van Breemen, 1998a) or signaled to do so by vasodilators (Boittin et al., 2003), but also to reload its Ca^{2+} store once depleted (Fameli, van Breemen, & Kuo, 2007; Spinelli & Trebak, 2016).

2.1.1 Refilling of the SR

As for all other cell types, refilling of the SR of smooth muscles is accomplished via either store depletion activated Ca^{2+} entry pathways (Ginsborg, House, & Mitchell, 1980a, 1980b; Putney, 1986; Soboloff, Rothberg, Madesh, & Gill, 2012) or a variety of receptor-operated mechanisms. This process appears to be facilitated, in part, by PM–SR nanojunctions that "funnel" Ca^{2+} influx from the extracellular space into the SR via SERCA during activating waves of SR Ca^{2+} release (Lee et al., 2001), and in doing so will provide necessary support for continued Ca^{2+} release from the SR via whichever nanojunction(s) is activated. Receptor-operated cation channels, such as the transient receptor potential channel TRPC6, may deliver Na^{+} to the cytoplasmic nanospaces formed by PM–SR junctions in a manner that promotes Ca^{2+} entry via reverse-mode Na^{+}/Ca^{2+} exchangers (NCX) at the PM, and may thus supply Ca^{2+} to SERCA on the adjacent, junctional SR membranes (Fameli, Kuo, & van Breemen, 2009; Fameli et al., 2007; Poburko et al., 2007). SR reloading via SERCA may also be facilitated by Ca^{2+} influx through voltage-gated Ca^{2+} channels (Takeda, Nystoriak, Nieves-Cintron, Santana, & Navedo, 2011), TRPCs (Albert, Saleh, & Large, 2009; Rosado, Diez, Smani, & Jardin, 2015; Shi, Miralles, Birnbaumer, Large, & Albert, 2016), and the stromal interaction molecule (STIM)/Orai system (Berra-Romani, Mazzocco-Spezzia, Pulina, & Golovina, 2008; Takahashi et al., 2007). The latter of these

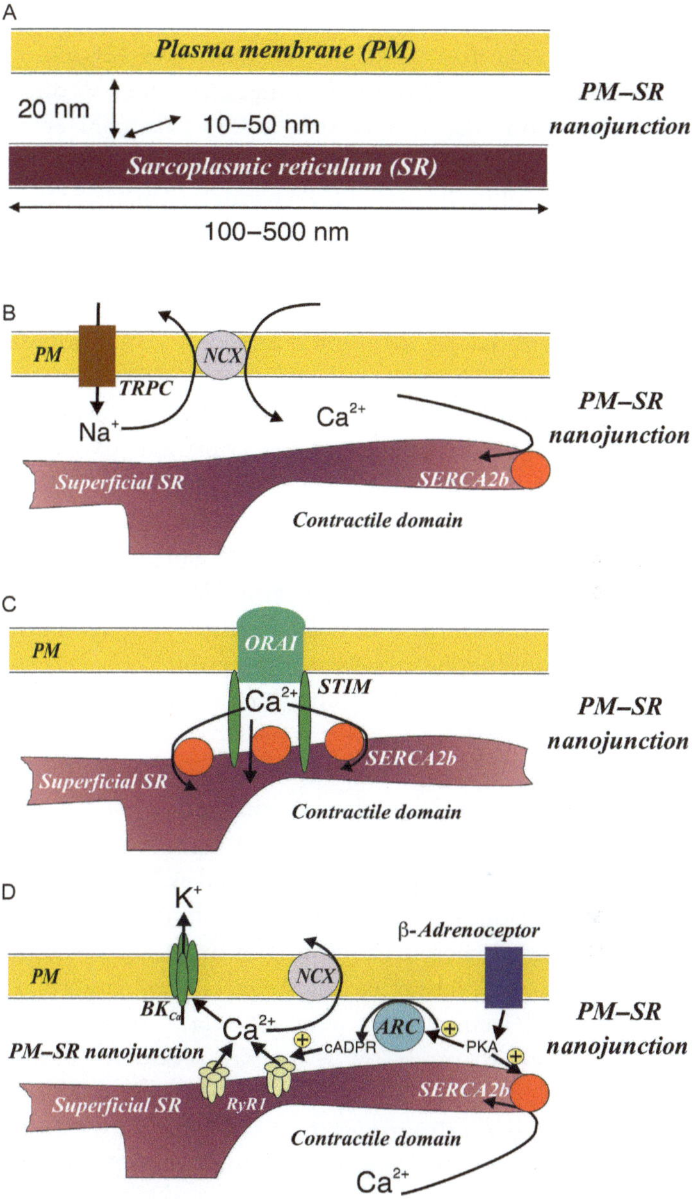

Fig. 1 Plasma membrane–SR nanojunctions are polymodal. Schematics show (A) dimensions of PM–SR junctions, refilling of the SR via PM–SR junctions by (B) reverse-mode activity of sodium–calcium exchangers (NCX), (C) STIM–Orai complexes, and (D) emptying of the SR through Ca^{2+} release via ryanodine receptor (RyR) subtype 1 and forward mode NCX activity.

processes of store refilling may well be aided by the shortening of junctional spaces and/or the formation of additional PM–SR nanojunctions through the assembly of STIM–Orai complexes (Berra-Romani et al., 2008; Lu, Wang, Shimoda, & Sylvester, 2008; Takahashi et al., 2007), which perhaps acts as a supplementary pathway over and above receptor-operated Na^{+}/Ca^{2+} influx through TRP channels (Albert et al., 2009; Shi, Ju, Large, & Albert, 2012) and allied reverse-mode activity of Na^{+}/Ca^{2+} exchangers (NCX) (Poburko, Fameli, Kuo, & van Breemen, 2008; Poburko et al., 2006; Syyong, Poburko, Fameli, & van Breemen, 2007). However, it is evident that STIM–Orai complexes may in some instances serve to replace other Ca^{2+} entry pathways (Soboloff et al., 2012), such as voltage-gated Ca^{2+} influx (Takeda et al., 2011). In this respect, it is therefore notable that all of these mechanisms may contribute to SR loading in a single type of vascular smooth muscle cell (Kato et al., 2013; Leblanc et al., 2015; Lu, Wang, Peng, Shimoda, & Sylvester, 2009; Lu et al., 2008; Ng et al., 2012; Ogawa, Firth, Smith, Maliakal, & Yuan, 2012; Snetkov et al., 2006). The precise functional designation and relative temporal characteristics for the activation of each store refilling pathway remain unclear, but distinctions will undoubtedly surface at some point.

2.1.2 Emptying of the SR

It is evident that PM–SR nanojunctions provide pathways for store emptying. In this instance, they likely aid the targeting of Ca^{2+} sparks, which arise from RyRs on the superficial SR (Nelson et al., 1995), and consequent recruitment of large conductance Ca^{2+}-activated K^{+} (BK_{Ca}) channels on the PM (Benham & Bolton, 1986; Benham, Bolton, Lang, & Takewaki, 1986), thus promoting hyperpolarization (Boittin et al., 2003), closure of voltage-gated channels (Nelson et al., 1995; Perez, Bonev, Patlak, & Nelson, 1999), and removal of released Ca^{2+} from the PM–SR nanojunction via forward mode NCX activity (Nazer & Van Breemen, 1998b). The resultant net loss of Ca^{2+} from the SR likely serves to promote, in part, myocyte relaxation (Boittin et al., 2003; Brayden & Nelson, 1992).

SERCA2b and RyR1 are preferentially targeted to the SR proximal to the PM in pulmonary arterial myocytes (Clark et al., 2010; Gilbert et al., 2014), and likely serve PM–SR nanojunctions in a manner required by the superficial buffer barrier hypothesis (Fig. 1). Consistent with this, we have shown that vasodilation in response to β-adrenoceptor activation may be inhibited by: (1) blocking SERCA with cyclopiazonic acid (Boittin et al., 2003); (2) blocking RyRs with ryanodine (Boittin et al., 2003); and (3) blocking BK_{Ca}

channels with iberiotoxin. In this cell type, therefore, SERCA2b may remove Ca^{2+} from the bulk cytoplasm to the superficial SR, from which Ca^{2+} sparks may be released into the PM–SR junction via RyR1 to drive hyperpolarization (Boittin et al., 2003), removal of Ca^{2+} from the cell by forward mode NCX activity and ultimately vasodilation (Boittin et al., 2003; Clark et al., 2010; Evans, 2009).

It should be noted, however, that when acting independently of RyR1, SERCA2b may be equally well placed to "refill" the superficial SR and support contraction by sequestering Ca^{2+} influx into the PM–SR junctions as indicated earlier. Indirect support for this view may be taken from the finding that in pulmonary arterial myocytes from patients with pulmonary arterial hypertension, cAMP appears to augment store-operated Ca^{2+} entry via a PKA-dependent pathway, yet inhibits store-operated calcium entry in pulmonary arterial myocytes from normotensive patients (Zhang et al., 2007). In short, there is a pathological swing in the balance from store emptying to store refilling at the PM–SR junctions, which may be further enhanced by increased expression of TRPC1, 3, 4, and 6 (Liu et al., 2012; Yu et al., 2009; Zhang et al., 2007, 2014).

These findings aside, further advances in our understanding of the functional organization of PM–SR nanojunctions are limited by their size alone, which renders nanojunctions beyond the resolution of current live cell experimental approaches. However, their importance to the coordination of ion exchange has now been visualized by 3D models of Ca^{2+} fluxes. These models incorporated dimensionally realistic intracellular architecture, Ca^{2+} pump, and release kinetics and transporter densities. Outcomes suggest that increases of one single Ca^{2+} ion may raise the local concentration from nanomolar to micromolar; i.e., there may be circumstances in which considerations on bulk concentration become irrelevant within cytoplasmic nanospaces (van Breemen et al., 2013). Even though these models only accounted for the stochastic element of diffusion, they highlighted that the functional integrity of PM–SR nanojunctions relies heavily on the close apposition of the two membranes, since various interrogations demonstrated that a separation of less than 50 nm was required to adequately provide for compartmentalized Ca^{2+} signaling. Of most importance perhaps was the observation that junctional integrity was lost when the separation of the PM and the junctional SR was raised above 50 nm (Fameli et al., 2014, 2007), because this indicates, as mentioned earlier, that nanojunctions have the capacity to restrict Ca^{2+} diffusion and thus "hold" Ca^{2+} within the junction.

The aforementioned studies apart, it could be argued that there is perhaps a lack of additional defining and quantifiable characteristics to extrapolate this argument across the cell. However, such a position is adequately countered by both qualitative and quantitative evidence in support of not only a requirement for junctional signaling beyond that originally envisaged at the superficial buffer barrier hypothesis, but also a growing body of evidence that highly localized signals arise in a manner constrained by closely apposed membranes in distant regions of the cell.

2.2 Multiple Releasable Pools of SR Ca^{2+}

The superficial buffer barrier hypothesis itself implies that smooth muscle contraction must be supported by Ca^{2+} release from the deeper, perinuclear SR (Nixon, Mignery, & Somlyo, 1994), i.e., from a subcompartment of the SR that is functionally segregated from, but continuous with (McCarron & Olson, 2008) the superficial SR that demarcates PM–SR nanojunctions (Poburko et al., 2004). Consistent with this view there is evidence to suggest that multiple releasable pools of SR Ca^{2+} exist (Golovina & Blaustein, 1997; Iino, Kobayashi, & Endo, 1988; Janiak, Wilson, Montague, & Hume, 2001; Yamaguchi, Kajita, & Madison, 1995). Significantly, in a variety of smooth muscle types the SERCA pump inhibitor cylcopiazonic acid has been shown to deplete one releasable pool of SR Ca^{2+}, while having little effect on Ca^{2+} release from another (Boittin et al., 2003; Clark et al., 2010; Dipp & Evans, 2001; Golovina & Blaustein, 1997; Janiak et al., 2001). In this respect it should be noted that cell-free assays on recombinant SERCA2a and SERCA2b have shown that cyclopiazonic acid is an equally effective inhibitor of both splice variants (Campbell, Kessler, Sagara, Inesi, & Fambrough, 1991; Verboomen, Wuytack, De Smedt, Himpens, & Casteels, 1992). The preferential depletion by cyclopiazonic acid of one SR compartment over another may therefore be due to its pharmacokinetics, with "selective access" to SERCA2b over SERCA2a perhaps being delivered by the combinatorial effects of the local cytoplasmic environment (e.g., pH and ATP concentration) (Hauser & Barth, 2007; Inesi, Lewis, Toyoshima, Hirata, & de Meis, 2008; Jensen, Sorensen, Olesen, Moller, & Nissen, 2006), the relative hydrophobicity of cyclopiazonic acid (K_p 3.1 compared to 4.9 for thapsigargin) and/or its relative affinity for SERCA (1000 times lower than that of thapsigargin) (Moncoq, Trieber, & Young, 2007). That said, it is possible that cell-specific, posttranslational modifications could confer a pharmacology distinct from that exhibited by recombinant SERCA, given that site-directed mutagenesis of

SERCA can selectively reduce the affinity of cyclopiazonic acid relative to thapsigargin (Ma et al., 1999).

If we accept their existence we must, therefore, ask what governs the segregation of multiple SR compartments and how might such segregation contribute to the multiplicity of Ca^{2+} signals required to appropriately regulate smooth muscle function, from contraction and dilation, to proliferation, migration, and ultimately programmed cell death. Clearly, the positioning of different SERCA, RyRs, and IP_3Rs in and around different nanojunctions of the SR could prove to be critical to both the segregation of functionally distinct "SR compartments" and, as required, the wider propagation of Ca^{2+} signals beyond the nanojunctions within which a given Ca^{2+} signal may arise. As mentioned previously, the capacity for this is evident given that arterial myocytes express multiple SERCA types (Clark et al., 2010; Eggermont, Wuytack, Verbist, & Casteels, 1990), up to three RyR subtypes (Herrmann-Frank et al., 1991; Kinnear et al., 2008; Neylon, Richards, Larsen, Agrotis, & Bobik, 1995; Westcott et al., 2012; Yang et al., 2005) and one or more IP_3R receptor subtypes (Westcott et al., 2012; Zhao et al., 2010).

2.2.1 *SERCA2a, RyR3, and RyR2 Are Resident in the Deep SR of Pulmonary Arterial Myocytes and Underpin Vasoconstriction*

It is now apparent that different types of SERCA may be located on the deep SR and the superficial SR proximal to PM–SR junctions, given that, in marked contrast to SERCA2b, SERCA2a is entirely restricted to the perinuclear SR of pulmonary arterial myocytes (Clark et al., 2010; Gilbert et al., 2014).

Importantly SERCA2b has a higher affinity for Ca^{2+} than SERCA2a, and this is allied to a low V_{max} (Verboomen et al., 1992). These differences in kinetics alone could conceivably allow activities of SERCA2b clusters on the superficial SR to determine, by withdrawal, the resting Ca^{2+} concentration of the bulk cytoplasm. However, these kinetic parameters will also serve to limit, due to saturation, the capacity for removal by SERCA2b of Ca^{2+} from the bulk cytoplasm during regenerative Ca^{2+} waves (Fig. 2). By contrast the high V_{max} and relatively low affinity for Ca^{2+} of SERCA2a (Verboomen et al., 1992; Verboomen, Wuytack, Van den Bosch, Mertens, & Casteels, 1994), allied to the fact that it is located on the perinuclear SR in pulmonary arterial myocytes (Clark et al., 2010; Gilbert et al., 2014), suggests that SERCA2a may function to recycle Ca^{2+} into the perinuclear SR (Clark et al., 2010; Evans, 2009) in support

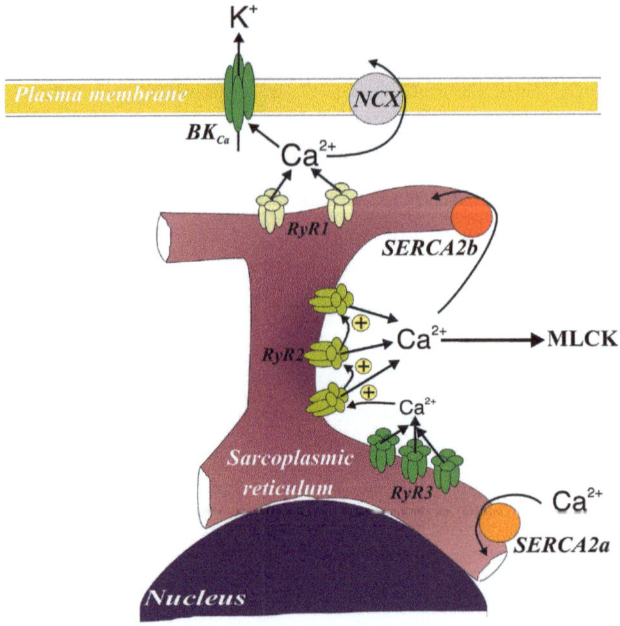

Fig. 2 The strategic positioning of SERCA and RyRs support myocyte contraction and relaxation. Schematic shows how different subtypes of SERCA and RyRs may be strategically positioned on the perinuclear, extraperinuclear, and superficial SR, in order to support or oppose vasoconstriction.

of those regenerative Ca^{2+} waves that maintain vasoconstriction (Asada, Yamazawa, Hirose, Takasaka, & Iino, 1999; Iino, Kasai, & Yamazawa, 1994; Perez & Sanderson, 2005; Ruehlmann, Lee, Poburko, & van Breemen, 2000). Indirect evidence for this comes from our finding that in pulmonary arterial myocytes cyclopiazonic acid does not block SR Ca^{2+} release via RyRs in response to hypoxia (Dipp & Evans, 2001; Dipp, Nye, & Evans, 2001) or endothelin-1 (Clark et al., 2010), despite the fact that Ca^{2+} release and constriction are blocked by ryanodine and the SERCA inhibitor thapsigargin (Clark et al., 2010; Kinnear et al., 2004).

Early functional studies on pulmonary arteries also raised the possibility that different populations of ryanodine receptors might be segregated from each other within pulmonary arterial myocytes and act in support of Ca^{2+} release from different SR compartments and thus regulate different cellular processes (Boittin et al., 2003; Dipp et al., 2001; Evans, 2009; Evans, Wyatt, Kinnear, Clark, & Blanco, 2005). This was first evident from the fact that blocking ryanodine receptors not only attenuated (~60%) isoprenaline-induced vasodilation of preconstricted pulmonary arteries

(Boittin et al., 2003) but also abolished hypoxic pulmonary vasoconstriction (Dipp et al., 2001); a process that serves to match lung perfusion to ventilation by diverting blood flow from oxygen-deprived to oxygen-rich areas of the lung (von Euler & Liljestrand, 1946).

That RyR subtypes were targeted to different aspects of the SR was confirmed by immunocytochemistry. The distribution of RyR3 was most notable in this respect given that, much like SERCA2a, this RyR subtype was mostly restricted to the perinuclear SR of pulmonary arterial myocytes (Clark et al., 2010; Gilbert et al., 2014; Kinnear et al., 2008). RyR3 is therefore very much separated from those clusters of RyR1 located on the superficial SR. The distribution of RyR2 was a little more puzzling, as this RyR subtype appears to radiate from the edge of the perinuclear region to the wider cell (Clark et al., 2010; Kinnear et al., 2008). RyR2 may therefore occupy and thus operate within the middle ground, that is the extraperinuclear space between the superficial and perinuclear regions of the cell in which, respectively, RyR1 and RyR3 are mostly located (Fig. 2). It was therefore proposed that perinuclear clusters of RyR3 might act, in some instances, as an initiation site for those propagating Ca^{2+} waves that evoke contraction (Dipp & Evans, 2001; Kinnear et al., 2008). Subsequent CICR via RyR2 might then serve to carry rapidly propagating Ca^{2+} waves away from the perinuclear region and across the myofilaments to enhance myocyte contraction (Clark et al., 2010; Evans, 2009; Kinnear et al., 2008).

2.2.2 Can Ca^{2+} Be Locked Within Junctional Complexes?

Intriguingly, data suggest that pulmonary arterial myocytes have the capacity to "lock" Ca^{2+} within the contractile domain. Briefly, vasoconstriction of pulmonary arteries without the endothelium can be repeatedly induced by hypoxia following preincubation with and in the continued presence of Ca^{2+} free extracellular medium containing 1 mM EGTA, and once induced vasoconstriction can be maintained for an hour or more (Dipp et al., 2001). We have no clear explanation for this observation, but one possible mechanism could be the inhibition by hypoxia of the process of Ca^{2+} removal via by SERCA2b at PM–SR nanojunctions, which appears to support vasodilation. There is some evidence to support this hypothesis, given that pulmonary vasodilation by β-adrenoceptor activation is abolished during hypoxia (McIntyre, Banerjee, Hahn, Agrafojo, & Fullerton, 1995) and HPV is enhanced by cyclopiazonic acid (Morio & McMurtry, 2002) yet abolished by thapsigargin (A.M. Evans, unpublished).

2.2.3 Multiple Paths to Smooth Muscle Contraction

There appear to be yet further complexities of oragnization over and above the hierarchical arrangement of the Ca^{2+} release pathways highlighted earlier, because of the apparent separation of SR resident IP_3Rs of pulmonary arterial myocytes from all RyRs that may be recruited by CICR (Boittin et al., 2002; Iino et al., 1988). This is evident from the fact that IP_3R activation by IP_3 elicits propagating, regenerative Ca^{2+} waves that arise in a manner entirely independent of RyR activation, given that they are abolished by the IP_3R antagonists xestospongin C but remain unaffected in the presence of ryanodine (Boittin et al., 2002; Iino et al., 1988). One can only conclude that there may be two distinct paths to myocyte contraction. Therefore, while the cytoplasmic Ca^{2+} signaling domain for contraction appears to be larger than the nanoscale, its distribution must be far from homogeneous. In support of this view, it is evident that there are separate PM regions for filament attachment and caveolae (Moore et al., 2004), that the density of myosin filaments is less in the cell periphery than central myoplasm (Lee, Poburko, Kuo, Seow, & van Breemen, 2002) and that the functional Ca^{2+}-binding protein calmodulin is tethered to the myofilaments rather than free in solution (Wilson, Sutherland, & Walsh, 2002). Consequently, the path length between the SR membrane pairs and calmodulin may be on the nanoscale and offer the capacity for SR–myofilament coupling along discrete routes of signal propagation, that might be selected for by way of receptor type and designation.

The aforementioned organizational structure is quite unlike that suggested by investigations on portal vein and mesenteric arterial smooth muscles, which appear to exhibit no such lines of segregation between IP_3Rs and RyRs. This is evident from the fact that in both of these myocyte types spontaneous Ca^{2+} release events are supported by IP_3Rs and RyRs (Gordienko & Bolton, 2002; Pucovsky & Bolton, 2006), and in portal vein myocytes CICR via RyRs is triggered by IP_3-induced Ca^{2+} release from IP_3Rs (Boittin, Coussin, Macrez, Mironneau, & Mironneau, 1998). Intriguingly, however, in these cell types multiple, perinuclear Ca^{2+} discharge regions (CDRs) have been identified, to which both IP_3Rs and RyRs appear to be targeted (Pucovsky & Bolton, 2006). It has been proposed that among these lies a primary frequent discharge site (FDR) that may at rest elicit periodic BK_{Ca} channel activations and thus oppose constriction. By contrast, in response to vasoconstrictor action and IP_3 accumulation it has been proposed that the FDR may deliver augmented Ca^{2+} release such that additional CDRs and Ca^{2+} entry pathways may be recruited in order to select for myocyte contraction (Pucovsky, Gordienko, & Bolton, 2002). The issue with this model is that, while it highlights the evident variability with respect to the organization of Ca^{2+} signaling machinery between smooth muscle types, we are restricted to a bimodal system, which cannot account for the versatility or plasticity of Ca^{2+} signaling observed within a given population of myocytes. Perhaps, the truth lies somewhere between the model proposed and one that incorporates nanojunctions of the SR.

2.3 Lysosome–SR Nanojunctions

Lysosome–SR nanojunctions are also evident in arterial myocytes (Fameli et al., 2014; Kinnear et al., 2004), and evidence from a variety of cell types has demonstrated that intracellular Ca^{2+} signaling may be supported by endolysosome targeted TPCs (*TPCN*1–3, gene name) (Brailoiu et al., 2009, 2010; Cai & Patel, 2010; Calcraft et al., 2009; Ishibashi, Suzuki, & Imai, 2000; Ruas et al., 2010; Zhu et al., 2010; Zong et al., 2009). It should be noted, however, that the role of TPCs in endolysosomal Ca^{2+} signaling remains controversial, on the grounds of ion selectivity and contrary findings with respect to the capacity for TPC channel gating by the Ca^{2+} mobilizing messenger NAADP (Jha, Ahuja, Patel, Brailoiu, & Muallem, 2014; Morgan & Galione, 2014; Pitt et al., 2010; Ruas et al., 2015; Schieder, Rotzer, Bruggemann, Biel, & Wahl-Schott, 2010; Wang et al., 2012). Nevertheless, there is substantial support for the view that NAADP may trigger intracellular Ca^{2+} release from acidic stores in a manner, at the very least, supported by all three subtypes of vertebrate TPCs. Of these, only the lysosome targeted TPC2 or TPC3 confer the level of lysosome–S/ER coupling necessary for subsequent amplification of Ca^{2+} bursts from acidic stores by CICR from the S/ER (Ogunbayo et al., 2015), and TPCN3 is absent in primates (including humans) and some rodents (e.g., mouse, rat) (Calcraft et al., 2009).

Our preliminary observations on pulmonary arterial myocytes are consistent with this view, in that global Ca^{2+} waves evoked by NAADP are abolished in myocytes from *Tpcn2* knockout mice (Ogunbayo, Zhu, et al., 2015). Yet in the context of this chapter, it is perhaps most significant that depletion of SR Ca^{2+} stores by inhibition of SERCA pumps with thapsigargin or block of RyRs with ryanodine, reveal spatially restricted bursts of Ca^{2+} release in response to NAADP that fail to propagate away from their point of initiation in the absence of either SR stores replete in

Ca^{2+} or functional RyRs (Boittin et al., 2002; Kinnear et al., 2004). In short, NAADP initiates global Ca^{2+} waves in an all-or-none manner by mobilizing acidic, lysosome-related Ca^{2+} stores that subsequently trigger CICR from the SR via RyRs (Kinnear et al., 2004). NAADP-induced Ca^{2+} bursts must therefore breach a given threshold in order to elicit a global Ca^{2+} wave by CICR via RyRs on the SR, and, it would appear, in a manner reminiscent of excitation–contraction coupling at the neuromuscular junction.

By contrast and as mentioned earlier, intracellular dialysis of IP_3 evokes regenerative Ca^{2+} waves in pulmonary arterial myocytes that remain unaffected following depletion of acidic stores with bafilomycin or block of RyRs with ryanodine. Moreover, IP_3- but not NAADP-evoked Ca^{2+} transients are blocked by the IP_3R antagonist xestospongin C (Boittin et al., 2002). Therefore, lysosomes couple to the SR of pulmonary arterial myocytes by CICR via RyRs but not IP_3Rs, i.e., there must be segregation of these receptors by region of location on the SR and perhaps by their specification to different nanojunctions. Further indirect support for this view may be taken from the findings of others (Janiak et al., 2001; Subedi, Paudel, & Sham, 2014). As discussed previously, however, we must always be mindful of the fact that in other smooth muscles, such as those of the portal vein, there may be no such lines of segregation between IP_3Rs and RyRs (Boittin et al., 1998; Gordienko & Bolton, 2002).

2.3.1 Lysosome–SR Junctions Form a Trigger Zone for Ca^{2+} Signaling by NAADP

LysoTracker red labeling in acutely isolated pulmonary arterial myocytes demonstrated that lysosomes form tight perinuclear clusters in a manner consistent with the spatially restricted nature of Ca^{2+} bursts triggered by NAADP (Boittin et al., 2002; Kinnear et al., 2004). Moreover lysosomal clusters associate with a subpopulation of RyRs labeled with Bodipy-Ryanodine and are separated from these RyRs by a narrow junction that is well beyond the resolution of deconvolution microscopy (i.e., $<0.2\ \mu m$) (Kinnear et al., 2004). It was therefore proposed that lysosomal clusters and RyRs form a "trigger zone," or intracellular synapse, for Ca^{2+} signaling by NAADP (Kinnear et al., 2004). Standard (2D) transmission electron microscopy has confirmed that lysosome–SR junctions do exist in pulmonary arterial myocytes and has allowed us to characterize these structures in greater detail. They are regularly occurring features of these cells and exhibit a distance of sepration between lysosomal and SR membranes of $\sim$16 nm and a lateral extension of $\sim$300 nm (Fameli et al., 2014).

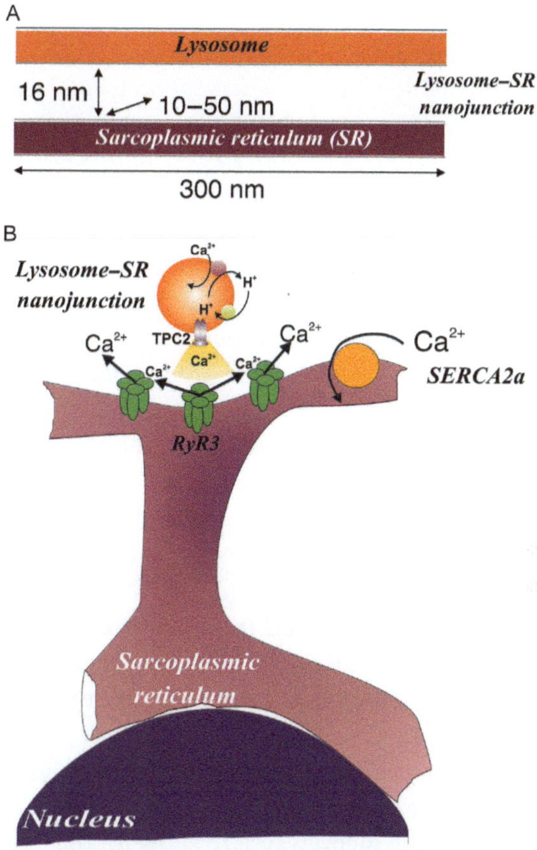

Fig. 3 The lysosome–SR nanojunction. Schematics show (A) the dimensions of lysosome–SR junctions and (B) the targeting of SERCA2a and RyR3 to the perinuclear SR proximal to these junctions.

Their architecture is, therefore, consistent with that of previously defined nanojunctions (Fig. 3A).

The presence of lysosome–SR nanojunctions may explain, in part, why Ca^{2+} bursts by NAADP induce global Ca^{2+} signals in an all-or-none manner by CICR from the SR via RyRs (Boittin et al., 2002), as this perhaps relies on the capacity for "quantal Ca^{2+} release" from lysosomes (Zhu et al., 2010). Moreover, from the signaling characteristics revealed it is evident that coupling of lysosomal Ca^{2+} stores to RyRs in pulmonary arterial myocytes could serve to provide a "margin of safety" with respect to the initiation of propagating, global Ca^{2+} waves, and contraction in response to certain stimuli, such as hypoxia. In short, lysosome-related Ca^{2+} release could conceivably contribute to the induction of hypoxic pulmonary vasoconstriction (Dipp & Evans, 2001).

2.3.2 Lysosomes Colocalize With RyR Subtype 3 to Form a Trigger Zone for Ca^{2+} Signaling in Pulmonary Arterial Smooth Muscle

Of the three RyR subtypes expressed in arterial smooth muscle, clusters of RyR3 (Fig. 3B) may be strategically positioned on those regions of the SR membrane incorporated within lysosome–SR nanojunctions of pulmonary arterial myocytes (Kinnear et al., 2008). This is evident from the fact that lysosomes and RyR3 share a common perinuclear designation and RyR3 labeling colocalizes with ~41% of the total volume of labeling for a lysosome-specific marker (αlgp120), compared to 13% and 14%, respectively, for RyR2 and RyR1 (Kinnear et al., 2008); in this respect, it is intriguing that Bolton and coworkers identified, as mentioned previously, perinuclear CDRs as the point of initiation of global Ca^{2+} waves in portal vein and mesenteric are myocytes (Pucovsky & Bolton, 2006).

2.3.3 Why Might RyR3 Be Targeted to Lysosome–SR Junctions?

When considering lysosome–SR junctions it is easy to forget that the classical functions of lysosomes, such as autophagy, are executed mostly independent of lysosome–SR interactions (Gozuacik & Kimchi, 2004), but are exquisitely sensitive to the luminal Ca^{2+} load of lysosomes and thus to Ca^{2+} fluxes across their limiting membranes. In short, lysosome–SR junctions must be engineered in such a way as to allow for the induction of highly localized Ca^{2+} signals that modulate lysosome-specific functions without engaging the SR, while at the same providing the capacity for initiation, as required, of propagating SR Ca^{2+} waves by way of CICR. Therefore, to understand the organizational hierarchy and the capacity for signal specification at lysosome–SR nanojunctions, or for that matter any other nanojunction, we must decipher the logic behind the choice of receptor targeted to each of the membrane pairs from which they are comprised.

One determining factor in this respect could be the relative sensitivity of each RyR subtype to CICR, the maximum gain in response to Ca^{2+} and the relative sensitivity of each receptor subtype to inactivation by Ca^{2+}. Cell-free assays suggest that the threshold for activation of RyR1, RyR2, and RyR3 is similar with channel activation evident at cytoplasmic Ca^{2+} concentrations >100 nM. By contrast, estimates of the EC_{50} are different with half maximal activation at >250 nM for RyR2 and >400 nM for RyR3 (Chen, Li, Ebisawa, & Zhang, 1997; Li & Chen, 2001). It should be noted, however, that such differences between subtypes could be greater in situ and might even differ across cell types due to, for example, cell-specific expression pattern for the wide array of RyR-binding partners (Song, Lee, Youn, Eom, & Kim, 2011). Nevertheless these are the best direct comparisons available for RyR subtypes and serve to highlight the role kinetics might play.

The higher EC_{50} exhibited by RyR3 could be significant, because this would provide for a higher operational "margin of safety" within which local lysosome-specific signals may arise in a manner filtered by the threshold that must be breached for all-or-none amplification of lysosomal Ca^{2+} bursts by CICR from the SR, i.e., the probability of false events being initiated would be lower for RyR3 than for RyR2. The P_o vs cytoplasmic Ca^{2+} concentration curves may also be significant, given that RyR3 (0–1) exhibits a higher gain in P_o than does RyR2 (0–0.9), while RyR1 (0–0.2) exhibits relatively little gain in P_o with increasing cytoplasmic Ca^{2+} concentration (Chen et al., 1997; Li & Chen, 2001). Therefore, once the threshold for activation of sufficient RyR3 clusters is breached these channels would offer greater amplification of Ca^{2+} bursts from lysosomal Ca^{2+} stores than would RyR2, while amplification via RyR1 would be marginal (see also Manunta et al., 2000; Yang et al., 2001). There is also marked variation in the relative sensitivity of each RyR subtype to inactivation by Ca^{2+}. RyR3 exhibits the lowest sensitivity to inactivation by Ca^{2+} with an IC_{50} of 3 mM while that for RyR2 is 2 mM; in each case channel activity may still be observed at concentrations >10 mM (Chen et al., 1997; Li & Chen, 2001). In marked contrast, RyR1 inactivation occurs within the micromolar range and full inactivation is achieved by 1 mM Ca^{2+}. Its sensitivity to inactivation by Ca^{2+} would therefore render RyR1 unsuitable for a role in the amplification of Ca^{2+} bursts at lysosome–SR junctions because the local Ca^{2+} concentration may exceed the operating range for RyR1. Therefore, the functional properties of RyR3 are likely best suited to a role in the amplification of Ca^{2+} bursts at lysosome–SR nanojunctions.

Taking these kinetic properties in to account, computational simulations estimate that within lysosome–SR nanojunctions half maximal SR Ca^{2+} release via RyR3 would likely occur at a junctional Ca^{2+} concentration of ~10 μM (Fameli et al., 2014). Applying the architecture of lysosome–sarcoplasmic junctions to computer modeling of the signaling complex they form, as determined by studies on pulmonary arterial myocytes, provided further powerful insights in to the mechanism of Ca^{2+} signaling within lysosome–SR junctions. Modeling outcomes showed that Ca^{2+} transients generated by junctional Ca^{2+} release from lysosomes were sufficient to reach those concentrations (~10 μM) necessary to breach the threshold for CICR via junctional RyR3s. However, increasing the separation of these nanojunctions decreases the maximum attainable junctional [Ca^{2+}] to values below those required for CICR via RyRs, and thus disrupts signaling across this junction. Perhaps most importantly, and consistent with previous studies on PM–SR nanojunctions, these analyses predicted a 30–50 nm functional operating limit for lysosome–SR nanojunctions, above which junctional integrity is lost due to inadequate control of ion movements within the cytoplasmic nanodomain between the lysosome–SR junctional membrane pair. In short, lysosome–SR nanojunctions appear necessary and sufficient for the generation of Ca^{2+} bursts and their subsequent amplification into propagating, global Ca^{2+} waves (Fameli et al., 2014).

It is evident that initiation of a propagating Ca^{2+} wave may not be governed by the action of a single junctional complex acting alone, and this too may contribute to the cells capacity to deliver lysosome-specific Ca^{2+} signals and even allow for highly restricted exchange of Ca^{2+} between these organelles and the SR. Assessment of lysosome–SR junction morphology in 3D using tomographic transmission electron microscopy identified many branch points on the SR that formed narrow cisternae, which rendered a single extension of the SR capable of coupling with multiple organelles. Furthermore, modeling outcomes predict that activation of between 60 and 100 lysosomes may be necessary to generate Ca^{2+} bursts of sufficient magnitude to yield the degree of RyR3 activation required to elicit a propagating Ca^{2+} wave; unless the rate of junctional coupling by CICR is faster than predicted. In other words, CICR engendered by a single lysosome–SR nanojunction would most likely be degenerate (Fameli et al., 2014), unless reinforced by a process of quantal Ca^{2+} release at other, proximal lysosome–SR junctions within a cluster (Zhu et al., 2010). Therefore, it seems most likely that multiple lysosome–SR junctions work in concert in a process characterized by both additive and regenerative elements, in order to provide the necessary threshold and margin of safety required to ensure all-or-none propagation by CICR of global Ca^{2+} waves beyond the confines of the cytoplasmic nanodomain demarcated by each lysosome–SR nanojunction. Summation will likely be governed by further lysosomal Ca^{2+} release, by SR Ca^{2+} release via junctional clusters of RyR3, and/or the recruitment of interjunctional clusters of RyR3.

We may also need to be mindful of the fact that the threshold for induction of CICR via RyRs may also be modulated by the luminal Ca^{2+} concentration of the SR (Beard, Sakowska, Dulhunty, & Laver, 2002; Ching,

Williams, & Sitsapesan, 2000; Gilchrist, Belcastro, & Katz, 1992; Gyorke & Gyorke, 1998; Tripathy & Meissner, 1996), which could in turn be primed by Ca^{2+} taken up via SERCA2a during lysosomal Ca^{2+} bursts that fail to breach the threshold for CICR from the SR.

Lysosome–SR signaling may also occur in reverse (Morgan et al., 2013). In short, Ca^{2+} release from SR compartments could increase lysosomal Ca^{2+} load, enhance the capacity for lysosomal Ca^{2+} release and thus augment coupling by CICR across lysosome–SR nanojunctions.

2.3.4 How May Ca^{2+} Signals Propagate Away From Lysosome–SR Junctions to the Wider Cell If RyR3 Is Restricted to the Perinuclear Region?

Significantly, the density of RyR3 labeling was found to decline markedly (between 4- and 14-fold by region) outside of the perinuclear region of the cell (Kinnear et al., 2008). It seems highly unlikely, therefore, that RyR3 functions to carry a propagating Ca^{2+} wave far beyond the point of initiation of CICR within the proposed trigger zone for Ca^{2+} signaling via lysosome–SR junctions. Given this finding it may be of significance that the density of labeling for RyR2 increases markedly in the extraperinuclear region of pulmonary arterial myocytes when compared to the perinuclear region and exhibits a ~3-fold greater density of labeling within this region than observed for either RyR3 or RyR1. This suggests that RyR2, but not RyR1, may function to receive Ca^{2+} from RyR3 clusters within extrajunctional spaces that lie at the interface of lysosome–SR nanojunctions (Fig. 4), and thereby allow for further propagation of the Ca^{2+} signal via CICR. Such a role would be supported by the lower EC_{50} for CICR via RyR2, which would also insure that once initiated a propagating Ca^{2+} wave would be less prone to failure.

If clusters of RyR3 do indeed sit within lysosome–SR nanojunctions and an array of RyR2 carries propagating Ca^{2+} signals away from these junctions, this may prove to be significant to our further understanding of the mechanisms that underpin hypoxic pulmonary vasoconstriction. This is evident from the fact that preincubation of pulmonary arteries with 8-bromo-cADPR, a competitive cADPR antagonist (Sethi, Empson, Bailey, Potter, & Galione, 1997; Wilson et al., 2001), blocks hypoxic pulmonary vasococnstriction in an all-or-none manner (Dipp & Evans, 2001), but once it is initiated 8-bromo-8-bromo-cADPR reverses hypoxic pulmonary vasoconstriction in a concentration-dependent manner. This is reminiscent of the block by alpha-bungarotoxin of transmission at the neuromuscular junction (Lee, Chen, & Katz, 1977). It seems plausible,

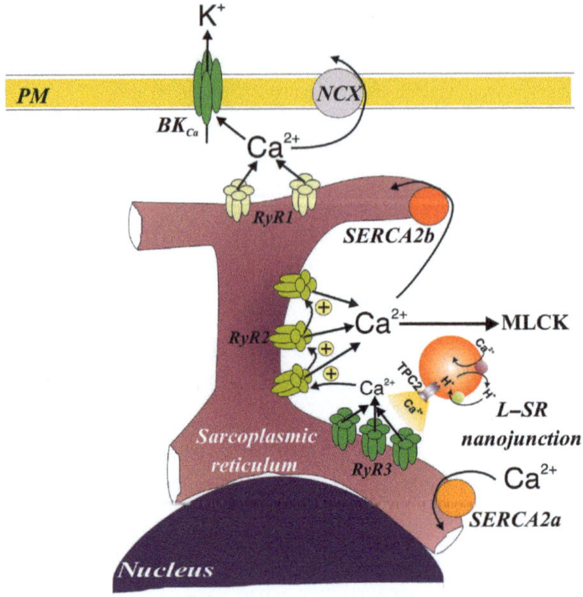

Fig. 4 RyR2 may support the propagation of Ca^{2+} waves away from lysosome–SR junctions. Schematic shows how RyR2 may be strategically positioned on the extra-perinuclear SR to carry a propagating Ca^{2+} wave away from perinuclear clusters of RyR3 at the interface of lysosome–SR junctions.

therefore, that 8-bromo-cADPR may deliver all-or-none block of HPV by inhibiting the activation by Ca^{2+} and/or cADPR of perinuclear clusters of RyR3 at lysosome–SR junctions. By contrast, the concentration-dependent reversal of HPV once it has been initiated could be explained if regenerative, propagating Ca^{2+} waves are carried by RyR2, in a manner maintained by increases in cADPR accumulation but independent of any further Ca^{2+} release from lysosome-related stores. In this respect, it is important to note that Ca^{2+} may also sensitize RyRs to activation by cADPR (Panfoli, Burlando, & Viarengo, 1999). The combinatorial effects of Ca^{2+} and cADPR are therefore of fundamental importance (Morgan & Galione, 2008), not least with regard to the threshold for activation of RyRs by either agent.

2.3.5 Tissue Specificity and Plasticity of Lysosome–SR Junctions

Identified limitations placed on coupling across lysosome–SR nanojunctions may be of greater significance because these nanojunctions exhibit perhaps the highest degree of plasticity of the family of nanojunctions described to date. Such heterogeneity and plasticity of function could be altered dramatically by variations in junctional distances and/or by altering signaling complexes targeted to lysosome–SR junctional membrane pairs. For example, in atrial myocytes it has been proposed that NAADP evokes Ca^{2+} release from lysosome-related stores in a manner which delivers increases in SR Ca^{2+} load (Collins, Bayliss, Churchill, Galione, & Terrar, 2011) and thus facilitates SR Ca^{2+} release by CICR via RyR2 clusters (Capel et al., 2015). This junctional variant may be provided by either: (1) an increase in junctional distance such that the lysosome-SR junctional Ca^{2+} concentration is insufficient to breach the threshold for activation of RyR2, yet sufficient to allow for SR Ca^{2+} uptake via apposing SERCA2a clusters or (2) the formation of lysosome–SR junctions between regions of the SR that possess dense SERCA2 clusters but are devoid of RyR2. Either way, in atrial myocytes the organization of lysosome–SR junctions likely insures, in contrast to outcomes for pulmonary arterial myocytes, the preeminence of PM–SR junctions in the context of cardiac excitation–contraction coupling.

Still greater plasticity of function must be supported in all cell types, given that under metabolic stresses lysosomal pathways support, in a manner influenced by TPC2 (Lin et al., 2015), the process of autophagy, which contributes to the maintenance of energy supply when cells are exposed to metabolic stresses, programmed cell death, and the recycling of organelles such as mitochondria (Gozuacik & Kimchi, 2004). Of particular importance to tonic smooth muscles may be autophagic support of energy supply during metabolic stresses, such as hypoxia, through glycogen metabolism via acid maltase, protein breakdown and, perhaps, free fatty acid supply for β-oxidation by mitochondria (Klionsky, Elazar, Seglen, & Rubinsztein, 2008; Kovsan, Bashan, Greenberg, & Rudich, 2010). Importantly, once engaged, autophagic processes will require separation of lysosome–SR junctions in support of vesicle trafficking and fusion events among lysosomes, endosomes, autophagosomes, and amphisomes (Klionsky et al., 2008), which might be initiated upon induction of either a global Ca^{2+} wave via lysosome–SR junctions or increased frequency of subthreshold lysosomal Ca^{2+} bursts. Thereafter, fusion events will be dependent on local Ca^{2+} signals for the effective formation of the SNARE complexes (Luzio et al., 2000), and the capacity for delivery of these will be enhanced by the uncoupling of lysosomal Ca^{2+} release from the process of excitation–contraction coupling. Of significance in this respect is the observation that mTOR may modulate autophagy, in part, by uncoupling from lysosome targeted TPC2 (Cang et al., 2013; Lu et al., 2013), because autophagy may in turn support hypoxic pulmonary vasoconstriction and, under certain circumstances, contribute to the development of hypoxic pulmonary hypertension (Evans et al., 2009; Goncharov et al., 2014). Dysfunction of these enzyme systems consequent to lysosome–SR junctional abnormalities could also conceivably contribute to pathologies associated with subclasses of lysosome storage disease such as Niemann–Pick disease type C1 (Lloyd-Evans et al., 2008; Tassoni, Fawaz, & Johnston, 1991), Pompe and Gaucher (Jmoudiak & Futerman, 2005; Noori et al., 2002), which may include hepatic portal (Tassoni et al., 1991) or pulmonary hypertension (Jmoudiak & Futerman, 2005; Noori et al., 2002), dysfunctions in cholesterol trafficking (Carstea et al., 1997), and consequent increases in plasma cholesterol levels, vascular lesion formation, atherosclerosis/thrombosis, and medial degradation (Ron & Horowitz, 2008; Tassoni et al., 1991).

2.4 Mitochondria–SR Nanojunctions

Investigations on cardiac ventricular myocytes first suggested that there may be a "structural proximity" between mitochondria and the SR, and highlighted the possibility of a role for mitochondria in Ca^{2+} sequestration (Ramesh et al., 1998). Subsequent investigations on neonatal cardiac myocytes confirmed that mitochondria buffer cytoplasmic Ca^{2+} signals without altering levels of ATP supply (Drago, De Stefani, Rizzuto, & Pozzan, 2012). Moreover studies on cell lines using Pericam-tagged linkers have identified discrete patches of mitochondria–ER junctions that exhibit pronounced heterogeneity with respect to their capacity for ER–mitochondrial Ca^{2+} transfer and estimated the optimal junctional width for efficient Ca^{2+} transfer to be on the nanoscale (Csordas et al., 2010). It is now clear that mitochondria–SR junctions have a diverse array of functions, and may contribute to the regulation of mitochondrial energy metabolism, lipid transport, apoptosis, and SR Ca^{2+} loading (Rowland & Voeltz, 2012).

In vascular smooth muscles, these mitochondria–SR junctional complexes are separated by cytoplasmic nanospaces of the order of 20 nm across (Tong et al., 2009), equivalent to that of PM–SR nanojunctions lysosome–SR nanojunctions. Although their precise functions in vascular smooth muscles remain to be determined, evidence suggests that Ca^{2+} may be bidirectionally exchanged across mitochondrial–SR junctions (Poburko et al., 2006), and that SR Ca^{2+} loading may be moderated by mitochondria during agonist-induced NCX reversal (Poburko et al., 2006). It is also evident that mitochondrial Ca^{2+} uptake can be observed following Ca^{2+} release from the SR via RyRs (Gilbert et al., 2014) or IP_3Rs (Olson, Chalmers, & McCarron, 2010), may occur quickly enough to influence the gating of IP_3Rs at the intracluster level, and may thus regulate both local and global Ca^{2+} signals (Olson et al., 2010). Such SR–mitochondrial Ca^{2+} transfer may deliver context-specific changes in mitochondrial energy metabolism (Chalmers & McCarron, 2008; Rowland & Voeltz, 2012) and this could perhaps support vasomotion (Poburko et al., 2006). As for other junctional complexes, mitochondria–SR nanojunctions also exhibit a high degree of plasticity that may be essential to phases of mitochondrial mobility during smooth muscle proliferation (Chalmers et al., 2012).

Given the above, it is intriguing to note that mitochondria–SR signaling and the processes it regulates may decline with aging, and thus contribute to age-related vascular dysfunction (Chalmers, Saunter, Girkin, & McCarron, 2016).

2.5 Nuclear Invaginations May Provide a Nanodomain Within Which Ca^{2+} Signals May Be Generated to Modulate Gene Expression

The aforementioned studies led us to perhaps the most intriguing nanodomain of all, namely the nuclear invaginations. Importantly, the SR is contiguous with the outer nuclear membrane (Lesh et al., 1998) and its tubular invaginations (Echevarria, Leite, Guerra, Zipfel, & Nathanson, 2003; Fricker, Hollinshead, White, & Vaux, 1997; Gerasimenko, Gerasimenko, Tepikin, & Petersen, 1995). However, despite the proposed role of SR Ca^{2+} release in excitation–transcription coupling and thus phenotypic modulation (Cartin, Lounsbury, & Nelson, 2000; Gomez, Stevenson, Bonev, Hill-Eubanks, & Nelson, 2002; Stevenson, Gomez, Hill-Eubanks, & Nelson, 2001; Wamhoff, Bowles, & Owens, 2006), there have been few detailed investigations on nuclear Ca^{2+} signaling in native smooth muscles (Wray & Burdyga, 2010). The current consensus (Bootman, Fearnley, Smyrnias, MacDonald, & Roderick, 2009; Queisser, Wiegert, & Bading, 2011) is that the nuclear envelope and its invaginations increase the surface area available for direct entry of Ca^{2+} into the nucleus via nuclear pores, driven either by Ca^{2+} influx across the PM or by activation of RyRs (Marius, Guerra, Nathanson, Ehrlich, & Leite, 2006) and IP_3Rs (Avedanian, Jacques, & Bkaily, 2011; Cardenas et al., 2004; Gerasimenko et al., 1995; Hirose, Stuyvers, Dun, Ter Keurs, & Boyden, 2008) resident in the deep, perinuclear S/ER.

However, early investigations on a variety of cell types, including arterial smooth muscles, strongly suggested that the nuclear membrane restricts direct Ca^{2+} flux into the nucleus (Himpens, De Smedt, & Casteels, 1992; Himpens, De Smedt, Droogmans, & Casteels, 1992; Neylon, Hoyland, Mason, & Irvine, 1990; Wamhoff et al., 2006; Waybill et al., 1991; Williams, Fogarty, Tsien, & Fay, 1985), raising the possibility that the nuclear membrane acts, much like PM–SR nanojunctions, as a buffer barrier that can restrict the free diffusion of Ca^{2+} to the nucleus. Our pilot studies on pulmonary arterial myocytes concur.

Consistent with prior studies on other cell types, electron microscopy has revealed that the boundary of nuclear invaginations is comprised of a double membrane, which is an extension of the nucleoplasmic reticulum (unpublished). The lumen of each invagination is continuous with the cytoplasm, demarcating a defined cytoplasmic nanospace with a radius of 10–100 nm; i.e., we are now considering a tubular system rather than the relatively simple junctional complexes formed at the interface between two membrane pairs. Our preliminary investigations suggest (Fig. 5) that a third subtype of SERCA pump, SERCA1, and RyR1 may be selectively targeted to the SR membrane within nuclear nvaginations of pulmonary arterial myocytes (Evans, 2013). Each nuclear invagination may thus demarcate a cytoplasmic nanodomain within which Ca^{2+} signals are delivered to selectively modulate gene expression, i.e., they may be functionally segregated from the bulk cytoplasm due to the restrictions placed on the free diffusion of Ca^{2+} by the dimensions of their "junctional" radii, length of the invagination in question, and the strategic positioning of different Ca^{2+} release channels and Ca^{2+} transporters.

2.5.1 How Might Nuclear Invaginations Coordinate Ca^{2+} Signals and Thus Gene Expression via Their Incorporated Cytoplasmic Nanodomains?

It seems likely that nuclear invaginations will allow, for example, vasoconstrictors to coordinate the release of Ca^{2+} to initiate myocyte contraction and gene expression, at the very least in a manner that may be opposed/modulated by vasodilators. In order to achieve this goal I envisage that gene expression will be coordinated, in part, by Ca^{2+} signals from nuclear invaginations of the SR (nSR).

Ca^{2+} signals arising via nSR resident RyR1 may be supported by Ca^{2+} uptake via SERCA1, and may thus be "informed" by the nature of Ca^{2+} signals arising within the bulk cytoplasm and perhaps, in part, via RyR1 activation driven by consequent increases in SR luminal Ca^{2+} concentration

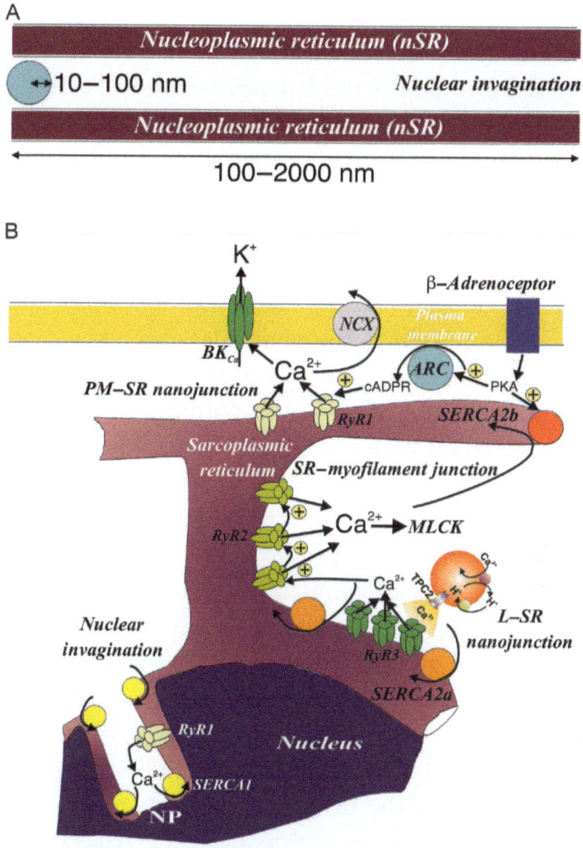

Fig. 5 Nanojunctions of the SR from PM to nucleus. Schematic shows (A) dimensions of nuclear invaginations and (B) the strategic positioning of SERCA and RyRs within nanojunctions of the SR that provide for independent, but informed, support of site- and function-specific calcium signals between the SR and the PM, lysosomes, and the nucleus.

(Beard et al., 2002; Ching et al., 2000; Gilchrist et al., 1992; Gyorke & Gyorke, 1998; Tripathy & Meissner, 1996). Importantly in this respect, SERCA1 has a high affinity for Ca^{2+}, which may be modulated by sarcolipin or phospholamban (Odermatt et al., 1998), while RyR1 provides limited support for signal propagation by CICR (Chen et al., 1997; Li & Chen, 2001; Yang et al., 2001). Therefore, SERCA1 may load the nSR Ca^{2+} store, and provide a "variable buffer barrier" that restricts free diffusion of Ca^{2+} from the perinuclear cytoplasm into the invaginations, limits loss of Ca^{2+} from the invaginations to the cytoplasm and may also act to limit free diffusion of Ca^{2+} into the nucleus via nuclear pores. Gating of RyR1 within invaginations, either directly by cADPR (Ogunbayo et al., 2011) and/or indirectly by local SR Ca^{2+} load consequent to modulation of calsequestrin–RyR1 interactions (Protasi, Paolini, Canato, Reggiani, & Quarta, 2012), may thereafter serve to adjust the balance between RyR1 and SERCA1 activities. The limited support provided by RyR1 for signal propagation allied to effective Ca^{2+} uptake via SERCA1, would insure that evoked Ca^{2+} release events do not propagate beyond the nuclear invaginations from which they arise. Moreover, the Ca^{2+} load of the nSR and thus the capacity for Ca^{2+} release by nuclear invaginations could be determined by the high affinity of SERCA1 for Ca^{2+}, through, for example, fractional uptake of Ca^{2+} mobilized from the perinuclear SR. The capacity for this is evident due to the fact that recycling of Ca^{2+} into the perinuclear SR will be conferred by SERCA2a, which has a low affinity for Ca^{2+} and a high V_{max}. Gating of RyR1s resident within nuclear invaginations could thus modulate gene expression in a manner informed by myocyte activation.

If we accept that invaginations of the nucleoplasmic reticulum could, as posited earlier, support site-specific Ca^{2+} signals with the capacity to orchestrate transcriptional cascades, how could this be achieved? Conceivably by providing for local modulation, for example, of the calcinuerin-NFAT3c pathway to guide the translocation of NFAT3c to the nucleus via nuclear pores, and this could explain, in part, previous contrary findings with respect to the regulation by differently evoked Ca^{2+} signals of calcineurin-NFAT3c signaling (Gomez et al., 2002). That aside, the entire boundary the IMN is lined by A-type LAMIN (Burke & Stewart, 2013) which associates with complexes that have been proposed to provide chromatin attachment points that confer transcriptional suppression (Burke & Stewart, 2013) via, for example, histone (Harr et al., 2015) and LINC complexes (Demmerle, Koch, & Holaska, 2013). Although assessing the mechanisms involved is beyond the optical resolution of current technologies, it has been suggested that transcriptional activities could be modulated by Ca^{2+}-dependent enzymes (Zuleger, Robson, & Schirmer, 2011) and may also be governed by conformational changes in these proteins consequent to the redistribution of electrostatic charge across the ONM and INM (Strelkov, Schumacher, Burkhard, Aebi, & Herrmann, 2004).

2.6 Could Nanojunctions of the SR Support Network Activity?

It is quite plausible that PM–SR junctions could deliver either a "gain" or "loss" of function at the perinuclear SR and thus the nSR, due to their capacity to support bidirectional adjustments of SR Ca^{2+} load. In short, nanojunctions of the SR could underpin a form of network activity.

As detailed earlier, SERCA2b is resident on the superficial SR proximal to PM–SR junctions and exhibits perhaps the highest affinity for Ca^{2+} (Dode et al., 2003; Odermatt, Kurzydlowski, & MacLennan, 1996; Verboomen et al., 1992). Its kinetics may thus support, as described earlier, a barrier to Ca^{2+} flux between the cytoplasmic nanodomains within PM–SR nanojunctions and the myofilaments incorporated within the contractile domains. Segregation of PM–SR nanojunctions could be further assisted by the limited support for signal propagation by CICR afforded by RyR1 (Chen et al., 1997; Li & Chen, 2001; Yang et al., 2001). Under resting conditions, the high affinity for Ca^{2+} of SERCA2b may therefore determine, in part, the Ca^{2+} concentration of the bulk cytoplasm, by removal of Ca^{2+} into PM–SR nanojunctions. Moreover, and as discussed previously, modulation of SERCA activity may alter the capacity for removal of Ca^{2+} to the peripheral SR, and this could impact in turn on the capacity for Ca^{2+} release at other junctional complexes.

2.6.1 Unloading of the SR, Loss of Function, and Vasodilation

In response to vasodilators (Gironacci, Valera, Yujnovsky, & Pena, 2004; Lara Lda et al., 2006; Liu, Oudit, Fang, Zhou, & Scholey, 2012), the V_{max} of SERCA2b may be increased through the phosphorylation of phospholamban by, for example, PKA or PKG (Raeymaekers, Eggermont, Wuytack, & Casteels, 1990), which results in the dissociation of this inhibitory protein from SERCA2b (Raeymaekers et al., 1990). This effectively increases the capacity for removal of cytoplasmic Ca^{2+} to the superficial SR. Unloading of the superficial SR and thus store emptying may then be facilitated by Ca^{2+} release into the PM–SR junctions due to PKA-dependent RyR1 activation, consequent BK_{Ca} channel activation, PM hyperpolarization, reductions in voltage-gated Ca^{2+} influx and accelerated Ca^{2+} removal from the cell by forward mode activity of NCX (Boittin et al., 2003; Nazer & Van Breemen, 1998b). This could conceivably result in a "net loss" of function within the perinuclear SR that will oppose vasoconstriction. It is equally plausible that an additional knock on effect of either a net reduction in luminal Ca^{2+} concentration of the perinuclear SR and/or reductions in Ca^{2+} release from the perinuclear SR, could be a decrease the Ca^{2+} load of the nSR that constitutes nuclear invaginations. At the very least this could depress Ca^{2+}-dependent pathways of transcriptional control.

2.6.2 Refilling of the SR and Gain of Function

By contrast with the above, refilling of the SR delivers gain of function. When myocyte contraction is supported by SR store refilling via SERCA2b through its receipt of Ca^{2+} from receptor-operated pathways (Albert et al., 2009; Shi et al., 2012), voltage-gated Ca^{2+} influx (Takeda et al., 2011), STIM–Orai complexes (Berra-Romani et al., 2008; Lu et al., 2008; Takahashi et al., 2007), and/or reverse-mode activity of NCX (Poburko et al., 2008, 2006; Syyong et al., 2007), the normally low V_{max} of SERCA2b would be self-limiting, due to saturation. Therefore removal of Ca^{2+} from the deeper cytoplasm during regenerative Ca^{2+} waves would be reduced and refilling of the deep SR facilitated. This would indirectly provide, perhaps by increased perinuclear Ca^{2+} release or load, a "gain of function" to the nSR within nuclear invaginations, and, at the very least, opposite effects in terms of transcriptional regulation when compared to unloading of the SR.

As mentioned previously, SERCA1 and SERCA2a may also be modulated by interactions with phospholamban, or for that matter by sarcolipin. Therefore, the extensive capacity for actively redirecting junctional Ca^{2+} fluxes across the SR network of arterial myocytes is clear.

2.7 Junctional Reorganization During the Switch From a Contractile to a Migratory and Proliferative Smooth Muscle Phenotype

We already know that among other changes to the Ca^{2+} signaling apparatus, increases in the expression of SERCA2a, RyR2, and RyR3 may be initiated during the transition from a contractile to a proliferating smooth muscle phenotype (Berra-Romani et al., 2008; Magnier et al., 1992), and increases in SERCA2b expression have been observed in myocytes of arteries following organ culture (Thorne & Paul, 2003). Moreover such changes have been linked to altered vascular reactivity (Thorne & Paul, 2003). Thus, it seems likely that the switch from a contractile to a proliferative–migratory phenotype(s) will lead to marked reorganization of cellular nanojunctions in a manner coupled to changes in SERCA and RyR expression at the very least. From the most cursory consideration of observations made thus far and the schematic model derived (Fig. 5) it is evident that reorganization of available nanojunctions and/or their allocated calcium signaling machinery could drastically alter outcomes in terms of regulated Ca^{2+} redistribution. As proposed previously (van Breemen et al., 2013), such reorganization could provide for the functional diversity noted not only across different types of vascular smooth muscle, but also during their differentiation and switch to secretory and/or proliferative phenotypes. Consistent with this view, recent investigations suggest that such reorganization of junctional complexes within pulmonary arterial myocytes may be triggered by hypoxia and contribute to the progression pulmonary hypertension (Gilbert et al., 2014), which is known to be driven, in part, by NFAT-dependent vascular remodeling (Bierer et al., 2011; Suzuki et al., 2002).

2.8 Couplons and Ca^{2+} Exchange Across and Between Cytoplasmic Nanodomains

The dimensions of membrane–membrane junctions will, in part, determine the cytoplasmic volume into which Ca^{2+} will be released, the degree to which diffusion may be restricted and thus the effective Ca^{2+} concentration. Moreover, the capacity for Ca^{2+} release within a given nanodomain will be dictated by, for example, the number of RyR clusters within this volume and the number of RyRs per cluster, which will likely determine the magnitude of "unitary" Ca^{2+} signals, e.g., Ca^{2+} sparks. Information on the spatial organization of RyRs and SERCA is therefore required in order to allow us to assess the capacity for signal propagation within and between nanodomains. The importance of such information is clearly demonstrated by studies on ventricular myocytes which have shown that intercluster distances of RyR2 are less than 1 μm on average, and that greater intercluster distances will compromise effective propagation of Ca^{2+} waves by CICR (Franzini-Armstrong & Protasi, 1997; Franzini-Armstrong, Protasi, & Ramesh, 1999; Soeller et al., 2007). Likewise, SERCA distribution and density will determine the capacity for Ca^{2+} removal per unit volume. In this respect, the concept of "couplons" may prove pivotal to our understanding of the path along which a given Ca^{2+} signal is allowed to propagate and those paths along which a signal may not pass. Couplons were originally defined in studies on skeletal and cardiac muscles (Stern, Pizarro, & Rios, 1997), as the functional grouping of RyRs, dihydropyridine receptors, and related junctional proteins that "cooperate" during excitation–contraction coupling. Further investigations on smooth muscles should ideally draw on similar considerations in relation to, for example, the organization of: (A) PM–SR junctions—SERCA2b, RyR1, BK_{Ca}, and NCX; (B) SR–myofilament junctions—SERCA2a, RyR3, RyR2, and calmodulin myofilaments; (C) lysosome–SR junctions—TPC2, RyR3, and SERCA2a; (D) mitochondria–SR junctions—RyRs, IP_3Rs, and MCU; and (E) nuclear invaginations—SERCA1, RyR1, and nuclear pore complexes. When allied to computer models of junctional signaling, this information could provide for unprecedented advances not only in our understanding of signaling on the nanoscale, but also of the capacity for novel pharmacological interventions.

3. CONCLUSION

All intracellular nanojunctions presently defined create a cytoplasmic nanodomain between each junctional membrane pair that is approximately 20 nm across, whether we consider the PM–SR junction, lysosome–SR junction, mitochondria–SR junction, or the nuclear invagination. The consistency of separation between membrane pairs that constitute each nanojunction is therefore evident, so why ~20 nm? Clearly, whichever nanojunction is modeled this degree of membrane separation ensures that Ca^{2+} transients within the corresponding cytoplasmic nanodomain are segregated from those in the bulk myoplasm, and may thus accurately confer site- and function-specific Ca^{2+} signals. In this respect, a number of factors act to confine Ca^{2+} transients within nanojunctions:

(1) The geometry of the junctions, especially the distance between membranes (Fameli et al., 2014, 2007).

(2) The relatively low diffusivity of cytoplasmic Ca^{2+} (free, nevermind buffered) (Allbritton, Meyer, & Stryer, 1992; Kushmerick & Podolsky, 1969) in combination with the restrictive junctional geometry provides enhanced capacity to limit diffusion of Ca^{2+} beyond a given nanodomain.

(3) The kinetics of Ca^{2+} pumps in the junctions, allied to the previous two factors, directs Ca^{2+} fluxes and may act as a further barrier to Ca^{2+} flux away from the junctional nanospace.

(4) Protein complexes that span junctions likely limit ion mobility within cytoplasmic nanodomains by increasing path tortuosity (Devine et al., 1972; Poburko et al., 2008).

The properties of a given nanojunction may, therefore, allow arterial myocytes to coordinate the delivery of Ca^{2+} signals in a manner that allows for the selective induction of, for example, vasoconstriction, vasodilation, autophagy, ATP supply, and/or gene expression. While the hypothesis presented is most likely an oversimplification of the Ca^{2+} signaling apparatus available to the cell, it provides us with a case to challenge. In doing so we must remain aware that the precise configuration of junctional complexes may be context specific, and vary not only between myocyte types but also throughout the path of cell differentiation, during repair, disease, and aging. This remains a relatively new horizon and it will be some time before we gain a true appreciation of the importance of nanojunctions and the complexity and versatility they afford cellular Ca^{2+} signaling.

CONFLICT OF INTEREST

The author has no conflict of interest to declare.

ACKNOWLEDGMENTS

The work author is presently funded by the British Heart Foundation Program Grant (12/14/29885).

REFERENCES

Albert, A. P., Saleh, S. N., & Large, W. A. (2009). Identification of canonical transient receptor potential (TRPC) channel proteins in native vascular smooth muscle cells. *Current Medicinal Chemistry*, *16*(9), 1158–1165.

Allbritton, N. L., Meyer, T., & Stryer, L. (1992). Range of messenger action of calcium ion and inositol 1,4,5-trisphosphate. *Science*, *258*(5089), 1812–1815.

Asada, Y., Yamazawa, T., Hirose, K., Takasaka, T., & Iino, M. (1999). Dynamic Ca2+ signalling in rat arterial smooth muscle cells under the control of local renin-angiotensin system. *The Journal of Physiology*, *521*(Pt. 2), 497–505. doi: PHY_9710 [pii].

Avedanian, L., Jacques, D., & Bkaily, G. (2011). Presence of tubular and reticular structures in the nucleus of human vascular smooth muscle cells. *Journal of Molecular and Cellular Cardiology*, *50*(1), 175–186. http://dx.doi.org/10.1016/j.yjmcc.2010.10.005. S0022-2828(10)00384-6 [pii].

Beard, N. A., Sakowska, M. M., Dulhunty, A. F., & Laver, D. R. (2002). Calsequestrin is an inhibitor of skeletal muscle ryanodine receptor calcium release channels. *Biophysical Journal*, *82*(1 Pt. 1), 310–320.

Benham, C. D., & Bolton, T. B. (1986). Spontaneous transient outward currents in single visceral and vascular smooth muscle cells of the rabbit. *The Journal of Physiology*, *381*, 385–406.

Benham, C. D., Bolton, T. B., Lang, R. J., & Takewaki, T. (1986). Calcium-activated potassium channels in single smooth muscle cells of rabbit jejunum and guinea-pig mesenteric artery. *The Journal of Physiology*, *371*, 45–67.

Berra-Romani, R., Mazzocco-Spezzia, A., Pulina, M. V., & Golovina, V. A. (2008). Ca2+ handling is altered when arterial myocytes progress from a contractile to a proliferative phenotype in culture. *American Journal of Physiology. Cell Physiology*, *295*(3), C779–C790. http://dx.doi.org/10.1152/ajpcell.00173.2008. 00173.2008 [pii].

Berridge, M. J. (2008a). Inositol trisphosphate and calcium signalling mechanisms. *Biochimica et Biophysica Acta*, *1793*, 933–940.

Berridge, M. J. (2008b). Smooth muscle cell calcium activation mechanisms. *The Journal of Physiology*, *586*(Pt. 21), 5047–5061.

Bierer, R., Nitta, C. H., Friedman, J., Codianni, S., de Frutos, S., Dominguez-Bautista, J. A., … Gonzalez Bosc, L. V. (2011). NFATc3 is required for chronic hypoxia-induced pulmonary hypertension in adult and neonatal mice. *American Journal of Physiology. Lung Cellular and Molecular Physiology*, *301*(6), L872–L880. http://dx.doi.org/10.1152/ajplung.00405.2010. ajplung.00405.2010 [pii].

Boittin, F. X., Coussin, F., Macrez, N., Mironneau, C., & Mironneau, J. (1998). Inositol 1,4,5-trisphosphate- and ryanodine-sensitive Ca2+ release channel-dependent Ca2+ signalling in rat portal vein myocytes. *Cell Calcium*, *23*(5), 303–311.

Boittin, F. X., Dipp, M., Kinnear, N. P., Galione, A., & Evans, A. M. (2003). Vasodilation by the calcium-mobilizing messenger cyclic ADP-ribose. *The Journal of Biological Chemistry*, *278*(11), 9602–9608.

Boittin, F. X., Galione, A., & Evans, A. M. (2002). Nicotinic acid adenine dinucleotide phosphate mediates Ca2+ signals and contraction in arterial smooth muscle via a two-pool mechanism. *Circulation Research*, *91*(12), 1168–1175.

Bootman, M. D., Fearnley, C., Smyrnias, I., MacDonald, F., & Roderick, H. L. (2009). An update on nuclear calcium signalling. *Journal of Cell Science*, *122*(Pt. 14), 2337–2350. http://dx.doi.org/10.1242/jcs.028100.122/14/2337 [pii].

Brailoiu, E., Churamani, D., Cai, X., Schrlau, M. G., Brailoiu, G. C., Gao, X., … Patel, S. (2009). Essential requirement for two-pore channel 1 in NAADP-mediated calcium signaling. *The Journal of Cell Biology*, *186*, 201–209.

Brailoiu, E., Hooper, R., Cai, X., Brailoiu, G. C., Keebler, M. V., Dun, N. J., … Patel, S. (2010). An ancestral deuterostome family of two-pore channels mediates nicotinic acid adenine dinucleotide phosphate-dependent calcium release from acidic organelles. *The Journal of Biological Chemistry*, *285*(5), 2897–2901.

Brayden, J. E., & Nelson, M. T. (1992). Regulation of arterial tone by activation of calcium-dependent potassium channels. *Science*, *256*(5056), 532–535.

Burke, B., & Stewart, C. L. (2013). The nuclear lamins: Flexibility in function. *Nature Reviews Molecular Cell Biology*, *14*(1), 13–24. http://dx.doi.org/10.1038/nrm3488.

Cai, X., & Patel, S. (2010). Degeneration of an intracellular ion channel in the primate lineage by relaxation of selective constraints. *Molecular Biology and Evolution*, *27*(10), 2352–2359. http://dx.doi.org/10.1093/molbev/msq122. msq122 [pii].

Calcraft, P. J., Ruas, M., Pan, Z., Cheng, X., Arredouani, A., Hao, X., … Zhu, M. X. (2009). NAADP mobilizes calcium from acidic organelles through two-pore channels. *Nature*, *459*(7246), 596–600.

Campbell, A. M., Kessler, P. D., Sagara, Y., Inesi, G., & Fambrough, D. M. (1991). Nucleotide sequences of avian cardiac and brain SR/ER Ca(2+)-ATPases and functional comparisons with fast twitch Ca(2+)-ATPase. Calcium affinities and inhibitor effects. *The Journal of Biological Chemistry*, *266*(24), 16050–16055.

Cang, C., Zhou, Y., Navarro, B., Seo, Y. J., Aranda, K., Shi, L., … Ren, D. (2013). mTOR regulates lysosomal ATP-sensitive two-pore Na(+) channels to adapt to metabolic state. *Cell*, *152*(4), 778–790. http://dx.doi.org/10.1016/j.cell.2013.01.023.

Capel, R. A., Bolton, E. L., Lin, W. K., Aston, D., Wang, Y., Liu, W., … Terrar, D. A. (2015). Two-pore channels (TPC2s) and nicotinic acid adenine dinucleotide phosphate (NAADP) at lysosomal-sarcoplasmic reticular junctions contribute to acute and chronic beta-adrenoceptor signaling in the heart. *The Journal of Biological Chemistry*, *290*(50), 30087–30098. http://dx.doi.org/10.1074/jbc.M115.684076.

Cardenas, C., Muller, M., Jaimovich, E., Perez, F., Buchuk, D., Quest, A. F., & Carrasco, M. A. (2004). Depolarization of skeletal muscle cells induces phosphorylation of cAMP response element binding protein via calcium and protein kinase Calpha. *The Journal of Biological Chemistry*, *279*(37), 39122–39131. http://dx.doi.org/10.1074/jbc.M401044200. M401044200 [pii].

Carstea, E. D., Morris, J. A., Coleman, K. G., Loftus, S. K., Zhang, D., Cummings, C., … Tagle, D. A. (1997). Niemann–Pick C1 disease gene: Homology to mediators of cholesterol homeostasis. *Science*, *277*(5323), 228–231.

Cartin, L., Lounsbury, K. M., & Nelson, M. T. (2000). Coupling of Ca(2+) to CREB activation and gene expression in intact cerebral arteries from mouse: Roles of ryanodine receptors and voltage-dependent Ca(2+) channels. *Circulation Research*, *86*(7), 760–767.

Chalmers, S., & McCarron, J. G. (2008). The mitochondrial membrane potential and Ca2+ oscillations in smooth muscle. *Journal of Cell Science, 121*(Pt. 1), 75–85. http://dx.doi.org/10.1242/jcs.014522.
Chalmers, S., Saunter, C., Girkin, J. M., & McCarron, J. G. (2016). Age decreases mitochondrial motility and increases mitochondrial size in vascular smooth muscle. *The Journal of Physiology, 594*, 4283–4295. http://dx.doi.org/10.1113/JP271942.
Chalmers, S., Saunter, C., Wilson, C., Coats, P., Girkin, J. M., & McCarron, J. G. (2012). Mitochondrial motility and vascular smooth muscle proliferation. *Arteriosclerosis, Thrombosis, and Vascular Biology, 32*(12), 3000–3011. http://dx.doi.org/10.1161/ATVBAHA.112.255174.
Chen, S. R., Li, X., Ebisawa, K., & Zhang, L. (1997). Functional characterization of the recombinant type 3 Ca2+ release channel (ryanodine receptor) expressed in HEK293 cells. *The Journal of Biological Chemistry, 272*(39), 24234–24246.
Cheng, H., Lederer, W. J., & Cannell, M. B. (1993). Calcium sparks: Elementary events underlying excitation-contraction coupling in heart muscle. *Science, 262*(5134), 740–744.
Ching, L. L., Williams, A. J., & Sitsapesan, R. (2000). Evidence for Ca(2+) activation and inactivation sites on the luminal side of the cardiac ryanodine receptor complex. *Circulation Research, 87*(3), 201–206.
Clark, J. H., Kinnear, N. P., Kalujnaiab, S., Cramb, G., Fleischer, S., Jeyakumar, L. H., … Evans, A. M. (2010). Identification of functionally segregated sarcoplasmic reticulum calcium stores in pulmonary arterial smooth muscle. *The Journal of Biological Chemistry, 285*, 13542–13549.
Collins, T. P., Bayliss, R., Churchill, G. C., Galione, A., & Terrar, D. A. (2011). NAADP influences excitation-contraction coupling by releasing calcium from lysosomes in atrial myocytes. *Cell Calcium, 50*(5), 449–458. http://dx.doi.org/10.1016/j.ceca.2011.07.007.
Csordas, G., Varnai, P., Golenar, T., Roy, S., Purkins, G., Schneider, T. G., … Hajnoczky, G. (2010). Imaging interorganelle contacts and local calcium dynamics at the ER-mitochondrial interface. *Molecular Cell, 39*(1), 121–132. http://dx.doi.org/10.1016/j.molcel.2010.06.029.
Dabertrand, F., Nelson, M. T., & Brayden, J. E. (2012). Acidosis dilates brain parenchymal arterioles by conversion of calcium waves to sparks to activate BK channels. *Circulation Research, 110*(2), 285–294. http://dx.doi.org/10.1161/CIRCRESAHA.111.258145.
Del Castillo, J., & Katz, B. (1954). Quantal components of the end-plate potential. *The Journal of Physiology, 124*(3), 560–573.
Del Castillo, J., & Katz, B. (1956). Localization of active spots within the neuromuscular junction of the frog. *The Journal of Physiology, 132*(3), 630–649.
Demmerle, J., Koch, A. J., & Holaska, J. M. (2013). Emerin and histone deacetylase 3 (HDAC3) cooperatively regulate expression and nuclear positions of MyoD, Myf5, and Pax7 genes during myogenesis. *Chromosome Research, 21*(8), 765–779. http://dx.doi.org/10.1007/s10577-013-9381-9.
Devine, C. E., Somlyo, A. V., & Somlyo, A. P. (1972). Sarcoplasmic reticulum and excitation-contraction coupling in mammalian smooth muscles. *The Journal of Cell Biology, 52*(3), 690–718.
Dipp, M., & Evans, A. M. (2001). Cyclic ADP-ribose is the primary trigger for hypoxic pulmonary vasoconstriction in the rat lung in situ. *Circulation Research, 89*(1), 77–83.
Dipp, M., Nye, P. C., & Evans, A. M. (2001). Hypoxic release of calcium from the sarcoplasmic reticulum of pulmonary artery smooth muscle. *American Journal of Physiology. Lung Cellular and Molecular Physiology, 281*(2), L318–L325.
Dode, L., Andersen, J. P., Leslie, N., Dhitavat, J., Vilsen, B., & Hovnanian, A. (2003). Dissection of the functional differences between sarco(endo)plasmic reticulum Ca2+-ATPase (SERCA) 1 and 2 isoforms and characterization of Darier disease (SERCA2) mutants by

steady-state and transient kinetic analyses. *The Journal of Biological Chemistry*, *278*(48), 47877–47889. http://dx.doi.org/10.1074/jbc.M306784200. M306784200 [pii].

Drago, I., De Stefani, D., Rizzuto, R., & Pozzan, T. (2012). Mitochondrial Ca2+ uptake contributes to buffering cytoplasmic Ca2+ peaks in cardiomyocytes. *Proceedings of the National Academy of Sciences of the United States of America*, *109*(32), 12986–12991. http://dx.doi.org/10.1073/pnas.1210718109.

Echevarria, W., Leite, M. F., Guerra, M. T., Zipfel, W. R., & Nathanson, M. H. (2003). Regulation of calcium signals in the nucleus by a nucleoplasmic reticulum. *Nature Cell Biology*, *5*(5), 440–446. http://dx.doi.org/10.1038/ncb980. ncb980 [pii].

Eggermont, J. A., Wuytack, F., Verbist, J., & Casteels, R. (1990). Expression of endoplasmic-reticulum Ca2(+)-pump isoforms and of phospholamban in pig smooth-muscle tissues. *The Biochemical Journal*, *271*(3), 649–653.

Evans, A. M. (2009). The role of intracellular ion channels in regulating cytoplasmic calcium in pulmonary arterial smooth muscle: Which store and where? *Advances in Experimental Medicine and Biology*, *661*, 57–76. http://dx.doi.org/10.1007/978-1-60761-500-2_4.

Evans, A. M. (2013). From contraction to gene expression: Function-specific calcium signals are delivered by the strategic positioning of calcium pumps and release channels within membrane–membrane nanojunctions of the sarcoplasmic reticulum. In *Proceedings of the Physiological Society, 37th Congress of IUPS (Birmingham, UK) SA402*.

Evans, A. M., Hardie, D. G., Peers, C., Wyatt, C. N., Viollet, B., Kumar, P., … Ogunbayo, O. (2009). Ion channel regulation by AMPK: The route of hypoxia-response coupling in thecarotid body and pulmonary artery. *Annals of the New York Academy of Sciences*, *1177*, 89–100.

Evans, A. M., Wyatt, C. N., Kinnear, N. P., Clark, J. H., & Blanco, E. A. (2005). Pyridine nucleotides and calcium signalling in arterial smooth muscle: From cell physiology to pharmacology. *Pharmacology & Therapeutics*, *107*(3), 286–313.

Fameli, N., Kuo, K. H., & van Breemen, C. (2009). A model for the generation of localized transient [Na+] elevations in vascular smooth muscle. *Biochemical and Biophysical Research Communications*, *389*(3), 461–465. http://dx.doi.org/10.1016/j.bbrc.2009.08.166. S0006-291X(09)01770-7 [pii].

Fameli, N., Ogunbayo, O. A., van Breemen, C., & Evans, A. M. (2014). Cytoplasmic nanojunctions between lysosomes and sarcoplasmic reticulum are required for specific calcium signaling. *F1000Research*, *3*, 93. http://dx.doi.org/10.12688/f1000research.3720.1.

Fameli, N., van Breemen, C., & Kuo, K. H. (2007). A quantitative model for linking Na+/Ca2+ exchanger to SERCA during refilling of the sarcoplasmic reticulum to sustain [Ca2+] oscillations in vascular smooth muscle. *Cell Calcium*, *42*(6), 565–575. http://dx.doi.org/10.1016/j.ceca.2007.02.001. S0143-4160(07)00028-0 [pii].

Fatt, P., & Katz, B. (1950). Membrane potentials at the motor end-plate. *The Journal of Physiology*, *111*(1–2), 46p–47p.

Fatt, P., & Katz, B. (1951). An analysis of the end-plate potential recorded with an intracellular electrode. *The Journal of Physiology*, *115*(3), 320–370.

Feldmeyer, D., Melzer, W., Pohl, B., & Zollner, P. (1990). Fast gating kinetics of the slow Ca2+ current in cut skeletal muscle fibres of the frog. *The Journal of Physiology*, *425*, 347–367.

Franzini-Armstrong, C., & Porter, K. R. (1964). Sarcolemmal invaginations constituting the T system in fish muscle fibers. *Journal of Cell Biology*, *2*, 675–696.

Franzini-Armstrong, C., & Protasi, F. (1997). Ryanodine receptors of striated muscles: A complex channel capable of multiple interactions. *Physiological Reviews*, 77(3), 699–729.

Franzini-Armstrong, C., Protasi, F., & Ramesh, V. (1999). Shape, size, and distribution of Ca(2+) release units and couplons in skeletal and cardiac muscles. *Biophysical Journal*,

77(3), 1528–1539. http://dx.doi.org/10.1016/S0006-3495(99)77000-1. S0006-3495 (99)77000-1 [pii].

Fricker, M., Hollinshead, M., White, N., & Vaux, D. (1997). Interphase nuclei of many mammalian cell types contain deep, dynamic, tubular membrane-bound invaginations of the nuclear envelope. *The Journal of Cell Biology*, *136*(3), 531–544.

Gabella, G. (1971). Caveolae intracellulares and sarcoplasmic reticulum in smooth muscle. *Journal of Cell Science*, *8*(3), 601–609.

Gerasimenko, O. V., Gerasimenko, J. V., Tepikin, A. V., & Petersen, O. H. (1995). ATP-dependent accumulation and inositol trisphosphate- or cyclic ADP-ribose-mediated release of Ca2+ from the nuclear envelope. *Cell*, *80*(3), 439–444. doi: 0092-8674(95) 90494-8 [pii].

Gilbert, G., Ducret, T., Marthan, R., Savineau, J. P., & Quignard, J. F. (2014). Stretch-induced Ca2+ signalling in vascular smooth muscle cells depends on Ca2+ store segregation. *Cardiovascular Research*, *103*(2), 313–323. http://dx.doi.org/10.1093/cvr/cvu069.

Gilchrist, J. S., Belcastro, A. N., & Katz, S. (1992). Intraluminal Ca2+ dependence of Ca2+ and ryanodine-mediated regulation of skeletal muscle sarcoplasmic reticulum Ca2+ release. *The Journal of Biological Chemistry*, *267*(29), 20850–20856.

Ginsborg, B. L., House, C. R., & Mitchell, M. R. (1980a). A calcium-readmission response recorded from Nauphoeta salivary gland acinar cells. *The Journal of Physiology*, *304*, 437–447.

Ginsborg, B. L., House, C. R., & Mitchell, M. R. (1980b). On the role of calcium in the electrical responses of cockroach salivary gland cells to dopamine. *The Journal of Physiology*, *303*, 325–335.

Gironacci, M. M., Valera, M. S., Yujnovsky, I., & Pena, C. (2004). Angiotensin-(1–7) inhibitory mechanism of norepinephrine release in hypertensive rats. *Hypertension*, *44*(5), 783–787. http://dx.doi.org/10.1161/01.HYP.0000143850.73831.9d. 01.HYP.0000143850.73831. 9d [pii].

Golovina, V. A., & Blaustein, M. P. (1997). Spatially and functionally distinct Ca2+ stores in sarcoplasmic and endoplasmic reticulum. *Science*, *275*(5306), 1643–1648.

Gomez, M. F., Stevenson, A. S., Bonev, A. D., Hill-Eubanks, D. C., & Nelson, M. T. (2002). Opposing actions of inositol 1,4,5-trisphosphate and ryanodine receptors on nuclear factor of activated T-cells regulation in smooth muscle. *The Journal of Biological Chemistry*, *277*(40), 37756–37764. http://dx.doi.org/10.1074/jbc.M203596200. M203596200 [pii].

Goncharov, D. A., Kudryashova, T. V., Ziai, H., Ihida-Stansbury, K., DeLisser, H., Krymskaya, V. P., ... Goncharova, E. A. (2014). Mammalian target of rapamycin complex 2 (mTORC2) coordinates pulmonary artery smooth muscle cell metabolism, proliferation, and survival in pulmonary arterial hypertension. *Circulation*, *129*(8), 864–874. http://dx.doi.org/10.1161/CIRCULATIONAHA.113.004581.

Gordienko, D. V., & Bolton, T. B. (2002). Crosstalk between ryanodine receptors and IP(3) receptors as a factor shaping spontaneous Ca(2+)-release events in rabbit portal vein myocytes. *The Journal of Physiology*, *542*(Pt. 3), 743–762.

Gozuacik, D., & Kimchi, A. (2004). Autophagy as a cell death and tumor suppressor mechanism. *Oncogene*, *23*(16), 2891–2906. http://dx.doi.org/10.1038/sj.onc.1207521.

Grayson, T. H., Haddock, R. E., Murray, T. P., Wojcikiewicz, R. J., & Hill, C. E. (2004). Inositol 1,4,5-trisphosphate receptor subtypes are differentially distributed between smooth muscle and endothelial layers of rat arteries. *Cell Calcium*, *36*(6), 447–458. http://dx.doi.org/10.1016/j.ceca.2004.04.005.

Gyorke, I., & Gyorke, S. (1998). Regulation of the cardiac ryanodine receptor channel by luminal Ca2+ involves luminal Ca2+ sensing sites. *Biophysical Journal*, *75*(6), 2801–2810.

Harr, J. C., Luperchio, T. R., Wong, X., Cohen, E., Wheelan, S. J., & Reddy, K. L. (2015). Directed targeting of chromatin to the nuclear lamina is mediated by chromatin state and

A-type lamins. *The Journal of Cell Biology, 208*(1), 33–52. http://dx.doi.org/10.1083/jcb.201405110.

Hauser, K., & Barth, A. (2007). Side-chain protonation and mobility in the sarcoplasmic reticulum Ca2+-ATPase: Implications for proton countertransport and Ca2+ release. *Biophysical Journal, 93*(9), 3259–3270.

Herrmann-Frank, A., Darling, E., & Meissner, G. (1991). Functional characterization of the Ca(2+)-gated Ca2+ release channel of vascular smooth muscle sarcoplasmic reticulum. *Pflügers Archiv, 418*(4), 353–359.

Hill-Eubanks, D. C., Werner, M. E., Heppner, T. J., & Nelson, M. T. (2011). Calcium signaling in smooth muscle. *Cold Spring Harbor Perspectives in Biology, 3*(9), a004549. http://dx.doi.org/10.1101/cshperspect.a004549. cshperspect.a004549 [pii].

Himpens, B., De Smedt, H., & Casteels, R. (1992a). Kinetics of nucleocytoplasmic Ca2+ transients in DDT1 MF-2 smooth muscle cells. *The American Journal of Physiology, 263*(5 Pt. 1), C978–C985.

Himpens, B., De Smedt, H., Droogmans, G., & Casteels, R. (1992b). Differences in regulation between nuclear and cytoplasmic Ca2+ in cultured smooth muscle cells. *The American Journal of Physiology, 263*(1 Pt. 1), C95–C105.

Hirose, K., Kadowaki, S., & Iino, M. (1998). Allosteric regulation by cytoplasmic Ca2+ and IP3 of the gating of IP3 receptors in permeabilized guinea pig vascular smooth muscle cells. *The Journal of Physiology, 506*(Pt. 2), 407–414.

Hirose, M., Stuyvers, B., Dun, W., Ter Keurs, H., & Boyden, P. A. (2008). Wide long lasting perinuclear Ca2+ release events generated by an interaction between ryanodine and IP3 receptors in canine Purkinje cells. *Journal of Molecular and Cellular Cardiology, 45*(2), 176–184. http://dx.doi.org/10.1016/j.yjmcc.2008.05.008. S0022-2828(08)00419-7 [pii].

Iino, M., Kasai, H., & Yamazawa, T. (1994). Visualization of neural control of intracellular Ca2+ concentration in single vascular smooth muscle cells in situ. *The EMBO Journal, 13*(21), 5026–5031.

Iino, M., Kobayashi, T., & Endo, M. (1988). Use of ryanodine for functional removal of the calcium store in smooth muscle cells of the guinea-pig. *Biochemical and Biophysical Research Communications, 152*(1), 417–422.

Inesi, G., Lewis, D., Toyoshima, C., Hirata, A., & de Meis, L. (2008). Conformational fluctuations of the Ca2+-ATPase in the native membrane environment. Effects of pH, temperature, catalytic substrates, and thapsigargin. *The Journal of Biological Chemistry, 283*(2), 1189–1196.

Ishibashi, K., Suzuki, M., & Imai, M. (2000). Molecular cloning of a novel form (two-repeat) protein related to voltage-gated sodium and calcium channels. *Biochemical and Biophysical Research Communications, 270*(2), 370–376.

Janiak, R., Wilson, S. M., Montague, S., & Hume, J. R. (2001). Heterogeneity of calcium stores and elementary release events in canine pulmonary arterial smooth muscle cells. *American Journal of Physiology. Cell Physiology, 280*(1), C22–C33.

Jensen, A. M., Sorensen, T. L., Olesen, C., Moller, J. V., & Nissen, P. (2006). Modulatory and catalytic modes of ATP binding by the calcium pump. *The EMBO Journal, 25*(11), 2305–2314.

Jha, A., Ahuja, M., Patel, S., Brailoiu, E., & Muallem, S. (2014). Convergent regulation of the lysosomal two-pore channel-2 by Mg2+, NAADP, PI(3,5)P2 and multiple protein kinases. *The EMBO Journal, 33*(5), 501–511. http://dx.doi.org/10.1002/embj.201387035.

Jmoudiak, M., & Futerman, A. H. (2005). Gaucher disease: Pathological mechanisms and modern management. *British Journal of Haematology, 129*(2), 178–188.

Kato, K., Okamura, K., Hatta, M., Morita, H., Kajioka, S., Naito, S., & Yamazaki, J. (2013). Involvement of IP3-receptor activation in endothelin-1-induced Ca(2+) influx in rat pulmonary small artery. *European Journal of Pharmacology, 720*(1–3), 255–263. http://dx.doi.org/10.1016/j.ejphar.2013.09.076.

Kinnear, N. P., Boittin, F. X., Thomas, J. M., Galione, A., & Evans, A. M. (2004). Lysosome-sarcoplasmic reticulum junctions. A trigger zone for calcium signaling by nicotinic acid adenine dinucleotide phosphate and endothelin-1. *The Journal of Biological Chemistry*, *279*(52), 54319–54326.

Kinnear, N. P., Wyatt, C. N., Clark, J. H., Calcraft, P. J., Fleischer, S., Jeyakumar, L. H., … Evans, A. M. (2008). Lysosomes co-localize with ryanodine receptor subtype 3 to form a trigger zone for calcium signalling by NAADP in rat pulmonary arterial smooth muscle. *Cell Calcium*, *44*(2), 190–201.

Klionsky, D. J., Elazar, Z., Seglen, P. O., & Rubinsztein, D. C. (2008). Does bafilomycin A1 block the fusion of autophagosomes with lysosomes? *Autophagy*, *4*(7), 849–850.

Kovsan, J., Bashan, N., Greenberg, A. S., & Rudich, A. (2010). Potential role of autophagy in modulation of lipid metabolism. *American Journal of Physiology. Endocrinology and Metabolism*, *298*(1), E1–E7. http://dx.doi.org/10.1152/ajpendo.00562.2009.

Kur, J., Bankhead, P., Scholfield, C. N., Curtis, T. M., & McGeown, J. G. (2013). Ca(2+) sparks promote myogenic tone in retinal arterioles. *British Journal of Pharmacology*, *168*(7), 1675–1686. http://dx.doi.org/10.1111/bph.12044.

Kushmerick, M. J., & Podolsky, R. J. (1969). Ionic mobility in muscle cells. *Science*, *166*(3910), 1297–1298.

Lara Lda, S., Cavalcante, F., Axelband, F., De Souza, A. M., Lopes, A. G., & Caruso-Neves, C. (2006). Involvement of the Gi/o/cGMP/PKG pathway in the AT2-mediated inhibition of outer cortex proximal tubule Na+-ATPase by Ang-(1–7). *The Biochemical Journal*, *395*(1), 183–190. http://dx.doi.org/10.1042/BJ20051455. BJ20051455 [pii].

Leblanc, N., Forrest, A. S., Ayon, R. J., Wiwchar, M., Angermann, J. E., Pritchard, H. A., … Greenwood, I. A. (2015). Molecular and functional significance of Ca(2+)-activated Cl(−) channels in pulmonary arterial smooth muscle. *Pulmonary Circulation*, *5*(2), 244–268. http://dx.doi.org/10.1086/680189.

Lee, H. C. (2004). Multiplicity of Ca2+ messengers and Ca2+ stores: A perspective from cyclic ADP-ribose and NAADP. *Current Molecular Medicine*, *4*(3), 227–237.

Lee, C., Chen, D., & Katz, R. L. (1977). Characteristics of nondepolarizing neuromuscular block: (I) post-junctional block by alpha-bungarotoxin. *Canadian Anaesthetists' Society Journal*, *24*(2), 212–219.

Lee, C. H., Poburko, D., Kuo, K. H., Seow, C. Y., & van Breemen, C. (2002). Ca(2+) oscillations, gradients, and homeostasis in vascular smooth muscle. *American Journal of Physiology. Heart and Circulatory Physiology*, *282*(5), H1571–H1583. http://dx.doi.org/10.1152/ajpheart.01035.2001.

Lee, C. H., Poburko, D., Sahota, P., Sandhu, J., Ruehlmann, D. O., & van Breemen, C. (2001). The mechanism of phenylephrine-mediated [Ca(2+)](i) oscillations underlying tonic contraction in the rabbit inferior vena cava. *The Journal of Physiology*, *534*(Pt. 3), 641–650. doi: PHY_12339 [pii].

Lesh, R. E., Nixon, G. F., Fleischer, S., Airey, J. A., Somlyo, A. P., & Somlyo, A. V. (1998). Localization of ryanodine receptors in smooth muscle. *Circulation Research*, *82*(2), 175–185.

Li, P., & Chen, S. R. (2001). Molecular basis of Ca(2)+activation of the mouse cardiac Ca(2)+release channel (ryanodine receptor). *The Journal of General Physiology*, *118*(1), 33–44.

Lifshitz, L. M., Carmichael, J. D., Lai, F. A., Sorrentino, V., Bellve, K., Fogarty, K. E., & ZhuGe, R. (2011). Spatial organization of RYRs and BK channels underlying the activation of STOCs by Ca(2+) sparks in airway myocytes. *The Journal of General Physiology*, *138*(2), 195–209. http://dx.doi.org/10.1085/jgp.201110626. jgp.201110626 [pii].

Lin, P. H., Duann, P., Komazaki, S., Park, K. H., Li, H., Sun, M., … Ma, J. (2015). Lysosomal two-pore channel subtype 2 (TPC2) regulates skeletal muscle autophagic signaling. *The Journal of Biological Chemistry, 290*(6), 3377–3389. http://dx.doi.org/10.1074/jbc.M114.608471.

Liu, G. C., Oudit, G. Y., Fang, F., Zhou, J., & Scholey, J. W. (2012). Angiotensin-(1–7)-induced activation of ERK1/2 is cAMP/protein kinase A-dependent in glomerular mesangial cells. *American Journal of Physiology. Renal Physiology, 302*(6), F784–F790. http://dx.doi.org/10.1152/ajprenal.00455.2011. ajprenal.00455.2011 [pii].

Liu, X. R., Zhang, M. F., Yang, N., Liu, Q., Wang, R. X., Cao, Y. N., ... Lin, M. J. (2012). Enhanced store-operated Ca(2) + entry and TRPC channel expression in pulmonary arteries of monocrotaline-induced pulmonary hypertensive rats. *American Journal of Physiology. Cell Physiology, 302*(1), C77–C87. http://dx.doi.org/10.1152/ajpcell.00247.2011.

Lloyd-Evans, E., Morgan, A. J., He, X., Smith, D. A., Elliot-Smith, E., Sillence, D. J., ... Platt, F. M. (2008). Niemann–Pick disease type C1 is a sphingosine storage disease that causes deregulation of lysosomal calcium. *Nature Medicine, 14*(11), 1247–1255.

Lu, Y., Hao, B. X., Graeff, R., Wong, C. W., Wu, W. T., & Yue, J. (2013). Two pore channel 2 (TPC2) inhibits autophagosomal-lysosomal fusion by alkalinizing lysosomal pH. *The Journal of Biological Chemistry, 288*(33), 24247–24263. http://dx.doi.org/10.1074/jbc.M113.484253.

Lu, W., Wang, J., Peng, G., Shimoda, L. A., & Sylvester, J. T. (2009). Knockdown of stromal interaction molecule 1 attenuates store-operated Ca2+ entry and Ca2+ responses to acute hypoxia in pulmonary arterial smooth muscle. *American Journal of Physiology. Lung Cellular and Molecular Physiology, 297*(1), L17–L25. http://dx.doi.org/10.1152/ajplung.00063.2009.

Lu, W., Wang, J., Shimoda, L. A., & Sylvester, J. T. (2008). Differences in STIM1 and TRPC expression in proximal and distal pulmonary arterial smooth muscle are associated with differences in Ca2+ responses to hypoxia. *American Journal of Physiology Lung Cellular and Molecular Physiology, 295*(1), L104–L113. http://dx.doi.org/10.1152/ajplung.00058.2008.00058.2008 [pii].

Luzio, J. P., Rous, B. A., Bright, N. A., Pryor, P. R., Mullock, B. M., & Piper, R. C. (2000). Lysosome-endosome fusion and lysosome biogenesis. *Journal of Cell Science, 113*(Pt. 9), 1515–1524.

Ma, J., Gonzalez, A., & Chen, R. (1996). Fast activation of dihydropyridine-sensitive calcium channels of skeletal muscle. Multiple pathways of channel gating. *The Journal of General Physiology, 108*(3), 221–232.

Ma, H., Zhong, L., Inesi, G., Fortea, I., Soler, F., & Fernandez-Belda, F. (1999). Overlapping effects of S3 stalk segment mutations on the affinity of Ca2+-ATPase (SERCA) for thapsigargin and cyclopiazonic acid. *Biochemistry, 38*(47), 15522–15527.

Magnier, C., Papp, B., Corvazier, E., Bredoux, R., Wuytack, F., Eggermont, J., ... Enouf, J. (1992). Regulation of sarco-endoplasmic reticulum Ca(2+)-ATPases during platelet-derived growth factor-induced smooth muscle cell proliferation. *The Journal of Biological Chemistry, 267*(22), 15808–15815.

Manunta, M., Rossi, D., Simeoni, I., Butelli, E., Romanin, C., Sorrentino, V., & Schindler, H. (2000). ATP-induced activation of expressed RyR3 at low free calcium. *FEBS Letters, 471*(2–3), 256–260.

Marius, P., Guerra, M. T., Nathanson, M. H., Ehrlich, B. E., & Leite, M. F. (2006). Calcium release from ryanodine receptors in the nucleoplasmic reticulum. *Cell Calcium, 39*(1), 65–73. http://dx.doi.org/10.1016/j.ceca.2005.09.010. S0143-4160(05)00186-7 [pii].

McCarron, J. G., & Olson, M. L. (2008). A single luminally continuous sarcoplasmic reticulum with apparently separate Ca2+ stores in smooth muscle. *The Journal of Biological Chemistry, 283*(11), 7206–7218. http://dx.doi.org/10.1074/jbc.M708923200.M708923200 [pii].

McIntyre, R. C., Jr., Banerjee, A., Hahn, A. R., Agrafojo, J., & Fullerton, D. A. (1995). Selective inhibition of cyclic adenosine monophosphate-mediated pulmonary vasodilation by acute hypoxia. *Surgery, 117*(3), 314–318.

Moncoq, K., Trieber, C. A., & Young, H. S. (2007). The molecular basis for cyclopiazonic acid inhibition of the sarcoplasmic reticulum calcium pump. *The Journal of Biological Chemistry*, *282*(13), 9748–9757. http://dx.doi.org/10.1074/jbc.M611653200.

Moore, E. D., Voigt, T., Kobayashi, Y. M., Isenberg, G., Fay, F. S., Gallitelli, M. F., & Franzini-Armstrong, C. (2004). Organization of Ca2+ release units in excitable smooth muscle of the guinea-pig urinary bladder. *Biophysical Journal*, *87*(3), 1836–1847. http://dx.doi.org/10.1529/biophysj.104.044123.

Morgan, A. J., Davis, L. C., Wagner, S. K., Lewis, A. M., Parrington, J., Churchill, G. C., & Galione, A. (2013). Bidirectional Ca(2)(+) signaling occurs between the endoplasmic reticulum and acidic organelles. *The Journal of Cell Biology*, *200*(6), 789–805. http://dx.doi.org/10.1083/jcb.201204078.

Morgan, A. J., & Galione, A. (2008). Investigating cADPR and NAADP in intact and broken cell preparations. *Methods*, *46*(3), 194–203.

Morgan, A. J., & Galione, A. (2014). Two-pore channels (TPCs): Current controversies. *Bioessays*, *36*(2), 173–183. http://dx.doi.org/10.1002/bies.201300118.

Morio, Y., & McMurtry, I. F. (2002). Ca(2+) release from ryanodine-sensitive store contributes to mechanism of hypoxic vasoconstriction in rat lungs. *Journal of Applied Physiology*, *92*(2), 527–534.

Nakai, J., Dirksen, R. T., Nguyen, H. T., Pessah, I. N., Beam, K. G., & Allen, P. D. (1996). Enhanced dihydropyridine receptor channel activity in the presence of ryanodine receptor. *Nature*, *380*(6569), 72–75. http://dx.doi.org/10.1038/380072a0.

Nazer, M. A., & van Breemen, C. (1998a). Functional linkage of Na(+)-Ca2+ exchange and sarcoplasmic reticulum Ca2+ release mediates Ca2+ cycling in vascular smooth muscle. *Cell Calcium*, *24*(4), 275–283. doi: S0143-4160(98)90051-3 [pii].

Nazer, M. A., & Van Breemen, C. (1998b). A role for the sarcoplasmic reticulum in Ca2+ extrusion from rabbit inferior vena cava smooth muscle. *The American Journal of Physiology*, *274*(1 Pt. 2), H123–H131.

Nelson, M. T., Cheng, H., Rubart, M., Santana, L. F., Bonev, A. D., Knot, H. J., & Lederer, W. J. (1995). Relaxation of arterial smooth muscle by calcium sparks. *Science*, *270*(5236), 633–637.

Neylon, C. B., Hoyland, J., Mason, W. T., & Irvine, R. F. (1990). Spatial dynamics of intracellular calcium in agonist-stimulated vascular smooth muscle cells. *The American Journal of Physiology*, *259*(4 Pt. 1), C675–C686.

Neylon, C. B., Richards, S. M., Larsen, M. A., Agrotis, A., & Bobik, A. (1995). Multiple types of ryanodine receptor/Ca2+ release channels are expressed in vascular smooth muscle. *Biochemical and Biophysical Research Communications*, *215*(3), 814–821.

Ng, L. C., O'Neill, K. G., French, D., Airey, J. A., Singer, C. A., Tian, H., … Hume, J. R. (2012). TRPC1 and Orai1 interact with STIM1 and mediate capacitative Ca(2+) entry caused by acute hypoxia in mouse pulmonary arterial smooth muscle cells. *American Journal of Physiology. Cell Physiology*, *303*(11), C1156–C1172. http://dx.doi.org/10.1152/ajpcell.00065.2012.

Nixon, G. F., Mignery, G. A., & Somlyo, A. V. (1994). Immunogold localization of inositol 1,4,5-trisphosphate receptors and characterization of ultrastructural features of the sarcoplasmic reticulum in phasic and tonic smooth muscle. *Journal of Muscle Research and Cell Motility*, *15*(6), 682–700.

Noori, S., Acherman, R., Siassi, B., Luna, C., Ebrahimi, M., Pavlova, Z., & Ramanathan, R. (2002). A rare presentation of Pompe disease with massive hypertrophic cardiomyopathy at birth. *Journal of Perinatal Medicine*, *30*(6), 517–521.

Odermatt, A., Becker, S., Khanna, V. K., Kurzydlowski, K., Leisner, E., Pette, D., & MacLennan, D. H. (1998). Sarcolipin regulates the activity of SERCA1, the fast-twitch skeletal muscle sarcoplasmic reticulum Ca2+-ATPase. *The Journal of Biological Chemistry*, *273*(20), 12360–12369.

Odermatt, A., Kurzydlowski, K., & MacLennan, D. H. (1996). The vmax of the Ca2+-ATPase of cardiac sarcoplasmic reticulum (SERCA2a) is not altered by Ca2+/calmodulin-dependent phosphorylation or by interaction with phospholamban. *The Journal of Biological Chemistry*, *271*(24), 14206–14213.

Ogawa, A., Firth, A. L., Smith, K. A., Maliakal, M. V., & Yuan, J. X. (2012). PDGF enhances store-operated Ca2+ entry by upregulating STIM1/Orai1 via activation of Akt/mTOR in human pulmonary arterial smooth muscle cells. *American Journal of Physiology. Cell Physiology*, *302*(2), C405–C411. http://dx.doi.org/10.1152/ajpcell.00337.2011.

Ogunbayo, O. A., Ma, J., Zhu, M. X., & Evans, A. M. (2015). Lysosome-ER coupling supported by two pore channel 2 is required for Nicotinic acid adenine dinucleotide phosphate-induced global calcium waves in pulmonary arterial myocytes. *Proceedings of the Physiological Society*, *34*, PC134.

Ogunbayo, O. A., Zhu, Y., Rossi, D., Sorrentino, V., Ma, J., Zhu, M. X., & Evans, A. M. (2011). Cyclic adenosine diphosphate ribose activates ryanodine receptors, whereas NAADP activates two-pore domain channels. *The Journal of Biological Chemistry*, *286*(11), 9136–9140. http://dx.doi.org/10.1074/jbc.M110.202002. M110.202002 [pii].

Ogunbayo, O. A., Zhu, Y., Shen, B., Agbani, E., Li, J., Ma, J., … Evans, A. M. (2015). Organelle-specific subunit interactions of the vertebrate two-pore channel family. *The Journal of Biological Chemistry*, *290*(2), 1086–1095. http://dx.doi.org/10.1074/jbc.M114.610493.

Olson, M. L., Chalmers, S., & McCarron, J. G. (2010). Mitochondrial Ca2+ uptake increases Ca2+ release from inositol 1,4,5-trisphosphate receptor clusters in smooth muscle cells. *The Journal of Biological Chemistry*, *285*(3), 2040–2050. http://dx.doi.org/10.1074/jbc.M109.027094.

Panfoli, I., Burlando, B., & Viarengo, A. (1999). Cyclic ADP-ribose-dependent Ca2+ release is modulated by free [Ca2+] in the scallop sarcoplasmic reticulum. *Biochemical and Biophysical Research Communications*, *257*(1), 57–62.

Perez, G. J., Bonev, A. D., Patlak, J. B., & Nelson, M. T. (1999). Functional coupling of ryanodine receptors to KCa channels in smooth muscle cells from rat cerebral arteries. *The Journal of General Physiology*, *113*(2), 229–238.

Perez, J. F., & Sanderson, M. J. (2005). The frequency of calcium oscillations induced by 5-HT, ACH, and KCl determine the contraction of smooth muscle cells of intrapulmonary bronchioles. *The Journal of General Physiology*, *125*(6), 535–553. http://dx.doi.org/10.1085/jgp.200409216. jgp.200409216 [pii].

Pitt, S. J., Funnell, T. M., Sitsapesan, M., Venturi, E., Rietdorf, K., Ruas, M., … Sitsapesan, R. (2010). TPC2 is a novel NAADP-sensitive Ca2+ release channel, operating as a dual sensor of luminal pH and Ca2+. *The Journal of Biological Chemistry*, *285*(45), 35039–35046. http://dx.doi.org/10.1074/jbc.M110.156927. M110.156927 [pii].

Poburko, D., Fameli, N., Kuo, K. H., & van Breemen, C. (2008). Ca2+ signaling in smooth muscle: TRPC6, NCX and LNats in nanodomains. *Channels (Austin, Tex)*, *2*(1), 10–12. doi: 6053 [pii].

Poburko, D., Kuo, K. H., Dai, J., Lee, C. H., & van Breemen, C. (2004). Organellar junctions promote targeted Ca2+ signaling in smooth muscle: Why two membranes are better than one. *Trends in Pharmacological Sciences*, *25*(1), 8–15. http://dx.doi.org/10.1016/j.tips.2003.10.011.S0165-6147(03)00353-5 [pii].

Poburko, D., Liao, C. H., Lemos, V. S., Lin, E., Maruyama, Y., Cole, W. C., & van Breemen, C. (2007). Transient receptor potential channel 6-mediated, localized cytosolic [Na+] transients drive Na+/Ca2+ exchanger-mediated Ca2+ entry in purinergically stimulated aorta smooth muscle cells. *Circulation Research*, *101*(10), 1030–1038. http://dx.doi.org/10.1161/CIRCRESAHA.107.155531.

Poburko, D., Potter, K., van Breemen, E., Fameli, N., Liao, C. H., Basset, O., … van Breemen, C. (2006). Mitochondria buffer NCX-mediated Ca2+-entry and limit its

diffusion into vascular smooth muscle cells. *Cell Calcium, 40*(4), 359–371. http://dx.doi.org/10.1016/j.ceca.2006.04.031. S0143-4160(06)00069-8 [pii].
Protasi, F., Paolini, C., Canato, M., Reggiani, C., & Quarta, M. (2012). Lessons from calsequestrin-1 ablation in vivo: Much more than a Ca(2+) buffer after all. *Journal of Muscle Research and Cell Motility, 32*(4–5), 257–270. http://dx.doi.org/10.1007/s10974-011-9277-2.
Pucovsky, V., & Bolton, T. B. (2006). Localisation, function and composition of primary Ca(2+) spark discharge region in isolated smooth muscle cells from guinea-pig mesenteric arteries. *Cell Calcium, 39*(2), 113–129. http://dx.doi.org/10.1016/j.ceca.2005.10.002.
Pucovsky, V., Gordienko, D. V., & Bolton, T. B. (2002). Effect of nitric oxide donors and noradrenaline on Ca2+ release sites and global intracellular Ca2+ in myocytes from guinea-pig small mesenteric arteries. *The Journal of Physiology, 539*(Pt. 1), 25–39.
Putney, J. W., Jr. (1986). A model for receptor-regulated calcium entry. *Cell Calcium,* 7(1), 1–12.
Queisser, G., Wiegert, S., & Bading, H. (2011). Structural dynamics of the cell nucleus: Basis for morphology modulation of nuclear calcium signaling and gene transcription. *Nucleus, 2*(2), 98–104. http://dx.doi.org/10.4161/nucl.2.2.15116.1949-1034-2-2-5 [pii].
Raeymaekers, L., Eggermont, J. A., Wuytack, F., & Casteels, R. (1990). Effects of cyclic nucleotide dependent protein kinases on the endoplasmic reticulum Ca2+ pump of bovine pulmonary artery. *Cell Calcium, 11*(4), 261–268. doi: 0143-4160(90)90002-C [pii].
Ramesh, V., Sharma, V. K., Sheu, S. S., & Franzini-Armstrong, C. (1998). Structural proximity of mitochondria to calcium release units in rat ventricular myocardium may suggest a role in Ca2+ sequestration. *Annals of the New York Academy of Sciences, 853*, 341–344.
Rembold, C. M., & Chen, X. L. (1998). The buffer barrier hypothesis, [Ca2+]i homogeneity, and sarcoplasmic reticulum function in swine carotid artery. *The Journal of Physiology, 513*(Pt. 2), 477–492.
Ron, I., & Horowitz, M. (2008). Intracellular cholesterol modifies the ERAD of glucocerebrosidase in Gaucher disease patients. *Molecular Genetics and Metabolism, 93*(4), 426–436.
Rosado, J. A., Diez, R., Smani, T., & Jardin, I. (2015). STIM and Orai1 variants in store-operated calcium entry. *Frontiers in Pharmacology, 6*, 325. http://dx.doi.org/10.3389/fphar.2015.00325.
Rosenbluth, J. (1962). Subsurface cisterns and their relationship to the neuronal plasma membrane. *Journal of Cell Biology, 13*, 405–421.
Rowland, A. A., & Voeltz, G. K. (2012). Endoplasmic reticulum-mitochondria contacts: Function of the junction. *Nature Reviews Molecular Cell Biology, 13*(10), 607–625. http://dx.doi.org/10.1038/nrm3440.
Ruas, M., Davis, L. C., Chen, C. C., Morgan, A. J., Chuang, K. T., Walseth, T. F., … Galione, A. (2015). Expression of Ca(2)(+)-permeable two-pore channels rescues NAADP signalling in TPC-deficient cells. *The EMBO Journal, 34*(13), 1743–1758. http://dx.doi.org/10.15252/embj.201490009.
Ruas, M., Rietdorf, K., Arredouani, A., Davis, L. C., Lloyd-Evans, E., Koegel, H., … Galione, A. (2010). Purified TPC isoforms form NAADP receptors with distinct roles for Ca(2+) signaling and endolysosomal trafficking. *Current Biology, 20*, 703–709.
Ruehlmann, D. O., Lee, C. H., Poburko, D., & van Breemen, C. (2000). Asynchronous Ca(2+) waves in intact venous smooth muscle. *Circulation Research, 86*(4), E72–E79.
Schieder, M., Rotzer, K., Bruggemann, A., Biel, M., & Wahl-Schott, C. A. (2010). Characterization of two-pore channel 2 (TPCN2)-mediated Ca2+ currents in isolated lysosomes. *The Journal of Biological Chemistry, 285*(28), 21219–21222. http://dx.doi.org/10.1074/jbc.C110.143123.
Sethi, J. K., Empson, R. M., Bailey, V. C., Potter, B. V., & Galione, A. (1997). 7-Deaza-8-bromo-cyclic ADP-ribose, the first membrane-permeant, hydrolysis-resistant cyclic ADP-ribose antagonist. *The Journal of Biological Chemistry, 272*(26), 16358–16363.

Shi, J., Ju, M., Large, W. A., & Albert, A. P. (2012). Pharmacological profile of phosphatidylinositol 3-kinases and related phosphoinositols mediating endothelin(A) receptor-operated native TRPC channels in rabbit coronary artery myocytes. *British Journal of Pharmacology, 166*, 2161–2175. http://dx.doi.org/10.1111/j.1476-5381.2012.01937.x.

Shi, J., Miralles, F., Birnbaumer, L., Large, W. A., & Albert, A. P. (2016). Store depletion induces Galphaq-mediated PLCbeta1 activity to stimulate TRPC1 channels in vascular smooth muscle cells. *The FASEB Journal, 30*(2), 702–715. http://dx.doi.org/10.1096/fj.15-280271.

Shmigol, A. V., Eisner, D. A., & Wray, S. (1999). The role of the sarcoplasmic reticulum as a Ca2+ sink in rat uterine smooth muscle cells. *The Journal of Physiology, 520*(Pt. 1), 153–163. doi: PHY_9587 [pii].

Snetkov, V. A., Knock, G. A., Baxter, L., Thomas, G. D., Ward, J. P., & Aaronson, P. I. (2006). Mechanisms of the prostaglandin F2alpha-induced rise in [Ca2+]i in rat intrapulmonary arteries. *The Journal of Physiology, 571*(Pt. 1), 147–163. http://dx.doi.org/10.1113/jphysiol.2005.101394.

Soboloff, J., Rothberg, B. S., Madesh, M., & Gill, D. L. (2012). STIM proteins: Dynamic calcium signal transducers. *Nature Reviews. Molecular Cell Biology, 13*(9), 549–565. http://dx.doi.org/10.1038/nrm3414.

Soeller, C., Crossman, D., Gilbert, R., & Cannell, M. B. (2007). Analysis of ryanodine receptor clusters in rat and human cardiac myocytes. *Proceedings of the National Academy of Sciences of the United States of America, 104*(38), 14958–14963. http://dx.doi.org/10.1073/pnas.0703016104. 0703016104 [pii].

Song, D. W., Lee, J. G., Youn, H. S., Eom, S. H., & Kim, D. H. (2011). Ryanodine receptor assembly: A novel systems biology approach to 3D mapping. *Progress in Biophysics and Molecular Biology, 105*(3), 145–161. http://dx.doi.org/10.1016/j.pbiomolbio.2010.09.021.

Spinelli, A. M., & Trebak, M. (2016). Orai channel-mediated Ca2+ signals in vascular and airway smooth muscle. *American Journal of Physiology Cell Physiology, 310*(6), C402–C413. http://dx.doi.org/10.1152/ajpcell.00355.2015.

Stern, M. D., Pizarro, G., & Rios, E. (1997). Local control model of excitation-contraction coupling in skeletal muscle. *The Journal of General Physiology, 110*(4), 415–440.

Stevenson, A. S., Gomez, M. F., Hill-Eubanks, D. C., & Nelson, M. T. (2001). NFAT4 movement in native smooth muscle. A role for differential Ca(2+) signaling. *The Journal of Biological Chemistry, 276*(18), 15018–15024. http://dx.doi.org/10.1074/jbc.M011684200M011684200 [pii].

Strelkov, S. V., Schumacher, J., Burkhard, P., Aebi, U., & Herrmann, H. (2004). Crystal structure of the human lamin A coil 2B dimer: Implications for the head-to-tail association of nuclear lamins. *Journal of Molecular Biology, 343*(4), 1067–1080. http://dx.doi.org/10.1016/j.jmb.2004.08.093.

Subedi, K. P., Paudel, O., & Sham, J. S. (2014). Detection of differentially regulated subsarcolemmal calcium signals activated by vasoactive agonists in rat pulmonary artery smooth muscle cells. *American Journal of Physiology. Cell Physiology, 306*(7), C659–C669. http://dx.doi.org/10.1152/ajpcell.00341.2013.

Suzuki, E., Nishimatsu, H., Satonaka, H., Walsh, K., Goto, A., Omata, M., … Hirata, Y. (2002). Angiotensin II induces myocyte enhancer factor 2- and calcineurin/nuclear factor of activated T cell-dependent transcriptional activation in vascular myocytes. *Circulation Research, 90*(9), 1004–1011.

Syyong, H. T., Poburko, D., Fameli, N., & van Breemen, C. (2007). ATP promotes NCX-reversal in aortic smooth muscle cells by DAG-activated Na+ entry. *Biochemical and Biophysical Research Communications, 357*(4), 1177–1182. http://dx.doi.org/10.1016/j.bbrc.2007.04.080. S0006-291X(07)00816-9 [pii].

Takahashi, Y., Watanabe, H., Murakami, M., Ono, K., Munehisa, Y., Koyama, T., … Ito, H. (2007). Functional role of stromal interaction molecule 1 (STIM1) in vascular smooth

muscle cells. *Biochemical and Biophysical Research Communications, 361*(4), 934–940. http://dx.doi.org/10.1016/j.bbrc.2007.07.096. S0006-291X(07)01578-1 [pii].
Takeda, Y., Nystoriak, M. A., Nieves-Cintron, M., Santana, L. F., & Navedo, M. F. (2011). Relationship between Ca2+ sparklets and sarcoplasmic reticulum Ca2+ load and release in rat cerebral arterial smooth muscle. *American Journal of Physiology. Heart and Circulatory Physiology, 301*(6), H2285–H2294. http://dx.doi.org/10.1152/ajpheart.00488.2011.
Tassoni, J. P., Jr., Fawaz, K. A., & Johnston, D. E. (1991). Cirrhosis and portal hypertension in a patient with adult Niemann–Pick disease. *Gastroenterology, 100*(2), 567–569.
Thorne, G. D., & Paul, R. J. (2003). Effects of organ culture on arterial gene expression and hypoxic relaxation: Role of the ryanodine receptor. *American Journal of Physiology. Cell Physiology, 284*(4), C999–C1005. http://dx.doi.org/10.1152/ajpcell.00158.200200158.2002 [pii].
Tong, W. C., Sweeney, M., Jones, C. J., Zhang, H., O'Neill, S. C., Prior, I., & Taggart, M. J. (2009). Three-dimensional electron microscopic reconstruction of intracellular organellar arrangements in vascular smooth muscle—Further evidence of nanospaces and contacts. *Journal of Cellular and Molecular Medicine, 13*(5), 995–998. http://dx.doi.org/10.1111/j.1582-4934.2009.00770.x.
Tripathy, A., & Meissner, G. (1996). Sarcoplasmic reticulum lumenal Ca2+ has access to cytosolic activation and inactivation sites of skeletal muscle Ca2+ release channel. *Biophysical Journal, 70*(6), 2600–2615.
Van Breemen, C. (1977). Calcium requirement for activation of intact aortic smooth muscle. *The Journal of Physiology, 272*(2), 317–329.
van Breemen, C., Chen, Q., & Laher, I. (1995). Superficial buffer barrier function of smooth muscle sarcoplasmic reticulum. *Trends in Pharmacological Sciences, 16*(3), 98–105. doi: S0165614700889907 [pii].
van Breemen, C., Fameli, N., & Evans, A. M. (2013). Pan-junctional sarcoplasmic reticulum in vascular smooth muscle: Nanospace Ca2+ transport for site- and function-specific Ca2+ signalling. *The Journal of Physiology, 591*(Pt. 8), 2043–2054. http://dx.doi.org/10.1113/jphysiol.2012.246348.
van Breemen, C., & Saida, K. (1989). Cellular mechanisms regulating [Ca2+]i smooth muscle. *Annual Review of Physiology, 51*, 315–329. http://dx.doi.org/10.1146/annurev.ph.51.030189.001531.
Verboomen, H., Wuytack, F., De Smedt, H., Himpens, B., & Casteels, R. (1992). Functional difference between SERCA2a and SERCA2b Ca2+ pumps and their modulation by phospholamban. *The Biochemical Journal, 286*(Pt. 2), 591–595.
Verboomen, H., Wuytack, F., Van den Bosch, L., Mertens, L., & Casteels, R. (1994). The functional importance of the extreme C-terminal tail in the gene 2 organellar Ca(2+)-transport ATPase (SERCA2a/b). *The Biochemical Journal, 303*(Pt. 3), 979–984.
von Euler, U. S., & Liljestrand, G. (1946). Observations on the pulmonary arterial blood pressure in the cat. *Acta Physiologica Scandinavica, 12*, 301–320.
Wamhoff, B. R., Bowles, D. K., & Owens, G. K. (2006). Excitation-transcription coupling in arterial smooth muscle. *Circulation Research, 98*(7), 868–878. http://dx.doi.org/10.1161/01.RES.0000216596.73005.3c. 98/7/868 [pii].
Wang, X., Zhang, X., Dong, X. P., Samie, M., Li, X., Cheng, X., … Xu, H. (2012). TPC proteins are phosphoinositide-activated sodium-selective ion channels in endosomes and lysosomes. *Cell, 151*(2), 372–383. http://dx.doi.org/10.1016/j.cell.2012.08.036.
Waybill, M. M., Yelamarty, R. V., Zhang, Y. L., Scaduto, R. C., Jr., LaNoue, K. F., Hsu, C. J., … Cheung, J. Y. (1991). Nuclear calcium gradients in cultured rat hepatocytes. *The American Journal of Physiology, 261*(1 Pt. 1), E49–E57.
Westcott, E. B., Goodwin, E. L., Segal, S. S., & Jackson, W. F. (2012). Function and expression of ryanodine receptors and inositol 1,4,5-trisphosphate receptors in smooth muscle cells of murine feed arteries and arterioles. *The Journal of Physiology, 590*(8), 1849–1869. http://dx.doi.org/10.1113/jphysiol.2011.222083.

White, C., & McGeown, J. G. (2000). Ca2+ uptake by the sarcoplasmic reticulum decreases the amplitude of depolarization-dependent [Ca2+]i transients in rat gastric myocytes. *Pflügers Archiv*, *440*(3), 488–495.
Williams, D. A., Fogarty, K. E., Tsien, R. Y., & Fay, F. S. (1985). Calcium gradients in single smooth muscle cells revealed by the digital imaging microscope using Fura-2. *Nature*, *318*(6046), 558–561.
Wilson, H. L., Dipp, M., Thomas, J. M., Lad, C., Galione, A., & Evans, A. M. (2001). ADP-ribosyl cyclase and cyclic ADP-ribose hydrolase act as a redox sensor. A primary role for cyclic ADP-ribose in hypoxic pulmonary vasoconstriction. *The Journal of Biological Chemistry*, *276*(14), 11180–11188.
Wilson, D. P., Sutherland, C., & Walsh, M. P. (2002). Ca2+ activation of smooth muscle contraction: Evidence for the involvement of calmodulin that is bound to the triton insoluble fraction even in the absence of Ca2+. *The Journal of Biological Chemistry*, *277*(3), 2186–2192. http://dx.doi.org/10.1074/jbc.M110056200.
Wray, S., & Burdyga, T. (2010). Sarcoplasmic reticulum function in smooth muscle. *Physiological Reviews*, *90*(1), 113–178. http://dx.doi.org/10.1152/physrev.00018.2008. 90/1/113 [pii].
Yamaguchi, H., Kajita, J., & Madison, J. M. (1995). Isoproterenol increases peripheral [Ca2+]i and decreases inner [Ca2+]i in single airway smooth muscle cells. *The American Journal of Physiology*, *268*(3 Pt. 1), C771–C779.
Yang, X. R., Lin, M. J., Yip, K. P., Jeyakumar, L. H., Fleischer, S., Leung, G. P., & Sham, J. S. (2005). Multiple ryanodine receptor subtypes and heterogeneous ryanodine receptor-gated Ca2+ stores in pulmonary arterial smooth muscle cells. *American Journal of Physiology. Lung Cellular and Molecular Physiology*, *289*(2), L338–L348. http://dx.doi.org/10.1152/ajplung.00328.2004.00328.2004 [pii].
Yang, D., Pan, Z., Takeshima, H., Wu, C., Nagaraj, R. Y., Ma, J., & Cheng, H. (2001). RyR3 amplifies RyR1-mediated Ca(2+)-induced Ca(2+) release in neonatal mammalian skeletal muscle. *The Journal of Biological Chemistry*, *276*(43), 40210–40214. http://dx.doi.org/10.1074/jbc.M106944200. M106944200 [pii].
Yoshikawa, A., van Breemen, C., & Isenberg, G. (1996). Buffering of plasmalemmal Ca2+ current by sarcoplasmic reticulum of guinea pig urinary bladder myocytes. *The American Journal of Physiology*, *271*(3 Pt. 1), C833–C841.
Young, R. C., Schumann, R., & Zhang, P. (2001). Intracellular calcium gradients in cultured human uterine smooth muscle: A functionally important subplasmalemmal space. *Cell Calcium*, *29*(3), 183–189. http://dx.doi.org/10.1054/ceca.2000.0182. S0143-4160(00)90182-9 [pii].
Yu, Y., Keller, S. H., Remillard, C. V., Safrina, O., Nicholson, A., Zhang, S. L., … Yuan, J. X. (2009). A functional single-nucleotide polymorphism in the TRPC6 gene promoter associated with idiopathic pulmonary arterial hypertension. *Circulation*, *119*(17), 2313–2322. http://dx.doi.org/10.1161/CIRCULATIONAHA.108.782458.
Zhang, S., Patel, H. H., Murray, F., Remillard, C. V., Schach, C., Thistlethwaite, P. A., … Yuan, J. X. (2007). Pulmonary artery smooth muscle cells from normal subjects and IPAH patients show divergent cAMP-mediated effects on TRPC expression and capacitative Ca2+ entry. *American Journal of Physiology. Lung Cellular and Molecular Physiology*, *292*(5), L1202–L1210. http://dx.doi.org/10.1152/ajplung.00214.2006.
Zhang, Y., Wang, Y., Yang, K., Tian, L., Fu, X., Wang, Y., … Wang, J. (2014). BMP4 increases the expression of TRPC and basal [Ca2+]i via the p38MAPK and ERK1/2 pathways independent of BMPRII in PASMCs. *PloS One*, *9*(12), e112695. http://dx.doi.org/10.1371/journal.pone.0112695.
Zhao, G., Neeb, Z. P., Leo, M. D., Pachuau, J., Adebiyi, A., Ouyang, K., … Jaggar, J. H. (2010). Type 1 IP3 receptors activate BKCa channels via local molecular coupling in

arterial smooth muscle cells. *The Journal of General Physiology*, *136*(3), 283–291. http://dx.doi.org/10.1085/jgp.201010453.

Zhu, M. X., Ma, J., Parrington, J., Calcraft, P. J., Galione, A., & Evans, A. M. (2010). Calcium signaling via two-pore channels: Local or global, that is the question. *American Journal of Physiology. Cell Physiology*, *298*(3), C430–C441.

Zong, X., Schieder, M., Cuny, H., Fenske, S., Gruner, C., Rotzer, K., … Wahl-Schott, C. (2009). The two-pore channel TPCN2 mediates NAADP-dependent Ca(2+)- release from lysosomal stores. *Pflügers Archiv*, *458*, 891–899.

Zuleger, N., Robson, M. I., & Schirmer, E. C. (2011). The nuclear envelope as a chromatin organizer. *Nucleus*, *2*(5), 339–349. http://dx.doi.org/10.4161/nucl.2.5.17846. http://dx.doi.org/10.4161/nucl.2.5.17846.

CHAPTER TWO

Calcium Channels in Vascular Smooth Muscle

D. Ghosh*, A.U. Syed*, M.P. Prada*, M.A. Nystoriak†, L.F. Santana*, M. Nieves-Cintrón*, M.F. Navedo*,[1]
*University of California, Davis, CA, United States
†Diabetes and Obesity Center, University of Louisville, Louisville, KY, United States
[1]Corresponding author: e-mail address: mfnavedo@ucdavis.edu

Contents

Abstract

Calcium (Ca^{2+}) plays a central role in excitation, contraction, transcription, and proliferation of vascular smooth muscle cells (VSMs). Precise regulation of intracellular Ca^{2+} concentration ($[Ca^{2+}]_i$) is crucial for proper physiological VSM function. Studies over the last several decades have revealed that VSMs express a variety of Ca^{2+}-permeable channels that orchestrate a dynamic, yet finely tuned regulation of $[Ca^{2+}]_i$. In this review, we discuss the major Ca^{2+}-permeable channels expressed in VSM and their contribution to vascular physiology and pathology.

ABBREVIATIONS

ANG II angiotensin II
2-APB 2-aminoethoxydiphenyl borate
BK_{Ca} large-conductance Ca^{2+}-activated potassium channel
Ca^{2+} calcium

Advances in Pharmacology, Volume 78
ISSN 1054-3589
http://dx.doi.org/10.1016/bs.apha.2016.08.002

$[Ca^{2+}]_i$ intracellular Ca^{2+} concentration
$[Ca^{2+}]_{mito}$ mitochondrial Ca^{2+} concentration
CaM calmodulin
ET-1 endothelin-1
GPCR G-protein-coupled receptor
IP_3R inositol-1,4,5,-trisphosphate receptor
LTCC L-type $Ca_V1.2$ channel
PKA protein kinase A
PKC protein kinase C
PKG protein kinase G
PLC phospholipase C
ROS reactive oxygen species
RyR ryanodine receptor
SM smooth muscle
SOCE store-operated Ca^{2+} entry
SR sarcoplasmic reticulum
STOC spontaneous transient outward current
TRP transient receptor potential
TTCC T-type Ca^{2+} channel
VDCC voltage-dependent Ca^{2+} channel
VSM vascular smooth muscle cells

1. INTRODUCTION

Highly coordinated control of VSM excitability is essential for proper vascular function and regulation of blood flow. Intracellular Ca^{2+} plays a pivotal role in this process. Accordingly, well-orchestrated and distinct signaling pathways allow tight regulation of $[Ca^{2+}]_i$, which, together with differential Ca^{2+} sensitivity of the contractile machinery, provide additional fine-tuning of VSM contractility. Studies over the last several decades have revealed the expression of multiple Ca^{2+}-permeable channels in VSM that coordinate a dynamic and precise control of $[Ca^{2+}]_i$, thereby playing a pivotal role in VSM physiology (Fig. 1). Changes in $[Ca^{2+}]_i$ are produced by Ca^{2+} influx through voltage-dependent and -independent plasmalemmal Ca^{2+}-permeable channels, as well as Ca^{2+} release from intracellular stores. L-type $Ca_V1.2$ channels (LTCCs) have long been considered the primary route of Ca^{2+} entry in VSM. Indeed, Ca^{2+} influx through LTCCs is the principal mediator of myogenic response, which is the intrinsic ability of VSM to contract/relax in response to changes in intraluminal pressure (Bayliss, 1902). Besides LTCC, T-type Ca^{2+} channels (TTCCs) are emerging as important

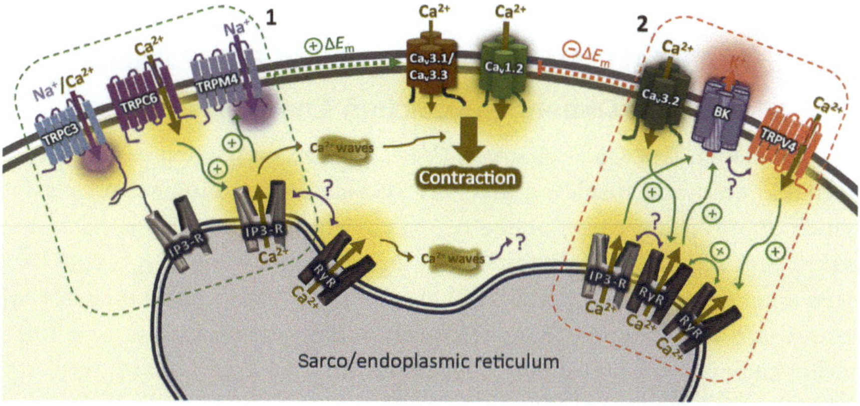

Fig. 1 Schematic representation of the interplay of major Ca^{2+}-permeable channels involved in the regulation of VSM $[Ca^{2+}]_i$ and contractility. Ca^{2+} influx predominantly through L-type $Ca_V1.2$ and to some extent T-type $Ca_V3.1/3.3$ channels promotes VSM contraction. L-type $Ca_V1.2$ and T-type $Ca_V3.1/3.3$ channel activity can be regulated, via membrane potential, by several Ca^{2+}-permeable channels serving (1) depolarizing and (2) hyperpolarizing roles, thus modulating the contractile state of VSM. The emerging role of mitochondria Ca^{2+} channels in regulation of VSM Ca^{2+} homeostasis and vascular reactivity is not depicted in this cartoon for simplicity. (+) denotes positive modulation, (−) represents negative modulation, and (?) indicates areas of uncertainty in the pathway.

contributors to myogenic tone. LTCCs also play a crucial role in excitation–transcription coupling in VSM (Amberg & Navedo, 2013). Members of the transient receptor potential (TRP) channel family, as well as Ca^{2+} release-activated channels (Orai/STIM), have also been found to contribute to regulation of VSM function. Moreover, Ca^{2+} release from intracellular stores through ryanodine receptors (RyRs) and inositol-1,4,5,-trisphosphate receptors (IP_3Rs) in the sarcoplasmic reticulum (SR) is an important contributor to $[Ca^{2+}]_i$ and VSM excitability. RyRs and IP_3Rs are also involved in VSM regulation via their communication with plasmalemmal ion channels. Recently, mitochondria have also been receiving substantial attention for their emerging relevance in VSM Ca^{2+} handling. This review presents an overview of the major Ca^{2+}-permeable channels that contribute to VSM Ca^{2+} handling and contractility, and vascular reactivity. More extensive reviews on individual channels can be found elsewhere (Amberg & Navedo, 2013; Earley & Brayden, 2015; Harraz & Welsh, 2013b; McCarron, Olson, Wilson, Sandison, & Chalmers, 2013; Narayanan, Adebiyi, & Jaggar, 2012; Navedo & Amberg, 2013).

2. PLASMALEMMAL Ca^{2+}-PERMEABLE CHANNELS

2.1 Voltage-Dependent Calcium Channels

Voltage-dependent Ca^{2+} channels (VDCCs) are widely expressed among excitable cells and display a diversity of electrophysiological properties, which allows them to influence many physiological functions (Catterall, 2011). Since their initial discovery in 1953, multiple VDCC subtypes have been characterized (Catterall, 2011). Of particular importance to this review are the LTCCs, which exhibit a large-conductance and long-lasting current with membrane depolarization, and TTCCs with tiny conductance and transient current at negative potentials. N-, P/Q-, and R-type Ca^{2+} channels have also been identified (Catterall, 2011). In the following section, we focus on LTCCs and TTCCs and their contribution to $[Ca^{2+}]_i$ in VSM.

2.1.1 L-Type $Ca_V1.2$ Channels

LTCCs form the fulcrum in Ca^{2+} dynamics of VSM. Ca^{2+} influx through LTCCs in these cells is a principal mediator of myogenic tone (Fig. 1) (Amberg & Navedo, 2013; Knot & Nelson, 1998; Nelson, Patlak, Worley, & Standen, 1990). The vascular LTCC was first sequenced from rabbit lungs in 1990 (Biel et al., 1990), showing 65% amino acid sequence homology with its skeletal muscle isoform. LTCCs are comprised of pore-forming α_{1c} and auxiliary β, $\alpha_2\delta$, and γ subunits that modulate channel function. The α_{1c}, which contains the voltage sensor, the gating apparatus, and the Ca^{2+}-permeable pore, is made up of four homologous domains (I, II, III, IV), each of which is composed of six transmembrane segments (S1–S6) and intracellular NH_2- and COOH-termini. The S5 and S6 of each homologous domain form the pore region of the channel. Two glutamate residues at the pore loop determine the Ca^{2+} selectivity. The S1–S4 forms the voltage sensor, which rotates to open the ion pore (Bezanilla, 2008).

The α_{1c} transcript undergoes extensive alternative spicing, which provides structural and functional diversity in cell-type selective expression patterns. For example, splice variation in rat α_{1c} exon-1 gives rise to arterial VSM specific α_{1c} subunit that has cysteine-rich NH_2-terminus (Cheng et al., 2007). This α_{1c} when coexpressed with only $\alpha_2\delta$ demonstrated more negative steady-state activation and deactivation kinetics, smaller whole-cell currents, and decreased plasmalemma incorporation. LTCCs containing a SM-selective 25 amino acid exon 9a in the I–II intracellular linker region exhibit more hyperpolarized window current and are a key regulator of cerebrovascular constriction (Liao et al., 2007; Nystoriak, Murakami, Penar, & Wellman, 2009). Other variations, which exclude exon 33, have window current closer to resting membrane potential and higher sensitivity to blockade by nifedipine (Liao et al., 2007).

The COOH-terminus of α_{1c} provides regulatory functions such as plasmalemma targeting, retention, and constitutive intracellular recycling of LTCCs (Catterall, 2011). It also contains the calmodulin (CaM)-binding site, which facilitates channel trafficking to the plasmalemma. In rat and human cerebral arteries, the COOH-terminus of α_{1c} is cleaved, producing a short LTCC and a 50-kDa COOH-terminus fragment (Bannister et al., 2013). This fragment can be detected in the cytosol and the nucleus of VSM and was shown to induce vasodilation by decreasing α_{1c} expression and shifting the channel voltage dependence of activation to more depolarized potentials. Such findings warrant the need for further in-depth research regarding the role of the α_{1c} COOH-terminal fragment in blood pressure regulation in physiological and pathological conditions.

In vitro and in vivo studies have established the critical role for α_{1c} in vascular function. For instance, dihydropyridine antagonists (e.g., nifedipine, isradipine, nicardipine) that selectively inhibit α_{1c} activity were found to abolish the pressure-induced increase in $[Ca^{2+}]_i$ and prevented the development of myogenic tone (Knot & Nelson, 1998). Conversely, dihydropyridine agonists (e.g., Bay K 8644) that specifically stimulate LTCC activity enhance the myogenic response (Hwa & Bevan, 1986). Furthermore, α_{1c} knockout (SMAKO) mice showed a dramatic drop in myogenic tone and mean arterial pressure (Moosmang et al., 2003). Swelling of VSM may also induce cell contraction through activation of LTCC. Reports from cerebral artery VSM, canine basilar artery VSM, and rat-tail artery VSM confirmed the involvement of LTCC α_{1c} in vasoconstriction to a hypo-osmotic challenge (Kimura et al., 2000; Welsh, Nelson, Eckman, & Brayden, 2000; Wijetunge & Hughes, 2007).

The β subunit is generally paired in a 1:1 stoichiometry with $Ca_V1.2\alpha_{1c}$. The β subunit is made up of two conserved core regions, akin to the Src-homology-3 (SH3) domain and the guanylate kinase domain. Four β subunits, which are encoded by four different genes with several known splice variants, have been identified. Both biophysical properties and plasmalemmal insertion of $Ca_V1.2\alpha_{1c}$ can be distinctively regulated by different β isoforms (Catterall, 2011). In VSM, β_3 is the predominant β subunit, and a recent study concluded that it plays a critical role in upregulating

LTCC activity and in the development of angiotensin II (ANG II)-induced hypertension (Kharade et al., 2013).

Initially thought to be distinct subunits, the $\alpha_2\delta$ subunit exists as a single subunit connected by a disulfide bond (Catterall, 2011). The δ portion is anchored to the plasmalemma, while the glycosylated extracellular α_2 domain interacts with the α_{1c} (Gurnett, De Waard, & Campbell, 1996). Three different $\alpha_2\delta$ isoforms have been identified ($\alpha_2\delta1$–$\alpha_2\delta3$). Heterologous coexpression of different $\alpha_2\delta$ isoforms with different α_{1c} and β subunits produced channels with distinct gating profiles and current densities, highlighting the important regulatory role of $\alpha_2\delta$ on channel function (Klugbauer, Lacinova, Marais, Hobom, & Hofmann, 1999). In cerebral VSM, $\alpha_2\delta$ is a crucial regulator of LTCC function as illustrated by decreased Ca^{2+} influx via LTCCs and vasodilation following $\alpha_2\delta1$ knockdown (Bannister et al., 2009). Furthermore, increased $\alpha_2\delta1$ mRNA and protein were observed in cerebral VSM from spontaneously hypertensive rats (Bannister et al., 2012). These data, and the β subunit data discussed earlier, suggest that increased $\alpha_2\delta1$ and $\beta3$ expression enhances LTCC expression and function, thus contributing to augmented vasoconstriction during hypertension. Therefore, targeting the $\alpha_2\delta1$ and $\beta3$ may be a viable therapeutic approach to reverse/ameliorate increased vasoconstriction during hypertension.

Eight γ subunits have been identified. These subunits contain four transmembrane regions with intracellular NH_2- and COOH-termini. The first extracellular loop contains the conserved region of the GLWXXC amino acid motif, the most distinct feature of all the γ subunits (Catterall, 2011). The γ subunits also regulate biophysical and trafficking properties of the α_{1c} (Arikkath & Campbell, 2003). However, little is known about the function of the γ subunit on the regulation of LTCCs in VSM.

LTCCs are major targets of second messenger/kinase signaling cascades such as protein kinase A (PKA) and protein kinase C (PKC). Modulation of LTCC activity by these kinases ultimately contributes to control VSM function and vascular reactivity. The modulation of vascular LTCC activity by PKA is controversial. Activation of PKA in VSM has been reported to inhibit, potentiate, or has no effect on LTCC activity (see review by Keef, Hume, & Zhong, 2001). This contrasts with abundant and consistent data, indicating that agonists that stimulate PKA activity typically trigger vasodilation. Surprisingly, it was recently reported that an elevation in extracellular D-glucose from 5 to 15–20 mM, which is similar to the glucose concentration typically observed in animal models of diabetes and human diabetic patients, potentiates LTCC activity in cerebral VSM via a mechanism that requires PKA (Navedo, Takeda, Nieves-Cintron, Molkentin, & Santana, 2010; Nystoriak et al., 2014). This increase in vascular LTCC activity was correlated with enhanced myogenic tone in response to elevated glucose, thus providing the first example, to our knowledge, of a PKA-mediated vasoconstriction. On the other hand, activation of PKC by phorbol esters and vasoconstrictors acting through G_q-coupled receptors (e.g., ANG II, endothelin-1) results in potentiation of vascular LTCC activity and vasoconstriction (Keef et al., 2001; Weiss & Dascal, 2015). Consistent with this, genetic ablation of PKC protected against ANG II-induced potentiation of LTCC activity and the development of hypertension (Nieves-Cintron, Amberg, Navedo, Molkentin, & Santana, 2008). These genetic studies found that PKC activity was also required for basal and persistent LTCC activity in some VSM (see later and Santana et al., 2008). An integrated view of the specific upstream pathways contributing to PKA- and PKC-mediated LTCC regulation remains to be fully elucidated. Further, work in this area may help to identify novel therapeutic targets to treat pathological conditions associated with LTCC dysfunction such as hypertension (Nieves-Cintron et al., 2008).

Studies combining classical electrophysiology with high-resolution total internal reflection fluorescence microscopy have provided important information regarding the spatial organization of functional LTCCs and resultant Ca^{2+} signal in VSM (Nystoriak, Nieves-Cintron, & Navedo, 2013). Using this approach, elementary Ca^{2+} influx events via LTCCs (i.e., LTCC sparklets) were imaged in VSM, and channel activity was found to occur through distinct loci of low and high activity (Navedo, Amberg, Votaw, & Santana, 2005). The molecular, biophysical, and regulatory properties as well as the functional role of LTCC sparklets in VSM have been extensively reviewed in recent papers (Navedo & Amberg, 2013; Navedo & Santana, 2013; Santana & Navedo, 2009).

LTCC sparklets are sensitive to dihydropyridines and extracellular Ca^{2+} concentration and insensitive to store depletion by thapsigargin. Noteworthy, LTCC sparklets are always associated with inward L-type Ca^{2+} currents, confirming that they are produced by Ca^{2+} influx via LTCC. Whereas low activity LTCC sparklets exhibit stochastic behavior, high activity LTCC sparklets are produced by prolonged channel openings and in many cases by the nonstochastic, coordinated opening of clustered LTCC channels (Navedo, Cheng, et al., 2010). The dual optical/electrical recording of Ca^{2+} influx via LTCC also provides critical information regarding functional regulation of these channels. Importantly, the structurally diverse scaffolding protein AKAP150 was shown to be critical in mediating LTCC regulation. By virtue of its ability to bind PKA, PKC, calcineurin, and the LTCC itself, AKAP150 facilitates LTCC regulation by these kinases and phosphatase (Navedo, Amberg, Nieves, Molkentin, & Santana, 2006; Navedo et al., 2008; Navedo & Santana, 2013; Navedo, Takeda, et al., 2010; Santana & Navedo, 2009). High activity LTCC sparklets were shown to require distinct PKC and calcineurin activity. Accordingly, high activity LTCC sparklets contribute to $[Ca^{2+}]_i$, and myogenic tone during physiological and pathological conditions (Amberg, Navedo, Nieves-Cintrón, Molkentin, & Santana, 2007; Takeda, Nystoriak, Nieves-Cintron, Santana, & Navedo, 2011). Indeed, exacerbated high LTCC sparklet activity that is dependent on PKC or PKA has been linked to increased myogenic tone and activation of prohypertensive signaling pathways in animal models of hypertension and diabetes, respectively (Navedo & Amberg, 2013; Navedo, Takeda, et al., 2010; Nieves-Cintron et al., 2008, 2015; Nystoriak et al., 2014).

In summary, splice variations for the α_{1c}, β, and $\alpha_2\delta$ subunits result in complex functional diversity of LTCCs. As the predominant Ca^{2+} entry pathway in VSM, LTCCs play a key role in modulating VSM contractility and myogenic tone. Thus, mechanisms regulating LTCC subunit composition, posttranslational modifications, and membrane organization have the potential to impact VSM function and vascular reactivity during physiological and pathological conditions.

2.1.2 T-Type Ca^{2+} Channels

TTCCs were first identified as a separate VDCC in guinea pig ventricular myocytes as transient conductance currents of ~8 pS with Ba^{2+} as the charge carrier (Catterall, 2011). T-type currents are activated at more hyperpolarized potentials (~-30 mV). These channels can be blocked by mibefradil, NNC 55-0396, pimozide, penfluridol, and nickel, although caution should be taken as many of these compounds have off-target effects (Gray & Macdonald, 2006). Dihydropyridines such as nifedipine (at nanomolar range) have generally minimal effects on TTCCs. However, it has been shown that nifedipine at micromolar concentrations (>1 μM) can suppress TTCC function (Akaike et al., 1989). The T-type conductance is similar with Ba^{2+} and Ca^{2+} as charge carriers, whereas L-type current is significantly larger in the presence of Ba^{2+} than with Ca^{2+} (Catterall, 2011). So far, no auxiliary β, $\alpha_2\delta$, or γ subunits have been purified for the TTCC. However, some studies suggest that LTCC auxiliary subunits may modulate TTCC functions (Perez-Reyes, 2006).

Several studies have now suggested a role for TTCCs in VSM physiology and vascular reactivity. At the molecular level, transcript and protein for $Ca_V3.1$ and $Ca_V3.2$ have been found in VSM from several vascular beds and in different species, including humans (Abd El-Rahman et al., 2013; Harraz, Abd El-Rahman, et al., 2014; Harraz, Visser, et al., 2015; Kuo, Ellis, Seymour, Sandow, & Hill, 2010). Studies on rat mesenteric arterioles indicate that TTCCs also contribute to vasoconstrictor responses (Gustafsson, Andreasen, Salomonsson, Jensen, & Holstein-Rathlou, 2001; Jensen, Salomonsson, Jensen, & Holstein-Rathlou, 2004). In skeletal muscle arteries, $Ca_V3.1$ and $Ca_V3.2$ are actively involved in maintenance of myogenic tone (VanBavel, Sorop, Andreasen, Pfaffendorf, & Jensen, 2002). In arteriolar SM from the retina, $Ca_V3.1$ activity has been reported, indicating a potentially important role in retinal microcirculation (Fernandez, McGahon, McGeown, & Curtis, 2015).

$Ca_V3.1$ and $Ca_V3.2$ channels have also been identified in rat and mouse cerebral VSM as the nifedipine-insensitive component of Ba^{2+} currents (Abd El-Rahman et al., 2013; Harraz, Abd El-Rahman, et al., 2014; Harraz, Visser, et al., 2015; Kuo et al., 2010; Nikitina et al., 2006). Interestingly, $Ca_V3.1$ seems to be replaced by $Ca_V3.3$ in human cells (Harraz, Visser, et al., 2015). $Ca_V3.1$/$Ca_V3.3$ and $Ca_V3.2$ contributions to the observed T-type current in these cells could be distinguished based on a 20-fold higher sensitivity of $Ca_V3.2$ to blockage by Ni^{2+} ($Ca_V3.2$ $EC_{50}=12\ \mu M$; $Ca_V3.1$ $EC_{50}=250\ \mu M$) (Lee, Gomora, Cribbs, & Perez-Reyes, 1999). By exploiting this selectivity and the use of genetically modified mice, the contribution of these channels to the regulation of the myogenic response was found to diverge (Fig. 1). Whereas $Ca_V3.1$/$Ca_V3.3$ seems to mediate pressure-induced constriction, $Ca_V3.2$ contributes to negative feedback regulation of pressure-induced tone by modulating the RyR—large-conductance Ca^{2+}-activated K^+ (BK_{Ca}) channel axis (Harraz, Abd El-Rahman, et al., 2014; Harraz, Brett, & Welsh, 2014; Harraz, Visser, et al., 2015). Another interesting detail is that TTCCs and LTCCs appear to respond to different intravascular pressures following their voltage dependence (Harraz, Abd El-Rahman, et al., 2014; Harraz, Visser, et al., 2015). Accordingly, $Ca_V3.1$/$Ca_V3.3$ channels predominantly contribute to myogenic tone at lower intraluminal pressures (e.g., 20–40 mmHg), in which membrane potential of VSM is ~–60 to –50 mV. Conversely, $Ca_V1.2$ function is prominent at more depolarized VSM membrane potentials (~–45 to –36 mV) observed at greater intraluminal pressures (Knot & Nelson, 1998). Regulation of TTCC activity by protein kinases may also contribute to modulate VSM function and the myogenic response. Accordingly, PKA and PKG activation has been shown to inhibit TTCC in VSM (Harraz, Brett, et al., 2014; Harraz & Welsh, 2013a). This TTCC suppression could limit extracellular Ca^{2+} entry, which may contribute to the well-known vasodilatory responses triggered by these kinases. Thus, along with LTCCs, TTCCs may contribute to precise maintenance of myogenic tone through their ability to activate at lower pressures.

2.2 TRP Channels

TRP channels are a superfamily of cationic channels with 28 encoding genes. Based on their sequence homology, these channels can be further categorized into six subfamilies: TRPC (canonical), TRPV (vanilloid), TRPM (melastatin), TRPP (polycystin), TRPA (ankyrin), and TRPML (mucolipin) (Earley & Brayden, 2015). Sequence analysis suggests that TRP channels consist of six membrane-spanning helices (S1–S6) with intracellular NH_2- and COOH-termini of variable lengths. While a crystal structure has not yet been resolved for any TRP channel, electron cryomicroscopy studies of the capsaicin-activated TRPV1 channel demonstrate four symmetrical subunits with S5 and S6 loop forming the ion pore. The structure also contains a selectivity filter, which is differentially regulated by endogenous and exogenous ligands (Liao, Cao, Julius, & Cheng, 2013). Functional TRP channels consist of four subunits and can be homomeric or heteromeric in nature. As most cells express multiple TRP channel isoforms, it is likely that these channels exist in heteromultimeric form. The NH_2- and COOH-termini have several domains that can modify and regulate channel function. For example, the number of ankyrin repeats on the TRPV1 and TRPA1 channels has been shown to regulate channel activity (Gaudet, 2008). Similarly, the COOH-terminus of some TRP channels contains a CaM/IP_3-binding domain, serine–threonine kinase target sequence, and PDZ protein–protein interaction domains, depending on transcript splicing patterns (Walker, Hume, & Horowitz, 2001).

Multiple TRP channels are expressed in VSM. In these cells, TRP channels contribute to regulation of membrane potential, contraction, and development of myogenic tone (Earley & Brayden, 2015). Additionally, certain TRP channels contribute to vascular mechanosensitivity via G-protein-coupled signaling in resistance arteries (Earley & Brayden, 2015). Almost all TRP channels are permeable to Ca^{2+} with the exception of TRPM4 and TRPM5, which are Ca^{2+} activated, but not Ca^{2+} permeable

(Earley & Brayden, 2015). Here, we review key issues on specific vascular TRP channels.

2.2.1 TRPV1

TRPV1 are nonselective cation channels with the preference for Ca^{2+} to Na^{+} ions (10:1) (Caterina et al., 1997). TRPV1 channels are primarily expressed in sensory neurons and play a crucial role in heat sensation and nociception (Meents, Neeb, & Reuter, 2010). TRPV1 can be endogenously activated by acidic $pH < 5.5$ and derivatives of arachidonic acid, and exogenously by capsaicin and resiniferatoxin. Furthermore, channel activity and trafficking are regulated by PKA and PKC (Efendiev, Bavencoffe, Hu, Zhu, & Dessauer, 2013; Koda et al., 2016) with potential important implications in the regulation of vascular tone (Earley & Brayden, 2015). Activation of these channels in sensory nerves results in release of vasodilatory neuropeptides and consequently, dilation of the blood vessels. In SM, TRPV1 activation elicits contraction. For example, skeletal muscle arterioles exhibit significant constriction in response to application of the TRPV1 activator capsaicin (Toth et al., 2014). TRPV1 expression has also been reported in arterioles from thermoregulatory tissues including dura, cremaster, skin, and ear (Cavanaugh et al., 2011). Yet, TRPV1 expression does not seem to be ubiquitous among vascular tissues (Baylie & Brayden, 2011).

2.2.2 TRPV2

TRPV2 channels have been reported in aortic, mesenteric, and basilar artery VSM (Earley & Brayden, 2015). In aortic myocytes, application of a hypotonic solution resulted in TRPV2 activation, increased Ca^{2+} influx, and constriction (Muraki et al., 2003). Yet, more studies are required to elucidate a functional role for TRPV2 in different vascular beds.

2.2.3 TRPV4

TRPV4 channels have been implicated in regulation of myogenic tone. This channel can be stimulated by mechanical stress, including sheer stress and cell swelling. In cerebral VSM, TRPV4 channels are regulated by PKC and their activation results in Ca^{2+} influx (i.e., TRPV4 sparklets) (Mercado et al., 2014). Unexpectedly, this activation was associated with vasodilation rather than vasoconstriction (Earley, Heppner, Nelson, & Brayden, 2005). Ca^{2+} entry during a single TRPV4 sparklet is ~100-fold larger than a $Ca_V1.2$ sparklet (Mercado et al., 2014). This highly localized TRPV4 sparklet may stimulate the activity of RyR in the SR resulting in generation of

Ca^{2+} sparks. The ensuing Ca^{2+} sparks activate BK_{Ca} channels leading to VSM membrane potential hyperpolarization and vasorelaxation (Fig. 1) (Earley et al., 2005, 2009). An intriguing possibility, based on the large and localized flux of Ca^{2+} through TRPV4, is that these channels could also directly activate nearby BK_{Ca} channels to regulate vascular reactivity. This hypothesis requires examination.

2.2.4 TRPC1

TRPC1 channels form functional heteromers with TRPC4 and/or TRPC5, which can be regulated by G_q-coupled signaling pathways (Sabourin et al., 2009; Strubing, Krapivinsky, Krapivinsky, & Clapham, 2001). The functional relevance of TRPC1 channels in VSM, however, is controversial. For instance, studies support (Bergdahl et al., 2003, 2005; Inoue et al., 2006) and refute (DeHaven et al., 2009; Dietrich et al., 2007; Varga-Szabo et al., 2008) a role for TRPC1 in store-operated Ca^{2+} entry (SOCE) in VSM. Moreover, Ca^{2+} influx via TRPC1 has been associated with BK_{Ca} channel activation and VSM relaxation (Kwan et al., 2009). This is somewhat paradoxical as a recent study suggested that (1) TRPC1 does not form functional homomeric channels and (2) TRPC1-containing heteromers exhibit decreased Ca^{2+} permeability (Storch et al., 2012). More research is necessary to define the mechanisms by which TRPC1 modulates BK_{Ca} channel activity and vascular reactivity.

2.2.5 TRPC3

TRPC3 in VSM has been linked to vascular tone regulation via stimulation of a variety of G-protein-coupled receptors (GPCRs) including ANG II and endothelin-1 (ET-1) receptors (Earley & Brayden, 2015). Accordingly, TRPC3 does not seem to have an effect on intrinsic vascular tone, suggesting that TRPC3 may not be essential in pressure-induced tone but rather receptor-mediated vasoconstriction of the arteries (Reading, Earley, Waldron, Welsh, & Brayden, 2005; Xi et al., 2008). Indeed, in cerebral VSM, the mechanism of TRPC3-induced vasoconstriction implicates IP_3 facilitated coupling of IP_3R and TRPC3 (Fig. 1) (Xi et al., 2008). The activation of TRPC3 results in cation influx (e.g., Na^+ and Ca^{2+}) and consequently membrane depolarization leading to opening of LTCCs and vasoconstriction.

2.2.6 TRPC4

A role for TRPC4 in aortic and mesenteric VSM has been suggested (Lindsey & Songu-Mize, 2010). Under prolonged cyclic stretch conditions, TRPC4 expression appears to decrease along with a decrease in SOCE. These studies suggest that downregulation of TRPC4 may be a protective mechanism against stretch-mediated increase in SOCE.

2.2.7 TRPC5

TRPC5 channels have been implicated in SOCE in VSM when coassembled with other TRPC subunits. For example, application of an anti-TRPC5 antibody was found to inhibit SOCE in VSM from cerebral arterioles in response to store depletion (Xu, Boulay, Flemming, & Beech, 2006). Likewise, currents evoked by cyclopiazonic acid (a SERCA pump inhibitor) were inhibited by an anti-TRPC5 antibody (Saleh, Albert, & Large, 2009). Altogether, these results suggest a role for TRPC5 in SOCE in VSM.

2.2.8 TRPC6

TRPC6-mediated Ca^{2+} mobilization in VSM has been associated with regulation of vasoconstriction. TRPC6 channels can be activated via mechanosensation, such as cell swelling and sheer stress, to promote vasoconstriction (Welsh, Morielli, Nelson, & Brayden, 2002). These channels can also be activated by GPCRs (Inoue et al., 2001). For example, application of 1 nM ANG II activates TRPC6 channels via a mechanism that is directly associated with diacylglycerol, and independent of PKC (Helliwell & Large, 1997; Saleh, Albert, Peppiatt, & Large, 2006). More recently, an exciting study proposed that Ca^{2+} entry via TRPC6 plays a critical role as part of a force-sensing complex that bolsters Ca^{2+} release through IP_3Rs on the SR to stimulate TRPM4 channel activity and myogenic tone (Fig. 1) (Gonzales et al., 2014). However, despite the wealth of information on this channel in VSM, additional studies are needed to establish their (1) contributions to local and global Ca^{2+} signals, (2) role in different vascular beds, and (3) mechanisms of activation and mechanosensation.

2.2.9 TRPM4

TRPM4 is one of two TRP channels that are Ca^{2+} activated, but not Ca^{2+} permeable. Nonetheless, TRPM4 channels in VSM play an essential role in development of myogenic tone (Earley, Waldron, & Brayden, 2004; Gonzales et al., 2014; Li & Brayden, 2015). In VSM from rat cerebral arteries, these channels are activated by local IP_3R-mediated increases in $[Ca^{2+}]_i$

(Fig. 1) (Gonzales, Amberg, & Earley, 2010). Their activity can be differentially regulated by specific signaling proteins depending on the vascular bed (Earley, Straub, & Brayden, 2007; Li & Brayden, 2015). More recently, RhoA/Rho-associated protein kinase has been demonstrated to potentiate TRPM4 in VSM of parenchymal arterioles (Li & Brayden, 2015). Other TRPM channels, such as TRPM7 and TRPM8, are also expressed in VSM. Whereas TRPM7 has been implicated mainly in Mg^{2+} homeostasis (He, Yao, Savoia, & Touyz, 2005), little is known about the functional role of TRPM8 in VSM.

2.2.10 TRPP2

TRPP are Ca^{2+}-permeable channels known to be mechanosensitive (Earley & Brayden, 2015). Recognized also as polycystic-1 and -2 proteins, TRPP1 and TRPP2 expression has been described in multiple vascular beds, including mesenteric and cerebral arteries (Narayanan et al., 2013; Sharif-Naeini et al., 2009). In these arterial beds, the activation of TRPP1 and TRPP2 seems to regulate the myogenic response, albeit via different mechanisms. TRPP2 activation is also associated with differential regulation of IP_3R and RyR in cells (presumably VSM) from cerebral arteries with important implications for modulation of vascular reactivity (Abdi et al., 2015). Yet, several issues still require further examination such as (1) contribution of TRPP2 to local and global Ca^{2+} signals, (2) how TRPP2 modulates IP_3R and RyR activity, (3) what role, if any, does interaction of TRPP2 with other ion channels (e.g., TRPP1) have on the control of vascular reactivity, and (4) how TRPP2 may contribute to VSM physiology and pathophysiology.

2.2.11 Intracellular TRP Channels

Accumulating evidence suggests that TRP channels in intracellular membranes may also play a critical role in regulating Ca^{2+} homeostasis and cell function (Dong, Wang, & Xu, 2010). For example, TRPV1 and TRPP2 localized to the endoplasmic reticulum are thought to be involved in intracellular Ca^{2+} regulation (Koulen et al., 2002; Olah et al., 2001). Furthermore, TRPM2 and TRPML may play a role in Ca^{2+} release from lysosomes, which may contribute to oxidative stress of the cell (Dong et al., 2010; Lange et al., 2009). In VSM, TRPM7-containing vesicles are quickly trafficked to the plasmalemma in response to shear stress (Oancea, Wolfe, & Clapham, 2006). This resulted in a significant increase in TRPM7-like currents, which may contribute to increased $[Ca^{2+}]_i$ and impaired VSM function during pathological conditions (Dong et al., 2010). Extensive studies, however, are still required to completely understand the role of intracellular TRP channels in VSM.

2.3 Orai and STIM

Depletion of Ca^{2+} from intracellular stores via activation of plasmalemmal phospholipase C (PLC)-coupled receptors and subsequent IP_3R-mediated Ca^{2+} release triggers Ca^{2+} influx from extracellular sources. This process of SOCE was first introduced as a mechanism of controlled and sustained Ca^{2+} entry following activation of surface membrane receptors (Putney, 1986). The physiological and pathological significance of the Ca^{2+} release-activated Ca^{2+} current (I_{CRAC}) was brought to light by rare cases of severe immunodeficiency in patients with inherited defects in components mediating SOCE that profoundly impairs immune cell function (Picard et al., 2009). A great deal of progress has been made in the field of SOCE and has highlighted an emerging importance of I_{CRAC} in multiple cell types, including VSM. Following decades-long investigations aimed at revealing the molecular identity of I_{CRAC}, the plasmalemmal Ca^{2+}-permeable channels and associated S/ER-localized channel activators responsible for SOCE were only recently identified as Orai and STIM, respectively (Roos et al., 2005; Zhang et al., 2005). It is now recognized that upon reduction of Ca^{2+} concentration in S/ER ($[Ca^{2+}]_{S/ER}$), the $[Ca^{2+}]_{S/ER}$ sensor STIM1 undergoes dynamic spatial reorganization into aggregate clusters that interact with Orai1 channels in the plasmalemma to facilitate I_{CRAC}. This section will briefly review the current state of knowledge regarding molecular and functional aspects of Orai-channel mediated Ca^{2+} signaling in VSM and its potential contribution to vascular disease states.

The Orai homologues are plasmalemmal ion channels encoded by three genes (i.e., Orai1, Orai2, and Orai3) with little genetic or structural similarity to that of other known Ca^{2+}-permeable channels. In mammals, alternative methionine translation initiation gives rise to two forms of Orai1: a 33 kDa long form (Oria1α) and a 23 kDa short form (Oria1β) (Fukushima, Tomita, Janoshazi, & Putney, 2012). Perhaps as a result of these proteomic and functional differences, Orai1α and Orai1β could display preference for certain binding partners (e.g., TRPC1 vs Orai3) to exhibit selective participation in distinct Ca^{2+} currents (Desai et al., 2015). Sequence analyses and crystallization studies have revealed that Orai channels are heteromeric structures, with each channel consisting of 4–6 Orai subunits

(Hou, Pedi, Diver, & Long, 2012). Each subunit consists of four highly conserved transmembrane helices (M1–M4). The side chains of amino acids in M1 helices of each Orai subunit form a 55 Å pore. The extracellular face of the Orai pore has a distinct ring of glutamate residues that form its selectivity filter (Hou et al., 2012). This feature is thought to render the highly selective nature of I_{CRAC} being almost exclusively carried by Ca^{2+} over Na^+ or K^+ ions (Hoth & Penner, 1993). Each Orai subunit consists of cytosolic NH_2 and COOH-termini, which contain sites for functional regulation by Ca^{2+}/CaM, PKC, and STIM proteins (Frischauf et al., 2009; Hooper et al., 2015; Mullins, Park, Dolmetsch, & Lewis, 2009).

It is now established that Orai channels are activated via physical molecular interaction with the stromal interaction molecules (STIM1 and STIM2), which function as Ca^{2+} sensors within the S/ER (Roos et al., 2005; Zhang et al., 2005). STIM proteins are single transmembrane proteins that are primarily located in the ER, although small populations of STIM1 that play a role in controlling Ca^{2+} entry have also been observed in the plasmalemma (Spassova et al., 2006). S/ER STIM senses alterations in luminal $[Ca^{2+}]_i$ via NH_2 terminal canonical EF-hand domains. Upon a depletion of S/ER Ca^{2+} and dissociation of Ca^{2+} ions from the NH_2 terminal EF-hand domains, STIM proteins undergo unfolding and aggregation at plasmalemmal-S/ER junctions where they activate Orai channels via direct physical interactions to induce conformational changes in the channel structure. Importantly, proper Orai/STIM communication requires COOH-terminus coiled-coil interaction domains of both Orai and STIM (Frischauf et al., 2009; Muik et al., 2008).

Determining the precise physiological role of the SOCE machinery in cardiovascular tissues has been hampered by a lack of selective pharmacological modulators of Orai channels and STIM proteins. The widely used nonselective cation channel inhibitor 2-aminoethoxydiphenyl borate (2-APB) has been shown to exhibit concentration-dependent and divergent effects on I_{CRAC} (Prakriya & Lewis, 2001). A new class of I_{CRAC} inhibitors was shown to have selective effects on Orai channels independent of STIM oligomerization or STIM/Orai interaction (Derler et al., 2013). For example, in VSM, the I_{CRAC} inhibitor S66 prevented Ca^{2+} influx following store depletion with nanomolar potency (Li et al., 2011). Thus, the emergence of novel compounds that can specifically target Orai–Orai/STIM interactions and STIM aggregation will significantly aid in future studies to investigate I_{CRAC}-related mechanisms in cardiovascular physiology and pathology.

Expression of Orai and STIM ranges from very low to nondetectable in quiescent VSM. Yet, store depletion by thapsigargin and subsequent SOCE was shown to be prominent in synthetic cultured rat aortic smooth muscle, but not in freshly dispersed VSM (Potier et al., 2009). In line with enhanced SOCE in dedifferentiated SM present in many disease states, transformation of contractile VSM to a noncontractile proliferative phenotype in culture is associated with substantial upregulation in the expression of Orai and STIM proteins (Berra-Romani, Mazzocco-Spezzia, Pulina, & Golovina, 2008; Potier et al., 2009). Consistent with in vitro findings, expression for both Orai1 and STIM1 were strongly upregulated in association with SM proliferation following balloon angioplasty-induced carotid injury in rats (Zhang et al., 2011). Lentivirus-mediated knockdown of Orai1 in injured vessels prevented conversion of SM to a proliferative phenotype and mitigated neointima formation, suggesting that Orai1-mediated Ca^{2+} entry may be an important determinant of vascular remodeling during injury such as in restenosis. Further, in vitro studies have confirmed that SOCE becomes a predominant source of Ca^{2+} influx in synthetic VSM that plays a pivotal role in proliferative and migratory processes. Thus, adaptive changes in Orai and STIM expression and function may drive phenotypic modulation during angiogenesis and vascular repair, as well as in disease.

The precise molecular determinants underlying Orai/STIM upregulation in phenotypic switching of quiescent to proliferative/migratory SM is still unresolved. A major driving factor of phenotypic switching of VSM is the polypeptide platelet-derived growth factor (PDGF). PDGF, via activation of the PDGFβ receptor and downstream PLCγ-mediated SR Ca^{2+} release, is an important activator of Orai1 channels in SM. Whether STIM-independent regulation of Orai occurring downstream of PDGF stimulation significantly contributes to Ca^{2+} influx in proliferating/migratory VSM is still unclear. However, in addition to the well-known role in activation of Orai channels, STIM1 clustered in ER/PM junctions also inhibits Ca^{2+} influx through $Ca_V1.2$ channels and leads to internalization of LTCC (Park, Shcheglovitov, & Dolmetsch, 2010; Wang et al., 2010). This mechanism, however, could not be confirmed in quiescent VSM (Takeda et al., 2011), perhaps reflecting distinct roles for STIM proteins in proliferative vs contractile cells.

It is also important to mention the role that Orai channels play in store-independent, ligand-activated Ca^{2+} entry in SM that is mediated by Orai1, Orai3, and STIM1. This nonstore-operated strongly inwardly rectifying current, termed I_{ARC}, is gated by arachidonic acid and its metabolite leukotriene C4 (Zhang et al., 2015). In VSM, Ca^{2+} entry requiring Orai1, Orai3, and STIM1 has been observed independent of sustained store depletion following application of the proinflammatory peptide thrombin. Like Orai1 and STIM1, Orai3 protein was also upregulated in an experimental model of carotid artery injury and in vivo knockdown of this subunit alone, blunted neointima formation. These findings suggest that heteromultimerization of Orai channels could give rise to store-dependent and store-independent Ca^{2+} entry pathways that could contribute to maintenance of the synthetic VSM phenotype in several disease states.

3. SR Ca^{2+} CHANNELS

Ca^{2+} release channels located on the SR membrane of VSM play pivotal roles in controlling cell excitability and vascular reactivity. The two major classes of Ca^{2+} release channels in VSM are RyR and IP_3R. In the following section, we describe their role in VSM.

3.1 Ryanodine Receptors

RyRs are intracellular Ca^{2+} channels that mediate Ca^{2+} release from the SR. Three RyR isoforms (RyR1–RyR3) encoded by three distinct genes have been identified (Lanner, Georgiou, Joshi, & Hamilton, 2010). Structural models predicted both NH_2- and COOH-termini to be cytosolic and the pore to have 4–12 membrane-spanning domains (Lanner et al., 2010). Single-particle electron cryomicroscopy studies have provided insights into channel gating and interaction with modulators (Samso, Wagenknecht, & Allen, 2005; Serysheva et al., 2008). Recently, 4.8 and 6.8 Å resolution structures of the RyR were described by two independent groups (Efremov, Leitner, Aebersold, & Raunser, 2015; Zalk et al., 2015). Both reports suggest that Ca^{2+} sensitivity of RyR1 is imparted by EF-hand domains in α-solenoid structures that connect the cytoplasmic region to the channel pore.

mRNA transcript and protein levels have been detected for all RyR isoforms in VSM. RyR1 and RyR2 mediate Ca^{2+} sparks (e.g., Ca^{2+} release from intracellular stores through RyRs) in portal vein SM (Coussin, Macrez, Morel, & Mironneau, 2000), whereas RyR2 has been found to be the predominant isoform in VSM from rat resistance vessels (Vaithianathan et al., 2010). Several studies suggest that in contrast to the tight coupling between LTCCs and RyRs in skeletal and cardiac muscle, a loose coupling mechanism may exist in VSM in which LTCCs indirectly modulate RyRs by contributing to global $[Ca^{2+}]_i$ and SR Ca^{2+} load (Collier, Ji, Wang, & Kotlikoff, 2000; Essin et al., 2007). Interestingly, recent studies demonstrate that application of Ni^{2+} at a concentration that selectively inhibits $Ca_V3.2$ reduced Ca^{2+} spark activity in VSM from WT mice, but had no effect in cells from $Ca_V3.2$ knockout mice (Harraz, Brett, et al., 2015). Consistent with a role for $Ca_V3.2$ in modulation of Ca^{2+} sparks, these channels were found juxtaposed with RyR in specific microdomains (Harraz, Abd El-Rahman, et al., 2014), suggesting that Ca^{2+} influx through TTCCs may contribute to RyR activation in VSM (Fig. 1). This may represent a novel mechanism for regulation of RyRs, VSM excitability, and vascular reactivity that requires further examination.

In VSM, RyR can be activated by caffeine, and depending on its concentration, it can induce massive Ca^{2+} release from intracellular stores or increase the frequency of Ca^{2+} sparks (Jaggar, Porter, Lederer, & Nelson, 2000). The receptor can be inhibited in a concentration-dependent manner by the alkaloid ryanodine. Accordingly, at low concentrations, ryanodine can activate RyR, while at higher concentrations, it inhibits the receptor. Other pharmacological agents such as tetracaine have been used to block RyR and examine their role in VSM function. However, their use is limited due to nonspecific effects. RyRs play a central role in excitation–contraction coupling in both skeletal and cardiac muscle where they contribute to the global increase in $[Ca^{2+}]_i$ necessary for contraction. However, RyRs influence VSM excitability indirectly by modulating the activity of plasmalemma ion channels (Fig. 1). In a landmark study, it was found that Ca^{2+} sparks could simultaneously activate multiple BK_{Ca} channels in the plasmalemma to produce spontaneous transient outward currents (STOCs) and promote hyperpolarization and relaxation of VSM in small resistance arteries (Nelson et al., 1995; Perez, Bonev, Patlak, & Nelson, 1999). The spatial proximity between the plasmalemma and the SR in VSM (Somlyo, 1985) permits Ca^{2+}sparks to activate BK_{Ca} channels with minimal effects on global $[Ca^{2+}]_i$. In rabbit portal vein, however, RyRs have been found to depolarize VSM through activation of Ca^{2+} sensitive chloride channels (Saleh & Greenwood, 2005; Wang, Hogg, & Large, 1992). This highlights the importance of RyR in fine-tuning VSM excitability among different vascular beds.

Regulation of the functional coupling between RyR and BK_{Ca} channels has profound implications for VSM function. For instance, the vasodilatory effects of nitric oxide and forskolin can be attributed, at least in part, to an increase in PKG and PKA activity that acts on RyR to stimulate Ca^{2+} sparks-mediated STOC frequency (Jaggar et al., 2000). Conversely, activators of PKC inhibit RyR activity, which reduces STOC frequency and could contribute to vasoconstriction (Amberg et al., 2007; Bonev, Jaggar, Rubart, & Nelson, 1997). In addition, any disturbance on BK_{Ca} channel Ca^{2+} sensitivity may impact functional RyR–BK_{Ca} coupling, and vascular contractility; a point well illustrated in animal models of hypertension. In these animals, decreased expression of BK_{Ca} channel β1 subunit reduces BK_{Ca} Ca^{2+} sensitivity resulting in impaired STOC activity, increased myogenic tone, and hypertension (Amberg, Bonev, Rossow, Nelson, & Santana, 2003; Nieves-Cintrón, Amberg, Nichols, Molkentin, & Santana, 2007). The coupling strength between RyR and BK_{Ca} channel is also affected in animal models of diabetes (Nystoriak et al., 2014). These examples highlight the relevance of the relationship between RyR and BK_{Ca} channels in VSM with impaired communication leading to vascular dysfunction.

3.2 Inositol-1,4,5,-Trisphosphate Receptors

IP_3R is a ubiquitously expressed Ca^{2+} release channel localized to the SR membrane (Narayanan et al., 2012; Nixon, Mignery, & Somlyo, 1994). These channels consist of four membrane-spanning subunits surrounding the central ion permeation pore. Each subunit contains six transmembrane domains, a luminal loop that forms the ion-conducting pore between transmembrane domains 5 and 6, and cytosolic NH_2- and COOH-termini. The NH_2-terminus is further subdivided into a suppression domain that inhibits IP_3 binding, an IP_3-binding core domain, binding sites for ATP and Ca^{2+}, and a coupling domain for physical interactions with TRPC channels. The COOH-terminus seems to contribute to IP_3R tetramerization, and recent electron cryomicroscopy studies with a resolved IP_3R structure at 4.7 Å implicated this domain in channel gating (Fan et al., 2015). Moreover, these studies also suggest that the gate for the Ca^{2+} conduction path includes several hydrophobic residues located closer to the cytosolic side of the SR membrane (Fan et al., 2015). Yet, further experiments will be needed to completely understand the permeation, gating, and regulatory mechanisms governing IP_3R function.

Three different isoforms of IP_3R (IP_3R1, IP_3R2, IP_3R3) have been reported (Narayanan et al., 2012). While expression of all three isoforms has been found in VSM from aorta, mesenteric, and cerebral arteries, IP_3R1 seems to be the predominant isoform in VSM from small resistance arteries (Grayson, Haddock, Murray, Wojcikiewicz, & Hill, 2004; Zhao, Adebiyi, Blaskova, Xi, & Jaggar, 2008; Zhou et al., 2008). Yet, the expression profile of the IP_3R isoforms can vary depending on the developmental stage. Accordingly, expression of IP_3R3 is high in neonatal SM, decreases during development, and is surpassed by increasing IP_3R1 expression in adult SM (Tasker, Michelangeli, & Nixon, 1999). High levels of IP_3R2 and IP_3R3 expression have also been found in proliferating SM (Tasker, Taylor, & Nixon, 2000). These IP_3R isoforms also differ in IP_3-binding affinities as follows: $IP_3R2 > IP_3R1 > IP_3R3$ (Newton, Mignery, & Sudhof, 1994; Wojcikiewicz & Luo, 1998). Physical localization of IP_3R on the SR membrane is also isoform and tissue-dependent, which may be important for distinct physiological functions such as gene expression and VSM excitability (Nixon et al., 1994; Tasker et al., 2000; Zhao et al., 2008).

IP_3R activation is stimulated by the second messenger IP_3, which results from the hydrolysis of phosphatidylinositol 4,5-bisphosphate by PLC in response to activation of $G_{q/11}$-coupled receptors. Indeed, IP_3R activity in VSM can be stimulated by many endogenous vasoactive molecules that act through $G_{q/11}$-coupled receptors to produce IP_3, including ET-1, acetylcholine, noradrenaline, and serotonin (Berridge, 2008). Pharmacological inhibition of IP_3R can be achieved with the application of widely used agents such as 2-APB and xestospongin C. IP_3R inhibition may have distinct effects on vascular reactivity. For instance, in mouse mesenteric arteries, IP_3R inhibition with xestospongin C did not affect the myogenic tone, but did prevent the phenylephrine-induced vasoconstriction (Mauban, Zacharia, Fairfax, & Wier, 2015). IP_3R activity in VSM can be modulated by $[Ca^{2+}]_i$, luminal SR Ca^{2+} load, ATP, several protein kinases (e.g., PKA, PKG), regulatory proteins (e.g., RACK, FKBP12), reactive oxygen species (ROS), and pH (Bezprozvanny, Watras, & Ehrlich, 1991; Iino, 1990; Narayanan et al., 2012).

The activation of IP_3R can produce multiple Ca^{2+} signals, including Ca^{2+} puffs and Ca^{2+} waves, with important implications for VSM function. Ca^{2+} puffs are elementary, localized Ca^{2+} release events produced by clusters of IP_3R (Parker & Smith, 2010; Tovey et al., 2001). Ca^{2+} puffs have been observed in colonic and ureteric SM (Boittin et al., 2000; Olson, Chalmers, & McCarron, 2010), but not in VSM, perhaps due to differences in IP_3R localization, distribution, and function. Yet, recent indirect evidence suggests that Ca^{2+} puffs could alter cerebral VSM function through modulation of plasmalemmal ion channel activity. Accordingly, localized Ca^{2+} release via IP_3R was shown to promote the opening of TRPM4 channels leading to pressure-induced membrane depolarization and cell contraction (Fig. 1) (Gonzales et al., 2010; Gonzales & Earley, 2012).

On the other hand, Ca^{2+} waves are propagating elevations in $[Ca^{2+}]_i$ resulting from Ca^{2+} release via IP_3R, RyR, or both due to electrical, mechanical, and receptor-mediated stimulation in VSM (Amberg & Navedo, 2013; Narayanan et al., 2012; Wray & Burdyga, 2010). The contributions of IP_3R and/or RyR to spontaneous and agonist-induced Ca^{2+} wave generation seem to differ in VSM according to the vascular bed (Fig. 1) (Boittin, Macrez, Halet, & Mironneau, 1999; Dabertrand, Nelson, & Brayden, 2012; Gordienko & Bolton, 2002; Jaggar, 2001; Wray & Burdyga, 2010; Zacharia, Zhang, & Wier, 2007; Zhao et al., 2008). For example, RyR but not IP_3R was found to play a prominent role in spontaneous Ca^{2+} wave generation and propagation in cerebral VSM (Jaggar, 2001; Jaggar & Nelson, 2000), whereas both RyR and IP_3R appear to be involved in Ca^{2+} waves in portal vein SM (Gordienko & Bolton, 2002). Conversely, IP_3Rs are involved in agonist-induced Ca^{2+} waves, which may propagate with involvement of RyR activation perhaps via a Ca^{2+}-induced Ca^{2+} release (CICR) mechanism (Boittin et al., 1999; Gordienko & Bolton, 2002; Wray & Burdyga, 2010; Zhao et al., 2008). The mechanisms of wave propagation, however, remain unclear. Future studies should comprehensively investigate the levels of RyR and IP_3R expression, subcellular distribution and activation, their regulation by cytosolic and SR Ca^{2+} concentration, as well as variations in these properties in VSM among different vascular beds.

IP_3R-mediated SR Ca^{2+} release-dependent and -independent mechanisms have been described to regulate VSM function. IP_3R-mediated Ca^{2+} waves have been suggested to contribute to agonist-induced VSM contraction and the myogenic response (Boittin et al., 1999; Lamont & Wier, 2004; Zacharia et al., 2007; Zhao et al., 2008). This response seems to involve an increase in the frequency of Ca^{2+} waves that could contribute, at least in part, to elevate global $[Ca^{2+}]_i$ to activate the contractile machinery (Hill-Eubanks, Werner, Heppner, & Nelson, 2011). Interestingly, recent studies have also revealed a significant contribution for IP_3R to VSM excitability that is independent of its ability to mediate SR Ca^{2+} release. At physiological intravascular pressures, PLC-coupled receptors promote vasoconstriction by activating a cation current (I_{Cat}) that requires physical coupling between TRPC3 and IP_3R (Adebiyi et al., 2010; Xi et al., 2008; Zhao et al., 2008).

IP_3R-mediated Ca^{2+} waves have also been proposed to contribute to pressure-induced vasoconstriction, at least at low intravascular pressures, in cerebral VSM (Adebiyi et al., 2010; Gonzales et al., 2014; Mufti et al., 2010; Xi et al., 2008). Two recent studies proposed a role for PLCγ1 in this process, albeit via distinct signaling pathways. One study implicates pressure-induced stimulation of PLCγ1 activity to TRPC6-mediated Ca^{2+} influx leading to activation (via CICR) of IP_3-sensitized IP_3R. The resulting localized rise in $[Ca^{2+}]_i$ activates neighboring TRPM4 channels to depolarize VSM and contribute to development of the myogenic response (Gonzales et al., 2014). A second study suggested the involvement of integrin $\alpha_v\beta_3$ in activation PLCγ1, IP_3 production, IP_3R activation, and Ca^{2+} waves generation in response to an increase in intravascular pressure (Mufti et al., 2015). The ensuing Ca^{2+} waves facilitate MLC_{20} phosphorylation and development of myogenic tone. In principle, these two IP_3R-mediated SR Ca^{2+} release-dependent mechanisms could synergize to contribute to the regulation of pressure-induced vasoconstriction. Future studies should be designed to test this possibility. Additionally, a somewhat counterintuitive role for IP_3, IP_3R, and IP_3R-mediated SR Ca^{2+} release on activation of BK_{Ca} channels in cerebral VSM has been described. IP_3R activation was found to increase BK_{Ca} Ca^{2+} sensitivity (Zhao et al., 2010). This was suggested to facilitate BK_{Ca} channel activity in response to IP_3R-mediated SR Ca^{2+} release to ameliorate agonist-induced vasoconstriction. Thus, multiple IP_3R-mediated SR Ca^{2+} release-dependent and -independent mechanisms can converge to regulate VSM function.

Proliferation of VSM seems to depend on IP_3R-mediated Ca^{2+} release, specifically increased frequency of Ca^{2+} waves (Wilkerson, Heppner, Bonev, & Nelson, 2006). Accordingly, suppression of IP_3R1 expression in A7r5 cell line prevented them from proliferating (Y. Wang et al., 2001). Furthermore, it was found that IP_3R-mediated Ca^{2+} waves are necessary for the dedifferentiation of native VSM from the contractile to proliferative state (Wilkerson et al., 2006), although the mechanisms require further examination.

IP_3R-mediated signaling has been proposed to contribute to vascular pathology. For example, mesenteric VSM of ANG II-induced hypertensive mice and spontaneously hypertensive rats display elevated mRNA and protein levels of IP_3R1 (Abou-Saleh et al., 2013). This increased expression was associated with sensitization of IP_3R-mediated Ca^{2+} release, resulting in augmented vasoconstriction in response to stimulation by vasoactive agents in ANG II-induced hypertensive mice. A different study found that increased TRPC3 expression and coupling between TRPC3 and IP_3R, but no changes in IP_3R expression, contributed to agonist-induced vasoconstriction during hypertension, and that this did not require IP_3R-mediated SR Ca^{2+} release (Adebiyi et al., 2012). Some of the disparities can be related to the use of different animal models of hypertension, evaluation of different proteins, and/or diverse experimental conditions and approaches. Impaired IP_3R expression, function, and IP_3R-mediated Ca^{2+} signals in VSM have also been documented to contribute to vascular dysfunction during diabetes and atherosclerosis (Massaeli, Austria, & Pierce, 1999; Searls, Loganathan, Smirnova, & Stehno-Bittel, 2010). Thus, IP_3R-mediated SR Ca^{2+} release-dependent and -independent mechanisms may also contribute to impaired VSM function during pathological conditions.

4. MITOCHONDRIAL Ca^{2+} CHANNELS

Intracellular organelles like mitochondria are emerging as important players in smooth muscle Ca^{2+} handling. Mitochondria harbor various Ca^{2+} channels executing mitochondrial Ca^{2+} turnover. Mitochondrial-calcium-uniporter (MCU), mitochondrial RyR, mitochondrial-Ca^{2+}-channel type-2, rapid mode of uptake, and H^+/Ca^{2+} exchanger (Letm1) play a major role in mitochondrial Ca^{2+} influx. The notable channels for mitochondrial Ca^{2+} extrusion include mitochondrial Na^+/Ca^{2+} exchanger, mitochondrial permeability transition pores (mPTPs), and Letm1 (which works as an efflux channel at high mitochondrial Ca^{2+} concentrations ($[Ca^{2+}]_{mito}$)) (Hoppe, 2010). These mitochondrial Ca^{2+} channels may play a physiologically relevant role in VSM, yet they are understudied. In fact, many recent studies centered on the role of the MCU in VSM contractility (McCarron et al., 2013).

In VSM, the mitochondria appear to be a relatively immobile organelle, localized in crucial intracellular regions to operate optimally (McCarron et al., 2013). They sequester cytosolic Ca^{2+} over a wide concentration range (200 nM–10 μM) through the highly Ca^{2+}-selective channel MCU. Mitochondrial Ca^{2+} buffering capacity lies in the substantial amount of phosphate inside the organelle and the electrochemical gradient created by expulsion of H^+ by electron transport chain complexes (McCarron et al., 2013). For instance, mitochondria localized to subplasmalemmal regions of VSM have been shown to buffer stretch-induced cytosolic Ca^{2+} elevation, thereby contributing to intracellular Ca^{2+} homeostasis (Gilbert, Ducret, Marthan, Savineau, & Quignard, 2014). Additionally, evidence suggests that mitochondria do not readily buffer the initial stage of Ca^{2+} influx through LTCC, but rather affect the declining phase of the LTCC-mediated Ca^{2+} signal. However, the organelle is far quicker to scavenge the Ca^{2+} release into the cytosol through IP_3R, thereby targeting the rising phase of Ca^{2+} transient produced by these receptors. Such buffering action may prevent Ca^{2+}-dependent deactivation of IP_3R in VSM, thereby allowing repeated occurrence of IP_3R-mediated Ca^{2+} oscillation and Ca^{2+} waves (McCarron et al., 2013). The cue to distinguish the mitochondrial buffering effect on LTCC Ca^{2+} signals vs IP_3R Ca^{2+} signals may lie in a seminal work in neurons that highlights the differential Ca^{2+} sequestering effect of mitochondria on Ca_V1 and Ca_V2 channels solely based on the positional/spatial aspect of the organelle with respect to the ion channel (Wheeler et al., 2012). Interestingly, Ca^{2+} uptake by mitochondria does not affect the ATP production by the organelle (Chalmers & McCarron, 2008).

Mitochondria can also regulate VSM function via its production of ROS and modulation of the activity of several ion channels. For example, Ca^{2+} intake by mitochondria residing at close proximity of the IP_3R causes depolarization of the organelle through an elevation in $[Ca^{2+}]_{mito}$, thereby culminating in increased production of ROS and NF-kB activation (Narayanan, Xi, Pfeffer, & Jaggar, 2010). NF-kB being a transcription modulator may regulate the expression of LTCC, thus influencing arterial contraction (Narayanan et al., 2010). Recent studies have also demonstrated that the vasoconstrictor ANG II couples with NADPH oxidase to produce discrete microdomains of ROS signaling (Amberg, Earley, & Glapa, 2010). These microdomains can be amplified by adjacent mitochondrial ROS-induced ROS release to promote oxidative activation of PKC resulting in local stimulation of LTCC activity, enhanced Ca^{2+} influx and vasoconstriction (Chaplin, Nieves-Cintron, Fresquez, Navedo, & Amberg, 2015). Notably, disruption of this pathway in vivo ameliorates vascular dysfunction associated with hypertension (Chaplin et al., 2015). Mitochondria-derived ROS can also modulate the activity of RyR and BK_{Ca} channels in VSM and therefore may contribute to vasodilation under certain conditions (Cheranov & Jaggar, 2004; Xi, Cheranov, & Jaggar, 2005). Thus, mitochondria can distinctly regulate VSM function. Future studies should further examine the expression, localization, and function of mitochondrial Ca^{2+} channels, as well as their interplay with other ion channels in modulating cellular Ca^{2+} signals and VSM function.

5. CONCLUSION

Intracellular Ca^{2+} VSM is controlled by an exquisite repertoire of Ca^{2+}-permeable channels to regulate cell excitability, vessel diameter, and ultimately, blood flow. Here, we have discussed our current understanding of the expression, structure, localization, regulation, and functional role of major Ca^{2+}-permeable channels in VSM. Altered regulation of these Ca^{2+}-permeable channels can have profound impact on cardiovascular physiology and pathology. Further research is still required to completely appreciate how all these Ca^{2+}-permeable channels contribute to Ca^{2+} handling, VSM excitability, and vascular reactivity. This can be accomplished with the employment of new, emerging technologies. For instance, the development of innovative imaging tools has made possible the recording of subcellular Ca^{2+} signals produced by a single or clusters of Ca^{2+}-permeable channels. As the superior spatiotemporal resolution afforded by these technologies has begun to refine our understanding of Ca^{2+} signaling, such technologies could be adapted to further examine elementary signals produced by distinct Ca^{2+}-permeable channels expressed in VSM. It is increasingly apparent that there is an intricate physical and functional relationship among many of these Ca^{2+}-permeable channels as well as with other ion channels and regulatory signaling proteins. Therefore, more comprehensive knowledge of the cellular distribution of these channels with interacting partners is required. The advent of super-resolution nanoscopy as well as proximity ligation assay technology should aid in this task. It will also be important to systematically examine sex- and tissue-specific variations in the expression, localization, regulation, functional role, and physiological significance of all Ca^{2+}-permeable channels in VSM. Finally, the role of Ca^{2+}-permeable channels in VSM from native human tissue should be examined. This translational approach may confirm mechanisms observed in animal models and, perhaps more importantly, may reveal new information regarding ion channel physiology and pharmacology specific to humans.

CONFLICT OF INTEREST

The authors have no conflicts of interest to declare.

ACKNOWLEDGMENTS

This work was supported by grants from the National Institute of Health R01-HL098200 (M.F.N.), R01-HL121059 (M.F.N.), R01-HL095870 (L.F.S.), and T32-HL086350 (A.U.S. and M.A.N.) and American Heart Association 14GRNT18730054 (M.F.N.) and 16SDG27260070 (M.A.N.).

REFERENCES

Abd El-Rahman, R. R., Harraz, O. F., Brett, S. E., Anfinogenova, Y., Mufti, R. E., Goldman, D., & Welsh, D. G. (2013). Identification of L- and T-type Ca2+ channels in rat cerebral arteries: Role in myogenic tone development. *American Journal of*

Physiology. Heart and Circulatory Physiology, *304*(1), H58–H71. http://dx.doi.org/10.1152/ajpheart.00476.2012.

Abdi, A., Mazzocco, C., Legeron, F. P., Yvert, B., Macrez, N., & Morel, J. L. (2015). TRPP2 modulates ryanodine- and inositol-1,4,5-trisphosphate receptors-dependent Ca^{2+} signals in opposite ways in cerebral arteries. *Cell Calcium*, *58*, 467–475.

Abou-Saleh, H., Pathan, A. R., Daalis, A., Hubrack, S., Abou-Jassoum, H., Al-Naeimi, H., … Machaca, K. (2013). Inositol 1,4,5-trisphosphate (IP3) receptor up-regulation in hypertension is associated with sensitization of Ca2+ release and vascular smooth muscle contractility. *Journal of Biological Chemistry*, *288*(46), 32941–32951. http://dx.doi.org/10.1074/jbc.M113.496802.

Adebiyi, A., Thomas-Gatewood, C. M., Leo, M. D., Kidd, M. W., Neeb, Z. P., & Jaggar, J. H. (2012). An elevation in physical coupling of type 1 inositol 1,4,5-trisphosphate (IP3) receptors to transient receptor potential 3 (TRPC3) channels constricts mesenteric arteries in genetic hypertension. *Hypertension*, *60*(5), 1213–1219. http://dx.doi.org/10.1161/HYPERTENSIONAHA.112.198820.

Adebiyi, A., Zhao, G., Narayanan, D., Thomas-Gatewood, C. M., Bannister, J. P., & Jaggar, J. H. (2010). Isoform-selective physical coupling of TRPC3 channels to IP3 receptors in smooth muscle cells regulates arterial contractility. *Circulation Research*, *106*(10), 1603–1612. http://dx.doi.org/10.1161/CIRCRESAHA.110.216804.

Akaike, N., Kanaide, H., Kuga, T., Nakamura, M., Sadoshima, J., & Tomoike, H. (1989). Low-voltage-activated calcium current in rat aorta smooth muscle cells in primary culture. *Journal of Physiology*, *416*, 141–160.

Amberg, G. C., Bonev, A. D., Rossow, C. F., Nelson, M. T., & Santana, L. F. (2003). Modulation of the molecular composition of large conductance, Ca^{2+} activated K^{+} channels in vascular smooth muscle during hypertension. *Journal of Clinical Investigation*, *112*(5), 717–724.

Amberg, G. C., Earley, S., & Glapa, S. A. (2010). Local regulation of arterial L-type calcium channels by reactive oxygen species. *Circulation Research*, *107*(8), 1002–1010. http://dx.doi.org/10.1161/CIRCRESAHA.110.217018.

Amberg, G. C., & Navedo, M. F. (2013). Calcium dynamics in vascular smooth muscle. *Microcirculation*, *20*(4), 281–289. http://dx.doi.org/10.1111/micc.12046.

Amberg, G. C., Navedo, M. F., Nieves-Cintrón, M., Molkentin, J. D., & Santana, L. F. (2007). Calcium sparklets regulate local and global calcium in murine arterial smooth muscle. *Journal of Physiology*, *579*(1), 187–201.

Arikkath, J., & Campbell, K. P. (2003). Auxiliary subunits: Essential components of the voltage-gated calcium channel complex. *Current Opinion in Neurobiology*, *13*(3), 298–307.

Bannister, J. P., Adebiyi, A., Zhao, G., Narayanan, D., Thomas, C. M., Feng, J. Y., & Jaggar, J. H. (2009). Smooth muscle cell alpha2delta-1 subunits are essential for vasoregulation by CaV1.2 channels. *Circulation Research*, *105*, 948–955.

Bannister, J. P., Bulley, S., Narayanan, D., Thomas-Gatewood, C., Luzny, P., Pachuau, J., & Jaggar, J. H. (2012). Transcriptional upregulation of alpha2delta-1 elevates arterial smooth muscle cell voltage-dependent Ca2+ channel surface expression and cerebrovascular constriction in genetic hypertension. *Hypertension*, *60*(4), 1006–1015. http://dx.doi.org/10.1161/HYPERTENSIONAHA.112.199661.

Bannister, J. P., Leo, M. D., Narayanan, D., Jangsangthong, W., Nair, A., Evanson, K. W., … Jaggar, J. H. (2013). The voltage-dependent L-type Ca2+ (CaV1.2) channel C-terminus fragment is a bi-modal vasodilator. *Journal of Physiology*, *591*(Pt. 12), 2987–2998. http://dx.doi.org/10.1113/jphysiol.2013.251926.

Baylie, R. L., & Brayden, J. E. (2011). TRPV channels and vascular function. *Acta Physiologica (Oxford, England)*, *203*(1), 99–116. http://dx.doi.org/10.1111/j.1748-1716.2010.02217.x.

Bayliss, W. M. (1902). On the local reaction of the arterial wall to changes in internal pressure. *Journal of Physiology, 28*, 220–231.

Bergdahl, A., Gomez, M. F., Dreja, K., Xu, S. Z., Adner, M., Beech, D. J., … Sward, K. (2003). Cholesterol depletion impairs vascular reactivity to endothelin-1 by reducing store-operated Ca2+ entry dependent on TRPC1. *Circulation Research, 93*(9), 839–847. http://dx.doi.org/10.1161/01.RES.0000100367.45446.A3.

Bergdahl, A., Gomez, M. F., Wihlborg, A. K., Erlinge, D., Eyjolfson, A., Xu, S. Z., … Hellstrand, P. (2005). Plasticity of TRPC expression in arterial smooth muscle: Correlation with store-operated Ca2+ entry. *American Journal of Physiology. Cell Physiology, 288*(4), C872–C880. http://dx.doi.org/10.1152/ajpcell.00334.2004.

Berra-Romani, R., Mazzocco-Spezzia, A., Pulina, M. V., & Golovina, V. A. (2008). Ca2+ handling is altered when arterial myocytes progress from a contractile to a proliferative phenotype in culture. *American Journal of Physiology. Cell Physiology, 295*(3), C779–C790. http://dx.doi.org/10.1152/ajpcell.00173.2008.

Berridge, M. J. (2008). Smooth muscle cell calcium activation mechanisms. *Journal of Physiology, 586*(21), 5047–5061. http://dx.doi.org/10.1113/jphysiol.2008.160440.

Bezanilla, F. (2008). How membrane proteins sense voltage. *Nature Reviews. Molecular Cell Biology, 9*(4), 323–332. http://dx.doi.org/10.1038/nrm2376.

Bezprozvanny, I., Watras, J., & Ehrlich, B. E. (1991). Bell-shaped calcium-response curves of Ins(1,4,5)P3- and calcium-gated channels from endoplasmic reticulum of cerebellum. *Nature, 351*(6329), 751–754. http://dx.doi.org/10.1038/351751a0.

Biel, M., Ruth, P., Bosse, E., Hullin, R., Stuhmer, W., Flockerzi, V., & Hofmann, F. (1990). Primary structure and functional expression of a high voltage activated calcium channel from rabbit lung. *FEBS Letters, 269*(2), 409–412.

Boittin, F. X., Coussin, F., Morel, J. L., Halet, G., Macrez, N., & Mironneau, J. (2000). Ca(2+) signals mediated by Ins(1,4,5)P(3)-gated channels in rat ureteric myocytes. *Biochemical Journal, 349*(Pt 1), 323–332.

Boittin, F. X., Macrez, N., Halet, G., & Mironneau, J. (1999). Norepinephrine-induced Ca(2+) waves depend on InsP(3) and ryanodine receptor activation in vascular myocytes. *American Journal of Physiology, 277*(1 Pt 1), C139–C151.

Bonev, A. D., Jaggar, J. H., Rubart, M., & Nelson, M. T. (1997). Activators of protein kinase C decrease Ca^{2+} spark frequency in smooth muscle cells from cerebral arteries. *American Journal of Physiology, 273*(6 Pt 1), C2090–C2095.

Caterina, M. J., Schumacher, M. A., Tominaga, M., Rosen, T. A., Levine, J. D., & Julius, D. (1997). The capsaicin receptor: A heat-activated ion channel in the pain pathway. *Nature, 389*(6653), 816–824. http://dx.doi.org/10.1038/39807.

Catterall, W. A. (2011). Voltage-gated calcium channels. *Cold Spring Harbor Perspectives in Biology, 3*(8), a003947. http://dx.doi.org/10.1101/cshperspect.a003947.

Cavanaugh, D. J., Chesler, A. T., Jackson, A. C., Sigal, Y. M., Yamanaka, H., Grant, R., … Basbaum, A. I. (2011). Trpv1 reporter mice reveal highly restricted brain distribution and functional expression in arteriolar smooth muscle cells. *Journal of Neuroscience, 31*(13), 5067–5077. http://dx.doi.org/10.1523/JNEUROSCI.6451-10.2011.

Chalmers, S., & McCarron, J. G. (2008). The mitochondrial membrane potential and Ca2+ oscillations in smooth muscle. *Journal of Cell Science, 121*(Pt 1), 75–85. http://dx.doi.org/10.1242/jcs.014522.

Chaplin, N. L., Nieves-Cintron, M., Fresquez, A. M., Navedo, M. F., & Amberg, G. C. (2015). Arterial smooth muscle mitochondria amplify hydrogen peroxide microdomains functionally coupled to L-type calcium channels. *Circulation Research, 117*(12), 1013–1023. http://dx.doi.org/10.1161/CIRCRESAHA.115.306996.

Cheng, X., Liu, J., Asuncion-Chin, M., Blaskova, E., Bannister, J. P., Dopico, A. M., & Jaggar, J. H. (2007). A novel Ca(V)1.2 N terminus expressed in smooth muscle cells of resistance size arteries modifies channel regulation by auxiliary subunits.

Journal of Biological Chemistry, 282(40), 29211–29221. http://dx.doi.org/10.1074/jbc.M610623200.

Cheranov, S. Y., & Jaggar, J. H. (2004). Mitochondrial modulation of Ca2+ sparks and transient KCa currents in smooth muscle cells of rat cerebral arteries. *Journal of Physiology, 556*(Pt 3), 755–771. http://dx.doi.org/10.1113/jphysiol.2003.059568.

Collier, M. L., Ji, G., Wang, Y., & Kotlikoff, M. I. (2000). Calcium-induced calcium release in smooth muscle: Loose coupling between the action potential and calcium release. *Journal of General Physiology, 115*(5), 653–662.

Coussin, F., Macrez, N., Morel, J. L., & Mironneau, J. (2000). Requirement of ryanodine receptor subtypes 1 and 2 for Ca(2+)-induced Ca(2+) release in vascular myocytes. *Journal of Biological Chemistry, 275*(13), 9596–9603.

Dabertrand, F., Nelson, M. T., & Brayden, J. E. (2012). Acidosis dilates brain parenchymal arterioles by conversion of calcium waves to sparks to activate BK channels. *Circulation Research, 110*(2), 285–294. http://dx.doi.org/10.1161/CIRCRESAHA.111.258145.

DeHaven, W. I., Jones, B. F., Petranka, J. G., Smyth, J. T., Tomita, T., Bird, G. S., & Putney, J. W., Jr. (2009). TRPC channels function independently of STIM1 and Orai1. *Journal of Physiology, 587*(Pt. 10), 2275–2298. http://dx.doi.org/10.1113/jphysiol.2009.170431.

Derler, I., Schindl, R., Fritsch, R., Heftberger, P., Riedl, M. C., Begg, M., … Romanin, C. (2013). The action of selective CRAC channel blockers is affected by the Orai pore geometry. *Cell Calcium, 53*(2), 139–151. http://dx.doi.org/10.1016/j.ceca.2012.11.005.

Desai, P. N., Zhang, X., Wu, S., Janoshazi, A., Bolimuntha, S., Putney, J. W., & Trebak, M. (2015). Multiple types of calcium channels arising from alternative translation initiation of the Orai1 message. *Science Signaling, 8*(387), ra74. http://dx.doi.org/10.1126/scisignal.aaa8323.

Dietrich, A., Kalwa, H., Storch, U., Mederos y Schnitzler, M., Salanova, B., Pinkenburg, O., … Gudermann, T. (2007). Pressure-induced and store-operated cation influx in vascular smooth muscle cells is independent of TRPC1. *Pflügers Archiv, 455*(3), 465–477. http://dx.doi.org/10.1007/s00424-007-0314-3.

Dong, X. P., Wang, X., & Xu, H. (2010). TRP channels of intracellular membranes. *Journal of Neurochemistry, 113*(2), 313–328. http://dx.doi.org/10.1111/j.1471-4159.2010.06626.x.

Earley, S., & Brayden, J. E. (2015). Transient receptor potential channels in the vasculature. *Physiological Reviews, 95*(2), 645–690. http://dx.doi.org/10.1152/physrev.00026.2014.

Earley, S., Heppner, T. J., Nelson, M. T., & Brayden, J. E. (2005). TRPV4 forms a novel Ca2+ signaling complex with ryanodine receptors and BKCa channels. *Circulation Research, 97*(12), 1270–1279.

Earley, S., Pauyo, T., Drapp, R., Tavares, M. J., Liedtke, W., & Brayden, J. E. (2009). TRPV4-dependent dilation of peripheral resistance arteries influences arterial pressure. *American Journal of Physiology. Heart and Circulatory Physiology, 297*(3), H1096–H1102. http://dx.doi.org/10.1152/ajpheart.00241.2009.

Earley, S., Straub, S. V., & Brayden, J. (2007). Protein kinase C regulates vascular myogenic tone through activation of TRPM4. *American Journal of Physiology, Heart and Circulatory Physiology, 292*(6), H2613–H2622.

Earley, S., Waldron, B. J., & Brayden, J. E. (2004). Critical role for transient receptor potential channel TRPM4 in myogenic constriction of cerebral arteries. *Circulation Research, 95*(9), 922–929.

Efendiev, R., Bavencoffe, A., Hu, H., Zhu, M. X., & Dessauer, C. W. (2013). Scaffolding by A-kinase anchoring protein enhances functional coupling between adenylyl cyclase and TRPV1 channel. *Journal of Biological Chemistry, 288*(6), 3929–3937. http://dx.doi.org/10.1074/jbc.M112.428144.

Efremov, R. G., Leitner, A., Aebersold, R., & Raunser, S. (2015). Architecture and conformational switch mechanism of the ryanodine receptor. *Nature, 517*(7532), 39–43. http://dx.doi.org/10.1038/nature13916.

Essin, K., Welling, A., Hofmann, F., Luft, F. C., Gollasch, M., & Moosmang, S. (2007). Indirect coupling between Cav1.2 channels and RyR to generate Ca^{2+} sparks in murine arterial smooth muscle cells. *Journal of Physiology*, *584*(Pt 1), 205–219.

Fan, G., Baker, M. L., Wang, Z., Baker, M. R., Sinyagovskiy, P. A., Chiu, W., … Serysheva, I. I. (2015). Gating machinery of InsP3R channels revealed by electron cryomicroscopy. *Nature*, *527*(7578), 336–341. http://dx.doi.org/10.1038/nature15249.

Fernandez, J. A., McGahon, M. K., McGeown, J. G., & Curtis, T. M. (2015). CaV3.1 T-type Ca2+ channels contribute to myogenic signaling in rat retinal arterioles. *Investigative Ophthalmology & Visual Science*, *56*(9), 5125–5132. http://dx.doi.org/10.1167/iovs.15-17299.

Frischauf, I., Muik, M., Derler, I., Bergsmann, J., Fahrner, M., Schindl, R., … Romanin, C. (2009). Molecular determinants of the coupling between STIM1 and Orai channels: Differential activation of Orai1-3 channels by a STIM1 coiled-coil mutant. *Journal of Biological Chemistry*, *284*(32), 21696–21706. http://dx.doi.org/10.1074/jbc.M109.018408.

Fukushima, M., Tomita, T., Janoshazi, A., & Putney, J. W. (2012). Alternative translation initiation gives rise to two isoforms of Orai1 with distinct plasma membrane mobilities. *Journal of Cell Science*, *125*(Pt. 18), 4354–4361. http://dx.doi.org/10.1242/jcs.104919.

Gaudet, R. (2008). A primer on ankyrin repeat function in TRP channels and beyond. *Molecular Biosystems*, *4*(5), 372–379. http://dx.doi.org/10.1039/b801481g.

Gilbert, G., Ducret, T., Marthan, R., Savineau, J. P., & Quignard, J. F. (2014). Stretch-induced Ca2+ signalling in vascular smooth muscle cells depends on Ca2+ store segregation. *Cardiovascular Research*, *103*(2), 313–323. http://dx.doi.org/10.1093/cvr/cvu069.

Gonzales, A. L., Amberg, G. C., & Earley, S. (2010). Ca2+ release from the sarcoplasmic reticulum is required for sustained TRPM4 activity in cerebral artery smooth muscle cells. *American Journal of Physiology. Cell Physiology*, *299*(2), C279–C288. http://dx.doi.org/10.1152/ajpcell.00550.2009.

Gonzales, A. L., & Earley, S. (2012). Endogenous cytosolic Ca(2+) buffering is necessary for TRPM4 activity in cerebral artery smooth muscle cells. *Cell Calcium*, *51*(1), 82–93. http://dx.doi.org/10.1016/j.ceca.2011.11.004.

Gonzales, A. L., Yang, Y., Sullivan, M. N., Sanders, L., Dabertrand, F., Hill-Eubanks, D. C., … Earley, S. (2014). A PLCgamma1-dependent, force-sensitive signaling network in the myogenic constriction of cerebral arteries. *Science Signaling*, 7(327), ra49. http://dx.doi.org/10.1126/scisignal.2004732.

Gordienko, D. V., & Bolton, T. B. (2002). Crosstalk between ryanodine receptors and IP(3) receptors as a factor shaping spontaneous Ca(2+)-release events in rabbit portal vein myocytes. *Journal of Physiology*, *542*(Pt. 3), 743–762.

Gray, L. S., & Macdonald, T. L. (2006). The pharmacology and regulation of T type calcium channels: New opportunities for unique therapeutics for cancer. *Cell Calcium*, *40*(2), 115–120. http://dx.doi.org/10.1016/j.ceca.2006.04.014.

Grayson, T. H., Haddock, R. E., Murray, T. P., Wojcikiewicz, R. J., & Hill, C. E. (2004). Inositol 1,4,5-trisphosphate receptor subtypes are differentially distributed between smooth muscle and endothelial layers of rat arteries. *Cell Calcium*, *36*(6), 447–458. http://dx.doi.org/10.1016/j.ceca.2004.04.005.

Gurnett, C. A., De Waard, M., & Campbell, K. P. (1996). Dual function of the voltage-dependent Ca2+ channel alpha 2 delta subunit in current stimulation and subunit interaction. *Neuron*, *16*(2), 431–440.

Gustafsson, F., Andreasen, D., Salomonsson, M., Jensen, B. L., & Holstein-Rathlou, N. (2001). Conducted vasoconstriction in rat mesenteric arterioles: Role for dihydropyridine-insensitive Ca(2+) channels. *American Journal of Physiology. Heart and Circulatory Physiology*, *280*(2), H582–H590.

Harraz, O. F., Abd El-Rahman, R. R., Bigdely-Shamloo, K., Wilson, S. M., Brett, S. E., Romero, M., … Welsh, D. G. (2014). Ca(V)3.2 channels and the induction of negative

feedback in cerebral arteries. *Circulation Research, 115*(7), 650–661. http://dx.doi.org/10.1161/CIRCRESAHA.114.304056.
Harraz, O. F., Brett, S. E., & Welsh, D. G. (2014). Nitric oxide suppresses vascular voltage-gated T-type Ca2+ channels through cGMP/PKG signaling. *American Journal of Physiology. Heart and Circulatory Physiology, 306*(2), H279–H285. http://dx.doi.org/10.1152/ajpheart.00743.2013.
Harraz, O. F., Brett, S. E., Zechariah, A., Romero, M., Puglisi, J. L., Wilson, S. M., & Welsh, D. G. (2015). Genetic ablation of CaV3.2 channels enhances the arterial myogenic response by modulating the RyR-BKCa axis. *Arteriosclerosis, Thrombosis, and Vascular Biology, 35*(8), 1843–1851. http://dx.doi.org/10.1161/ATVBAHA.115.305736.
Harraz, O. F., Visser, F., Brett, S. E., Goldman, D., Zechariah, A., Hashad, A. M., … Welsh, D. G. (2015). CaV1.2/CaV3.x channels mediate divergent vasomotor responses in human cerebral arteries. *Journal of General Physiology, 145*(5), 405–418. http://dx.doi.org/10.1085/jgp.201511361.
Harraz, O. F., & Welsh, D. G. (2013a). Protein kinase A regulation of T-type Ca2+ channels in rat cerebral arterial smooth muscle. *Journal of Cell Science, 126*(Pt. 13), 2944–2954. http://dx.doi.org/10.1242/jcs.128363.
Harraz, O. F., & Welsh, D. G. (2013b). T-type Ca(2)(+) channels in cerebral arteries: Approaches, hypotheses, and speculation. *Microcirculation, 20*(4), 299–306. http://dx.doi.org/10.1111/micc.12038.
He, Y., Yao, G., Savoia, C., & Touyz, R. M. (2005). Transient receptor potential melastatin 7 ion channels regulate magnesium homeostasis in vascular smooth muscle cells: Role of angiotensin II. *Circulation Research, 96*(2), 207–215. http://dx.doi.org/10.1161/01.RES.0000152967.88472.3e.
Helliwell, R. M., & Large, W. A. (1997). Alpha 1-adrenoceptor activation of a non-selective cation current in rabbit portal vein by 1,2-diacyl-sn-glycerol. *Journal of Physiology, 499*(Pt. 2), 417–428.
Hill-Eubanks, D. C., Werner, M. E., Heppner, T. J., & Nelson, M. T. (2011). Calcium signaling in smooth muscle. *Cold Spring Harbor Perspectives in Biology, 3*(9), a004549. http://dx.doi.org/10.1101/cshperspect.a004549.
Hooper, R., Zhang, X., Webster, M., Go, C., Kedra, J., Marchbank, K., … Soboloff, J. (2015). Novel protein kinase C-mediated control of Orai1 function in invasive melanoma. *Molecular and Cellular Biology, 35*(16), 2790–2798. http://dx.doi.org/10.1128/MCB.01500-14.
Hoppe, U. C. (2010). Mitochondrial calcium channels. *FEBS Letters, 584*(10), 1975–1981. http://dx.doi.org/10.1016/j.febslet.2010.04.017.
Hoth, M., & Penner, R. (1993). Calcium release-activated calcium current in rat mast cells. *Journal of Physiology, 465*, 359–386.
Hou, X., Pedi, L., Diver, M. M., & Long, S. B. (2012). Crystal structure of the calcium release-activated calcium channel Orai. *Science, 338*(6112), 1308–1313. http://dx.doi.org/10.1126/science.1228757.
Hwa, J. J., & Bevan, J. A. (1986). Dihydropyridine calcium agonists selectively enhance resistance artery myogenic tone. *European Journal of Pharmacology, 126*(3), 231–238.
Iino, M. (1990). Biphasic Ca2+ dependence of inositol 1,4,5-trisphosphate-induced Ca release in smooth muscle cells of the guinea pig taenia caeci. *Journal of General Physiology, 95*(6), 1103–1122.
Inoue, R., Jensen, L. J., Shi, J., Morita, H., Nishida, M., Honda, A., & Ito, Y. (2006). Transient receptor potential channels in cardiovascular function and disease. *Circulation Research, 99*(2), 119–131. http://dx.doi.org/10.1161/01.RES.0000233356.10630.8a.
Inoue, R., Okada, T., Onoue, H., Hara, Y., Shimizu, S., Naitoh, S., … Mori, Y. (2001). The transient receptor potential protein homologue TRP6 is the essential component of

vascular alpha(1)-adrenoceptor-activated Ca(2+)-permeable cation channel. *Circulation Research, 88*(3), 325–332.
Jaggar, J. H. (2001). Intravascular pressure regulates local and global Ca(2+) signaling in cerebral artery smooth muscle cells. *American Journal of Physiology Cell Physiology, 281*(2), C439–C448.
Jaggar, J. H., & Nelson, M. T. (2000). Differential regulation of Ca^{2+} sparks and Ca^{2+} waves by UTP in rat cerebral artery smooth muscle cells. *American Journal of Physiology. Cell Physiology, 279*(5), C1528–C1539.
Jaggar, J. H., Porter, V. A., Lederer, W. J., & Nelson, M. T. (2000). Calcium sparks in smooth muscle. *American Journal of Physiology. Cell Physiology, 278*(2), C235–C256.
Jensen, L. J., Salomonsson, M., Jensen, B. L., & Holstein-Rathlou, N. H. (2004). Depolarization-induced calcium influx in rat mesenteric small arterioles is mediated exclusively via mibefradil-sensitive calcium channels. *British Journal of Pharmacology, 142*(4), 709–718. http://dx.doi.org/10.1038/sj.bjp.0705841.
Keef, K. D., Hume, J. R., & Zhong, J. (2001). Regulation of cardiac and smooth muscle Ca(2+) channels (Ca(V)1.2a, b) by protein kinases. *American Journal of Physiology. Cell Physiology, 281*(6), C1743–C1756.
Kharade, S. V., Sonkusare, S. K., Srivastava, A. K., Thakali, K. M., Fletcher, T. W., Rhee, S. W., & Rusch, N. J. (2013). The beta3 subunit contributes to vascular calcium channel upregulation and hypertension in angiotensin II-infused C57BL/6 mice. *Hypertension, 61*(1), 137–142. http://dx.doi.org/10.1161/HYPERTENSIONAHA.112.197863.
Kimura, M., Obara, K., Sasase, T., Ishikawa, T., Tanabe, Y., & Nakayama, K. (2000). Specific inhibition of stretch-induced increase in L-type calcium channel currents by herbimycin A in canine basilar arterial myocytes. *British Journal of Pharmacology, 130*(4), 923–931. http://dx.doi.org/10.1038/sj.bjp.0703360.
Klugbauer, N., Lacinova, L., Marais, E., Hobom, M., & Hofmann, F. (1999). Molecular diversity of the calcium channel alpha2delta subunit. *Journal of Neuroscience, 19*(2), 684–691.
Knot, H. J., & Nelson, M. T. (1998). Regulation of arterial diameter and wall [Ca^{2+}] in cerebral arteries of rat by membrane potential and intravascular pressure. *Journal of Physiology, 508*(Pt. 1), 199–209.
Koda, K., Hyakkoku, K., Ogawa, K., Takasu, K., Imai, S., Sakurai, Y., … Morioka, Y. (2016). Sensitization of TRPV1 by protein kinase C in rats with mono-iodoacetate-induced joint pain. *Osteoarthritis and Cartilage, 24*(7), 1254–1262. http://dx.doi.org/10.1016/j.joca.2016.02.010.
Koulen, P., Cai, Y., Geng, L., Maeda, Y., Nishimura, S., Witzgall, R., … Somlo, S. (2002). Polycystin-2 is an intracellular calcium release channel. *Nature Cell Biology, 4*(3), 191–197. http://dx.doi.org/10.1038/ncb754.
Kuo, I. Y., Ellis, A., Seymour, V. A., Sandow, S. L., & Hill, C. E. (2010). Dihydropyridine-insensitive calcium currents contribute to function of small cerebral arteries. *Journal of Cerebral Blood Flow and Metabolism, 30*(6), 1226–1239. http://dx.doi.org/10.1038/jcbfm.2010.11.
Kwan, H. Y., Shen, B., Ma, X., Kwok, Y. C., Huang, Y., Man, Y. B., … Yao, X. (2009). TRPC1 associates with BK(Ca) channel to form a signal complex in vascular smooth muscle cells. *Circulation Research, 104*(5), 670–678. http://dx.doi.org/10.1161/CIRCRESAHA.108.188748.
Lamont, C., & Wier, W. G. (2004). Different roles of ryanodine receptors and inositol (1,4,5)-trisphosphate receptors in adrenergically stimulated contractions of small arteries. *American Journal of Physiology. Heart and Circulatory Physiology, 287*(2), H617–H625. http://dx.doi.org/10.1152/ajpheart.00708.2003.
Lange, I., Yamamoto, S., Partida-Sanchez, S., Mori, Y., Fleig, A., & Penner, R. (2009). TRPM2 functions as a lysosomal Ca2+-release channel in beta cells. *Science Signaling, 2*(71), ra23. http://dx.doi.org/10.1126/scisignal.2000278.

Lanner, J. T., Georgiou, D. K., Joshi, A. D., & Hamilton, S. L. (2010). Ryanodine receptors: Structure, expression, molecular details, and function in calcium release. *Cold Spring Harbor Perspectives in Biology*, *2*(11), a003996. http://dx.doi.org/10.1101/cshperspect.a003996.

Lee, J. H., Gomora, J. C., Cribbs, L. L., & Perez-Reyes, E. (1999). Nickel block of three cloned T-type calcium channels: Low concentrations selectively block alpha1H. *Biophysical Journal*, *77*(6), 3034–3042.

Li, Y., & Brayden, J. E. (2015). Rho kinase activity governs arteriolar myogenic depolarization. *Journal of Cerebral Blood Flow and Metabolism*. http://dx.doi.org/10.1177/0271678X15621069, Epub ahead of print.

Li, J., McKeown, L., Ojelabi, O., Stacey, M., Foster, R., O'Regan, D., … Beech, D. J. (2011). Nanomolar potency and selectivity of a Ca(2)(+) release-activated Ca(2)(+) channel inhibitor against store-operated Ca(2)(+) entry and migration of vascular smooth muscle cells. *British Journal of Pharmacology*, *164*(2), 382–393. http://dx.doi.org/10.1111/j.1476-5381.2011.01368.x.

Liao, M., Cao, E., Julius, D., & Cheng, Y. (2013). Structure of the TRPV1 ion channel determined by electron cryo-microscopy. *Nature*, *504*(7478), 107–112. http://dx.doi.org/10.1038/nature12822.

Liao, P., Yu, D., Li, G., Yong, T. F., Soon, J. L., Chua, Y. L., & Soong, T. W. (2007). A smooth muscle Cav1.2 calcium channel splice variant underlies hyperpolarized window current and enhanced state-dependent inhibition by nifedipine. *Journal of Biological Chemistry*, *282*(48), 35133–35142. http://dx.doi.org/10.1074/jbc.M705478200.

Lindsey, S. H., & Songu-Mize, E. (2010). Stretch-induced TRPC4 downregulation is accompanied by reduced capacitative Ca2+ entry in WKY but not SHR mesenteric smooth muscle cells. *Clinical and Experimental Hypertension*, *32*(5), 288–292. http://dx.doi.org/10.3109/10641960903443525.

Massaeli, H., Austria, J. A., & Pierce, G. N. (1999). Chronic exposure of smooth muscle cells to minimally oxidized LDL results in depressed inositol 1,4,5-trisphosphate receptor density and Ca(2+) transients. *Circulation Research*, *85*(6), 515–523.

Mauban, J. R., Zacharia, J., Fairfax, S., & Wier, W. G. (2015). PC-PLC/sphingomyelin synthase activity plays a central role in the development of myogenic tone in murine resistance arteries. *American Journal of Physiology. Heart and Circulatory Physiology*, *308*(12), H1517–H1524. http://dx.doi.org/10.1152/ajpheart.00594.2014.

McCarron, J. G., Olson, M. L., Wilson, C., Sandison, M. E., & Chalmers, S. (2013). Examining the role of mitochondria in Ca(2)(+) signaling in native vascular smooth muscle. *Microcirculation*, *20*(4), 317–329. http://dx.doi.org/10.1111/micc.12039.

Meents, J. E., Neeb, L., & Reuter, U. (2010). TRPV1 in migraine pathophysiology. *Trends in Molecular Medicine*, *16*(4), 153–159. http://dx.doi.org/10.1016/j.molmed.2010.02.004.

Mercado, J., Baylie, R., Navedo, M. F., Yuan, C., Scott, J. D., Nelson, M. T., … Santana, L. F. (2014). Local control of TRPV4 channels by AKAP150-targeted PKC in arterial smooth muscle. *Journal of General Physiology*, *143*(5), 559–575. http://dx.doi.org/10.1085/jgp.201311050.

Moosmang, S., Schulla, V., Welling, A., Feil, R., Feil, S., Wegener, J. W., … Klugbauer, N. (2003). Dominant role of smooth muscle L-type calcium channel Cav1.2 for blood pressure regulation. *EMBO Journal*, *22*(22), 6027–6034.

Mufti, R. E., Brett, S. E., Tran, C. H., Abd El-Rahman, R., Anfinogenova, Y., El-Yazbi, A., … Welsh, D. G. (2010). Intravascular pressure augments cerebral arterial constriction by inducing voltage-insensitive Ca2+ waves. *Journal of Physiology*, *588*(Pt. 20), 3983–4005. http://dx.doi.org/10.1113/jphysiol.2010.193300.

Mufti, R. E., Zechariah, A., Sancho, M., Mazumdar, N., Brett, S. E., & Welsh, D. G. (2015). Implications of alphavbeta3 integrin signaling in the regulation of Ca2+ waves and myogenic tone in cerebral arteries. *Arteriosclerosis, Thrombosis, and Vascular Biology*, *35*(12), 2571–2578. http://dx.doi.org/10.1161/ATVBAHA.115.305619.

Muik, M., Frischauf, I., Derler, I., Fahrner, M., Bergsmann, J., Eder, P., … Romanin, C. (2008). Dynamic coupling of the putative coiled-coil domain of ORAI1 with STIM1 mediates ORAI1 channel activation. *Journal of Biological Chemistry*, *283*(12), 8014–8022. http://dx.doi.org/10.1074/jbc.M708898200.

Mullins, F. M., Park, C. Y., Dolmetsch, R. E., & Lewis, R. S. (2009). STIM1 and calmodulin interact with Orai1 to induce Ca2+-dependent inactivation of CRAC channels. *Proceedings of the National Academy of Sciences of the United States of America*, *106*(36), 15495–15500. http://dx.doi.org/10.1073/pnas.0906781106.

Muraki, K., Iwata, Y., Katanosaka, Y., Ito, T., Ohya, S., Shigekawa, M., & Imaizumi, Y. (2003). TRPV2 is a component of osmotically sensitive cation channels in murine aortic myocytes. *Circulation Research*, *93*(9), 829–838. http://dx.doi.org/10.1161/01.RES.0000097263.10220.0C.

Narayanan, D., Adebiyi, A., & Jaggar, J. H. (2012). Inositol trisphosphate receptors in smooth muscle cells. *American Journal of Physiology. Heart and Circulatory Physiology*, *302*(11), H2190–H2210. http://dx.doi.org/10.1152/ajpheart.01146.2011.

Narayanan, D., Bulley, S., Leo, M. D., Burris, S. K., Gabrick, K. S., Boop, F. A., & Jaggar, J. H. (2013). Smooth muscle cell transient receptor potential polycystin-2 (TRPP2) channels contribute to the myogenic response in cerebral arteries. *Journal of Physiology*, *591*(Pt. 20), 5031–5046. http://dx.doi.org/10.1113/jphysiol.2013.258319.

Narayanan, D., Xi, Q., Pfeffer, L. M., & Jaggar, J. H. (2010). Mitochondria control functional CaV1.2 expression in smooth muscle cells of cerebral arteries. *Circulation Research*, *107*(5), 631–641. http://dx.doi.org/10.1161/CIRCRESAHA.110.224345.

Navedo, M. F., & Amberg, G. C. (2013). Local regulation of L-type ca(2+) channel sparklets in arterial smooth muscle. *Microcirculation*, *20*(4), 290–298. http://dx.doi.org/10.1111/micc.12021.

Navedo, M. F., Amberg, G. C., Nieves, M., Molkentin, J. D., & Santana, L. F. (2006). Mechanisms underlying heterogeneous Ca^{2+} sparklet activity in arterial smooth muscle. *Journal of General Physiology*, *127*(6), 611–622.

Navedo, M. F., Amberg, G., Votaw, S. V., & Santana, L. F. (2005). Constitutively active L-type Ca^{2+} channels. *Proceedings of the National Academy of Sciences of the United States of America*, *102*(31), 11112–11117.

Navedo, M. F., Cheng, E. P., Yuan, C., Votaw, S., Molkentin, J. D., Scott, J. D., & Santana, L. F. (2010). Increased coupled gating of L-type Ca2+ channels during hypertension and Timothy syndrome. *Circulation Research*, *106*(4), 748–756. http://dx.doi.org/10.1161/CIRCRESAHA.109.213363.

Navedo, M. F., Nieves-Cintron, M., Amberg, G. C., Yuan, C., Votaw, V. S., Lederer, W. J., … Santana, L. F. (2008). AKAP150 is required for stuttering persistent Ca2+ sparklets and angiotensin II-induced hypertension. *Circulation Research*, *102*(2), e1–e11. http://dx.doi.org/10.1161/CIRCRESAHA.107.167809. CIRCRESAHA.107.167809 [pii].

Navedo, M. F., & Santana, L. F. (2013). CaV1.2 sparklets in heart and vascular smooth muscle. *Journal of Molecular and Cellular Cardiology*, *58*, 67–76. http://dx.doi.org/10.1016/j.yjmcc.2012.11.018.

Navedo, M. F., Takeda, Y., Nieves-Cintron, M., Molkentin, J. D., & Santana, L. F. (2010). Elevated Ca2+ sparklet activity during acute hyperglycemia and diabetes in cerebral arterial smooth muscle cells. *American Journal of Physiology. Cell Physiology*, *298*(2), C211–C220. http://dx.doi.org/10.1152/ajpcell.00267.2009.

Nelson, M. T., Cheng, H., Rubart, M., Santana, L. F., Bonev, A. D., Knot, H. J., & Lederer, W. J. (1995). Relaxation of arterial smooth muscle by calcium sparks. *Science*, *270*(5236), 633–637.

Nelson, M. T., Patlak, J. B., Worley, J. F., & Standen, N. B. (1990). Calcium channels, potassium channels, and voltage dependence of arterial smooth muscle tone. *American Journal of Physiology*, *259*(1 Pt. 1), C3–C18.

Newton, C. L., Mignery, G. A., & Sudhof, T. C. (1994). Co-expression in vertebrate tissues and cell lines of multiple inositol 1,4,5-trisphosphate (InsP3) receptors with distinct affinities for InsP3. *Journal of Biological Chemistry*, *269*(46), 28613–28619.

Nieves-Cintron, M., Amberg, G. C., Navedo, M. F., Molkentin, J. D., & Santana, L. F. (2008). The control of Ca2+ influx and NFATc3 signaling in arterial smooth muscle during hypertension. *Proceedings of the National Academy of Sciences of the United States of America*, *105*(40), 15623–15628. http://dx.doi.org/10.1073/pnas.0808759105. 0808759105 [pii].

Nieves-Cintrón, M., Amberg, G. C., Nichols, C. B., Molkentin, J. D., & Santana, L. F. (2007). Activation of NFATc3 down-regulates the β1 subunit of large conductance, calcium-activated K^+ channels in arterial smooth muscle and contributes to hypertension. *Journal of Biological Chemistry*, *282*(5), 3231–3240.

Nieves-Cintron, M., Nystoriak, M. A., Prada, M. P., Johnson, K., Fayer, W., Dell'Acqua, M. L., … Navedo, M. F. (2015). Selective downregulation of Kv2.1 function contributes to enhanced arterial tone during diabetes. *Journal of Biological Chemistry*, *290*(12), 7918–7929.

Nikitina, E., Jahromi, B. S., Bouryi, V. A., Takahashi, M., Xie, A., Zhang, Z. D., & Macdonald, R. L. (2006). Voltage-dependent calcium channels of dog basilar artery. *The Journal of Physiology*, *580*, 523–541.

Nixon, G. F., Mignery, G. A., & Somlyo, A. V. (1994). Immunogold localization of inositol 1,4,5-trisphosphate receptors and characterization of ultrastructural features of the sarcoplasmic reticulum in phasic and tonic smooth muscle. *Journal of Muscle Research and Cell Motility*, *15*(6), 682–700.

Nystoriak, M. A., Murakami, K., Penar, P. L., & Wellman, G. C. (2009). Ca(v)1.2 splice variant with exon 9* is critical for regulation of cerebral artery diameter. *American Journal of Physiology. Heart and Circulatory Physiology*, *297*(5), H1820–H1828. http://dx.doi.org/10.1152/ajpheart.00326.2009.

Nystoriak, M. A., Nieves-Cintron, M., & Navedo, M. F. (2013). Capturing single L-type Ca(2+) channel function with optics. *Biochimica et Biophysica Acta*, *1833*(7), 1657–1664. http://dx.doi.org/10.1016/j.bbamcr.2012.10.027.

Nystoriak, M. A., Nieves-Cintron, M., Nygren, P. J., Hinke, S. A., Nichols, C. B., Chen, C. Y., … Navedo, M. F. (2014). AKAP150 contributes to enhanced vascular tone by facilitating large-conductance Ca2+-activated K+ channel remodeling in hyperglycemia and diabetes mellitus. *Circulation Research*, *114*(4), 607–615. http://dx.doi.org/10.1161/CIRCRESAHA.114.302168.

Oancea, E., Wolfe, J. T., & Clapham, D. E. (2006). Functional TRPM7 channels accumulate at the plasma membrane in response to fluid flow. *Circulation Research*, *98*(2), 245–253. http://dx.doi.org/10.1161/01.RES.0000200179.29375.cc.

Olah, Z., Szabo, T., Karai, L., Hough, C., Fields, R. D., Caudle, R. M., … Iadarola, M. J. (2001). Ligand-induced dynamic membrane changes and cell deletion conferred by vanilloid receptor 1. *Journal of Biological Chemistry*, *276*(14), 11021–11030. http://dx.doi.org/10.1074/jbc.M008392200.

Olson, M. L., Chalmers, S., & McCarron, J. G. (2010). Mitochondrial Ca2+ uptake increases Ca2+ release from inositol 1,4,5-trisphosphate receptor clusters in smooth muscle cells. *Journal of Biological Chemistry*, *285*(3), 2040–2050. http://dx.doi.org/10.1074/jbc.M109.027094.

Park, C. Y., Shcheglovitov, A., & Dolmetsch, R. (2010). The CRAC channel activator STIM1 binds and inhibits L-type voltage-gated calcium channels. *Science*, *330*(6000), 101–105. http://dx.doi.org/10.1126/science.1191027.

Parker, I., & Smith, I. F. (2010). Recording single-channel activity of inositol trisphosphate receptors in intact cells with a microscope, not a patch clamp. *Journal of General Physiology*, *136*(2), 119–127. http://dx.doi.org/10.1085/jgp.200910390.

Perez, G. J., Bonev, A. D., Patlak, J. B., & Nelson, M. T. (1999). Functional coupling of ryanodine receptors to KCa channels in smooth muscle cells from rat cerebral arteries. *Journal of General Physiology*, *113*(2), 229–238.

Perez-Reyes, E. (2006). Molecular characterization of T-type calcium channels. *Cell Calcium*, *40*(2), 89–96. http://dx.doi.org/10.1016/j.ceca.2006.04.012.

Picard, C., McCarl, C. A., Papolos, A., Khalil, S., Luthy, K., Hivroz, C., … Feske, S. (2009). STIM1 mutation associated with a syndrome of immunodeficiency and autoimmunity. *New England Journal of Medicine*, *360*(19), 1971–1980. http://dx.doi.org/10.1056/NEJMoa0900082.

Potier, M., Gonzalez, J. C., Motiani, R. K., Abdullaev, I. F., Bisaillon, J. M., Singer, H. A., & Trebak, M. (2009). Evidence for STIM1- and Orai1-dependent store-operated calcium influx through ICRAC in vascular smooth muscle cells: Role in proliferation and migration. *FASEB Journal*, *23*(8), 2425–2437. http://dx.doi.org/10.1096/fj.09-131128.

Prakriya, M., & Lewis, R. S. (2001). Potentiation and inhibition of Ca(2+) release-activated Ca(2+) channels by 2-aminoethyldiphenyl borate (2-APB) occurs independently of IP(3) receptors. *Journal of Physiology*, *536*(Pt. 1), 3–19.

Putney, J. W., Jr. (1986). A model for receptor-regulated calcium entry. *Cell Calcium*, 7(1), 1–12.

Reading, S. A., Earley, S., Waldron, B. J., Welsh, D. G., & Brayden, J. E. (2005). TRPC3 mediates pyrimidine receptor-induced depolarization of cerebral arteries. *American Journal of Physiology. Heart and Circulatory Physiology*, *288*(5), H2055–H2061.

Roos, J., DiGregorio, P. J., Yeromin, A. V., Ohlsen, K., Lioudyno, M., Zhang, S., … Stauderman, K. A. (2005). STIM1, an essential and conserved component of store-operated Ca2+ channel function. *Journal of Cell Biology*, *169*(3), 435–445. http://dx.doi.org/10.1083/jcb.200502019.

Sabourin, J., Lamiche, C., Vandebrouck, A., Magaud, C., Rivet, J., Cognard, C., … Constantin, B. (2009). Regulation of TRPC1 and TRPC4 cation channels requires an alpha1-syntrophin-dependent complex in skeletal mouse myotubes. *Journal of Biological Chemistry*, *284*(52), 36248–36261. http://dx.doi.org/10.1074/jbc.M109.012872.

Saleh, S. N., Albert, A. P., & Large, W. A. (2009). Activation of native TRPC1/C5/C6 channels by endothelin-1 is mediated by both PIP3 and PIP2 in rabbit coronary artery myocytes. *Journal of Physiology*, *587*(Pt. 22), 5361–5375. http://dx.doi.org/10.1113/jphysiol.2009.180331.

Saleh, S. N., Albert, A. P., Peppiatt, C. M., & Large, W. A. (2006). Angiotensin II activates two cation conductances with distinct TRPC1 and TRPC6 channel properties in rabbit mesenteric artery myocytes. *Journal of Physiology*, *577*(Pt. 2), 479–495. http://dx.doi.org/10.1113/jphysiol.2006.119305.

Saleh, S. N., & Greenwood, I. A. (2005). Activation of chloride currents in murine portal vein smooth muscle cells by membrane depolarization involves intracellular calcium release. *American Journal of Physiology. Cell Physiology*, *288*(1), C122–C131. http://dx.doi.org/10.1152/ajpcell.00384.2004.

Samso, M., Wagenknecht, T., & Allen, P. D. (2005). Internal structure and visualization of transmembrane domains of the RyR1 calcium release channel by cryo-EM. *Nature Structural & Molecular Biology*, *12*(6), 539–544. http://dx.doi.org/10.1038/nsmb938.

Santana, L. F., & Navedo, M. F. (2009). Molecular and biophysical mechanisms of Ca2+ sparklets in smooth muscle. *Journal of Molecular and Cellular Cardiology*, *47*(4), 436–444. http://dx.doi.org/10.1016/j.yjmcc.2009.07.008. S0022-2828(09)00279-X [pii].

Santana, L. F., Navedo, M. F., Amberg, G. C., Nieves-Cintron, M., Votaw, V. S., & Ufret-Vincenty, C. A. (2008). Calcium sparklets in arterial smooth muscle. *Clinical and Experimental Pharmacology & Physiology*, *35*(9), 1121–1126. http://dx.doi.org/10.1111/j.1440-1681.2007.04867.x. CEP4867 [pii].

Searls, Y. M., Loganathan, R., Smirnova, I. V., & Stehno-Bittel, L. (2010). Intracellular Ca2 + regulating proteins in vascular smooth muscle cells are altered with type 1 diabetes due to the direct effects of hyperglycemia. *Cardiovascular Diabetology*, *9*, 8. http://dx.doi.org/10.1186/1475-2840-9-8.

Serysheva, I. I., Ludtke, S. J., Baker, M. L., Cong, Y., Topf, M., Eramian, D., … Chiu, W. (2008). Subnanometer-resolution electron cryomicroscopy-based domain models for the cytoplasmic region of skeletal muscle RyR channel. *Proceedings of the National Academy of Sciences of the United States of America*, *105*(28), 9610–9615. http://dx.doi.org/10.1073/pnas.0803189105.

Sharif-Naeini, R., Folgering, J. H., Bichet, D., Duprat, F., Lauritzen, I., Arhatte, M., … Honore, E. (2009). Polycystin-1 and -2 dosage regulates pressure sensing. *Cell*, *139*(3), 587–596. http://dx.doi.org/10.1016/j.cell.2009.08.045.

Somlyo, A. P. (1985). Excitation-contraction coupling and the ultrastructure of smooth muscle. *Circulation Research*, *57*(4), 497–507.

Spassova, M. A., Soboloff, J., He, L. P., Xu, W., Dziadek, M. A., & Gill, D. L. (2006). STIM1 has a plasma membrane role in the activation of store-operated Ca(2+) channels. *Proceedings of the National Academy of Sciences of the United States of America*, *103*(11), 4040–4045. http://dx.doi.org/10.1073/pnas.0510050103.

Storch, U., Forst, A. L., Philipp, M., Gudermann, T., & Mederos y Schnitzler, M. (2012). Transient receptor potential channel 1 (TRPC1) reduces calcium permeability in heteromeric channel complexes. *Journal of Biological Chemistry*, *287*(5), 3530–3540. http://dx.doi.org/10.1074/jbc.M111.283218.

Strubing, C., Krapivinsky, G., Krapivinsky, L., & Clapham, D. E. (2001). TRPC1 and TRPC5 form a novel cation channel in mammalian brain. *Neuron*, *29*(3), 645–655.

Takeda, Y., Nystoriak, M. A., Nieves-Cintron, M., Santana, L. F., & Navedo, M. F. (2011). Relationship between Ca2+ sparklets and sarcoplasmic reticulum Ca2+ load and release in rat cerebral arterial smooth muscle. *American Journal of Physiology. Heart and Circulatory Physiology*, *301*(6), H2285–H2294. http://dx.doi.org/10.1152/ajpheart.00488.2011.

Tasker, P. N., Michelangeli, F., & Nixon, G. F. (1999). Expression and distribution of the type 1 and type 3 inositol 1,4,5-trisphosphate receptor in developing vascular smooth muscle. *Circulation Research*, *84*(5), 536–542.

Tasker, P. N., Taylor, C. W., & Nixon, G. F. (2000). Expression and distribution of InsP (3) receptor subtypes in proliferating vascular smooth muscle cells. *Biochemical and Biophysical Research Communications*, *273*(3), 907–912. http://dx.doi.org/10.1006/bbrc.2000.3036.

Toth, A., Czikora, A., Pasztor, E. T., Dienes, B., Bai, P., Csernoch, L., … Boczan, J. (2014). Vanilloid receptor-1 (TRPV1) expression and function in the vasculature of the rat. *Journal of Histochemistry and Cytochemistry*, *62*(2), 129–144. http://dx.doi.org/10.1369/0022155413513589.

Tovey, S. C., de Smet, P., Lipp, P., Thomas, D., Young, K. W., Missiaen, L., … Bootman, M. D. (2001). Calcium puffs are generic InsP(3)-activated elementary calcium signals and are downregulated by prolonged hormonal stimulation to inhibit cellular calcium responses. *Journal of Cell Science*, *114*(Pt. 22), 3979–3989.

Vaithianathan, T., Narayanan, D., Asuncion-Chin, M. T., Jeyakumar, L. H., Liu, J., Fleischer, S., … Dopico, A. M. (2010). Subtype identification and functional characterization of ryanodine receptors in rat cerebral artery myocytes. *American Journal of Physiology. Cell Physiology*, *299*(2), C264–C278. http://dx.doi.org/10.1152/ajpcell.00318.2009.

VanBavel, E., Sorop, O., Andreasen, D., Pfaffendorf, M., & Jensen, B. L. (2002). Role of T-type calcium channels in myogenic tone of skeletal muscle resistance arteries. *American Journal of Physiology. Heart and Circulatory Physiology*, *283*(6), H2239–H2243. http://dx.doi.org/10.1152/ajpheart.00531.2002.

Varga-Szabo, D., Authi, K. S., Braun, A., Bender, M., Ambily, A., Hassock, S. R., … Nieswandt, B. (2008). Store-operated Ca(2+) entry in platelets occurs independently of transient receptor potential (TRP) C1. *Pflügers Archiv*, *457*(2), 377–387. http://dx.doi.org/10.1007/s00424-008-0531-4.

Walker, R. L., Hume, J. R., & Horowitz, B. (2001). Differential expression and alternative splicing of TRP channel genes in smooth muscles. *American Journal of Physiology. Cell Physiology*, *280*(5), C1184–C1192.

Wang, Y., Chen, J., Wang, Y., Taylor, C. W., Hirata, Y., Hagiwara, H., … Sakaki, Y. (2001). Crucial role of type 1, but not type 3, inositol 1,4,5-trisphosphate (IP(3)) receptors in IP(3)-induced Ca(2+) release, capacitative Ca(2+) entry, and proliferation of A7r5 vascular smooth muscle cells. *Circulation Research*, *88*(2), 202–209.

Wang, Y., Deng, X., Mancarella, S., Hendron, E., Eguchi, S., Soboloff, J., … Gill, D. L. (2010). The calcium store sensor, STIM1, reciprocally controls Orai and CaV1.2 channels. *Science*, *330*(6000), 105–109. http://dx.doi.org/10.1126/science.1191086.

Wang, Q., Hogg, R. C., & Large, W. A. (1992). Properties of spontaneous inward currents recorded in smooth muscle cells isolated from the rabbit portal vein. *Journal of Physiology*, *451*, 525–537.

Weiss, S., & Dascal, N. (2015). Molecular aspects of modulation of L-type calcium channels by protein kinase C. *Current Molecular Pharmacology*, *8*(1), 43–53.

Welsh, D. G., Morielli, A. D., Nelson, M. T., & Brayden, J. E. (2002). Transient receptor potential channels regulate myogenic tone of resistance arteries. *Circulation Research*, *90*(3), 248–250.

Welsh, D. G., Nelson, M. T., Eckman, D. M., & Brayden, J. E. (2000). Swelling-activated cation channels mediate depolarization of rat cerebrovascular smooth muscle by hyposmolarity and intravascular pressure. *Journal of Physiology*, *527*(Pt. 1), 139–148.

Wheeler, D. G., Groth, R. D., Ma, H., Barrett, C. F., Owen, S. F., Safa, P., & Tsien, R. W. (2012). Ca(V)1 and Ca(V)2 channels engage distinct modes of Ca(2+) signaling to control CREB-dependent gene expression. *Cell*, *149*(5), 1112–1124. http://dx.doi.org/10.1016/j.cell.2012.03.041.

Wijetunge, S., & Hughes, A. D. (2007). Src family tyrosine kinases mediate contraction of rat isolated tail arteries in response to a hyposmotic stimulus. *Journal of Hypertension*, *25*(9), 1871–1878. http://dx.doi.org/10.1097/HJH.0b013e328255e8f0.

Wilkerson, M. K., Heppner, T. J., Bonev, A. D., & Nelson, M. T. (2006). Inositol trisphosphate receptor calcium release is required for cerebral artery smooth muscle cell proliferation. *American Journal of Physiology. Heart and Circulatory Physiology*, *290*(1), H240–H247. http://dx.doi.org/10.1152/ajpheart.01191.2004.

Wojcikiewicz, R. J., & Luo, S. G. (1998). Differences among type I, II, and III inositol-1,4,5-trisphosphate receptors in ligand-binding affinity influence the sensitivity of calcium stores to inositol-1,4,5-trisphosphate. *Molecular Pharmacology*, *53*(4), 656–662.

Wray, S., & Burdyga, T. (2010). Sarcoplasmic reticulum function in smooth muscle. *Physiological Reviews*, *90*(1), 113–178. http://dx.doi.org/10.1152/physrev.00018.2008.

Xi, Q., Adebiyi, A., Zhao, G., Chapman, K. E., Waters, C. M., Hassid, A., & Jaggar, J. H. (2008). IP3 constricts cerebral arteries via IP3 receptor-mediated TRPC3 channel activation and independently of sarcoplasmic reticulum Ca2+ release. *Circulation Research*, *102*(9), 1118–1126. http://dx.doi.org/10.1161/CIRCRESAHA.108.173948.

Xi, Q., Cheranov, S. Y., & Jaggar, J. H. (2005). Mitochondria-derived reactive oxygen species dilate cerebral arteries by activating Ca2+ sparks. *Circulation Research*, *97*(4), 354–362. http://dx.doi.org/10.1161/01.RES.0000177669.29525.78.

Xu, S. Z., Boulay, G., Flemming, R., & Beech, D. J. (2006). E3-targeted anti-TRPC5 antibody inhibits store-operated calcium entry in freshly isolated pial arterioles. *American Journal of Physiology. Heart and Circulatory Physiology*, *291*(6), H2653–H2659. http://dx.doi.org/10.1152/ajpheart.00495.2006.

Zacharia, J., Zhang, J., & Wier, W. G. (2007). Ca2+ signaling in mouse mesenteric small arteries: Myogenic tone and adrenergic vasoconstriction. *American Journal of Physiology. Heart and Circulatory Physiology*, *292*(3), H1523–H1532. http://dx.doi.org/10.1152/ajpheart.00670.2006.
Zalk, R., Clarke, O. B., des Georges, A., Grassucci, R. A., Reiken, S., Mancia, F., … Marks, A. R. (2015). Structure of a mammalian ryanodine receptor. *Nature*, *517*(7532), 44–49. http://dx.doi.org/10.1038/nature13950.
Zhang, W., Halligan, K. E., Zhang, X., Bisaillon, J. M., Gonzalez-Cobos, J. C., Motiani, R. K., … Trebak, M. (2011). Orai1-mediated I (CRAC) is essential for neointima formation after vascular injury. *Circulation Research*, *109*(5), 534–542. http://dx.doi.org/10.1161/CIRCRESAHA.111.246777.
Zhang, S. L., Yu, Y., Roos, J., Kozak, J. A., Deerinck, T. J., Ellisman, M. H., … Cahalan, M. D. (2005). STIM1 is a Ca2+ sensor that activates CRAC channels and migrates from the Ca2+ store to the plasma membrane. *Nature*, *437*(7060), 902–905. http://dx.doi.org/10.1038/nature04147.
Zhang, W., Zhang, X., Gonzalez-Cobos, J. C., Stolwijk, J. A., Matrougui, K., & Trebak, M. (2015). Leukotriene-C4 synthase, a critical enzyme in the activation of store-independent Orai1/Orai3 channels, is required for neointimal hyperplasia. *Journal of Biological Chemistry*, *290*(8), 5015–5027. http://dx.doi.org/10.1074/jbc.M114.625822.
Zhao, G., Adebiyi, A., Blaskova, E., Xi, Q., & Jaggar, J. H. (2008). Type 1 inositol 1,4,5-trisphosphate receptors mediate UTP-induced cation currents, Ca2+ signals, and vasoconstriction in cerebral arteries. *American Journal of Physiology. Cell Physiology*, *295*(5), C1376–C1384. http://dx.doi.org/10.1152/ajpcell.00362.2008.
Zhao, G., Neeb, Z. P., Leo, M. D., Pachuau, J., Adebiyi, A., Ouyang, K., … Jaggar, J. H. (2010). Type 1 IP3 receptors activate BKCa channels via local molecular coupling in arterial smooth muscle cells. *Journal of General Physiology*, *136*(3), 283–291. http://dx.doi.org/10.1085/jgp.201010453.
Zhou, H., Nakamura, T., Matsumoto, N., Hisatsune, C., Mizutani, A., Iesaki, T., … Mikoshiba, K. (2008). Predominant role of type 1 IP3 receptor in aortic vascular muscle contraction. *Biochemical and Biophysical Research Communications*, *369*(1), 213–219. http://dx.doi.org/10.1016/j.bbrc.2007.12.194.

CHAPTER THREE

Potassium Channels in Regulation of Vascular Smooth Muscle Contraction and Growth

W.F. Jackson[1]
Michigan State University, East Lansing, MI, United States
[1]Corresponding author: e-mail address: jacks783@msu.edu

Contents

Abstract

Potassium channels importantly contribute to the regulation of vascular smooth muscle (VSM) contraction and growth. They are the dominant ion conductance of the VSM cell membrane and importantly determine and regulate membrane potential. Membrane potential, in turn, regulates the open-state probability of voltage-gated Ca^{2+} channels (VGCC), Ca^{2+} influx through VGCC, intracellular Ca^{2+}, and VSM contraction. Membrane potential also affects release of Ca^{2+} from internal stores and the Ca^{2+} sensitivity of the contractile machinery such that K^+ channels participate in all aspects of regulation of VSM contraction. Potassium channels also regulate proliferation of VSM cells through membrane potential-dependent and membrane potential-independent

Advances in Pharmacology, Volume 78
ISSN 1054-3589
http://dx.doi.org/10.1016/bs.apha.2016.07.001

mechanisms. VSM cells express multiple isoforms of at least five classes of K^+ channels that contribute to the regulation of contraction and cell proliferation (growth). This review will examine the structure, expression, and function of large conductance, Ca^{2+}-activated K^+ (BK_{Ca}) channels, intermediate-conductance Ca^{2+}-activated K^+ ($K_{Ca}3.1$) channels, multiple isoforms of voltage-gated K^+ (K_V) channels, ATP-sensitive K^+ (K_{ATP}) channels, and inward-rectifier K^+ (K_{IR}) channels in both contractile and proliferating VSM cells.

ABBREVIATIONS

AKAP A-kinase anchoring protein
BK_{Ca} large-conductance, Ca^{2+}-activated K^+
$Ca_V1.2$ L-type voltage-gated Ca^{2+} channels
CO carbon monoxide
COOH carboxy
CREB cAMP-response element-binding protein
EBIO 1-ethyl-2-benzimidazolinone
EC electrochemical
EETs epoxyeicosatrienoic acids
E_K K^+ equilibrium potential
ERK extracellular signal-regulated kinase
grad gradient
H_2S hydrogen sulfide
IP_3 inositol-1,4,5-trisphosphate
IP_3R IP_3 receptors
K_{ATP} ATP-sensitive K^+
$K_{Ca}3.1$, sK4, IK1 intermediate-conductance Ca^{2+}-activated K^+ channel
K_{IR} inward-rectifier K^+
K_V voltage-gated K^+
LRRCs leucine-rich-repeat-containing proteins
MEK mitogen-activated protein kinase kinase
mTOR mammalian target of rapamycin
MuRF1 muscle RING finger protein 1
NF nuclear factor
NFATC3 nuclear factor of activated T-cells, cytoplasmic 3
NH_2 amino
NO nitric oxide
NOR-1 neuron-derived orphan receptor-1
NOX5 NADPH oxidase 5
NS1619 1,3-dihydro-1-[2-hydroxy-5-(trifluoromethyl)phenyl]-5-(trifluoromethyl)-2H-benzimidazol-2-one
NS309 6,7-dichloro-1H-indole-2,3-dione 3-oxime
PDGF platelet-derived growth factor
PKA protein kinase A
PKC protein kinase C
PKG protein kinase G
RCK regulator of K^+ conductance

REST repressor element 1-silencing transcription factor
ROC receptor-operated channels
RyR ryanodine receptors
SAC stretch-activated channels
SKA-31 naphtho[1,2-d]thiazol-2-ylamine
SOC store-operated channels
SUR sulphonylurea receptors
TM transmembrane
TRP transient receptor potential
VGCC voltage-gated Ca^{2+} channels
VSM vascular smooth muscle

1. INTRODUCTION

Potassium channels importantly contribute to the regulation of vascular smooth muscle (VSM) contraction and growth. They are the dominant ion conductance of the VSM cell membrane and importantly determine and regulate VSM cell membrane potential (Jackson, 2000, 2005). Membrane potential, in turn, regulates the open-state probability of voltage-gated Ca^{2+} channels (VGCC), Ca^{2+} influx through these channels, intracellular Ca^{2+} and VSM contraction (Jackson, 2000, 2005). Membrane potential also affects release of Ca^{2+} from internal stores and the Ca^{2+} sensitivity of the contractile machinery such that K^+ channels participate in all aspects of regulation of VSM contraction (del Valle-Rodriguez, Lopez-Barneo, & Urena, 2003; Fernández-Tenorio et al., 2010, 2011; Ganitkevich & Isenberg, 1993; Kukuljan, Rojas, Catt, & Stojilkovic, 1994; Liu et al., 2009; Mahaut-Smith, Martinez-Pinna, & Gurung, 2008; Okada, Yanagisawa, & Taira, 1993; Urena, del Valle-Rodriguez, & Lopez-Barneo, 2007; Yamagishi, Yanagisawa, & Taira, 1992; Yamamura, Ohya, Muraki, & Imaizumi, 2012; Yanagisawa, Yamagishi, & Okada, 1993). Potassium channels also contribute to the regulation of proliferation of VSM cells through membrane potential-dependent (Bi et al., 2013; Miguel-Velado et al., 2005, 2010) and membrane potential-independent mechanisms (Cidad et al., 2012, 2015; Jimenez-Perez et al., 2016).

VSM cells express multiple isoforms of at least five classes of K^+ channels that participate in the regulation of contraction and cell proliferation (growth). These include large conductance, Ca^{2+}-activated K^+ (BK_{Ca}) channels, intermediate-conductance Ca^{2+}-activated K^+ ($K_{Ca}3.1$) channels,

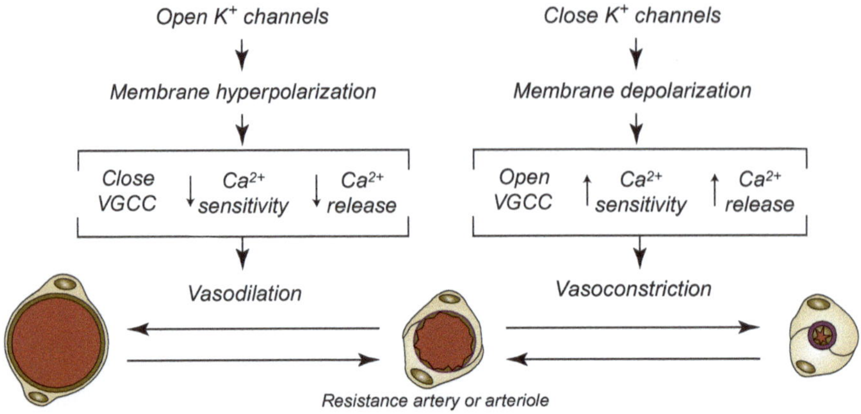

Fig. 1 Potassium channels regulate vascular smooth muscle contraction. Schematic diagram outlining the effects of K^+ channel opening and closing on the membrane potential of VSM cells, which, in turn, affects processes that lead to relaxation or contraction of VSM leading to vasodilation or vasoconstriction. Voltage-gated Ca^{2+} channels (VGCC). See text for more information. *Modified from Jackson, W. F. (2005). Potassium channels in the peripheral microcirculation.* Microcirculation, 12*(1), 113–127. doi:10.1080/10739680590896072.*

multiple isoforms of voltage-gated K^+ (K_V) channels, ATP-sensitive K^+ (K_{ATP}) channels, inward-rectifier K^+ (K_{IR}) channels, and members of the two-pore K^+ (K_{2P}) channel family of K^+ channels. Subsequent sections of this review will examine the function of K^+ channels in the regulation of VSM cell contraction and proliferation (Fig. 1). The expression and function of K_{2P} channels in VSM cells will not be addressed and the reader is referred to the literature for information on these channels (Feliciangeli, Chatelain, Bichet, & Lesage, 2015; Gurney & Manoury, 2009; O'Connell, Morton, & Hunter, 2002; Renigunta, Schlichthorl, & Daut, 2015; Sepulveda, Pablo Cid, Teulon, & Niemeyer, 2015).

2. POTASSIUM CHANNELS AND REGULATION OF VSM CONTRACTION

2.1 Setting the Stage

VSM cells, in small arteries and arterioles that develop myogenic tone when pressurized, are relatively depolarized, with membrane potentials on the order of −45 to −30 mV (Burns, Cohen, & Jackson, 2004; Emerson & Segal, 2000; Knot & Nelson, 1998; Siegl, Koeppen, Wolfle, Pohl, & de

Wit, 2005; Welsh, Jackson, & Segal, 1998). At physiological ion concentrations (3–5 mM K^+ extracellular, 140 mM K^+ intracellular), the electrochemical (EC) gradient (grad) for K^+ (the driving force for movement of K^+ through a K^+ channel) is outward. This means that opening of K^+ channels will lead to K^+ diffusion out of the cell, loss of positive charge, and membrane hyperpolarization (Nelson, Patlak, Worley, & Standen, 1990). Conversely, closure of open K^+ channels will result in a decrease in this hyperpolarizing current and membrane depolarization. The resistance of the plasma membrane of VSM cells to current flow is very high, on the order of 1–10 GΩ (Nelson, Patlak, et al., 1990). This means that very small currents produced by only a few active K^+ channels can have very large effects on membrane potential (Nelson, Patlak, et al., 1990). VGCC contribute substantially to the regulation of intracellular Ca^{2+} and contraction of VSM cells in differentiated, contractile VSM cells, particularly in resistance arteries and arterioles (Nelson, Patlak, et al., 1990). Voltage-dependent activation (depolarization) and deactivation (hyperpolarization) of these channels importantly regulates VSM contraction (Nelson, Patlak, et al., 1990).

The structure of the ion-conducting pore of K^+ channels is thought to be similar across all of the channels based on studies of KcsA, a two transmembrane (TM) domain K^+ channel from *Streptomyces lividans* (Kuang, Purhonen, & Hebert, 2015) (Fig. 2). As shown in Fig. 2B and C, the pore is formed by TM2 and the P-loop, which connects TM1 and TM2. Conserved residues in the P-loop (Thr-Val-Gly-Tyr-Gly) comprise the K^+ selectivity filter (red highlighted segment of the P-loop in Fig. 2B) (Kuang et al., 2015). In BK_{Ca} channels and K_V channels, segment 6 (S6) and the P-loop between S5 and S6 form the channels' pores (Figs. 3 and 4) (Kuang et al., 2015).

2.2 BK_{Ca} Channels and VSM Contraction

VSM cells typically express a high density of BK_{Ca} channels in their plasma membranes that importantly contribute to the negative feedback regulation of vascular tone (Fig. 3). A homotetramer of α-subunits encoded by the KCNMA1 gene composes the channels (Butler, Tsunoda, McCobb, Wei, & Salkoff, 1993; Pallanck & Ganetzky, 1994) (Fig. 3A). Segment 6 and the P-loop connecting S5 and S6 form the ion-conducting pore (Meera, Wallner, Song, & Toro, 1997) (Fig. 3A). Two regulators of K^+ conductance (RCK) domains (RCK1 and RCK2) in the C-terminus of the α-subunits contain the channel's Ca^{2+} sensors (Hoshi, Pantazis, &

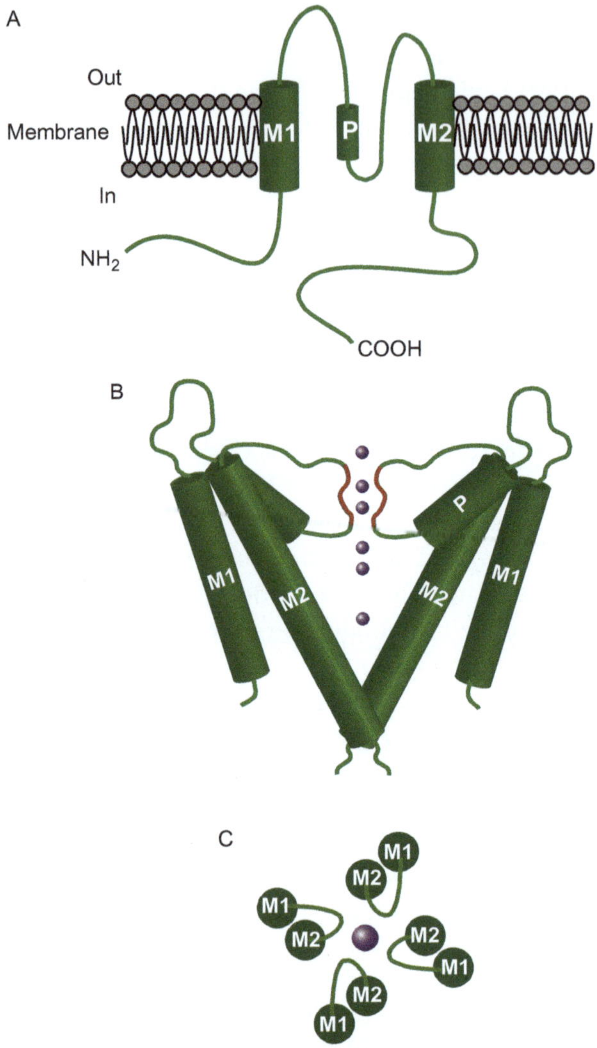

Fig. 2 Structure of the ion-conducting pore of K^+ channels. Top panel (A) shows a schematic representation of a two membrane-spanning domain (M1 and M2) K^+ channel. Functional channels are formed from a tetramer of these units, with the ion-conducting pore being formed by M2 and the P-loop domain that connects M1 and M2 (P in figure refers to the pore helical domain). Middle panel (B) shows approximate orientation of two sets of the M1, M2, and P-loops forming the channel. The *blue spheres* represent K^+ ions, and the *red highlighted regions* of the P-loops represent the selectivity filter of the channel's pore. The bottom panel (C) shows a top view of the channel subunits and the P-loop forming the K^+ ion-conducting pore. See text for references.

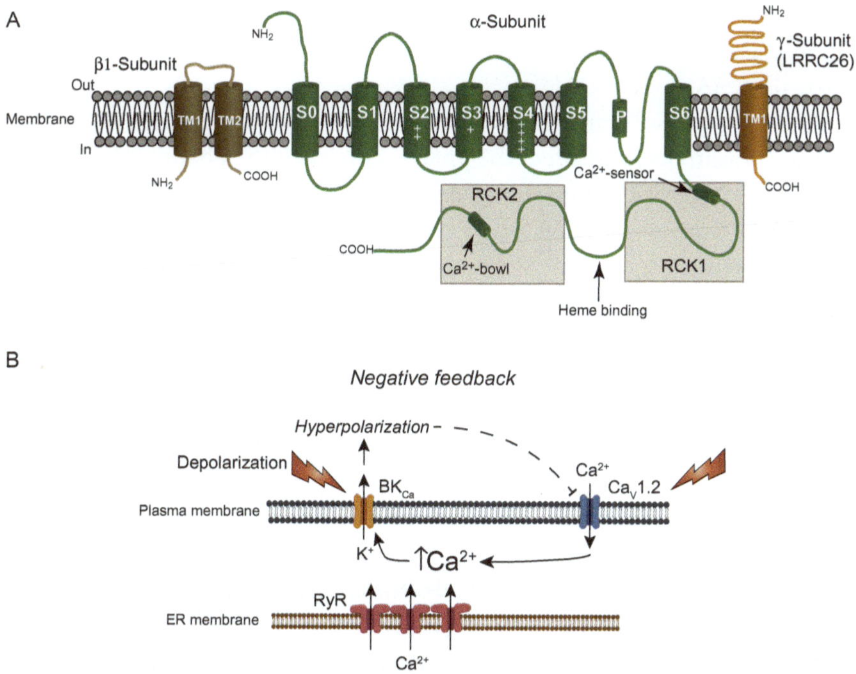

Fig. 3 Vascular BK_{Ca} channels. Panel (A) shows a β_1-subunit with two membrane-spanning domains, one pore-forming α-subunit with seven membrane-spanning domains and a γ-subunit (LRRC26, for example) with one membrane-spanning domain. Panel (B) shows a schematic of the primary negative feedback role for BK_{Ca} channels in contractile VSM. Membrane depolarization (due to activation of other membrane channels, not shown), or increases in intracellular Ca^{2+} in the vicinity of BK_{Ca} channels due to release of Ca^{2+} from ryanodine receptors (RyR, Ca^{2+} sparks), or influx of Ca^{2+} through L-type voltage-gated Ca^{2+} channels ($Ca_V1.2$), results in activation and opening of BK_{Ca} channels. The efflux of K^+ through these channels leads to membrane hyperpolarization and closure of $Ca_V1.2$ channels, negative feedback regulation of VSM excitability.

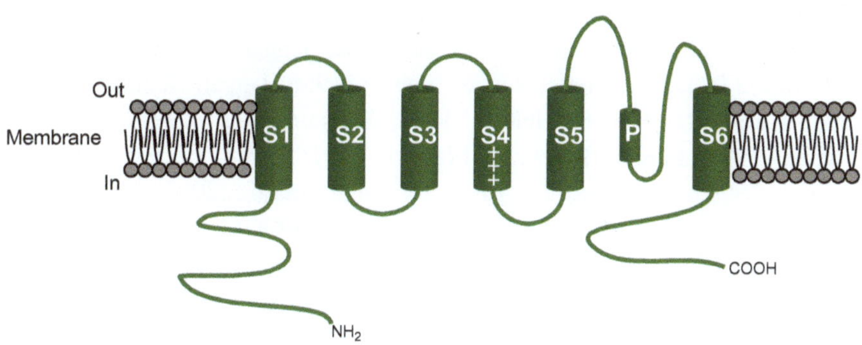

Fig. 4 The pore-forming α-subunit of K_V channels. Shown is a schematic of the six membrane-spanning domains of a typical K_V channel. The length and composition of the carboxy (COOH) and amino (NH_2) varies among the large number of K_V channel isoforms expressed in VSM cells. See text for more information.

Olcese, 2013) (Fig. 3A). Positively charged residues in TM domains S2, S3, and S4 serve as the channel's voltage sensors (Hoshi, Pantazis, et al., 2013) (Fig. 3A).

Accessory β_1-subunits (locus: KCNMB1) slow gating kinetics, increase the Ca^{2+} sensitivity, and affect the pharmacology of the channels (McManus et al., 1995; Meera, Wallner, Jiang, & Toro, 1996; Tseng-Crank et al., 1996) (Fig. 2). Activation of VSM BK_{Ca} channels by 17β-estradiol, lithocholate, dehydroepiandrosterone, dehydrosoyasaponin-I, and docosahexaenoic acid requires the presence of β_1-subunits (Hoshi, Tian, Xu, Heinemann, & Hou, 2013; Hou, Heinemann, & Hoshi, 2009). These subunits also contribute to dynamic trafficking of α-subunits to plasma membrane (Leo et al., 2014). The degree of coupling between the α-subunits and the β_1-subunits may account for the high Ca^{2+}-setpoint observed in arteriolar BK_{Ca} channels (Yang et al., 2009; Yang, Sohma, et al., 2013).

In addition to the β_1-subunits, leucine-rich-repeat-containing proteins (LRRCs), such as LRRC26, have been proposed as γ-subunits of VSM BK_{Ca} channels (Evanson, Bannister, Leo, & Jaggar, 2014). These subunits interact with BK_{Ca} channels, increasing both their voltage sensitivity and channel activation by agents such as NS1619 (Evanson et al., 2014).

Membrane depolarization and increases in intracellular Ca^{2+} activate BK_{Ca} channels (Fig. 3B). In resistance arteries and arterioles that develop pressure-induced myogenic tone, ex vivo, BK_{Ca} channels are active and contribute to resting VSM membrane potential. However, the source of Ca^{2+} responsible for BK_{Ca} channel activation may differ dependent on the anatomical origin of the vessel. Calcium released from groups of ryanodine receptors (RyR) in the subplasma membrane endoplasmic reticulum, in the form of Ca^{2+} sparks, control the activity of overlying BK_{Ca} channels in many resistance arteries (Brenner et al., 2000; Bychkov, Gollasch, Ried, Luft, & Haller, 1997; Furstenau et al., 2000; Gollasch et al., 2000; Jaggar, Porter, Lederer, & Nelson, 2000; Jaggar, Stevenson, & Nelson, 1998; Jaggar, Wellman, et al., 1998; Knot, Standen, & Nelson, 1998; Nelson et al., 1995; Nelson & Quayle, 1995; Perez, Bonev, & Nelson, 2001; Perez, Bonev, Patlak, & Nelson, 1999; Porter et al., 1998; Wellman et al., 2002; Wellman & Nelson, 2003). In contrast, Ca^{2+} influx through VGCC may activate BK_{Ca} channels in other vessels including hamster and mouse cremaster arterioles (Westcott, Goodwin, Segal, & Jackson, 2012; Westcott & Jackson, 2011), rabbit coronary arteries (Guia, Wan, Courtemanche, & Leblanc, 1999), and mouse mesenteric arteries (Suzuki, Yamamura, Ohya, & Imaizumi, 2013).

In striated muscle resistance arteries, both RyR-based Ca^{2+} sparks and VGCC Ca^{2+}-influx contribute to activation of BK_{Ca} channels (Westcott et al., 2012; Westcott & Jackson, 2011) suggesting that both mechanisms may be active in some cells. In cerebral VSM cells, Ca^{2+} influx through T-type, CaV3.2 VGCC stimulates RyR-based Ca^{2+} sparks contributing to the negative feedback regulation of myogenic tone (Harraz et al., 2014, 2015).

Vasoconstrictors have been reported to both activate (Berczi, Stekiel, Contney, & Rusch, 1992; Brayden & Nelson, 1992; Ganitkevich & Isenberg, 1990; Hashemzadeh-Gargari & Rembold, 1992; Jackson & Blair, 1998; Nelson et al., 1995; Nelson & Quayle, 1995; Rusch & Liu, 1997; Wakatsuki, Nakaya, & Inoue, 1992) and inhibit (Lange, Gebremedhin, Narayanan, & Harder, 1997; Scornik & Toro, 1992; Toro, Amador, & Stefani, 1990; Wesselman, Schubert, VanBavel, Nilsson, & Mulvany, 1997) BK_{Ca} channels. Activation of BK_{Ca} channels would tend to hyperpolarize VSM cells, deactivate VGCC, and limit VSM contraction, essentially preventing vasospasm (Fig. 3B). This activation results from both vasoconstrictor-induced depolarization and increases in intracellular Ca^{2+} (Fig. 3B). In contrast, inhibition of BK_{Ca} channels would promote depolarization and would enhance VSM contraction, in a positive feedback fashion. Protein kinase C (PKC), which is commonly activated by vasoconstrictors that activate $G_{q/11}$-coupled, heptahelical receptors, may be involved in this process in some blood vessels (Lange et al., 1997; Minami, Fukuzawa, & Nakaya, 1993), and may involve internalization and degradation of BK_{Ca} channels (Leo et al., 2015). Despite the evidence for inhibition of BK_{Ca} channel activity, the dominant effect of vasoconstrictors in most blood vessels is to activate BK_{Ca} channels.

Vasodilators that act through receptors coupled to the guanine nucleotide-binding protein, $G\alpha_s$, and formation of cAMP activate BK_{Ca} channels, as part of their mechanism of action (Kume, Graziano, & Kotlikoff, 1992; Kume, Takai, Tokuno, & Tomita, 1989; Sadoshima, Akaike, Kanaide, & Nakamura, 1988). Activation of BK_{Ca} channels may result from a number of mechanisms. Interaction of $G\alpha_s$ with BK_{Ca} channels, independent from cAMP and protein kinase A (PKA), may increase channel activity (Kume et al., 1992; Kume, Hall, Washabau, Takagi, & Kotlikoff, 1994; Scornik, Codina, Birnbaumer, & Toro, 1993). PKA-dependent phosphorylation of the α-subunits also can activate BK_{Ca} channels (Nara, Dhulipala, Wang, & Kotlikoff, 1998; Tian et al., 2004, 2001). Increased trafficking of the β1-subunits to the plasma membrane may also contribute to the mechanism of action of cAMP-related agonists

(Matsumoto, Szasz, Tostes, & Webb, 2012). Vasodilators that act via cAMP and PKA increase BK_{Ca} channel activity by increasing Ca^{2+} spark activity (Porter et al., 1998; Wellman, Santana, Bonev, & Nelson, 2001; Yamaguchi, Kajita, & Madison, 1995). In addition, exchange proteins activated by cAMP (EPACs) participate in cAMP-related activation of BK_{Ca} channels in VSM (Roberts, Kamishima, Barrett-Jolley, Quayle, & Dart, 2013). Thus, cAMP-related vasodilators may activate BK_{Ca} channels by a number of mechanisms.

Endothelium-derived nitric oxide (NO), nitrovasodilators that release NO and other vasodilators that act through the cGMP-protein kinase G (PKG) signaling cascade also have been proposed to activate BK_{Ca} channels (Fujino et al., 1991; Li, Zou, & Campbell, 1997; Robertson, Schubert, Hescheler, & Nelson, 1993; Taniguchi, Furukawa, & Shigekawa, 1993; Williams, Katz, Roy-Contancin, & Reuben, 1988; Winquist et al., 1985; Winquist, Faison, & Nutt, 1984). This may occur through modulation of Ca^{2+} sparks (Jewell, Saundry, Bonev, Tranmer, & Wellman, 2004; Mandala, Heppner, Bonev, & Nelson, 2007; Yuill, McNeish, Kansui, Garland, & Dora, 2010), by phosphorylation of the channels by PKG (Alioua, Huggins, & Rousseau, 1995; Swayze & Braun, 2001) or altered channel trafficking (Leo et al., 2014). Activation of BK_{Ca} channels by NO, independent from cGMP has also been proposed (Ahern, Hsu, & Jackson, 1999; Bolotina, Najibi, Palacino, Pagano, & Cohen, 1994; Buxton, Kaiser, Malmquist, & Tichenor, 2001; Lang, Harvey, McPhee, & Klemm, 2000; Lang, Harvey, & Mulholland, 2003; Li, Jin, & Campbell, 1998; Lovren & Triggle, 2000; Mistry & Garland, 1998; Plane, Hurrell, Jeremy, & Garland, 1996; Plane, Sampson, Smith, & Garland, 2001; Yu, Sun, Maier, Harder, & Roman, 2002; Zhang, Tazzeo, Chu, & Janssen, 2006). In contrast, there are also a number of studies that have failed to demonstrate participation of BK_{Ca} channels in the mechanism of action of NO (Armstead, 1997; Bialecki & Stinson-Fisher, 1995; Brayden, 1990; Cooke, Rossitch, Andon, Loscalzo, & Dzau, 1991; Dong, Waldron, Galipeau, Cole, & Triggle, 1997; Fukami et al., 1998; Garland & McPherson, 1992; Ghisdal, Gomez, & Morel, 2000; Hansen & Olesen, 1997; Hernanz et al., 1999; Kilpatrick & Cocks, 1994; Plane & Garland, 1993; Plane, Wiley, Jeremy, Cohen, & Garland, 1998; Taguchi, Heistad, Kitazono, & Faraci, 1995; Wellman & Bevan, 1995; Zhu, Beny, Flammer, Luscher, & Haefliger, 1997). Thus, there may be regional or species differences that account for the presence or lack of effect of NO on BK_{Ca} channel activity.

Carbon monoxide (CO) (Abraham & Kappas, 2008; Jaggar et al., 2002, 2005; Li et al., 2008; Wang, Wang, & Wu, 1997; Wang & Wu, 1997; Wang, Wu, & Wang, 1997; Xi et al., 2004, 2010), epoxyeicosatrienoic acids (EETs) (Campbell, Gebremedhin, Pratt, & Harder, 1996; Earley, Heppner, Nelson, & Brayden, 2005; Eckman, Hopkins, McBride, & Keef, 1998), H_2O_2 (Barlow & White, 1998; Cheranov & Jaggar, 2006; Thengchaisri & Kuo, 2003), and hydrogen sulfide (H_2S) (Jackson-Weaver et al., 2013; Liang, Xi, Leffler, & Jaggar, 2012) all may activate BK_{Ca} channels. Effects of CO on BK_{Ca} channels may be direct via associated heme proteins that interact with the C-terminal domain of the α-subunits, between RCK1 and RCK2 (Jaggar et al., 2005) or by interaction with histidine residues in RCK1 (Hou et al., 2009) (Fig. 3A). In addition, CO augments Ca^{2+} spark frequency and their coupling to BK_{Ca} channels (Jaggar et al., 2002; Li et al., 2008; Xi et al., 2010). EETs also may stimulate Ca^{2+} sparks to activate BK_{Ca} channels through actions of EETs on transient receptor potential (TRP) V4 channels (Earley et al., 2005). Stimulation of Ca^{2+} sparks also may underlie the activation of BK_{Ca} channels by H_2S (Jackson-Weaver et al., 2013; Liang et al., 2012).

2.3 Diseases and VSM BK_{Ca} Channels

The effects of disease states on BK_{Ca} channel expression and function are complex. The activity of BK_{Ca} channels appears to be depressed in obesity (Borbouse et al., 2009, 2010; Frisbee, Maier, & Stepp, 2002; Nystoriak et al., 2014; Ozkor et al., 2011; Rusch, 2009), diabetes (Dong et al., 2009; Fernandez-Velasco, Ruiz-Hurtado, Gomez, & Rueda, 2014; Liu & Gutterman, 2002; McGahon et al., 2007; Mokelke, Dietz, Eckman, Nelson, & Sturek, 2005; Nystoriak et al., 2014; Wang et al., 2010; Yi et al., 2014; Zhou, Wang, Lamping, & Lee, 2006), and some models of aging (Albarwani, Al-Siyabi, Baomar, & Hassan, 2010; Marijic et al., 2001). In diabetes, the impaired function of BK_{Ca} channels involves reduced expression and function of the β_1-subunits (McGahon et al., 2007; Nystoriak et al., 2014; Yi et al., 2014). Diabetes and hyperglycemia results in increased proteolytic degradation of β_1-subunits via nuclear factor (NF)-kB-dependent expression and function of the muscle RING finger protein 1 (MuRF1) ubiquitin ligase (Yi et al., 2014). The downregulation of the β_1-subunit in diabetes also may involve activation of calcineurin/nuclear factor of activated T-cells, cytoplasmic 3 (NFATC3) signaling that is facilitated by A-kinase anchoring protein 150 (AKAP150) (Nystoriak et al., 2014).

However, the effects of hypertension on BK_{Ca} channel function are not clear, because both increased (Asano, Masuzawa-Ito, & Matsuda, 1993; Asano, Matsuda, Hayakawa, Ito, & Ito, 1993; Liu, Pleyte, Knaus, & Rusch, 1997; Paterno, Heistad, & Faraci, 1997; Rusch, Delucena, Wooldridge, England, & Cowley, 1992; Rusch & Liu, 1997; Zhang, Gao, Zuo, Lee, & Janssen, 2005) and reduced (Amberg, Bonev, Rossow, Nelson, & Santana, 2003; Amberg & Santana, 2003; Ambroisine et al., 2007; Bratz, Dick, Partridge, & Kanagy, 2005; Bratz, Swafford, Kanagy, & Dick, 2005; Callera, Yogi, Tostes, Rossoni, & Bendhack, 2004; Li, Lu, & Shi, 2014; Moreno-Dominguez, Cidad, Miguel-Velado, Lopez-Lopez, & Perez-Garcia, 2009; Nieves-Cintron, Amberg, Nichols, Molkentin, & Santana, 2007; Yang, Li, et al., 2013) functions have been reported. Differences in hypertension models, duration of hypertension, type of blood vessel, and the species studied may account for the lack of consensus on the effects of hypertension on BK_{Ca} channel function.

2.4 K_V Channels and VSM Contraction

VSM cells express a diverse array of K_V channels that include members of the K_V1 (loci: KCNA2–6) (Cox, 2005), K_V2 (loci: KCNB2–3) (Cox, 2005), K_V3 (loci: KCNC2–4) (Cox, 2005), K_V4 (loci: KCND1–3) (Cox, 2005), $K_V6.3$ (locus: KCNG3) (Moreno-Dominguez et al., 2009), K_V7 (loci: KCNQ1, KCNQ4–5) (Greenwood & Ohya, 2009; Jepps, Olesen, & Greenwood, 2013; Mackie & Byron, 2008), and $K_V9.3$ (locus: KCNS3) (Cox, 2005) families of K_V channels. They consist of homo- or heterotetramers of α-subunits (Fig. 4) (Kuang et al., 2015). Segment 6 (S6) and the P-loop between S5 and S6 form the channel's pore, as noted earlier (Kuang et al., 2015). Positively charged residues in S4 confer voltage sensitivity to the channels (Fig. 4) (Kuang et al., 2015). Modulatory accessory subunits accompany many K_V channels, affecting channel membrane expression, gating kinetics, and voltage sensitivity (Gutman et al., 2005).

Membrane depolarization activates K_V channels, and, in general, they participate in the negative feedback regulation of VSM contraction along with BK_{Ca} channels. Consistent with this negative feedback role, block of K_V channels potentiates VSM contraction induced by vasoconstrictors (Chadha et al., 2014; Cheong, Dedman, & Beech, 2001; Cheong, Dedman, Xu, & Beech, 2001; Cook, 1989; Hald et al., 2012; Martinez et al., 2009; Pagan et al., 2009; Shimizu, Yokoshiki, Sperelakis, & Paul,

2000). K_V channels are active at the resting membrane potential of VSM cells in blood vessels displaying myogenic tone; closure of these channels leads to membrane depolarization and vasoconstriction (Cox, 2005; Jackson, 2000, 2005; Nelson & Quayle, 1995).

Vasoconstrictors, including phenylephrine (Mistry & Garland, 1999), 5-HT (Bae et al., 2006; Ko, Park, Firth, Hong, et al., 2010; Sung et al., 2013), and angiotensin II (Clement-Chomienne, Walsh, & Cole, 1996) all inhibit K_V channels, probably acting through PKC (Clement-Chomienne et al., 1996; Hayabuchi, Standen, & Davies, 2001; Ko, Park, Firth, Hong, et al., 2010), Src tyrosine kinase (Sung et al., 2013), Rho kinase (Luykenaar, Brett, Wu, Wiehler, & Welsh, 2004; Luykenaar, El-Rahman, Walsh, & Welsh, 2009), and/or increased intracellular Ca^+ (Cox & Petrou, 1999; Gelband, Ishikawa, Post, Keef, & Hume, 1993; Ishikawa, Hume, & Keef, 1993). This closure of K_V channels may contribute to vasoconstrictor-induced VSM cell membrane depolarization and the mechanism of action of vasoconstrictors.

Vasodilators acting through the cAMP-PKA pathway activate K_V channels and contribute to their mechanism of action (Aiello, Malcolm, Walsh, & Cole, 1998; Aiello, Walsh, & Cole, 1994, 1995; Berwick et al., 2010; Chadha et al., 2014, 2012; Dick et al., 2008; Dong, Waldron, Cole, & Triggle, 1998; Heaps & Bowles, 2002; Heaps, Tharp, & Bowles, 2005; Khanamiri et al., 2013; Li, Chai, Gutterman, & Liu, 2003; Moore, Nelson, Parelkar, Rusch, & Rhee, 2014; Satake, Shibata, & Shibata, 1996). NO (Dick et al., 2008; Sobey & Faraci, 1999; Stott, Barrese, Jepps, Leighton, & Greenwood, 2015; Tanaka et al., 2006), H_2S (Cheang et al., 2010; Martelli et al., 2013; Rogers, Chilian, Bratz, Bryan, & Dick, 2007; Schleifenbaum et al., 2010), hypoxia (Hedegaard et al., 2014), acidosis (Berger, Vandier, Bonnet, Jackson, & Rusch, 1998), and anticontractile substances release by perivascular adipose tissue (Tano, Schleifenbaum, & Gollasch, 2014; Zavaritskaya et al., 2013) also may activate VSM K_V channels in some blood vessels.

2.5 Disease and VSM K_V Channels

The expression and function of K_V channels are reduced in diabetes (Bubolz, Li, Wu, & Liu, 2005; Chai, Liu, & Chen, 2005; Chai et al., 2007; Ko, Park, Firth, Kim, et al., 2010; Li et al., 2003) that may be mediated by elevated glucose (Li, Gutterman, Rusch, Bubolz, & Liu, 2004; Liu, Terata, Rusch, & Gutterman, 2001) and channel nitration (Li et al., 2004). The reduced K_V channel function may contribute to the increased VSM contractile function that is observed in diabetes.

However, in hypertension and obesity, the impact on K_V channels is not as clear. Increased (Cox, Folander, & Swanson, 2001; Cox, Fromme, Folander, & Swanson, 2008), decreased (Bratz, Dick, et al., 2005; Bratz, Swafford, et al., 2005; Cox, 1996; Cox, Lozinskaya, & Dietz, 2001; Liu, Hudetz, Knaus, & Rusch, 1998; Martens & Gelband, 1996; Tobin et al., 2009), or no change (Liu, Jones, & Sturek, 1994; Liu et al., 1997) in K_V channel function in hypertension has been reported. There also is no clear effect of obesity on K_V channel function, with decreased (Berwick et al., 2012; Dick & Tune, 2010; Nieves-Cintron et al., 2015; Yang, Jones, Thomas, & Rubin, 2007) and increased (Jiang, Thoren, Caligiuri, Hansson, & Pernow, 1999; Ko, Park, Firth, Hong, et al., 2010) function reported. Specific effects may depend on the vascular bed studied, the duration and severity of the pathology, and the species studied.

2.6 K_{ATP} Channels and VSM Contraction

VSM cells express K_{ATP} channels that consist of tetramers of pore-forming $K_{IR}6.1$ subunits (locus: KCNJ8) (Aziz et al., 2014; Li et al., 2013; Miki et al., 2002; Miura et al., 2003; Suzuki et al., 2001; Yamada et al., 1997), associated with an equal number of accessory sulphonylurea receptors (SUR) 2B (locus: ABCC9) (Adebiyi, McNally, & Jaggar, 2011; Miura et al., 2003; Quayle, Nelson, & Standen, 1997) (Fig. 5). These channels were named because millimolar intracellular ATP closes the channels (Foster & Coetzee, 2016). However, their modulation by vasodilator substances is probably more important for their physiological function. Sensitivity to ATP is conferred by the $K_{IR}6.1$ subunit, whereas sensitivity to channel blockade by sulphonylureas, such as glibenclamide, and activation by agonists such pinacidil and cromakalim resides in the SUR2B subunit (Foster & Coetzee, 2016).

Vascular K_{ATP} channels appear to be active under resting conditions in coronary (Berwick et al., 2010; Dankelman, Van der Ploeg, & Spaan, 1994; Duncker, van Zon, Pavek, Herrlinger, & Bache, 1995; Farouque & Meredith, 2007; Farouque, Worthley, & Meredith, 2004; Farouque, Worthley, Meredith, Skyrme-Jones, & Zhang, 2002; Imamura et al., 1992; Jackson, Konig, Dambacher, & Busse, 1993; Merkus et al., 2003; Merkus, Sorop, Houweling, Hoogteijling, & Duncker, 2006; Mori et al., 1995; Randall, 1995; Richmond, Tune, Gorman, & Feigl, 1999, 2000;

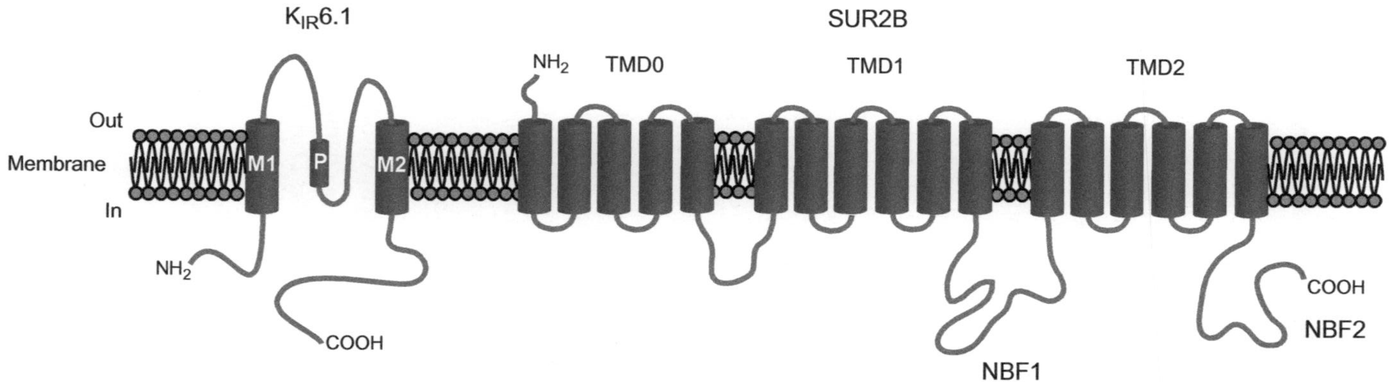

Fig. 5 Subunits of K_{ATP} channels. Shown are the $K_{IR}6.1$ and SUR2B subunits that are thought to comprise VSM K_{ATP} channels. The $K_{IR}6.1$ subunits have two membrane-spanning domains, whereas the SUR2B subunits have 17 membrane-spanning domains clustered into three groups (TMD0, TMD1, and TMD2), as shown. Functional channels are formed from a heterooctamer of these two subunits. See text for more information and references.

Samaha, Heineman, Ince, Fleming, & Balaban, 1992; Sharifi-Sanjani et al., 2013; Stepp, Kroll, & Feigl, 1997; Zhou, Teng, Tilley, Ledent, & Mustafa, 2014), skin (Abbink et al., 2002; Cankar & Strucl, 2008; Hojs, Strucl, & Cankar, 2009), and renal (Duncker, Oei, Hu, Stubenitsky, & Verdouw, 2001; Holdsworth et al., 2015) circulations, at rest. The resting activity of K_{ATP} channels is not as clear in skeletal muscle, because there is evidence both for (Jackson, 1993; Kosmas, Levy, & Hussain, 1995; Saito, McKay, Eraslan, & Hester, 1996; Vanelli, Chang, Gatensby, & Hussain, 1994; Vanelli & Hussain, 1994) and against (Banitt, Smits, Williams, Ganz, & Creager, 1996; Bank, Sih, Mullen, Osayamwen, & Lee, 2000; Bijlstra et al., 1996; Duncker et al., 2001; Farouque & Meredith, 2003a, 2003b, 2003c; Hammer, Ligon, & Hester, 2001; Holdsworth et al., 2015; Murrant & Sarelius, 2002) resting activity of these channels. In the cerebral circulation, K_{ATP} channels appear to be closed at rest (Faraci & Heistad, 1998; Horinaka et al., 1997; Leffler et al., 2011; Lindauer, Vogt, Schuh-Hofer, Dreier, & Dirnagl, 2003; Nnorom et al., 2014; Toyoda et al., 1997; Wei & Kontos, 1999).

Vasoconstrictors that activate PKC, close VSM K_{ATP} channels (Bonev & Nelson, 1996; Chrissobolis & Sobey, 2002; Cole, Malcolm, Walsh, & Light, 2000; Hayabuchi, Davies, & Standen, 2001; Quinn, Cui, Giblin, Clapp, & Tinker, 2003; Sampson, Davies, Barrett-Jolley, Standen, & Dart, 2007) and cause channel internalization (Jiao, Garg, Yang, Elton, & Hu, 2008). Vasoconstrictor-induced increases in intracellular Ca^{2+} also lead to K_{ATP} channel closure through activation of protein phosphatase 2b (calcineurin) (Wilson, Jabr, & Clapp, 2000). Vasoconstrictor-induced activation of $G_{i/o}$ signaling also inhibits K_{ATP} channels through inhibition of adenylate cyclase, reduced cAMP and decreased channel phosphorylation (Hayabuchi, Davies, et al., 2001).

As with BK_{Ca} channels and K_V channels, vasodilators that signal through the cAMP signaling pathway activate K_{ATP} channels (Akatsuka et al., 1994; Bouchard, Dumont, & Lamontagne, 1994; Dart & Standen, 1993; Eguchi et al., 2007; Jackson, 1993; Kitazono, Heistad, & Faraci, 1993b; Kleppisch & Nelson, 1995; Ming, Parent, & Lavallee, 1997; Nakashima & Vanhoutte, 1995; Nelson, Huang, Brayden, Hescheler, & Standen, 1990; Nelson et al., 2011; Quayle, Bonev, Brayden, & Nelson, 1994; Randall, 1995; Sawmiller, Ashtari, Urueta, Leschinsky, & Henning, 2006; Wellman, Quayle, & Standen, 1998; Yang et al., 2008). This may involve phosphorylation of both $K_{IR}6.1$ (Quinn, Giblin, & Tinker, 2004) and SUR2B (Shi et al., 2007, 2008) subunits. H_2S (Cheng, Ndisang, Tang, Cao, & Wang,

2004; Leffler et al., 2011; Liang et al., 2011; Mustafa et al., 2011; Zhao, Zhang, Lu, & Wang, 2001), acidosis (Faraci, Breese, & Heistad, 1994; Heintz, Damm, Brand, Koch, & Deussen, 2008; Lindauer et al., 2003), and hypoxia (Daut et al., 1990; Loutzenhiser & Parker, 1994; Marshall, Thomas, & Turner, 1993; Nakhostine & Lamontagne, 1993, 1994; Taguchi, Heistad, Kitazono, & Faraci, 1994; Tomiyama, Brian, & Todd, 1999; von Beckerath, Cyrys, Dischner, & Daut, 1991) may act, in part, in some vascular beds, by activation of VSM K_{ATP} channels.

2.7 Disease and VSM K_{ATP} Channels

The function of VSM K_{ATP} channels appears to be decreased in obesity (Erdos, Miller, & Busija, 2002; Erdos, Simandle, Snipes, Miller, & Busija, 2004; Hodnett, Xiang, Dearman, Carter, & Hester, 2008; Irat, Aslamaci, Karasu, & Ari, 2006; Lu et al., 2013; Miller, Tulbert, Puskar, & Busija, 2002; Spallarossa et al., 2001) and diabetes (Bouchard, Dumont, & Lamontagne, 1999; Kamata, Miyata, & Kasuya, 1989; Kinoshita et al., 2006; Li et al., 2015; Mayhan, 1994; Mayhan & Faraci, 1993; Miura et al., 2003). However, in hypertension K_{ATP} channel function has been reported to be decreased (Ghosh, Hanna, Wang, & McNeill, 2004; Kalliovalkama et al., 1999; Kam, Pfaffendorf, & van Zwieten, 1994; Kawata et al., 1998; Kitazono, Heistad, & Faraci, 1993a; Ohya et al., 1996; Tajada, Cidad, Moreno-Dominguez, Perez-Garcia, & Lopez-Lopez, 2012; Takaba et al., 1996; Van de Voorde, Vanheel, & Leusen, 1992), increased (Furspan & Webb, 1993; Miyata, Tsuchida, & Otomo, 1990), or not changed (Blanco-Rivero et al., 2008; Hutri-Kahonen et al., 1999; Kolias, Chai, & Webb, 1993).

2.8 K_{IR} Channels and VSM Contraction

VSM cells, particularly those in small resistance arteries and arterioles, also express one or more members of the strong K_{IR} channels, with $K_{IR}2.1$ (locus: KCNJ2) being the dominant isoform expressed (Longden & Nelson, 2015). These channels are formed from a tetramer of two membrane-spanning domain K_{IR} channel subunits (Kuang et al., 2015) (Fig. 6A). Block of the channel pore by intracellular polyamines and Mg^{2+} is responsible for the strong, voltage-dependent inward current rectification that is characteristic of these channels (Kuang et al., 2015) (Fig. 6B).

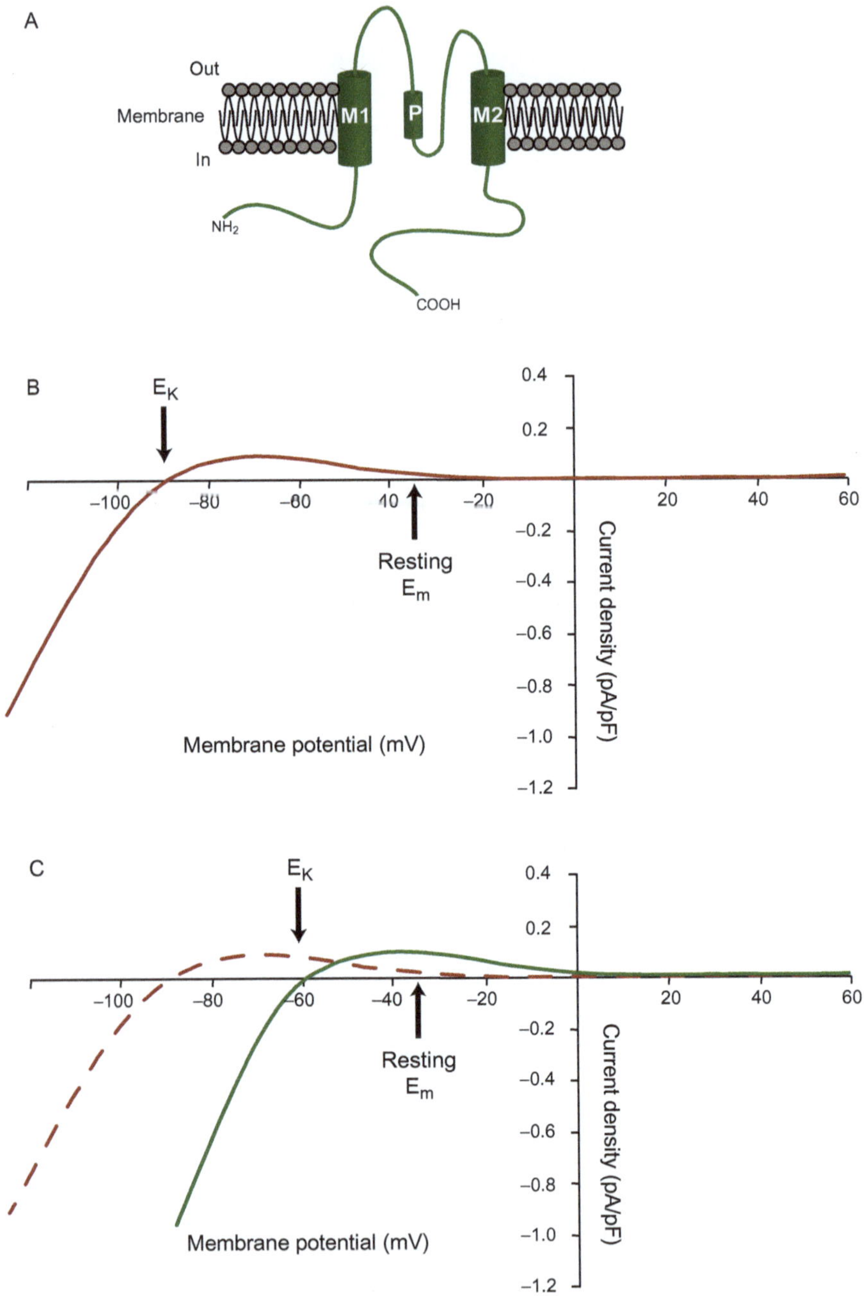

Fig. 6 Vascular K_{IR} channels and their currents. Panel (A) shows the topology of K_{IR} channels with two membrane-spanning domains. Functional channels are composed of a tetramer these subunits (see Fig. 2). Panel (B) shows a schematic of the current–voltage

While these channels derive their name from the inward currents that they conduct at membrane potentials more negative than the K^+ equilibrium potential (E_K), it is the small, outward "hump" in the current–voltage relationship at potentials positive to E_K that contributes to their physiology (Longden & Nelson, 2015; Quayle et al., 1997) (Fig. 6B).

Current through K_{IR} channels contributes to the resting membrane potential in a number of vascular beds (Burns et al., 2004; Chilton & Loutzenhiser, 2001; Chilton et al., 2008; Chilton, Smirnov, Loutzenhiser, Wang, & Loutzenhiser, 2011; Edwards & Hirst, 1988; Edwards, Hirst, & Silverberg, 1988; Jantzi et al., 2006; Jiang, Si, Lasarev, & Nuttall, 2001; Johnson, Marrelli, Steenberg, Childres, & Bryan, 1998; McCarron & Halpern, 1990; Smith et al., 2008; Troncoso Brindeiro, Fallet, Lane, & Carmines, 2008; Wu et al., 2007). Importantly, because of the shape of the current–voltage relationship, anything that hyperpolarizes the membrane will recruit outward current through K_{IR} channels (see Fig. 6B). Thus, these channels act to amplify hyperpolarization induced by opening of other K^+ channels or other cellular processes, such as the Na^+/K^+ ATPase, and may contribute to the mechanism of action of a number of vasodilators (Jackson, 2005; Jantzi et al., 2006; Longden & Nelson, 2015; Smith et al., 2008; Sonkusare, Dalsgaard, Bonev, & Nelson, 2016).

Increases in extracellular K^+ also activate K_{IR} channels, allowing these channels to contribute to functional hyperemia in electrically active tissues such as the brain (Filosa et al., 2006; Girouard et al., 2010; Paisansathan, Xu, Vetri, Hernandez, & Pelligrino, 2010; Vetri, Xu, Paisansathan, &

relationship for VSM K_{IR} channels for a cell with 5 mM K^+ in the extracellular solution (140 mM K^+ intracellular) and is based on data from Filosa et al. (2006). At membrane potentials more negative than the K^+ equilibrium potential (E_K, approximately −90 mV in 5 mM K^+), the channel conducts K^+ into the cells, as shown by the negative current density values. At potentials more positive than E_K up to approximately −30 mV, K_{IR} channels conduct K^+ ions out of the cell, and contribute to the resting membrane potential as denoted by the small positive currents at the assumed resting membrane potential of −35 mV. Note that anything that hyperpolarizes the membrane will recruit outward, positive current through K_{IR} channels, effectively amplifying the initial hyperpolarization. Panel (C) demonstrates the effects of increasing extracellular K^+ from 5 mM (*red dashed curve*) to 15 mM K^+ (*green solid curve*): increased extracellular K^+ shifts the E_K from −90 mV to approximately −60 mV. Note that there is now an elevated outward K^+ current at the original resting membrane potential. This enhanced outward K^+ current will hyperpolarize the VSM cell membrane from its resting value toward E_K, leading to vasodilation.

Pelligrino, 2012) and skeletal muscle (Armstrong, Dua, & Murrant, 2007; Crecelius, Kirby, Luckasen, Larson, & Dinenno, 2013; Crecelius, Luckasen, Larson, & Dinenno, 2014) (Fig. 6C). These channels may also be activated by K^+ released through other VSM or endothelial cell K^+ channels, another means by which K_{IR} channels can amplify the effects of vasodilators (Busse et al., 2002; Haddy, Vanhoutte, & Feletou, 2006; Longden & Nelson, 2015).

Vasoconstrictors may close K_{IR} channels through mechanisms involving PKC (Henry, Pearson, & Nichols, 1996; Park, Han, Kim, Youm, et al., 2005; Park et al., 2006; Zitron et al., 2004) or tyrosine kinases (Wischmeyer, Doring, & Karschin, 1998; Zitron et al., 2008), although this has not been well studied in blood vessels. Vasodilators that act through cAMP signaling may activate K_{IR} channels in some blood vessels (Paisansathan et al., 2010; Park, Han, Kim, Ko, et al., 2005; Son et al., 2005). However, it is unclear whether this is due to PKA-dependent phosphorylation of K_{IR} channels, or due to activation of other K^+ channels and amplification of hyperpolarization initiated by the opening of the other channel, as noted earlier.

2.9 Diseases and VSM K_{IR} Channels

The effects of hypertension on K_{IR} channel function are not clear; increases (Nakahata et al., 2006), decreases (Seo et al., 2014), or no change in function (Tajada et al., 2012) have been reported. Similarly, diabetes has been reported to increase (Troncoso Brindeiro et al., 2008; Troncoso Brindeiro, Lane, & Carmines, 2012) or decrease (Matsushita & Puro, 2006; Mayhan, Mayhan, Sun, & Patel, 2004; Vetri et al., 2012) K_{IR} channel function in different models. Regional, species, or model-dependent differences could be responsible for this heterogeneity.

Obesity (de Kreutzenberg et al., 2003; Haddock et al., 2011; Vigili de Kreutzenberg, Kiwanuka, Tiengo, & Avogaro, 2003), stress (Longden, Dabertrand, Hill-Eubanks, Hammack, & Nelson, 2014), and ischemia (Bastide et al., 1999, 2003; Marrelli, Johnson, Khorovets, Childres, & Bryan, 1998; Povlsen, Longden, Bonev, Hill-Eubanks, & Nelson, 2016) are associated with decreased K_{IR} channel function. Membrane cholesterol and hypercholesterolemia strongly suppress K_{IR} channel function in other systems (Fang et al., 2006). However, the effects of hypercholesterolemia on VSM K_{IR} channel expression and function have not been directly studied.

3. K^+ CHANNELS AND VSM PROLIFERATION

Remodeling of blood vessels after injury or due to diseases, such as atherosclerosis, results in phenotypic modulation of VSM cells from a quiescent, nondividing, contractile phenotype into proliferating cells. Potassium channels importantly contribute to the proliferative phenotype in VSM cells. An increase in K^+ channel expression and function are required for cells to proliferate (Neylon, 2002; Pardo, 2004; Urrego, Tomczak, Zahed, Stuhmer, & Pardo, 2014; Wonderlin & Strobl, 1996). Inhibition of K^+ channel function attenuates proliferation of VSM (Kohler et al., 2003; Miguel-Velado et al., 2005; Neylon, 2002; Tharp & Bowles, 2009; Tharp, Wamhoff, Turk, & Bowles, 2006) and other cells (Pardo, 2004; Urrego et al., 2014; Wonderlin & Strobl, 1996). Potassium channels are required for cells to progress through the cell cycle, as during proliferation (Urrego et al., 2014). They participate in this process by several mechanisms including membrane potential regulation, cell volume regulation, and ion-permeation-independent mechanisms (Urrego et al., 2014). Potassium channels also participate in apoptosis, a required component of vascular remodeling after injury (Kondratskyi, Kondratska, Skryma, & Prevarskaya, 2015).

In quiescent, contractile VSM cells, Ca^{2+} influx through high-voltage activated, L-type VGCC importantly contributes to cell Ca^{2+} regulation and contractile function (Jackson, 2000, 2005) (Figs. 1 and 3B). In this setting, activation of K^+ channels leads to membrane hyperpolarization, closure of VGCC, and decreases in intracellular Ca^{2+} (Jackson, 2000, 2005) (Figs. 1 and 3B). These cells also express a number of members of the TRP family of ion channels including TRPC1, TRPC3, TRPC4, TRPC5, TRPC6, TRPM4, and TRPV4 (Earley & Brayden, 2015). These channels serve as store-operated channels (SOC; TRPC1, TRPC4, TRPC5), receptor-operated channels (ROC; TRPC3, TRPC6, TRPM4, TRPV4), and stretch-activated channels (SAC; TRPC6, TRPM4) that contribute to agonist and pressure-induced contraction of native VSM cells (Earley & Brayden, 2015). However, in proliferating VSM cells, there is significant ion channel remodeling: expression of L-type VGCC is reduced, whereas expression of T-type VGCC, TRPC1, TRPC6, and SOC composed of ORAI and the endoplasmic reticulum Ca^{2+}-sensing protein, STIM is increased (Beech, 2007; House, Potier, Bisaillon, Singer, & Trebak, 2008; Munoz et al., 2013; Trebak, 2012; Tzeng et al., 2012) (Fig. 7). Importantly, there are also

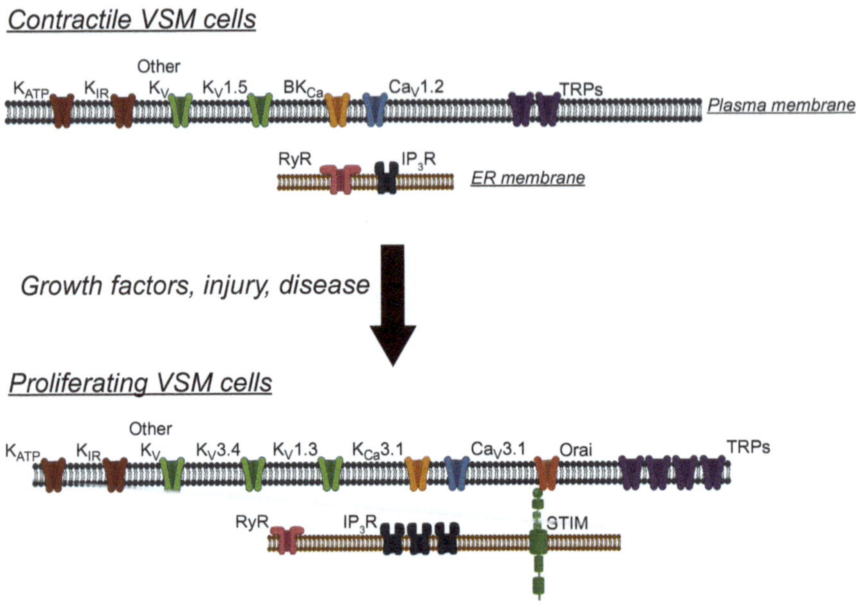

Fig. 7 Ion channel remodeling in proliferating VSM cells. Schematic summary of the ion channels expressed in contractile VSM cells and those expressed in proliferating VSM cells. Contractile VSM cells express predominantly L-type voltage-gated Ca^{2+} channels ($Ca_V1.2$), BK_{Ca} channels, and $K_V1.5$, in addition to K_{ATP}, K_{IR}, and several additional types of K_V channels. A number of transient receptor potential channels (TRPs) also are expressed that contribute to store-operated, receptor-operated, and stretch-activated cation channels. Intracellular ryanodine receptors (RyR) and inositol-1,4,5-trisphosphate (IP_3) receptors (IP_3R) also are expressed and functional in these cells. In response to growth factors, injury or disease, the pattern of expression of ion channels is remodeled. Proliferating VSM cells lose expression of $Ca_V1.2$, BK_{Ca}, and $K_V1.5$ channels. In their place, T-type Ca^{2+} ($Ca_V3.1$), $K_{Ca}3.1$, $K_V1.3$, and $K_V3.4$ channels are expressed. Proliferating cells also upregulate expression and function of store-operated channels (Orai/STIM), TRP channels, and IP_3R. See text for more information.

changes in K^+ channel expression that are essential for VSM cells to progress through the cell cycle and proliferate (Fig. 7 and Sections 3.1 and 3.2). During proliferation, increased expression and activation of a K^+ channel will either hyperpolarize the membrane to increase or maintain the EC gradient for Ca^{2+} entry through TRP channels and ORAI/STIM-based SOC (e.g., Fig. 8), which will increase or sustain Ca^{2+} influx through these channels (Bi et al., 2013; Munoz et al., 2013; Urrego et al., 2014). Increased intracellular Ca^{2+} concentration is an important signal for cell proliferation (Bi et al., 2013; Munoz et al., 2013; Urrego et al., 2014). As noted earlier, other roles for K^+ channels are also possible (Cidad et al., 2012, 2015; Urrego et al., 2014).

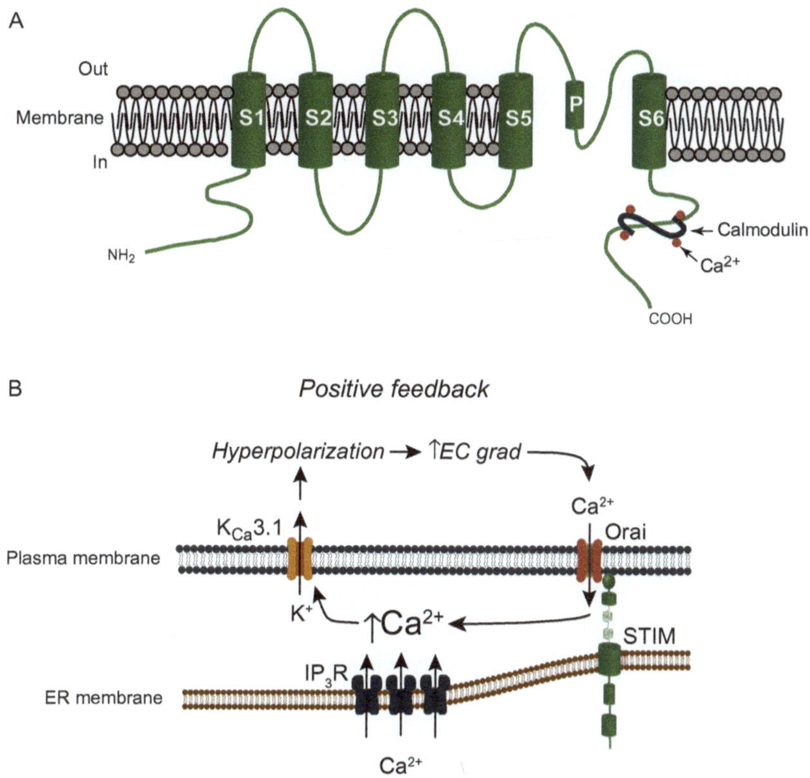

Fig. 8 Vascular $K_{Ca}3.1$ channels and their function in proliferating VSM cells. Panel (A) shows a schematic of an α-subunit demonstrating the typical six membrane-spanning domain structure of $K_{Ca}3.1$ channels. Calcium sensitivity is conferred by the Ca^{2+}-binding protein, calmodulin, that binds to the channel's C-terminus. Panel (B) shows a schematic of the role played by $K_{Ca}3.1$ channels in proliferating VSM cells. In proliferating cells, store-operated Ca^{2+} channels, composed of Orai proteins and the Ca^{2+}-sensing protein, STIM, are upregulated, as are IP_3R receptors. Increases in intracellular Ca^{2+} produced by increased activity of Orai/STIM and IP_3R, activates $K_{Ca}3.1$ channels, leading membrane hyperpolarization. This hyperpolarization increases the electrochemical (EC) gradient (grad) for Ca^{2+} diffusion into the cells, augmenting Ca^{2+} influx through Orai/STIM and other nonvoltage-gated Ca^{2+} channels in proliferating cells, leading to an increase in intracellular Ca^{2+}. This is a positive feedback system, in contrast to the negative feedback system that is found in contractile VSM cells (see Fig. 3B).

3.1 $K_{Ca}3.1$ and VSM Proliferation

The intermediate-conductance Ca^{2+}-activated K^+ channel, $K_{Ca}3.1$ (sK4, IK1, locus: *KCNN4*) has consistently been shown to play an important role in proliferation of VSM (Gole, Tharp, & Bowles, 2014; Kohler et al., 2003; Neylon, 2002; Tharp et al., 2006; Toyama et al., 2008) and other cells (Urrego et al., 2014) (Fig. 8). These K^+ channels are voltage insensitive and use calmodulin as the Ca^{2+} sensor (Fanger et al., 1999). Calmodulin interacts with the intracellular C-terminus of the channel to gate channel opening (Fanger et al., 1999) (Fig. 8A). The concentration of free Ca^{2+} required for 50% of maximal activation of $K_{Ca}3.1$ is on the order of 300 nM, with the threshold for activity at approximately 100 nM and maximal activity at 1 μM (Ishii et al., 1997).

Growth factors upregulate expression of $K_{Ca}3.1$ in cultured VSM cells (Gole et al., 2014; Kohler et al., 2003; Neylon, 2002; Tharp et al., 2006; Toyama et al., 2008). In porcine coronary artery VSM cells, NADPH oxidase 5 (NOX5)-related increases in reactive oxygen species appear to mediate the upregulation of $K_{Ca}3.1$ expression that results from simulation by basic fibroblast growth factor (Gole et al., 2014). Growth factors regulate expression of $K_{Ca}3.1$ both by decreasing the activity of repressor element 1-silencing transcription factor (REST) (Cheong et al., 2005) and by increasing activity of the AP-1 transcription factor (Bi et al., 2013; Ghanshani et al., 2000). Nucleoside diphosphate kinase B-dependent phosphorylation of $K_{Ca}3.1$ may contribute to activation of these channels in a mouse model of vascular injury (Zhou et al., 2015).

Selective inhibition of $K_{Ca}3.1$ reduces VSM cell growth and proliferation (Kohler et al., 2003; Neylon, 2002; Tharp et al., 2006; Toyama et al., 2008). Importantly, inhibition of these channels lessens restenosis after balloon injury in rat (Kohler et al., 2003) and pig (Tharp et al., 2008), and limits VSM proliferation in a mouse model of atherosclerosis (Toyama et al., 2008). These ion channels also have been implicated in the VSM proliferation that occurs after organ transplantation (Chen, Lam, Gregory, Schrepfer, & Wulff, 2013) and in chronic kidney disease (Huang, Pollock, & Chen, 2015). The proproliferative effect of $K_{Ca}3.1$ is mediated by increases in intracellular Ca^{2+} (Bi et al., 2013), likely due to increased influx of extracellular Ca^{2+} driven by $K_{Ca}3.1$-induced hyperpolarization and the increased EC gradient for Ca^{2+} influx through TRP channels and ORAI/STIM-based SOC (Fig. 8B). The proproliferative effect of increases in intracellular Ca^{2+} involves phosphorylated cAMP-response element-binding protein (CREB), c-Fos, and neuron-derived orphan receptor-1 (NOR-1) in human coronary VSM cells (Bi et al., 2013). Unexpectedly, activators of $K_{ca}3.1$ such as EBIO, SKA-31, and NS309 attenuated platelet-derived growth factor (PDGF)-induced proliferation of these cells (Bi et al., 2013). The inhibition of PDGF-induced proliferation appeared to arise from strong suppression of $K_{Ca}3.1$ expression by the activators, and the resultant reduction in intracellular Ca^{2+} signaling (Bi et al., 2013).

3.2 K_V Channels and VSM Proliferation

In addition to $K_{Ca}3.1$, there are also K_V channels that contribute to VSM cell proliferation. The K_V channel, $K_V3.4$ (locus: *KCNC4*) is upregulated during proliferation of human uterine artery VSM cells, and selective inhibition of $K_V3.4$ blocks proliferation (Miguel-Velado et al., 2005, 2010) and prevents progression of the cells through the G1 phase of the cell cycle (Miguel-Velado et al., 2010). The inhibitory effects of $K_v3.4$ blockade on proliferation could be mimicked by incubation of cells with elevated extracellular K^+ to produce depolarization equivalent to that produced by $K_V3.4$ blockade (Miguel-Velado et al., 2010). These data suggest that the proliferative effects of $K_V3.4$ may be related to the channel's impact on membrane potential (hyperpolarization), similar to the proposed mechanism for $K_{Ca}3.1$ channel stimulation of Ca^{2+} influx (Fig. 8B).

In contrast, balloon injury of mouse arteries results in upregulation of $K_V1.3$ (locus: *KCNE3*) (Cidad et al., 2010) and downregulation of $K_V1.5$ (Cidad et al., 2012, 2014, 2015). Proliferation and migration of VSM cells in this model can be attenuated by selective blockade of $K_V1.3$ channels (Cheong et al., 2011; Cidad et al., 2010). Studies in human VSM cells also confirm a role for $K_V1.3$ in proliferation (Cheong et al., 2011; Cidad et al., 2015). Interestingly, $K_V1.3$ may contribute to VSM proliferation by ion-permeation-independent mechanisms (Cidad et al., 2012, 2015; Jimenez-Perez et al., 2016). The proproliferative effects of $K_V1.3$ are mediated by voltage-dependent exposure of key residues in the channel's C-terminus (Tyr-447 and Ser-459) (Jimenez-Perez et al., 2016). These K_V channels may act as scaffolding proteins that recruit signaling proteins into signalplexes to promote the proliferative phenotype, independent from K^+ diffusion through the pore of the channel, and changes in membrane potential (Cahalan & Chandy, 2009; Jimenez-Perez et al., 2016; Schwab, Hanley, Fabian, & Stock, 2008). In human coronary VSM cells, this may involve mitogen-activated protein kinase kinase (MEK)/extracellular signal-regulated kinase (ERK) and phospholipase Cγ signaling pathways (Cidad et al., 2015). This may provide additional targets to combat vascular proliferative diseases, in addition to the phosphatidylinositol-4,5-bisphosphate

3-kinase/mammalian target of rapamycin (mTOR) pathway targeted by current therapies (Cidad et al., 2015).

4. CONCLUSION

While we have learned much about the expression and function of K^+ channels in the regulation of VSM contraction and proliferation in the past 30 years, there remain several outstanding questions. First, why do VSM cells express so many different K_V channels? Is this simply a matter of redundancy, or does the pattern of expression of these channels tune the electrophysiology of VSM cells in different vascular beds in ways that are not yet clear (Zhong et al., 2010)? Second, while it is clear that K^+ channels, like all ion channels, exist in multiprotein signaling domains (Abriel, Rougier, & Jalife, 2015; Kim & Oh, 2016; Levitan, 2006), our understanding of the regional heterogeneity in the nature and composition of these signaling domains in different vascular beds is incomplete. Finally, our understanding of the regulation of expression and function of K^+ channels in major cardiovascular disease states also remains incomplete, particularly as they relate to different vascular beds around the body. These are research areas where single cell transcriptome studies, high-resolution proteomics and informatics along with detailed electrophysiology and mechanical studies would aid in providing a clearer picture of the expression and function of the diverse array of K^+ channels that contribute to the regulation of VSM contraction and proliferation in health and disease.

CONFLICT OF INTEREST

The author has no conflicts of interest to declare.

ACKNOWLEDGMENTS

Supported by the National Heart, Lung and Blood Institute of the National Institutes of Health, under award numbers RO1 HL32469 and P01 HL070687. The content is solely the responsibility of the author and does not necessarily represent the official views of the National Institutes of Health.

REFERENCES

Abbink, E. J., Wollersheim, H., Netten, P. M., Russel, F. G., Lutterman, J. A., & Smits, P. (2002). Microcirculatory effects of KATP channel blockade by sulphonylurea derivatives in humans. *European Journal of Clinical Investigation*, *32*(3), 163–171.

Abraham, N. G., & Kappas, A. (2008). Pharmacological and clinical aspects of heme oxygenase. *Pharmacological Reviews*, *60*(1), 79–127. http://dx.doi.org/10.1124/pr.107.07104.

Abriel, H., Rougier, J. S., & Jalife, J. (2015). Ion channel macromolecular complexes in cardiomyocytes: Roles in sudden cardiac death. *Circulation Research*, *116*(12), 1971–1988. http://dx.doi.org/10.1161/CIRCRESAHA.116.305017.

Adebiyi, A., McNally, E. M., & Jaggar, J. H. (2011). Vasodilation induced by oxygen/glucose deprivation is attenuated in cerebral arteries of SUR2 null mice. *American Journal of Physiology. Heart and Circulatory Physiology*, *301*(4), H1360–H1368. http://dx.doi.org/10.1152/ajpheart.00406.2011.

Ahern, G. P., Hsu, S. F., & Jackson, M. B. (1999). Direct actions of nitric oxide on rat neurohypophysial K^+ channels. *The Journal of Physiology*, *520*(Pt. 1), 165–176.

Aiello, E. A., Malcolm, A. T., Walsh, M. P., & Cole, W. C. (1998). Beta-adrenoceptor activation and PKA regulate delayed rectifier K^+ channels of vascular smooth muscle cells. *The American Journal of Physiology*, *275*(2 Pt. 2), H448–H459.

Aiello, E. A., Walsh, M. P., & Cole, W. C. (1994). Isoproterenol and forskolin increase and PKI inhibits delayed rectifier K^+ current in vascular myocytes isolated from rabbit coronary artery and portal vein. *Canadian Journal of Physiology and Pharmacology*, *72*, 47.

Aiello, E. A., Walsh, M. P., & Cole, W. C. (1995). Phosphorylation by protein kinase A enhances delayed rectifier K^+ current in rabbit vascular smooth muscle cells. *The American Journal of Physiology*, *268*(2 Pt. 2), H926–H934.

Akatsuka, Y., Egashira, K., Katsuda, Y., Narishige, T., Ueno, H., Shimokawa, H., & Takeshita, A. (1994). ATP sensitive potassium channels are involved in adenosine A2 receptor mediated coronary vasodilatation in the dog. *Cardiovascular Research*, *28*(6), 906–911.

Albarwani, S., Al-Siyabi, S., Baomar, H., & Hassan, M. O. (2010). Exercise training attenuates ageing-induced BKCa channel downregulation in rat coronary arteries. *Experimental Physiology*, *95*(6), 746–755. http://dx.doi.org/10.1113/expphysiol.2009.051250.

Alioua, A., Huggins, J. P., & Rousseau, E. (1995). PKG-I alpha phosphorylates the alpha-subunit and upregulates reconstituted GKCa channels from tracheal smooth muscle. *The American Journal of Physiology*, *268*(6 Pt. 1), L1057–L1063.

Amberg, G. C., Bonev, A. D., Rossow, C. F., Nelson, M. T., & Santana, L. F. (2003). Modulation of the molecular composition of large conductance, Ca(2+) activated K(+) channels in vascular smooth muscle during hypertension. *The Journal of Clinical Investigation*, *112*(5), 717–724. http://dx.doi.org/10.1172/JCI18684.

Amberg, G. C., & Santana, L. F. (2003). Downregulation of the BK channel beta1 subunit in genetic hypertension. *Circulation Research*, *93*(10), 965–971. http://dx.doi.org/10.1161/01.RES.0000100068.43006.36.

Ambroisine, M. L., Favre, J., Oliviero, P., Rodriguez, C., Gao, J., Thuillez, C., … Delcayre, C. (2007). Aldosterone-induced coronary dysfunction in transgenic mice involves the calcium-activated potassium (BKCa) channels of vascular smooth muscle cells. *Circulation*, *116*(21), 2435–2443. http://dx.doi.org/10.1161/CIRCULATIONAHA.107.722009.

Armstead, W. M. (1997). Role of activation of calcium-sensitive K^+ channels in NO- and hypoxia-induced pial artery vasodilation. *American Journal of Physiology. Heart and Circulatory Physiology*, *272*(4), H1785–H1790.

Armstrong, M. L., Dua, A. K., & Murrant, C. L. (2007). Potassium initiates vasodilatation induced by a single skeletal muscle contraction in hamster cremaster muscle. *The Journal of Physiology*, *581*(Pt. 2), 841–852. http://dx.doi.org/10.1113/jphysiol.2007.130013.

Asano, M., Masuzawa-Ito, K., & Matsuda, T. (1993). Charybdotoxin-sensitive K^+ channels regulate the myogenic tone in the resting state of arteries from spontaneously hypertensive rats. *British Journal of Pharmacology*, *108*(1), 214–222.

Asano, M., Matsuda, T., Hayakawa, M., Ito, K. M., & Ito, K. (1993). Increased resting Ca^{2+} maintains the myogenic tone and activates K^+ channels in arteries from young spontaneously hypertensive rats. *European Journal of Pharmacology*, *247*(3), 295–304.

Aziz, Q., Thomas, A. M., Gomes, J., Ang, R., Sones, W. R., Li, Y., … Tinker, A. (2014). The ATP-sensitive potassium channel subunit, Kir6.1, in vascular smooth muscle plays a major role in blood pressure control. *Hypertension*, *64*(3), 523–529. http://dx.doi.org/10.1161/HYPERTENSIONAHA.114.03116.

Bae, Y. M., Kim, A., Kim, J., Park, S. W., Kim, T. K., Lee, Y. R., … Cho, S. I. (2006). Serotonin depolarizes the membrane potential in rat mesenteric artery myocytes by decreasing voltage-gated K^+ currents. *Biochemical and Biophysical Research Communications*, *347*(2), 468–476. http://dx.doi.org/10.1016/j.bbrc.2006.06.116.

Banitt, P. F., Smits, P., Williams, S. B., Ganz, P., & Creager, M. A. (1996). Activation of ATP-sensitive potassium channels contributes to reactive hyperemia in humans. *The American Journal of Physiology*, *271*(4 Pt. 2), H1594–H1598.

Bank, A. J., Sih, R., Mullen, K., Osayamwen, M., & Lee, P. C. (2000). Vascular ATP-dependent potassium channels, nitric oxide, and human forearm reactive hyperemia. *Cardiovascular Drugs and Therapy*, *14*(1), 23–29.

Barlow, R. S., & White, R. E. (1998). Hydrogen peroxide relaxes porcine coronary arteries by stimulating BKCa channel activity. *The American Journal of Physiology*, *275*(4 Pt. 2), H1283–H1289.

Bastide, M., Bordet, R., Pu, Q., Robin, E., Puisieux, F., & Dupuis, B. (1999). Relationship between inward rectifier potassium current impairment and brain injury after cerebral ischemia/reperfusion. *Journal of Cerebral Blood Flow and Metabolism*, *19*(12), 1309–1315. http://dx.doi.org/10.1097/00004647-199912000-00003.

Bastide, M., Gele, P., Petrault, O., Pu, Q., Caliez, A., Robin, E., … Bordet, R. (2003). Delayed cerebrovascular protective effect of lipopolysaccharide in parallel to brain ischemic tolerance. *Journal of Cerebral Blood Flow and Metabolism*, *23*(4), 399–405.

Beech, D. J. (2007). Ion channel switching and activation in smooth-muscle cells of occlusive vascular diseases. *Biochemical Society Transactions*, *35*(Pt. 5), 890–894. http://dx.doi.org/10.1042/BST0350890.

Berczi, V., Stekiel, W. J., Contney, S. J., & Rusch, N. J. (1992). Pressure-induced activation of membrane K^+ current in rat saphenous artery. *Hypertension*, *19*(6 Pt. 2), 725–729.

Berger, M. G., Vandier, C., Bonnet, P., Jackson, W. F., & Rusch, N. J. (1998). Intracellular acidosis differentially regulates KV channels in coronary and pulmonary vascular muscle. *The American Journal of Physiology*, *275*(4 Pt. 2), H1351–H1359.

Berwick, Z. C., Dick, G. M., Moberly, S. P., Kohr, M. C., Sturek, M., & Tune, J. D. (2012). Contribution of voltage-dependent K(+) channels to metabolic control of coronary blood flow. *Journal of Molecular and Cellular Cardiology*, *52*(4), 912–919. http://dx.doi.org/10.1016/j.yjmcc.2011.07.004.

Berwick, Z. C., Payne, G. A., Lynch, B., Dick, G. M., Sturek, M., & Tune, J. D. (2010). Contribution of adenosine A(2A) and A(2B) receptors to ischemic coronary dilation: Role of K(V) and K(ATP) channels. *Microcirculation*, *17*(8), 600–607. http://dx.doi.org/10.1111/j.1549-8719.2010.00054.x.

Bi, D., Toyama, K., Lemaitre, V., Takai, J., Fan, F., Jenkins, D. P., … Miura, H. (2013). The intermediate conductance calcium-activated potassium channel KCa3.1 regulates vascular smooth muscle cell proliferation via controlling calcium-dependent signaling. *The Journal of Biological Chemistry*, *288*(22), 15843–15853. http://dx.doi.org/10.1074/jbc.M112.427187.

Bialecki, R. A., & Stinson-Fisher, C. (1995). KCa channel antagonists reduce NO donor-mediated relaxation of vascular and tracheal smooth muscle. *The American Journal of Physiology*, *268*(1 Pt. 1), L152–L159.

Bijlstra, P. J., den Arend, J. A., Lutterman, J. A., Russel, F. G., Thien, T., & Smits, P. (1996). Blockade of vascular ATP-sensitive potassium channels reduces the vasodilator response to ischaemia in humans. *Diabetologia*, *39*(12), 1562–1568.

Blanco-Rivero, J., Gamallo, C., Aras-Lopez, R., Cobeno, L., Cogolludo, A., Perez-Vizcaino, F., ... Balfagon, G. (2008). Decreased expression of aortic KIR6.1 and SUR2B in hypertension does not correlate with changes in the functional role of K(ATP) channels. *European Journal of Pharmacology*, *587*(1–3), 204–208. http://dx.doi.org/10.1016/j.ejphar.2008.03.039.

Bolotina, V. M., Najibi, S., Palacino, J. J., Pagano, P. J., & Cohen, R. A. (1994). Nitric oxide directly activates calcium-dependent potassium channels in vascular smooth muscle. *Nature*, *368*(6474), 850–853. http://dx.doi.org/10.1038/368850a0.

Bonev, A. D., & Nelson, M. T. (1996). Vasoconstrictors inhibit ATP-sensitive K^+ channels in arterial smooth muscle through protein kinase C. *Journal of General Physiology*, *108*(4), 315–323. http://dx.doi.org/10.1085/jgp.108.4.315.

Borbouse, L., Dick, G. M., Asano, S., Bender, S. B., Dincer, U. D., Payne, G. A., ... Tune, J. D. (2009). Impaired function of coronary BK(Ca) channels in metabolic syndrome. *American Journal of Physiology. Heart and Circulatory Physiology*, *297*(5), H1629–H1637. http://dx.doi.org/10.1152/ajpheart.00466.2009.

Borbouse, L., Dick, G. M., Payne, G. A., Payne, B. D., Svendsen, M. C., Neeb, Z. P., ... Tune, J. D. (2010). Contribution of BK(Ca) channels to local metabolic coronary vasodilation: Effects of metabolic syndrome. *American Journal of Physiology. Heart and Circulatory Physiology*, *298*(3), H966–H973. http://dx.doi.org/10.1152/ajpheart.00876.2009.

Bouchard, J. F., Dumont, E., & Lamontagne, D. (1994). Evidence that prostaglandins I2, E2, and D2 may activate ATP sensitive potassium channels in the isolated rat heart. *Cardiovascular Research*, *28*(6), 901–905.

Bouchard, J. F., Dumont, E. C., & Lamontagne, D. (1999). Modification of vasodilator response in streptozotocin-induced diabetic rat. *Canadian Journal of Physiology and Pharmacology*, 77(12), 980–985.

Bratz, I. N., Dick, G. M., Partridge, L. D., & Kanagy, N. L. (2005). Reduced molecular expression of K(+) channel proteins in vascular smooth muscle from rats made hypertensive with N{omega}-nitro-L-arginine. *American Journal of Physiology. Heart and Circulatory Physiology*, *289*(3), H1277–H1283. http://dx.doi.org/10.1152/ajpheart.01052.2004.

Bratz, I. N., Swafford, A. N., Jr., Kanagy, N. L., & Dick, G. M. (2005). Reduced functional expression of K(+) channels in vascular smooth muscle cells from rats made hypertensive with N{omega}-nitro-L-arginine. *American Journal of Physiology. Heart and Circulatory Physiology*, *289*(3), H1284–H1290. http://dx.doi.org/10.1152/ajpheart.01053.2004.

Brayden, J. E. (1990). Membrane hyperpolarization is a mechanism of endothelium-dependent cerebral vasodilation. *American Journal of Physiology*, *259*(3), H668–H673.

Brayden, J. E., & Nelson, M. T. (1992). Regulation of arterial tone by activation of calcium-dependent potassium channels. *Science*, *256*(5056), 532–535.

Brenner, R., Perez, G. J., Bonev, A. D., Eckman, D. M., Kosek, J. C., Wiler, S. W., ... Aldrich, R. W. (2000). Vasoregulation by the beta1 subunit of the calcium-activated potassium channel. *Nature*, *407*(6806), 870–876. http://dx.doi.org/10.1038/35038011.

Bubolz, A. H., Li, H., Wu, Q., & Liu, Y. (2005). Enhanced oxidative stress impairs cAMP-mediated dilation by reducing Kv channel function in small coronary arteries of diabetic rats. *American Journal of Physiology. Heart and Circulatory Physiology*, *289*(5), H1873–H1880. http://dx.doi.org/10.1152/ajpheart.00357.2005.

Burns, W. R., Cohen, K. D., & Jackson, W. F. (2004). K^+-induced dilation of hamster cremasteric arterioles involves both the $Na+/K^+$-ATPase and inward-rectifier K^+ channels. *Microcirculation*, *11*(3), 279–293. http://dx.doi.org/10.1080/10739680490425985.

Busse, R., Edwards, G., Feletou, M., Fleming, I., Vanhoutte, P. M., & Weston, A. H. (2002). EDHF: Bringing the concepts together. *Trends in Pharmacological Sciences*, *23*(8), 374–380.

Butler, A., Tsunoda, S., McCobb, D. P., Wei, A., & Salkoff, L. (1993). *mSlo*, a complex mouse gene encoding "maxi" calcium-activated potassium channels. *Science*, *261*(5118), 221–224.

Buxton, I. L., Kaiser, R. A., Malmquist, N. A., & Tichenor, S. (2001). NO-induced relaxation of labouring and non-labouring human myometrium is not mediated by cyclic GMP. *British Journal of Pharmacology*, *134*(1), 206–214. http://dx.doi.org/10.1038/sj.bjp.0704226.

Bychkov, R., Gollasch, M., Ried, C., Luft, F. C., & Haller, H. (1997). Regulation of spontaneous transient outward potassium currents in human coronary arteries. *Circulation*, *95*(2), 503–510.

Cahalan, M. D., & Chandy, K. G. (2009). The functional network of ion channels in T lymphocytes. *Immunological Reviews*, *231*(1), 59–87. http://dx.doi.org/10.1111/j.1600-065X.2009.00816.x.

Callera, G. E., Yogi, A., Tostes, R. C., Rossoni, L. V., & Bendhack, L. M. (2004). Ca^{2+}-activated K^+ channels underlying the impaired acetylcholine-induced vasodilation in 2K-1C hypertensive rats. *The Journal of Pharmacology and Experimental Therapeutics*, *309*(3), 1036–1042. http://dx.doi.org/10.1124/jpet.103.062810.

Campbell, W. B., Gebremedhin, D., Pratt, P. F., & Harder, D. R. (1996). Identification of epoxyeicosatrienoic acids as endothelium-derived hyperpolarizing factors. *Circulation Research*, *78*(3), 415–423.

Cankar, K., & Strucl, M. (2008). The effect of glibenclamide on cutaneous laser-Doppler flux. *Microvascular Research*, *75*(1), 97–103. http://dx.doi.org/10.1016/j.mvr.2007.06.005.

Chadha, P. S., Jepps, T. A., Carr, G., Stott, J. B., Zhu, H. L., Cole, W. C., & Greenwood, I. A. (2014). Contribution of kv7.4/kv7.5 heteromers to intrinsic and calcitonin gene-related peptide-induced cerebral reactivity. *Arteriosclerosis, Thrombosis, and Vascular Biology*, *34*(4), 887–893. http://dx.doi.org/10.1161/ATVBAHA.114.303405.

Chadha, P. S., Zunke, F., Zhu, H. L., Davis, A. J., Jepps, T. A., Olesen, S. P., … Greenwood, I. A. (2012). Reduced KCNQ4-encoded voltage-dependent potassium channel activity underlies impaired beta-adrenoceptor-mediated relaxation of renal arteries in hypertension. *Hypertension*, *59*(4), 877–884. http://dx.doi.org/10.1161/HYPERTENSIONAHA.111.187427.

Chai, Q., Liu, Z., & Chen, L. (2005). Effects of streptozotocin-induced diabetes on Kv channels in rat small coronary smooth muscle cells. *The Chinese Journal of Physiology*, *48*(1), 57–63.

Chai, Q., Xu, X., Jia, Q., Dong, Q., Liu, Z., Zhang, W., & Chen, L. (2007). Molecular basis of dysfunctional Kv channels in small coronary artery smooth muscle cells of streptozotocin-induced diabetic rats. *The Chinese Journal of Physiology*, *50*(4), 171–177.

Cheang, W. S., Wong, W. T., Shen, B., Lau, C. W., Tian, X. Y., Tsang, S. Y., … Huang, Y. (2010). 4-Aminopyridine-sensitive K^+ channels contributes to NaHS-induced membrane hyperpolarization and relaxation in the rat coronary artery. *Vascular Pharmacology*, *53*(3–4), 94–98. http://dx.doi.org/10.1016/j.vph.2010.04.004.

Chen, Y. J., Lam, J., Gregory, C. R., Schrepfer, S., & Wulff, H. (2013). The Ca(2)(+)-activated K(+) channel KCa3.1 as a potential new target for the prevention of allograft vasculopathy. *PLoS One*, *8*(11), e81006. http://dx.doi.org/10.1371/journal.pone.0081006.

Cheng, Y., Ndisang, J. F., Tang, G., Cao, K., & Wang, R. (2004). Hydrogen sulfide-induced relaxation of resistance mesenteric artery beds of rats. *American Journal of Physiology. Heart and Circulatory Physiology*, *287*(5), H2316–H2323. http://dx.doi.org/10.1152/ajpheart.00331.2004.

Cheong, A., Bingham, A. J., Li, J., Kumar, B., Sukumar, P., Munsch, C., … Wood, I. C. (2005). Downregulated REST transcription factor is a switch enabling critical potassium channel expression and cell proliferation. *Molecular Cell*, *20*(1), 45–52. http://dx.doi.org/10.1016/j.molcel.2005.08.030.

Cheong, A., Dedman, A. M., & Beech, D. J. (2001). Expression and function of native potassium channel [K(V)alpha1] subunits in terminal arterioles of rabbit. *The Journal of Physiology, 534*(Pt. 3), 691–700.
Cheong, A., Dedman, A. M., Xu, S. Z., & Beech, D. J. (2001). K(V)alpha1 channels in murine arterioles: Differential cellular expression and regulation of diameter. *American Journal of Physiology. Heart and Circulatory Physiology, 281*(3), H1057–H1065.
Cheong, A., Li, J., Sukumar, P., Kumar, B., Zeng, F., Riches, K., … Beech, D. J. (2011). Potent suppression of vascular smooth muscle cell migration and human neointimal hyperplasia by KV1.3 channel blockers. *Cardiovascular Research, 89*(2), 282–289. http://dx.doi.org/10.1093/cvr/cvq305.
Cheranov, S. Y., & Jaggar, J. H. (2006). TNF-alpha dilates cerebral arteries via NAD(P)H oxidase-dependent Ca^{2+} spark activation. *American Journal of Physiology. Cell Physiology, 290*(4), C964–C971. http://dx.doi.org/10.1152/ajpcell.00499.2005.
Chilton, L., & Loutzenhiser, R. (2001). Functional evidence for an inward rectifier potassium current in rat renal afferent arterioles. *Circulation Research, 88*(2), 152–158.
Chilton, L., Loutzenhiser, K., Morales, E., Breaks, J., Kargacin, G. J., & Loutzenhiser, R. (2008). Inward rectifier K(+) currents and Kir2.1 expression in renal afferent and efferent arterioles. *Journal of the American Society of Nephrology, 19*(1), 69–76. http://dx.doi.org/10.1681/ASN.2007010039.
Chilton, L., Smirnov, S. V., Loutzenhiser, K., Wang, X., & Loutzenhiser, R. (2011). Segment-specific differences in the inward rectifier K(+) current along the renal interlobular artery. *Cardiovascular Research, 92*(1), 169–177. http://dx.doi.org/10.1093/cvr/cvr179.
Chrissobolis, S., & Sobey, C. G. (2002). Inhibitory effects of protein kinase C on inwardly rectifying K^+- and ATP-sensitive K^+ channel-mediated responses of the basilar artery. *Stroke, 33*(6), 1692–1697.
Cidad, P., Jimenez-Perez, L., Garcia-Arribas, D., Miguel-Velado, E., Tajada, S., Ruiz-McDavitt, C., … Perez-Garcia, M. T. (2012). Kv1.3 channels can modulate cell proliferation during phenotypic switch by an ion-flux independent mechanism. *Arteriosclerosis, Thrombosis, and Vascular Biology, 32*(5), 1299–1307. http://dx.doi.org/10.1161/ATVBAHA.111.242727.
Cidad, P., Miguel-Velado, E., Ruiz-McDavitt, C., Alonso, E., Jimenez-Perez, L., Asuaje, A., … Lopez-Lopez, J. R. (2015). Kv1.3 channels modulate human vascular smooth muscle cells proliferation independently of mTOR signaling pathway. *Pflügers Archiv, 467*(8), 1711–1722. http://dx.doi.org/10.1007/s00424-014-1607-y.
Cidad, P., Moreno-Dominguez, A., Novensa, L., Roque, M., Barquin, L., Heras, M., … Lopez-Lopez, J. R. (2010). Characterization of ion channels involved in the proliferative response of femoral artery smooth muscle cells. *Arteriosclerosis, Thrombosis, and Vascular Biology, 30*(6), 1203–1211. http://dx.doi.org/10.1161/ATVBAHA.110.205187.
Cidad, P., Novensa, L., Garabito, M., Batlle, M., Dantas, A. P., Heras, M., … Roque, M. (2014). K^+ channels expression in hypertension after arterial injury, and effect of selective Kv1.3 blockade with PAP-1 on intimal hyperplasia formation. *Cardiovascular Drugs and Therapy, 28*(6), 501–511. http://dx.doi.org/10.1007/s10557-014-6554-5.
Clement-Chomienne, O., Walsh, M. P., & Cole, W. C. (1996). Angiotensin II activation of protein kinase C decreases delayed rectifier K^+ current in rabbit vascular myocytes. *The Journal of Physiology, 495*(Pt. 3), 689–700.
Cole, W. C., Malcolm, T., Walsh, M. P., & Light, P. E. (2000). Inhibition by protein kinase C of the K(NDP) subtype of vascular smooth muscle ATP-sensitive potassium channel. *Circulation Research, 87*(2), 112–117.
Cook, N. S. (1989). Effect of some potassium channel blockers on contractile responses of the rabbit aorta. *Journal of Cardiovascular Pharmacology, 13*(2), 299–306. http://dx.doi.org/10.1097/00005344-198902000-00019.

Cooke, J. P., Rossitch, E., Jr., Andon, N. A., Loscalzo, J., & Dzau, V. J. (1991). Flow activates an endothelial potassium channel to release an endogenous nitrovasodilator. *The Journal of Clinical Investigation*, *88*(5), 1663–1671. http://dx.doi.org/10.1172/JCI115481.

Cox, R. H. (1996). Comparison of K^+ channel properties in freshly isolated myocytes from thoracic aorta of WKY and SHR. *American Journal of Hypertension*, *9*(9), 884–894.

Cox, R. H. (2005). Molecular determinants of voltage-gated potassium currents in vascular smooth muscle. *Cell Biochemistry and Biophysics*, *42*(2), 167–195. http://dx.doi.org/10.1385/CBB:42:2:167.

Cox, R. H., Folander, K., & Swanson, R. (2001). Differential expression of voltage-gated K(+) channel genes in arteries from spontaneously hypertensive and Wistar-Kyoto rats. *Hypertension*, *37*(5), 1315–1322.

Cox, R. H., Fromme, S. J., Folander, K. L., & Swanson, R. J. (2008). Voltage gated K^+ channel expression in arteries of Wistar–Kyoto and spontaneously hypertensive rats. *American Journal of Hypertension*, *21*(2), 213–218. http://dx.doi.org/10.1038/ajh.2007.44.

Cox, R. H., Lozinskaya, I., & Dietz, N. J. (2001). Differences in K^+ current components in mesenteric artery myocytes from WKY and SHR. *American Journal of Hypertension*, *14*(9 Pt. 1), 897–907.

Cox, R. H., & Petrou, S. (1999). Ca(2+) influx inhibits voltage-dependent and augments Ca(2+)-dependent K(+) currents in arterial myocytes. *The American Journal of Physiology*, *277*(1 Pt. 1), C51–C63.

Crecelius, A. R., Kirby, B. S., Luckasen, G. J., Larson, D. G., & Dinenno, F. A. (2013). Mechanisms of rapid vasodilation after a brief contraction in human skeletal muscle. *American Journal of Physiology. Heart and Circulatory Physiology*, *305*(1), H29–H40. http://dx.doi.org/10.1152/ajpheart.00298.2013.

Crecelius, A. R., Luckasen, G. J., Larson, D. G., & Dinenno, F. A. (2014). KIR channel activation contributes to onset and steady-state exercise hyperemia in humans. *American Journal of Physiology. Heart and Circulatory Physiology*, *307*(5), H782–H791. http://dx.doi.org/10.1152/ajpheart.00212.2014.

Dankelman, J., Van der Ploeg, C. P., & Spaan, J. A. (1994). Glibenclamide decelerates the responses of coronary regulation in the goat. *The American Journal of Physiology*, *266*(5 Pt. 2), H1715–H1721.

Dart, C., & Standen, N. B. (1993). Adenosine-activated potassium current in smooth muscle cells isolated from the pig coronary artery. *The Journal of Physiology*, *471*, 767–786.

Daut, J., Maierrudolph, W., Vonbeckerath, N., Mehrke, G., Gunther, K., & Goedelmeinen, L. (1990). Hypoxic dilation of coronary-arteries is mediated by ATP-sensitive potassium channels. *Science*, *247*(4948), 1341–1344. http://dx.doi.org/10.1126/science.2107575.

de Kreutzenberg, S. V., Puato, M., Kiwanuka, E., Del Prato, S., Pauletto, P., Pasini, L., … Avogaro, A. (2003). Elevated non-esterified fatty acids impair nitric oxide independent vasodilation, in humans: Evidence for a role of inwardly rectifying potassium channels. *Atherosclerosis*, *169*(1), 147–153.

del Valle-Rodriguez, A., Lopez-Barneo, J., & Urena, J. (2003). Ca^{2+} channel-sarcoplasmic reticulum coupling: A mechanism of arterial myocyte contraction without Ca^{2+} influx. *The EMBO Journal*, *22*(17), 4337–4345. http://dx.doi.org/10.1093/emboj/cdg432.

Dick, G. M., Bratz, I. N., Borbouse, L., Payne, G. A., Dincer, U. D., Knudson, J. D., … Tune, J. D. (2008). Voltage-dependent K^+ channels regulate the duration of reactive hyperemia in the canine coronary circulation. *American Journal of Physiology. Heart and Circulatory Physiology*, *294*(5), H2371–H2381. http://dx.doi.org/10.1152/ajpheart.01279.2007.

Dick, G. M., & Tune, J. D. (2010). Role of potassium channels in coronary vasodilation. *Experimental Biology and Medicine (Maywood, N.J.)*, *235*(1), 10–22. http://dx.doi.org/10.1258/ebm.2009.009201.

Dong, H., Waldron, G. J., Cole, W. C., & Triggle, C. R. (1998). Roles of calcium-activated and voltage-gated delayed rectifier potassium channels in endothelium-dependent vasorelaxation of the rabbit middle cerebral artery. *British Journal of Pharmacology*, *123*(5), 821–832. http://dx.doi.org/10.1038/sj.bjp.0701680.

Dong, H., Waldron, G. J., Galipeau, D., Cole, W. C., & Triggle, C. R. (1997). NO/PGI2-independent vasorelaxation and the cytochrome P450 pathway in rabbit carotid artery. *British Journal of Pharmacology*, *120*(4), 695–701. http://dx.doi.org/10.1038/sj.bjp.0700945.

Dong, L., Xie, M. J., Zhang, P., Ji, L. L., Liu, W. C., Dong, M. Q., & Gao, F. (2009). Rotenone partially reverses decreased BK Ca currents in cerebral artery smooth muscle cells from streptozotocin-induced diabetic mice. *Clinical and Experimental Pharmacology & Physiology*, *36*(10), e57–e64. http://dx.doi.org/10.1111/j.1440-1681.2009.05222.x.

Duncker, D. J., Oei, H. H., Hu, F., Stubenitsky, R., & Verdouw, P. D. (2001). Role of K(ATP)(+) channels in regulation of systemic, pulmonary, and coronary vasomotor tone in exercising swine. *American Journal of Physiology. Heart and Circulatory Physiology*, *280*(1), H22–H33.

Duncker, D. J., van Zon, N. S., Pavek, T. J., Herrlinger, S. K., & Bache, R. J. (1995). Endogenous adenosine mediates coronary vasodilation during exercise after K(ATP)+ channel blockade. *The Journal of Clinical Investigation*, *95*(1), 285–295. http://dx.doi.org/10.1172/JCI117653.

Earley, S., & Brayden, J. E. (2015). Transient receptor potential channels in the vasculature. *Physiological Reviews*, *95*(2), 645–690. http://dx.doi.org/10.1152/physrev.00026.2014.

Earley, S., Heppner, T. J., Nelson, M. T., & Brayden, J. E. (2005). TRPV4 forms a novel Ca^{2+} signaling complex with ryanodine receptors and BKCa channels. *Circulation Research*, *97*(12), 1270–1279. http://dx.doi.org/10.1161/01.RES.0000194321.60300.d6.

Eckman, D. M., Hopkins, N., McBride, C., & Keef, K. D. (1998). Endothelium-dependent relaxation and hyperpolarization in guinea-pig coronary artery: Role of epoxyeicosatrienoic acid. *British Journal of Pharmacology*, *124*(1), 181–189. http://dx.doi.org/10.1038/sj.bjp.0701778.

Edwards, F. R., & Hirst, G. D. (1988). Inward rectification in submucosal arterioles of guinea-pig ileum. *The Journal of Physiology*, *404*, 437–454.

Edwards, F. R., Hirst, G. D., & Silverberg, G. D. (1988). Inward rectification in rat cerebral arterioles; involvement of potassium ions in autoregulation. *The Journal of Physiology*, *404*, 455–466.

Eguchi, S., Kawano, T., Yinhua, Tanaka, K., Yasui, S., Mawatari, K., … Nakajo, N. (2007). Effects of prostaglandin E1 on vascular ATP-sensitive potassium channels. *Journal of Cardiovascular Pharmacology*, *50*(6), 686–691. http://dx.doi.org/10.1097/FJC.0b013e3181583d9b.

Emerson, G. G., & Segal, S. S. (2000). Electrical coupling between endothelial cells and smooth muscle cells in hamster feed arteries: Role in vasomotor control. *Circulation Research*, *87*(6), 474–479.

Erdos, B., Miller, A. W., & Busija, D. W. (2002). Alterations in KATP and KCa channel function in cerebral arteries of insulin-resistant rats. *American Journal of Physiology. Heart and Circulatory Physiology*, *283*(6), H2472–H2477. http://dx.doi.org/10.1152/ajpheart.00516.2002.

Erdos, B., Simandle, S. A., Snipes, J. A., Miller, A. W., & Busija, D. W. (2004). Potassium channel dysfunction in cerebral arteries of insulin-resistant rats is mediated by reactive oxygen species. *Stroke*, *35*(4), 964–969. http://dx.doi.org/10.1161/01.STR.0000119753.05670.F1.

Evanson, K. W., Bannister, J. P., Leo, M. D., & Jaggar, J. H. (2014). LRRC26 is a functional BK channel auxiliary gamma subunit in arterial smooth muscle cells. *Circulation Research*, *115*(4), 423–431. http://dx.doi.org/10.1161/CIRCRESAHA.115.303407.

Fang, Y., Mohler, E. R., 3rd., Hsieh, E., Osman, H., Hashemi, S. M., Davies, P. F., … Levitan, I. (2006). Hypercholesterolemia suppresses inwardly rectifying K^+ channels in aortic endothelium in vitro and in vivo. *Circulation Research*, *98*(8), 1064–1071. http://dx.doi.org/10.1161/01.RES.0000218776.87842.43.

Fanger, C. M., Ghanshani, S., Logsdon, N. J., Rauer, H., Kalman, K., Zhou, J., … Aiyar, J. (1999). Calmodulin mediates calcium-dependent activation of the intermediate conductance KCa channel, IKCa1. *The Journal of Biological Chemistry*, *274*(9), 5746–5754.

Faraci, F. M., Breese, K. R., & Heistad, D. D. (1994). Cerebral vasodilation during hypercapnia. Role of glibenclamide-sensitive potassium channels and nitric oxide. *Stroke*, *25*(8), 1679–1683.

Faraci, F. M., & Heistad, D. D. (1998). Regulation of the cerebral circulation: Role of endothelium and potassium channels. *Physiological Reviews*, *78*(1), 53–97.

Farouque, H. M., & Meredith, I. T. (2003a). Effects of inhibition of ATP-sensitive potassium channels on metabolic vasodilation in the human forearm. *Clinical Science (London, England)*, *104*(1), 39–46. http://dx.doi.org/10.1042/.

Farouque, H. M., & Meredith, I. T. (2003b). Inhibition of vascular ATP-sensitive K^+ channels does not affect reactive hyperemia in human forearm. *American Journal of Physiology. Heart and Circulatory Physiology*, *284*(2), H711–H718. http://dx.doi.org/10.1152/ajpheart.00315.2002.

Farouque, H. M., & Meredith, I. T. (2003c). Relative contribution of vasodilator prostanoids, NO, and KATP channels to human forearm metabolic vasodilation. *American Journal of Physiology. Heart and Circulatory Physiology*, *284*(6), H2405–H2411. http://dx.doi.org/10.1152/ajpheart.00879.2002.

Farouque, H. M., & Meredith, I. T. (2007). Effect of adenosine triphosphate-sensitive potassium channel inhibitors on coronary metabolic vasodilation. *Trends in Cardiovascular Medicine*, *17*(2), 63–68. http://dx.doi.org/10.1016/j.tcm.2006.12.003.

Farouque, H. M., Worthley, S. G., & Meredith, I. T. (2004). Effect of ATP-sensitive potassium channel inhibition on coronary metabolic vasodilation in humans. *Arteriosclerosis, Thrombosis, and Vascular Biology*, *24*(5), 905–910. http://dx.doi.org/10.1161/01.ATV.0000125701.18648.48.

Farouque, H. M., Worthley, S. G., Meredith, I. T., Skyrme-Jones, R. A., & Zhang, M. J. (2002). Effect of ATP-sensitive potassium channel inhibition on resting coronary vascular responses in humans. *Circulation Research*, *90*(2), 231–236.

Feliciangeli, S., Chatelain, F. C., Bichet, D., & Lesage, F. (2015). The family of K2P channels: Salient structural and functional properties. *The Journal of Physiology*, *593*(12), 2587–2603. http://dx.doi.org/10.1113/jphysiol.2014.287268.

Fernández-Tenorio, M., González-Rodríguez, P., Porras, C., Castellano, A., Moosmang, S., Hofmann, F., … López-Barneo, J. (2010). Short communication: Genetic ablation of L-type Ca^{2+} channels abolishes depolarization-induced Ca^{2+} release in arterial smooth muscle. *Circulation Research*, *106*(7), 1285–1289. http://dx.doi.org/10.1161/circresaha.109.213967.

Fernández-Tenorio, M., Porras-Gonzalez, C., Castellano, A., del Valle-Rodriguez, A., Lopez-Barneo, J., & Urena, J. (2011). Metabotropic regulation of RhoA/Rho-associated kinase by L-type Ca^{2+} channels. *Circulation Research*, *108*, 1348–1357. http://dx.doi.org/10.1161/circresaha.111.240127.

Fernandez-Velasco, M., Ruiz-Hurtado, G., Gomez, A. M., & Rueda, A. (2014). Ca(2+) handling alterations and vascular dysfunction in diabetes. *Cell Calcium*, *56*(5), 397–407. http://dx.doi.org/10.1016/j.ceca.2014.08.007.

Filosa, J. A., Bonev, A. D., Straub, S. V., Meredith, A. L., Wilkerson, M. K., Aldrich, R. W., & Nelson, M. T. (2006). Local potassium signaling couples neuronal activity to vasodilation in the brain. *Nature Neuroscience*, *9*(11), 1397–1403. http://dx.doi.org/10.1038/nn1779.

Foster, M. N., & Coetzee, W. A. (2016). KATP channels in the cardiovascular system. *Physiological Reviews*, *96*(1), 177–252. http://dx.doi.org/10.1152/physrev.00003.2015.

Frisbee, J. C., Maier, K. G., & Stepp, D. W. (2002). Oxidant stress-induced increase in myogenic activation of skeletal muscle resistance arteries in obese Zucker rats. *American Journal of Physiology. Heart and Circulatory Physiology*, *283*(6), H2160–H2168. http://dx.doi.org/10.1152/ajpheart.00379.2002.

Fujino, K., Nakaya, S., Wakatsuki, T., Miyoshi, Y., Nakaya, Y., Mori, H., & Inoue, I. (1991). Effects of nitroglycerin on ATP-induced Ca(++)-mobilization, Ca(++)-activated K channels and contraction of cultured smooth muscle cells of porcine coronary artery. *The Journal of Pharmacology and Experimental Therapeutics*, *256*(1), 371–377.

Fukami, Y., Toki, Y., Numaguchi, Y., Nakashima, Y., Mukawa, H., Matsui, H., … Ito, T. (1998). Nitroglycerin-induced aortic relaxation mediated by calcium-activated potassium channel is markedly diminished in hypertensive rats. *Life Sciences*, *63*(12), 1047–1055.

Furspan, P. B., & Webb, R. C. (1993). Decreased ATP sensitivity of a K^+ channel and enhanced vascular smooth muscle relaxation in genetically hypertensive rats. *Journal of Hypertension*, *11*(10), 1067–1072.

Furstenau, M., Lohn, M., Ried, C., Luft, F. C., Haller, H., & Gollasch, M. (2000). Calcium sparks in human coronary artery smooth muscle cells resolved by confocal imaging. *Journal of Hypertension*, *18*(9), 1215–1222.

Ganitkevich, V., & Isenberg, G. (1990). Isolated guinea-pig coronary smooth-muscle cells. Acetylcholine induces hyperpolarization due to sarcoplasmic-reticulum calcium release activating potassium channels. *Circulation Research*, *67*(2), 525–528.

Ganitkevich, V. Y., & Isenberg, G. (1993). Membrane potential modulates inositol 1,4,5-trisphosphate-mediated Ca^{2+} transients in guinea-pig coronary myocytes. *Journal of Physiology (London)*, *470*, 35–44.

Garland, C. J., & McPherson, G. A. (1992). Evidence that nitric oxide does not mediate the hyperpolarization and relaxation to acetylcholine in the rat small mesenteric artery. *British Journal of Pharmacology*, *105*(2), 429–435.

Gelband, C. H., Ishikawa, T., Post, J. M., Keef, K. D., & Hume, J. R. (1993). Intracellular divalent cations block smooth muscle K^+ channels. *Circulation Research*, *73*(1), 24–34.

Ghanshani, S., Wulff, H., Miller, M. J., Rohm, H., Neben, A., Gutman, G. A., … Chandy, K. G. (2000). Up-regulation of the IKCa1 potassium channel during T-cell activation. Molecular mechanism and functional consequences. *The Journal of Biological Chemistry*, *275*(47), 37137–37149. http://dx.doi.org/10.1074/jbc.M003941200.

Ghisdal, P., Gomez, J. P., & Morel, N. (2000). Action of a NO donor on the excitation-contraction pathway activated by noradrenaline in rat superior mesenteric artery. *The Journal of Physiology*, *522*(Pt. 1), 83–96.

Ghosh, M., Hanna, S. T., Wang, R., & McNeill, J. R. (2004). Altered vascular reactivity and KATP channel currents in vascular smooth muscle cells from deoxycorticosterone acetate (DOCA)-salt hypertensive rats. *Journal of Cardiovascular Pharmacology*, *44*(5), 525–531.

Girouard, H., Bonev, A. D., Hannah, R. M., Meredith, A., Aldrich, R. W., & Nelson, M. T. (2010). Astrocytic endfoot Ca^{2+} and BK channels determine both arteriolar dilation and constriction. *Proceedings of the National Academy of Sciences of the United States of America*, *107*(8), 3811–3816. http://dx.doi.org/10.1073/pnas.0914722107.

Gole, H. K., Tharp, D. L., & Bowles, D. K. (2014). Upregulation of intermediate-conductance Ca^{2+}-activated K^+ channels (KCNN4) in porcine coronary smooth muscle requires NADPH oxidase 5 (NOX5). *PLoS One, 9*(8), e105337. http://dx.doi.org/10.1371/journal.pone.0105337.

Gollasch, M., Lohn, M., Furstenau, M., Nelson, M. T., Luft, F. C., & Haller, H. (2000). Ca^{2+} channels, Ca^{2+} sparks, and regulation of arterial smooth muscle function. *Zeitschrift für Kardiologie, 89*(Suppl. 2), 15–19.

Greenwood, I. A., & Ohya, S. (2009). New tricks for old dogs: KCNQ expression and role in smooth muscle. *British Journal of Pharmacology, 156*(8), 1196–1203. http://dx.doi.org/10.1111/j.1476-5381.2009.00131.x.

Guia, A., Wan, X., Courtemanche, M., & Leblanc, N. (1999). Local Ca^{2+} entry through L-type Ca^{2+} channels activates Ca^{2+}-dependent K^+ channels in rabbit coronary myocytes. *Circulation Research, 84*(9), 1032–1042.

Gurney, A., & Manoury, B. (2009). Two-pore potassium channels in the cardiovascular system. *European Biophysics Journal, 38*(3), 305–318. http://dx.doi.org/10.1007/s00249-008-0326-8.

Gutman, G. A., Chandy, K. G., Grissmer, S., Lazdunski, M., Mckinnon, D., Pardo, L. A., … Wang, X. (2005). International Union of Pharmacology. LIII. Nomenclature and molecular relationships of voltage-gated potassium channels. *Pharmacological Reviews, 57*(4), 473–508. http://dx.doi.org/10.1124/pr.57.4.10.

Haddock, R. E., Grayson, T. H., Morris, M. J., Howitt, L., Chadha, P. S., & Sandow, S. L. (2011). Diet-induced obesity impairs endothelium-derived hyperpolarization via altered potassium channel signaling mechanisms. *PLoS One, 6*(1), e16423. http://dx.doi.org/10.1371/journal.pone.0016423.

Haddy, F. J., Vanhoutte, P. M., & Feletou, M. (2006). Role of potassium in regulating blood flow and blood pressure. *American Journal of Physiology. Regulatory, Integrative and Comparative Physiology, 290*(3), R546–R552. http://dx.doi.org/10.1152/ajpregu.00491.2005.

Hald, B. O., Jacobsen, J. C., Braunstein, T. H., Inoue, R., Ito, Y., Sorensen, P. G., … Jensen, L. J. (2012). BKCa and KV channels limit conducted vasomotor responses in rat mesenteric terminal arterioles. *Pflügers Archiv, 463*(2), 279–295. http://dx.doi.org/10.1007/s00424-011-1049-8.

Hammer, L. W., Ligon, A. L., & Hester, R. L. (2001). Differential inhibition of functional dilation of small arterioles by indomethacin and glibenclamide. *Hypertension, 37*(2 Pt. 2), 599–603.

Hansen, P. R., & Olesen, S. P. (1997). Relaxation of rat resistance arteries by acetylcholine involves a dual mechanism: Activation of K^+ channels and formation of nitric oxide. *Pharmacology & Toxicology, 80*(6), 280–285.

Harraz, O. F., Abd El-Rahman, R. R., Bigdely-Shamloo, K., Wilson, S. M., Brett, S. E., Romero, M., … Welsh, D. G. (2014). Ca(V)3.2 channels and the induction of negative feedback in cerebral arteries. *Circulation Research, 115*(7), 650–661. http://dx.doi.org/10.1161/CIRCRESAHA.114.304056.

Harraz, O. F., Brett, S. E., Zechariah, A., Romero, M., Puglisi, J. L., Wilson, S. M., & Welsh, D. G. (2015). Genetic ablation of CaV3.2 channels enhances the arterial myogenic response by modulating the RyR-BKCa axis. *Arteriosclerosis, Thrombosis, and Vascular Biology, 35*(8), 1843–1851. http://dx.doi.org/10.1161/ATVBAHA.115.305736.

Hashemzadeh-Gargari, H., & Rembold, C. M. (1992). Histamine activates whole cell K^+ currents in swine carotid arterial smooth muscle cell. *Comparative Biochemistry and Physiology. C, 102*(1), 33–37.

Hayabuchi, Y., Davies, N. W., & Standen, N. B. (2001). Angiotensin II inhibits rat arterial KATP channels by inhibiting steady-state protein kinase A activity and activating protein kinase Ce. *The Journal of Physiology, 530*(Pt. 2), 193–205.

Hayabuchi, Y., Standen, N. B., & Davies, N. W. (2001). Angiotensin II inhibits and alters kinetics of voltage-gated K(+) channels of rat arterial smooth muscle. *American Journal of Physiology. Heart and Circulatory Physiology, 281*(6), H2480–H2489.

Heaps, C. L., & Bowles, D. K. (2002). Gender-specific K(+)-channel contribution to adenosine-induced relaxation in coronary arterioles. *Journal of Applied Physiology (Bethesda, Md.: 1985)*, *92*(2), 550–558. http://dx.doi.org/10.1152/japplphysiol.00566.2001.

Heaps, C. L., Tharp, D. L., & Bowles, D. K. (2005). Hypercholesterolemia abolishes voltage-dependent K^+ channel contribution to adenosine-mediated relaxation in porcine coronary arterioles. *American Journal of Physiology. Heart and Circulatory Physiology*, *288*(2), H568–H576. http://dx.doi.org/10.1152/ajpheart.00157.2004.

Hedegaard, E. R., Nielsen, B. D., Kun, A., Hughes, A. D., Kroigaard, C., Mogensen, S., … Simonsen, U. (2014). KV 7 channels are involved in hypoxia-induced vasodilatation of porcine coronary arteries. *British Journal of Pharmacology*, *171*(1), 69–82. http://dx.doi.org/10.1111/bph.12424.

Heintz, A., Damm, M., Brand, M., Koch, T., & Deussen, A. (2008). Coronary flow regulation in mouse heart during hypercapnic acidosis: Role of NO and its compensation during eNOS impairment. *Cardiovascular Research*, 77(1), 188–196. http://dx.doi.org/10.1093/cvr/cvm014.

Henry, P., Pearson, W. L., & Nichols, C. G. (1996). Protein kinase C inhibition of cloned inward rectifier (HRK1/KIR2.3) K^+ channels expressed in Xenopus oocytes. *The Journal of Physiology*, *495*(Pt. 3), 681–688.

Hernanz, R., Alonso, M. J., Baena, A. B., Salaices, M., Alvarez, L., Castillo-Olivares, J. L., & Marin, J. (1999). Mechanisms involved in relaxation induced by exogenous nitric oxide in pig coronary arteries. *Methods and Findings in Experimental and Clinical Pharmacology*, *21*(3), 155–160.

Hodnett, B. L., Xiang, L., Dearman, J. A., Carter, C. B., & Hester, R. L. (2008). K(ATP)-mediated vasodilation is impaired in obese Zucker rats. *Microcirculation*, *15*(6), 485–494. http://dx.doi.org/10.1080/10739680801942240.

Hojs, N., Strucl, M., & Cankar, K. (2009). The effect of glibenclamide on acetylcholine and sodium nitroprusside induced vasodilatation in human cutaneous microcirculation. *Clinical Physiology and Functional Imaging*, *29*(1), 38–44. http://dx.doi.org/10.1111/j.1475-097X.2008.00833.x.

Holdsworth, C. T., Copp, S. W., Ferguson, S. K., Sims, G. E., Poole, D. C., & Musch, T. I. (2015). Acute inhibition of ATP-sensitive K^+ channels impairs skeletal muscle vascular control in rats during treadmill exercise. *American Journal of Physiology. Heart and Circulatory Physiology*, *308*(11), H1434–H1442. http://dx.doi.org/10.1152/ajpheart.00772.2014.

Horinaka, N., Kuang, T. Y., Pak, H., Wang, R., Jehle, J., Kennedy, C., & Sokoloff, L. (1997). Blockade of cerebral blood flow response to insulin-induced hypoglycemia by caffeine and glibenclamide in conscious rats. *Journal of Cerebral Blood Flow and Metabolism*, *17*(12), 1309–1318. http://dx.doi.org/10.1097/00004647-199712000-00006.

Hoshi, T., Pantazis, A., & Olcese, R. (2013). Transduction of voltage and Ca^{2+} signals by Slo1 BK channels. *Physiology (Bethesda)*, *28*(3), 172–189. http://dx.doi.org/10.1152/physiol.00055.2012.

Hoshi, T., Tian, Y., Xu, R., Heinemann, S. H., & Hou, S. (2013). Mechanism of the modulation of BK potassium channel complexes with different auxiliary subunit compositions by the omega-3 fatty acid DHA. *Proceedings of the National Academy of Sciences of the United States of America*, *110*(12), 4822–4827. http://dx.doi.org/10.1073/pnas.1222003110.

Hou, S., Heinemann, S. H., & Hoshi, T. (2009). Modulation of BKCa channel gating by endogenous signaling molecules. *Physiology (Bethesda)*, *24*, 26–35. http://dx.doi.org/10.1152/physiol.00032.2008.

House, S. J., Potier, M., Bisaillon, J., Singer, H. A., & Trebak, M. (2008). The non-excitable smooth muscle: Calcium signaling and phenotypic switching during vascular disease. *Pflügers Archiv*, *456*(5), 769–785. http://dx.doi.org/10.1007/s00424-008-0491-8.

Huang, C., Pollock, C. A., & Chen, X. M. (2015). KCa3.1: A new player in progressive kidney disease. *Current Opinion in Nephrology and Hypertension*, *24*(1), 61–66. http://dx.doi.org/10.1097/MNH.0000000000000083.

Hutri-Kahonen, N., Kahonen, M., Wu, X., Sand, J., Nordback, I., Taurio, J., & Porsti, I. (1999). Control of vascular tone in isolated mesenteric arterial segments from hypertensive patients. *British Journal of Pharmacology*, *127*(7), 1735–1743. http://dx.doi.org/10.1038/sj.bjp.0702716.

Imamura, Y., Tomoike, H., Narishige, T., Takahashi, T., Kasuya, H., & Takeshita, A. (1992). Glibenclamide decreases basal coronary blood flow in anesthetized dogs. *The American Journal of Physiology*, *263*(2 Pt. 2), H399–H404.

Irat, A. M., Aslamaci, S., Karasu, C., & Ari, N. (2006). Alteration of vascular reactivity in diabetic human mammary artery and the effects of thiazolidinediones. *The Journal of Pharmacy and Pharmacology*, *58*(12), 1647–1653. http://dx.doi.org/10.1211/jpp.58.12.0012.

Ishii, T. M., Silvia, C., Hirschberg, B., Bond, C. T., Adelman, J. P., & Maylie, J. (1997). A human intermediate conductance calcium-activated potassium channel. *Proceedings of the National Academy of Sciences of the United States of America*, *94*(21), 11651–11656.

Ishikawa, T., Hume, J. R., & Keef, K. D. (1993). Modulation of K^+ and Ca^{2+} channels by histamine H_1-receptor stimulation in rabbit coronary artery cells. *The Journal of Physiology*, *468*, 379–400.

Jackson, W. F. (1993). Arteriolar tone is determined by activity of ATP-sensitive potassium channels. *American Journal of Physiology*, *265*(5), H1797–H1803.

Jackson, W. F. (2000). Ion channels and vascular tone. *Hypertension*, *35*(1 Pt. 2), 173–178.

Jackson, W. F. (2005). Potassium channels in the peripheral microcirculation. *Microcirculation*, *12*(1), 113–127. http://dx.doi.org/10.1080/10739680590896072.

Jackson, W. F., & Blair, K. L. (1998). Characterization and function of Ca(2+)-activated K^+ channels in arteriolar muscle cells. *The American Journal of Physiology*, *274*(1 Pt. 2), H27–H34.

Jackson, W. F., Konig, A., Dambacher, T., & Busse, R. (1993). Prostacyclin-induced vasodilation in rabbit heart is mediated by ATP-sensitive potassium channels. *The American Journal of Physiology*, *264*(1 Pt. 2), H238–H243.

Jackson-Weaver, O., Osmond, J. M., Riddle, M. A., Naik, J. S., Gonzalez Bosc, L. V., Walker, B. R., & Kanagy, N. L. (2013). Hydrogen sulfide dilates rat mesenteric arteries by activating endothelial large-conductance Ca(2)(+)-activated K(+) channels and smooth muscle Ca(2)(+) sparks. *American Journal of Physiology. Heart and Circulatory Physiology*, *304*(11), H1446–H1454. http://dx.doi.org/10.1152/ajpheart.00506.2012.

Jaggar, J. H., Leffler, C. W., Cheranov, S. Y., Tcheranova, D., Shuyu, E., & Cheng, X. (2002). Carbon monoxide dilates cerebral arterioles by enhancing the coupling of Ca^{2+} sparks to Ca^{2+}-activated K^+ channels. *Circulation Research*, *91*(7), 610–617.

Jaggar, J. H., Li, A. L., Parfenova, H., Liu, J. X., Umstot, E. S., Dopico, A. M., & Leffler, C. W. (2005). Heme is a carbon monoxide receptor for large-conductance Ca^{2+}-activated K^+ channels. *Circulation Research*, *97*(8), 805–812. http://dx.doi.org/10.1161/RES.000018618.47148.7b.

Jaggar, J. H., Porter, V. A., Lederer, W. J., & Nelson, M. T. (2000). Calcium sparks in smooth muscle. *American Journal of Physiology. Cell Physiology*, *278*(2), C235–C256.

Jaggar, J. H., Stevenson, A. S., & Nelson, M. T. (1998). Voltage dependence of Ca^{2+} sparks in intact cerebral arteries. *The American Journal of Physiology*, *274*(6 Pt. 1), C1755–C1761.

Jaggar, J. H., Wellman, G. C., Heppner, T. J., Porter, V. A., Perez, G. J., Gollasch, M., … Nelson, M. T. (1998). Ca^{2+} channels, ryanodine receptors and Ca(2+)-activated K^+ channels: A functional unit for regulating arterial tone. *Acta Physiologica Scandinavica*, *164*(4), 577–587. http://dx.doi.org/10.1046/j.1365-201X.1998.00462.x.

Jantzi, M. C., Brett, S. E., Jackson, W. F., Corteling, R., Vigmond, E. J., & Welsh, D. G. (2006). Inward rectifying potassium channels facilitate cell-to-cell communication in hamster retractor muscle feed arteries. *American Journal of Physiology. Heart and Circulatory Physiology*, *291*(3), H1319–H1328. http://dx.doi.org/10.1152/ajpheart.00217.2006.

Jepps, T. A., Olesen, S. P., & Greenwood, I. A. (2013). One man's side effect is another man's therapeutic opportunity: Targeting Kv7 channels in smooth muscle disorders. *British Journal of Pharmacology*, *168*(1), 19–27. http://dx.doi.org/10.1111/j.1476-5381.2012.02133.x.

Jewell, R. P., Saundry, C. M., Bonev, A. D., Tranmer, B. I., & Wellman, G. C. (2004). Inhibition of Ca++ sparks by oxyhemoglobin in rabbit cerebral arteries. *Journal of Neurosurgery*, *100*(2), 295–302. http://dx.doi.org/10.3171/jns.2004.100.2.0295.

Jiang, Z. G., Si, J. Q., Lasarev, M. R., & Nuttall, A. L. (2001). Two resting potential levels regulated by the inward-rectifier potassium channel in the guinea-pig spiral modiolar artery. *The Journal of Physiology*, *537*(Pt. 3), 829–842.

Jiang, J., Thoren, P., Caligiuri, G., Hansson, G. K., & Pernow, J. (1999). Enhanced phenylephrine-induced rhythmic activity in the atherosclerotic mouse aorta via an increase in opening of KCa channels: Relation to Kv channels and nitric oxide. *British Journal of Pharmacology*, *128*(3), 637–646. http://dx.doi.org/10.1038/sj.bjp.0702855.

Jiao, J., Garg, V., Yang, B., Elton, T. S., & Hu, K. (2008). Protein kinase C-epsilon induces caveolin-dependent internalization of vascular adenosine 5′-triphosphate-sensitive K^+ channels. *Hypertension*, *52*(3), 499–506. http://dx.doi.org/10.1161/HYPERTENSIONAHA.108.110817.

Jimenez-Perez, L., Cidad, P., Alvarez-Miguel, I., Santos-Hipolito, A., Torres-Merino, R., Alonso, E., … Perez-Garcia, M. T. (2016). Molecular determinants of Kv1.3 potassium channels-induced proliferation. *The Journal of Biological Chemistry*, *291*(7), 3569–3580. http://dx.doi.org/10.1074/jbc.M115.678995.

Johnson, T. D., Marrelli, S. P., Steenberg, M. L., Childres, W. F., & Bryan, R. M., Jr. (1998). Inward rectifier potassium channels in the rat middle cerebral artery. *The American Journal of Physiology*, *274*(2 Pt. 2), R541–R547.

Kalliovalkama, J., Jolma, P., Tolvanen, J. P., Kahonen, M., Hutri-Kahonen, N., Wu, X., … Porsti, I. (1999). Arterial function in nitric oxide-deficient hypertension: Influence of long-term angiotensin II receptor antagonism. *Cardiovascular Research*, *42*(3), 773–782.

Kam, K. L., Pfaffendorf, M., & van Zwieten, P. A. (1994). Drug-induced endothelium-dependent and -independent relaxations in isolated resistance vessels taken from simultaneously hypertensive and streptozotocin-diabetic rats. *Blood Pressure*, *3*(6), 418–427.

Kamata, K., Miyata, N., & Kasuya, Y. (1989). Functional changes in potassium channels in aortas from rats with streptozotocin-induced diabetes. *European Journal of Pharmacology*, *166*(2), 319–323. http://dx.doi.org/10.1016/0014-2999(89)90076-9.

Kawata, T., Mimuro, T., Onuki, T., Tsuchiya, K., Nihei, H., & Koike, T. (1998). The K(ATP) channel opener nicorandil: Effect on renal hemodynamics in spontaneously hypertensive and Wistar Kyoto rats. *Kidney International. Supplement*, *67*, S231–S233.

Khanamiri, S., Soltysinska, E., Jepps, T. A., Bentzen, B. H., Chadha, P. S., Schmitt, N., … Olesen, S. P. (2013). Contribution of Kv7 channels to basal coronary flow and active response to ischemia. *Hypertension*, *62*(6), 1090–1097. http://dx.doi.org/10.1161/HYPERTENSIONAHA.113.01244.

Kilpatrick, E. V., & Cocks, T. M. (1994). Evidence for differential roles of nitric-oxide (No) and hyperpolarization in endothelium-dependent relaxation of pig isolated coronary-artery. *British Journal of Pharmacology*, *112*(2), 557–565.

Kim, H., & Oh, K. H. (2016). Protein network interacting with BK channels. *International Review of Neurobiology*, *128*, 127–161. http://dx.doi.org/10.1016/bs.irn.2016.03.003.

Kinoshita, H., Azma, T., Iranami, H., Nakahata, K., Kimoto, Y., Dojo, M., … Hatano, Y. (2006). Synthetic peroxisome proliferator-activated receptor-gamma agonists restore impaired vasorelaxation via ATP-sensitive K^+ channels by high glucose. *The Journal of*

Pharmacology and Experimental Therapeutics, *318*(1), 312–318. http://dx.doi.org/10.1124/jpet.106.100958.
Kitazono, T., Heistad, D. D., & Faraci, F. M. (1993a). ATP-sensitive potassium channels in the basilar artery during chronic hypertension. *Hypertension*, *22*(5), 677–681.
Kitazono, T., Heistad, D. D., & Faraci, F. M. (1993b). Role of ATP-sensitive K^+ channels in CGRP-induced dilatation of basilar artery in vivo. *The American Journal of Physiology*, *265*(2 Pt. 2), H581–H585.
Kleppisch, T., & Nelson, M. T. (1995). Adenosine activates ATP-sensitive potassium channels in arterial myocytes via A_2 receptors and cAMP-dependent protein kinase. *Proceedings of the National Academy of Sciences of the United States of America*, *92*(26), 12441–12445.
Knot, H. J., & Nelson, M. T. (1998). Regulation of arterial diameter and wall $[Ca^{2+}]$ in cerebral arteries of rat by membrane potential and intravascular pressure. *The Journal of Physiology*, *508*(Pt. 1), 199–209.
Knot, H. J., Standen, N. B., & Nelson, M. T. (1998). Ryanodine receptors regulate arterial diameter and wall $[Ca^{2+}]$ in cerebral arteries of rat via Ca^{2+}-dependent K^+ channels. *The Journal of Physiology*, *508*(Pt. 1), 211–221.
Ko, E. A., Park, W. S., Firth, A. L., Hong, D. H., Choi, S. W., Heo, H. J., … Han, J. (2010a). Increased sensitivity of serotonin on the voltage-dependent K^+ channels in mesenteric arterial smooth muscle cells of OLETF rats. *Progress in Biophysics and Molecular Biology*, *103*(1), 88–94. http://dx.doi.org/10.1016/j.pbiomolbio.2010.02.003.
Ko, E. A., Park, W. S., Firth, A. L., Kim, N., Yuan, J. X., & Han, J. (2010b). Pathophysiology of voltage-gated K^+ channels in vascular smooth muscle cells: Modulation by protein kinases. *Progress in Biophysics and Molecular Biology*, *103*(1), 95–101. http://dx.doi.org/10.1016/j.pbiomolbio.2009.10.001.
Kohler, R., Wulff, H., Eichler, I., Kneifel, M., Neumann, D., Knorr, A., … Hoyer, J. (2003). Blockade of the intermediate-conductance calcium-activated potassium channel as a new therapeutic strategy for restenosis. *Circulation*, *108*(9), 1119–1125. http://dx.doi.org/10.1161/01.CIR.0000086464.04719.DD.
Kolias, T. J., Chai, S., & Webb, R. C. (1993). Potassium channel antagonists and vascular reactivity in stroke-prone spontaneously hypertensive rats. *American Journal of Hypertension*, *6*(6 Pt. 1), 528–533.
Kondratskyi, A., Kondratska, K., Skryma, R., & Prevarskaya, N. (2015). Ion channels in the regulation of apoptosis. *Biochimica et Biophysica Acta*, *1848*(10 Pt. B), 2532–2546. http://dx.doi.org/10.1016/j.bbamem.2014.10.030.
Kosmas, E. N., Levy, R. D., & Hussain, S. N. (1995). Acute effects of glyburide on the regulation of peripheral blood flow in normal humans. *European Journal of Pharmacology*, *274*(1–3), 193–199.
Kuang, Q., Purhonen, P., & Hebert, H. (2015). Structure of potassium channels. *Cellular and Molecular Life Sciences*, *72*(19), 3677–3693. http://dx.doi.org/10.1007/s00018-015-1948-5.
Kukuljan, M., Rojas, E., Catt, K. J., & Stojilkovic, S. S. (1994). Membrane potential regulates inositol 1,4,5-trisphosphate-controlled cytoplasmic Ca^{2+} oscillations in pituitary gonadotrophs. *The Journal of Biological Chemistry*, *269*(7), 4860–4865.
Kume, H., Graziano, M. P., & Kotlikoff, M. I. (1992). Stimulatory and inhibitory regulation of calcium-activated potassium channels by guanine nucleotide-binding proteins. *Proceedings of the National Academy of Sciences of the United States of America*, *89*(22), 11051–11055.
Kume, H., Hall, I. P., Washabau, R. J., Takagi, K., & Kotlikoff, M. I. (1994). Beta-adrenergic agonists regulate KCa channels in airway smooth muscle by cAMP-dependent and -independent mechanisms. *The Journal of Clinical Investigation*, *93*(1), 371–379. http://dx.doi.org/10.1172/JCI116969.

Kume, H., Takai, A., Tokuno, H., & Tomita, T. (1989). Regulation of Ca^{2+}-dependent K^+-channel activity in tracheal myocytes by phosphorylation. *Nature*, *341*(6238), 152–154. http://dx.doi.org/10.1038/341152a0.

Lang, R. J., Harvey, J. R., McPhee, G. J., & Klemm, M. F. (2000). Nitric oxide and thiol reagent modulation of Ca^{2+}-activated K^+ (BKCa) channels in myocytes of the guinea-pig taenia caeci. *The Journal of Physiology*, *525*(Pt. 2), 363–376.

Lang, R. J., Harvey, J. R., & Mulholland, E. L. (2003). Sodium (2-sulfonatoethyl) methanethiosulfonate prevents S-nitroso-L-cysteine activation of Ca^{2+}-activated K^+ (BKCa) channels in myocytes of the guinea-pig taenia caeca. *British Journal of Pharmacology*, *139*(6), 1153–1163. http://dx.doi.org/10.1038/sj.bjp.0705349.

Lange, A., Gebremedhin, D., Narayanan, J., & Harder, D. (1997). 20-Hydroxyeicosatetraenoic acid-induced vasoconstriction and inhibition of potassium current in cerebral vascular smooth muscle is dependent on activation of protein kinase C. *Journal of Biological Chemistry*, *272*(43), 27345–27352. http://dx.doi.org/10.1074/jbc.272.43.27345.

Leffler, C. W., Parfenova, H., Basuroy, S., Jaggar, J. H., Umstot, E. S., & Fedinec, A. L. (2011). Hydrogen sulfide and cerebral microvascular tone in newborn pigs. *American Journal of Physiology. Heart and Circulatory Physiology*, *300*(2), H440–H447. http://dx.doi.org/10.1152/ajpheart.00722.2010.

Leo, M. D., Bannister, J. P., Narayanan, D., Nair, A., Grubbs, J. E., Gabrick, K. S., … Jaggar, J. H. (2014). Dynamic regulation of beta1 subunit trafficking controls vascular contractility. *Proceedings of the National Academy of Sciences of the United States of America*, *111*(6), 2361–2366. http://dx.doi.org/10.1073/pnas.1317527111.

Leo, M. D., Bulley, S., Bannister, J. P., Kuruvilla, K. P., Narayanan, D., & Jaggar, J. H. (2015). Angiotensin II stimulates internalization and degradation of arterial myocyte plasma membrane BK channels to induce vasoconstriction. *American Journal of Physiology. Cell Physiology*, *309*(6), C392–C402. http://dx.doi.org/10.1152/ajpcell.00127.2015.

Levitan, I. B. (2006). Signaling protein complexes associated with neuronal ion channels. *Nature Neuroscience*, *9*(3), 305–310. http://dx.doi.org/10.1038/nn1647.

Li, H., Chai, Q., Gutterman, D. D., & Liu, Y. (2003). Elevated glucose impairs cAMP-mediated dilation by reducing Kv channel activity in rat small coronary smooth muscle cells. *American Journal of Physiology. Heart and Circulatory Physiology*, *285*(3), H1213–H1219. http://dx.doi.org/10.1152/ajpheart.00226.2003.

Li, S. S., Cui, N., Yang, Y., Trower, T. C., Wei, Y. M., Wu, Y., … Jiang, C. (2015). Impairment of the vascular KATP channel imposes fatal susceptibility to experimental diabetes due to multi-organ injuries. *Journal of Cellular Physiology*, *230*(12), 2915–2926. http://dx.doi.org/10.1002/jcp.25003.

Li, H., Gutterman, D. D., Rusch, N. J., Bubolz, A., & Liu, Y. (2004). Nitration and functional loss of voltage-gated K^+ channels in rat coronary microvessels exposed to high glucose. *Diabetes*, *53*(9), 2436–2442.

Li, P. L., Jin, M. W., & Campbell, W. B. (1998). Effect of selective inhibition of soluble guanylyl cyclase on the K(Ca) channel activity in coronary artery smooth muscle. *Hypertension*, *31*(1 Pt. 2), 303–308.

Li, A., Knutsen, R. H., Zhang, H., Osei-Owusu, P., Moreno-Dominguez, A., Harter, T. M., … Nichols, C. G. (2013). Hypotension due to Kir6.1 gain-of-function in vascular smooth muscle. *Journal of the American Heart Association*, *2*(4), e000365. http://dx.doi.org/10.1161/JAHA.113.000365.

Li, Z., Lu, N., & Shi, L. (2014). Exercise training reverses alterations in Kv and BKCa channel molecular expression in thoracic aorta smooth muscle cells from spontaneously hypertensive rats. *Journal of Vascular Research*, *51*(6), 447–457. http://dx.doi.org/10.1159/000369928.

Li, A., Xi, Q., Umstot, E. S., Bellner, L., Schwartzman, M. L., Jaggar, J. H., & Leffler, C. W. (2008). Astrocyte-derived CO is a diffusible messenger that mediates glutamate-induced cerebral arteriolar dilation by activating smooth muscle cell KCa channels. *Circulation Research*, *102*(2), 234–241. http://dx.doi.org/10.1161/CIRCRESAHA.107.164145.

Li, P. L., Zou, A. P., & Campbell, W. B. (1997). Regulation of potassium channels in coronary arterial smooth muscle by endothelium-derived vasodilators. *Hypertension*, *29*(1 Pt. 2), 262–267.

Liang, G. H., Adebiyi, A., Leo, M. D., McNally, E. M., Leffler, C. W., & Jaggar, J. H. (2011). Hydrogen sulfide dilates cerebral arterioles by activating smooth muscle cell plasma membrane KATP channels. *American Journal of Physiology. Heart and Circulatory Physiology*, *300*(6), H2088–H2095. http://dx.doi.org/10.1152/ajpheart.01290.2010.

Liang, G. H., Xi, Q., Leffler, C. W., & Jaggar, J. H. (2012). Hydrogen sulfide activates Ca(2)(+) sparks to induce cerebral arteriole dilatation. *The Journal of Physiology*, *590*(11), 2709–2720. http://dx.doi.org/10.1113/jphysiol.2011.225128.

Lindauer, U., Vogt, J., Schuh-Hofer, S., Dreier, J. P., & Dirnagl, U. (2003). Cerebrovascular vasodilation to extraluminal acidosis occurs via combined activation of ATP-sensitive and Ca^{2+}-activated potassium channels. *Journal of Cerebral Blood Flow and Metabolism*, *23*(10), 1227–1238. http://dx.doi.org/10.1097/01.WCB.0000088764.02615.B7.

Liu, Y., & Gutterman, D. D. (2002). The coronary circulation in diabetes: Influence of reactive oxygen species on K^+ channel-mediated vasodilation. *Vascular Pharmacology*, *38*(1), 43–49.

Liu, Y., Hudetz, A. G., Knaus, H. G., & Rusch, N. J. (1998). Increased expression of Ca^{2+}-sensitive K^+ channels in the cerebral microcirculation of genetically hypertensive rats: Evidence for their protection against cerebral vasospasm. *Circulation Research*, *82*(6), 729–737.

Liu, Y., Jones, A. W., & Sturek, M. (1994). Increased barium influx and potassium current in stroke-prone spontaneously hypertensive rats. *Hypertension*, *23*(6 Pt. 2), 1091–1095.

Liu, Y., Pleyte, K., Knaus, H. G., & Rusch, N. J. (1997). Increased expression of Ca^{2+}-sensitive K^+ channels in aorta of hypertensive rats. *Hypertension*, *30*(6), 1403–1409.

Liu, Y., Terata, K., Rusch, N. J., & Gutterman, D. D. (2001). High glucose impairs voltage-gated K(+) channel current in rat small coronary arteries. *Circulation Research*, *89*(2), 146–152.

Liu, Q. H., Zheng, Y. M., Korde, A. S., Yadav, V. R., Rathore, R., Wess, J., & Wang, Y. X. (2009). Membrane depolarization causes a direct activation of G protein-coupled receptors leading to local Ca^{2+} release in smooth muscle. *Proceedings of the National Academy of Sciences of the United States of America*, *106*(27), 11418–11423. http://dx.doi.org/10.1073/pnas.0813307106.

Longden, T. A., Dabertrand, F., Hill-Eubanks, D. C., Hammack, S. E., & Nelson, M. T. (2014). Stress-induced glucocorticoid signaling remodels neurovascular coupling through impairment of cerebrovascular inwardly rectifying K^+ channel function. *Proceedings of the National Academy of Sciences of the United States of America*, *111*(20), 7462–7467. http://dx.doi.org/10.1073/pnas.1401811111.

Longden, T. A., & Nelson, M. T. (2015). Vascular inward rectifier K^+ channels as external K^+ sensors in the control of cerebral blood flow. *Microcirculation*, *22*(3), 183–196. http://dx.doi.org/10.1111/micc.12190.

Loutzenhiser, R. D., & Parker, M. J. (1994). Hypoxia inhibits myogenic reactivity of renal afferent arterioles by activating ATP-sensitive K^+ channels. *Circulation Research*, *74*(5), 861–869.

Lovren, F., & Triggle, C. (2000). Nitric oxide and sodium nitroprusside-induced relaxation of the human umbilical artery. *British Journal of Pharmacology*, *131*(3), 521–529. http://dx.doi.org/10.1038/sj.bjp.0703588.

Lu, S., Xiang, L., Clemmer, J. S., Gowdey, A. R., Mittwede, P. N., & Hester, R. L. (2013). Impaired vascular KATP function attenuates exercise capacity in obese zucker rats. *Microcirculation*, *20*(7), 662–669. http://dx.doi.org/10.1111/micc.12065.

Luykenaar, K. D., Brett, S. E., Wu, B. N., Wiehler, W. B., & Welsh, D. G. (2004). Pyrimidine nucleotides suppress KDR currents and depolarize rat cerebral arteries by activating Rho kinase. *American Journal of Physiology. Heart and Circulatory Physiology*, *286*(3), H1088–H1100. http://dx.doi.org/10.1152/ajpheart.00903.2003.

Luykenaar, K. D., El-Rahman, R. A., Walsh, M. P., & Welsh, D. G. (2009). Rho-kinase-mediated suppression of KDR current in cerebral arteries requires an intact actin cytoskeleton. *American Journal of Physiology. Heart and Circulatory Physiology*, *296*(4), H917–H926. http://dx.doi.org/10.1152/ajpheart.01206.2008.

Mackie, A. R., & Byron, K. L. (2008). Cardiovascular KCNQ (Kv7) potassium channels: Physiological regulators and new targets for therapeutic intervention. *Molecular Pharmacology*, *74*(5), 1171–1179. http://dx.doi.org/10.1124/mol.108.049825.

Mahaut-Smith, M. P., Martinez-Pinna, J., & Gurung, I. S. (2008). A role for membrane potential in regulating GPCRs? *Trends in Pharmacological Sciences*, *29*(8), 421–429. http://dx.doi.org/10.1016/j.tips.2008.05.007.

Mandala, M., Heppner, T. J., Bonev, A. D., & Nelson, M. T. (2007). Effect of endogenous and exogenous nitric oxide on calcium sparks as targets for vasodilation in rat cerebral artery. *Nitric Oxide*, *16*(1), 104–109. http://dx.doi.org/10.1016/j.niox.2006.06.007.

Marijic, J., Li, Q., Song, M., Nishimaru, K., Stefani, E., & Toro, L. (2001). Decreased expression of voltage- and Ca(2+)-activated K(+) channels in coronary smooth muscle during aging. *Circulation Research*, *88*(2), 210–216.

Marrelli, S. P., Johnson, T. D., Khorovets, A., Childres, W. F., & Bryan, R. M., Jr. (1998). Altered function of inward rectifier potassium channels in cerebrovascular smooth muscle after ischemia/reperfusion. *Stroke*, *29*(7), 1469–1474.

Marshall, J. M., Thomas, T., & Turner, L. (1993). A link between adenosine, ATP-sensitive K^+ channels, potassium and muscle vasodilatation in the rat in systemic hypoxia. *The Journal of Physiology*, *472*, 1–9.

Martelli, A., Testai, L., Breschi, M. C., Lawson, K., McKay, N. G., Miceli, F., … Calderone, V. (2013). Vasorelaxation by hydrogen sulphide involves activation of Kv7 potassium channels. *Pharmacological Research*, *70*(1), 27–34. http://dx.doi.org/10.1016/j.phrs.2012.12.005.

Martens, J. R., & Gelband, C. H. (1996). Alterations in rat interlobar artery membrane potential and K^+ channels in genetic and nongenetic hypertension. *Circulation Research*, *79*(2), 295–301.

Martinez, A. C., Pagan, R. M., Prieto, D., Recio, P., Garcia-Sacristan, A., Hernandez, M., & Benedito, S. (2009). Modulation of noradrenergic neurotransmission in isolated rat radial artery. *Journal of Pharmacological Sciences*, *111*(3), 299–311.

Matsumoto, T., Szasz, T., Tostes, R. C., & Webb, R. C. (2012). Impaired beta-adrenoceptor-induced relaxation in small mesenteric arteries from DOCA-salt hypertensive rats is due to reduced K(Ca) channel activity. *Pharmacological Research*, *65*(5), 537–545. http://dx.doi.org/10.1016/j.phrs.2012.02.004.

Matsushita, K., & Puro, D. G. (2006). Topographical heterogeneity of K(IR) currents in pericyte-containing microvessels of the rat retina: Effect of diabetes. *The Journal of Physiology*, *573*(Pt. 2), 483–495. http://dx.doi.org/10.1113/jphysiol.2006.107102.

Mayhan, W. G. (1994). Effect of diabetes mellitus on response of the basilar artery to activation of ATP-sensitive potassium channels. *Brain Research*, *636*(1), 35–39.

Mayhan, W. G., & Faraci, F. M. (1993). Responses of cerebral arterioles in diabetic rats to activation of ATP-sensitive potassium channels. *The American Journal of Physiology*, *265*(1 Pt. 2), H152–H157.

Mayhan, W. G., Mayhan, J. F., Sun, H., & Patel, K. P. (2004). In vivo properties of potassium channels in cerebral blood vessels during diabetes mellitus. *Microcirculation*, *11*(7), 605–613. http://dx.doi.org/10.1080/10739680490503410.

McCarron, J. G., & Halpern, W. (1990). Potassium dilates rat cerebral arteries by two independent mechanisms. *The American Journal of Physiology*, *259*(3 Pt. 2), H902–H908.

McGahon, M. K., Dash, D. P., Arora, A., Wall, N., Dawicki, J., Simpson, D. A., … Curtis, T. M. (2007). Diabetes downregulates large-conductance Ca^{2+}-activated potassium beta 1 channel subunit in retinal arteriolar smooth muscle. *Circulation Research*, *100*(5), 703–711. http://dx.doi.org/10.1161/01.RES.0000260182.36481.c9.

McManus, O. B., Helms, L. M., Pallanck, L., Ganetzky, B., Swanson, R., & Leonard, R. J. (1995). Functional role of the beta subunit of high conductance calcium-activated potassium channels. *Neuron*, *14*(3), 645–650.

Meera, P., Wallner, M., Jiang, Z., & Toro, L. (1996). A calcium switch for the functional coupling between alpha (hslo) and beta subunits (KV, Ca beta) of maxi K channels. *FEBS Letters*, *382*, 84–88.

Meera, P., Wallner, M., Song, M., & Toro, L. (1997). Large conductance voltage- and calcium-dependent K^+ channel, a distinct member of voltage-dependent ion channels with seven N-terminal transmembrane segments (S0-S6), an extracellular N terminus, and an intracellular (S9-S10) C terminus. *Proceedings of the National Academy of Sciences of the United States of America*, *94*(25), 14066–14071.

Merkus, D., Haitsma, D. B., Fung, T. Y., Assen, Y. J., Verdouw, P. D., & Duncker, D. J. (2003). Coronary blood flow regulation in exercising swine involves parallel rather than redundant vasodilator pathways. *American Journal of Physiology. Heart and Circulatory Physiology*, *285*(1), H424–H433. http://dx.doi.org/10.1152/ajpheart.00916.2002.

Merkus, D., Sorop, O., Houweling, B., Hoogteijling, B. A., & Duncker, D. J. (2006). KCa+ channels contribute to exercise-induced coronary vasodilation in swine. *American Journal of Physiology. Heart and Circulatory Physiology*, *291*(5), H2090–H2097. http://dx.doi.org/10.1152/ajpheart.00315.2006.

Miguel-Velado, E., Moreno-Dominguez, A., Colinas, O., Cidad, P., Heras, M., Perez-Garcia, M. T., & Lopez-Lopez, J. R. (2005). Contribution of Kv channels to phenotypic remodeling of human uterine artery smooth muscle cells. *Circulation Research*, *97*(12), 1280–1287. http://dx.doi.org/10.1161/01.RES.0000194322.91255.13.

Miguel-Velado, E., Perez-Carretero, F. D., Colinas, O., Cidad, P., Heras, M., Lopez-Lopez, J. R., & Perez-Garcia, M. T. (2010). Cell cycle-dependent expression of Kv3.4 channels modulates proliferation of human uterine artery smooth muscle cells. *Cardiovascular Research*, *86*(3), 383–391. http://dx.doi.org/10.1093/cvr/cvq011.

Miki, T., Suzuki, M., Shibasaki, T., Uemura, H., Sato, T., Yamaguchi, K., … Seino, S. (2002). Mouse model of Prinzmetal angina by disruption of the inward rectifier Kir6.1.. *Nature Medicine*, *8*(5), 466–472. http://dx.doi.org/10.1038/nm0502-466.

Miller, A. W., Tulbert, C., Puskar, M., & Busija, D. W. (2002). Enhanced endothelin activity prevents vasodilation to insulin in insulin resistance. *Hypertension*, *40*(1), 78–82.

Minami, K., Fukuzawa, K., & Nakaya, Y. (1993). Protein kinase C inhibits the Ca(2+)-activated K^+ channel of cultured porcine coronary artery smooth muscle cells. *Biochemical and Biophysical Research Communications*, *190*(1), 263–269. http://dx.doi.org/10.1006/bbrc.1993.1040.

Ming, Z., Parent, R., & Lavallee, M. (1997). Beta 2-adrenergic dilation of resistance coronary vessels involves KATP channels and nitric oxide in conscious dogs. *Circulation*, *95*(6), 1568–1576.

Mistry, D. K., & Garland, C. J. (1998). Nitric oxide (NO)-induced activation of large conductance Ca^{2+}-dependent K^+ channels (BK(Ca)) in smooth muscle cells isolated from the rat mesenteric artery. *British Journal of Pharmacology*, *124*(6), 1131–1140. http://dx.doi.org/10.1038/sj.bjp.0701940.

Mistry, D. K., & Garland, C. J. (1999). The influence of phenylephrine outward potassium currents in single smooth muscle cells from the rabbit mesenteric artery. *General Pharmacology, 33*(5), 389–399.

Miura, H., Wachtel, R. E., Loberiza, F. R., Jr., Saito, T., Miura, M., Nicolosi, A. C., & Gutterman, D. D. (2003). Diabetes mellitus impairs vasodilation to hypoxia in human coronary arterioles: Reduced activity of ATP-sensitive potassium channels. *Circulation Research, 92*(2), 151–158.

Miyata, N., Tsuchida, K., & Otomo, S. (1990). Functional changes in potassium channels in carotid arteries from stroke-prone spontaneously hypertensive rats. *European Journal of Pharmacology, 182*(1), 209–210.

Mokelke, E. A., Dietz, N. J., Eckman, D. M., Nelson, M. T., & Sturek, M. (2005). Diabetic dyslipidemia and exercise affect coronary tone and differential regulation of conduit and microvessel K^+ current. *American Journal of Physiology. Heart and Circulatory Physiology, 288*(3), H1233–H1241. http://dx.doi.org/10.1152/ajpheart.00732.2004.

Moore, C. L., Nelson, P. L., Parelkar, N. K., Rusch, N. J., & Rhee, S. W. (2014). Protein kinase A-phosphorylated KV1 channels in PSD95 signaling complex contribute to the resting membrane potential and diameter of cerebral arteries. *Circulation Research, 114*(8), 1258–1267. http://dx.doi.org/10.1161/CIRCRESAHA.114.303167.

Moreno-Dominguez, A., Cidad, P., Miguel-Velado, E., Lopez-Lopez, J. R., & Perez-Garcia, M. T. (2009). De novo expression of Kv6.3 contributes to changes in vascular smooth muscle cell excitability in a hypertensive mice strain. *The Journal of Physiology, 587*(3), 625–640. http://dx.doi.org/10.1113/jphysiol.2008.165217.

Mori, H., Chujo, M., Tanaka, E., Yamakawa, A., Shinozaki, Y., Mohamed, M. U., & Nakazawa, H. (1995). Modulation of adrenergic coronary vasoconstriction via ATP-sensitive potassium channel. *The American Journal of Physiology, 268*(3 Pt. 2), H1077–H1085.

Munoz, E., Hernandez-Morales, M., Sobradillo, D., Rocher, A., Nunez, L., & Villalobos, C. (2013). Intracellular Ca(2+) remodeling during the phenotypic journey of human coronary smooth muscle cells. *Cell Calcium, 54*(5), 375–385. http://dx.doi.org/10.1016/j.ceca.2013.08.006.

Murrant, C. L., & Sarelius, I. H. (2002). Multiple dilator pathways in skeletal muscle contraction-induced arteriolar dilations. *American Journal of Physiology. Regulatory, Integrative and Comparative Physiology, 282*(4), R969–R978. http://dx.doi.org/10.1152/ajpregu.00405.2001.

Mustafa, A. K., Sikka, G., Gazi, S. K., Steppan, J., Jung, S. M., Bhunia, A. K., … Snyder, S. H. (2011). Hydrogen sulfide as endothelium-derived hyperpolarizing factor sulfhydrates potassium channels. *Circulation Research, 109*(11), 1259–1268. http://dx.doi.org/10.1161/CIRCRESAHA.111.240242.

Nakahata, K., Kinoshita, H., Tokinaga, Y., Ishida, Y., Kimoto, Y., Dojo, M., … Hatano, Y. (2006). Vasodilation mediated by inward rectifier K^+ channels in cerebral microvessels of hypertensive and normotensive rats. *Anesthésie et Analgésie, 102*(2), 571–576. http://dx.doi.org/10.1213/01.ane.0000194303.00844.5e.

Nakashima, M., & Vanhoutte, P. M. (1995). Isoproterenol causes hyperpolarization through opening of ATP-sensitive potassium channels in vascular smooth-muscle of the canine saphenous-vein. *Journal of Pharmacology and Experimental Therapeutics, 272*(1), 379–384.

Nakhostine, N., & Lamontagne, D. (1993). Adenosine contributes to hypoxia-induced vasodilation through ATP-sensitive K^+ channel activation. *American Journal of Physiology, 265*(4), H1289–H1293.

Nakhostine, N., & Lamontagne, D. (1994). Contribution of prostaglandins in hypoxia-induced vasodilation in isolated rabbit hearts. Relation to adenosine and KATP channels. *Pflügers Archiv, 428*(5–6), 526–532.

Nara, M., Dhulipala, P. D., Wang, Y. X., & Kotlikoff, M. I. (1998). Reconstitution of beta-adrenergic modulation of large conductance, calcium-activated potassium (maxi-K) channels in Xenopus oocytes. Identification of the camp-dependent protein kinase phosphorylation site. *The Journal of Biological Chemistry*, *273*(24), 14920–14924.

Nelson, M. T., Cheng, H., Rubart, M., Santana, L. F., Bonev, A. D., Knot, H. J., & Lederer, W. J. (1995). Relaxation of arterial smooth muscle by calcium sparks [see comments]. *Science*, *270*, 633–637.

Nelson, M. T., Huang, Y., Brayden, J. E., Hescheler, J., & Standen, N. B. (1990). Arterial dilations in response to calcitonin gene-related peptide involve activation of K^+ channels. *Nature*, *344*(6268), 770–773. http://dx.doi.org/10.1038/344770a0.

Nelson, M. T., Patlak, J. B., Worley, J. F., & Standen, N. B. (1990). Calcium channels, potassium channels, and voltage dependence of arterial smooth muscle tone. *The American Journal of Physiology*, *259*(1 Pt. 1), C3–C18.

Nelson, M. T., & Quayle, J. M. (1995). Physiological roles and properties of potassium channels in arterial smooth muscle. *The American Journal of Physiology*, *268*(4 Pt. 1), C799–C822.

Nelson, C. P., Rainbow, R. D., Brignell, J. L., Perry, M. D., Willets, J. M., Davies, N. W., … Challiss, R. A. (2011). Principal role of adenylyl cyclase 6 in K(+) channel regulation and vasodilator signalling in vascular smooth muscle cells. *Cardiovascular Research*, *91*(4), 694–702. http://dx.doi.org/10.1093/cvr/cvr137.

Neylon, C. B. (2002). Potassium channels and vascular proliferation. *Vascular Pharmacology*, *38*(1), 35–41.

Nieves-Cintron, M., Amberg, G. C., Nichols, C. B., Molkentin, J. D., & Santana, L. F. (2007). Activation of NFATc3 down-regulates the beta1 subunit of large conductance, calcium-activated K^+ channels in arterial smooth muscle and contributes to hypertension. *The Journal of Biological Chemistry*, *282*(5), 3231–3240. http://dx.doi.org/10.1074/jbc.M608822200.

Nieves-Cintron, M., Nystoriak, M. A., Prada, M. P., Johnson, K., Fayer, W., Dell'Acqua, M. L., … Navedo, M. F. (2015). Selective down-regulation of KV2.1 function contributes to enhanced arterial tone during diabetes. *The Journal of Biological Chemistry*, *290*(12), 7918–7929. http://dx.doi.org/10.1074/jbc.M114.622811.

Nnorom, C. C., Davis, C., Fedinec, A. L., Howell, K., Jaggar, J. H., Parfenova, H., … Leffler, C. W. (2014). Contributions of KATP and KCa channels to cerebral arteriolar dilation to hypercapnia in neonatal brain. *Physiological Reports*, *2*(8), e12127. http://dx.doi.org/10.14814/phy2.12127.

Nystoriak, M. A., Nieves-Cintron, M., Nygren, P. J., Hinke, S. A., Nichols, C. B., Chen, C. Y., … Navedo, M. F. (2014). AKAP150 contributes to enhanced vascular tone by facilitating large-conductance Ca^{2+}-activated K^+ channel remodeling in hyperglycemia and diabetes mellitus. *Circulation Research*, *114*(4), 607–615. http://dx.doi.org/10.1161/CIRCRESAHA.114.302168.

O'Connell, A. D., Morton, M. J., & Hunter, M. (2002). Two-pore domain K^+ channels-molecular sensors. *Biochimica et Biophysica Acta*, *1566*(1–2), 152–161.

Ohya, Y., Setoguchi, M., Fujii, K., Nagao, T., Abe, I., & Fujishima, M. (1996). Impaired action of levcromakalim on ATP-sensitive K^+ channels in mesenteric artery cells from spontaneously hypertensive rats. *Hypertension*, *27*(6), 1234–1239.

Okada, Y., Yanagisawa, T., & Taira, N. (1993). BRL 38227 (levcromakalim)-induced hyperpolarization reduces the sensitivity to Ca^{2+} of contractile elements in canine coronary artery. *Naunyn-Schmiedeberg's Archives of Pharmacology*, *347*, 438–444.

Ozkor, M. A., Murrow, J. R., Rahman, A. M., Kavtaradze, N., Lin, J., Manatunga, A., & Quyyumi, A. A. (2011). Endothelium-derived hyperpolarizing factor determines resting and stimulated forearm vasodilator tone in health and in disease. *Circulation*, *123*(20), 2244–2253. http://dx.doi.org/10.1161/CIRCULATIONAHA.110.990317.

Pagan, R. M., Martinez, A. C., Martinez, M. P., Hernandez, M., Garcia-Sacristan, A., Correa, C., … Benedito, S. (2009). Endothelial and potassium channel dependent modulation of noradrenergic vasoconstriction in the pig radial artery. *European Journal of Pharmacology, 616*(1–3), 166–174. http://dx.doi.org/10.1016/j.ejphar.2009.06.002.

Paisansathan, C., Xu, H., Vetri, F., Hernandez, M., & Pelligrino, D. A. (2010). Interactions between adenosine and K^+ channel-related pathways in the coupling of somatosensory activation and pial arteriolar dilation. *American Journal of Physiology. Heart and Circulatory Physiology, 299*(6), H2009–H2017. http://dx.doi.org/10.1152/ajpheart.00702.2010.

Pallanck, L., & Ganetzky, B. (1994). Cloning and characterization of human and mouse homologs of the *Drosophila* calcium-activated potassium channel gene, *slowpoke*. *Human Molecular Genetics, 3*(8), 1239–1243.

Pardo, L. A. (2004). Voltage-gated potassium channels in cell proliferation. *Physiology (Bethesda), 19*, 285–292. http://dx.doi.org/10.1152/physiol.00011.2004.

Park, W. S., Han, J., Kim, N., Ko, J. H., Kim, S. J., & Earm, Y. E. (2005a). Activation of inward rectifier K^+ channels by hypoxia in rabbit coronary arterial smooth muscle cells. *American Journal of Physiology. Heart and Circulatory Physiology, 289*(6), H2461–H2467. http://dx.doi.org/10.1152/ajpheart.00331.2005.

Park, W. S., Han, J., Kim, N., Youm, J. B., Joo, H., Kim, H. K., … Earm, Y. E. (2005b). Endothelin-1 inhibits inward rectifier K^+ channels in rabbit coronary arterial smooth muscle cells through protein kinase C. *Journal of Cardiovascular Pharmacology, 46*(5), 681–689.

Park, W. S., Kim, N., Youm, J. B., Warda, M., Ko, J. H., Kim, S. J., … Han, J. (2006). Angiotensin II inhibits inward rectifier K^+ channels in rabbit coronary arterial smooth muscle cells through protein kinase Calpha. *Biochemical and Biophysical Research Communications, 341*(3), 728–735. http://dx.doi.org/10.1016/j.bbrc.2006.01.026.

Paterno, R., Heistad, D. D., & Faraci, F. M. (1997). Functional activity of Ca^{2+}-dependent K^+ channels is increased in basilar artery during chronic hypertension. *American Journal of Physiology. Heart and Circulatory Physiology, 272*(3), H1287–H1291.

Perez, G. J., Bonev, A. D., & Nelson, M. T. (2001). Micromolar Ca(2+) from sparks activates Ca(2+)-sensitive K(+) channels in rat cerebral artery smooth muscle. *American Journal of Physiology. Cell Physiology, 281*(6), C1769–C1775.

Perez, G. J., Bonev, A. D., Patlak, J. B., & Nelson, M. T. (1999). Functional coupling of ryanodine receptors to KCa channels in smooth muscle cells from rat cerebral arteries. *The Journal of General Physiology, 113*(2), 229–238.

Plane, F., & Garland, C. J. (1993). Differential effects of acetylcholine, nitric oxide and levcromakalim on smooth muscle membrane potential and tone in the rabbit basilar artery. *British Journal of Pharmacology, 110*(2), 651–656.

Plane, F., Hurrell, A., Jeremy, J. Y., & Garland, C. J. (1996). Evidence that potassium channels make a major contribution to SIN-1-evoked relaxation of rat isolated mesenteric artery. *British Journal of Pharmacology, 119*(8), 1557–1562.

Plane, F., Sampson, L. J., Smith, J. J., & Garland, C. J. (2001). Relaxation to authentic nitric oxide and SIN-1 in rat isolated mesenteric arteries: Variable role for smooth muscle hyperpolarization. *British Journal of Pharmacology, 133*(5), 665–672. http://dx.doi.org/10.1038/sj.bjp.0704127.

Plane, F., Wiley, K. E., Jeremy, J. Y., Cohen, R. A., & Garland, C. J. (1998). Evidence that different mechanisms underlie smooth muscle relaxation to nitric oxide and nitric oxide donors in the rabbit isolated carotid artery. *British Journal of Pharmacology, 123*(7), 1351–1358. http://dx.doi.org/10.1038/sj.bjp.0701746.

Porter, V. A., Bonev, A. D., Knot, H. J., Heppner, T. J., Stevenson, A. S., Kleppisch, T., … Nelson, M. T. (1998). Frequency modulation of Ca^{2+} sparks is involved in regulation of arterial diameter by cyclic nucleotides. *American Journal of Physiology-Cell Physiology, 274*(5), C1346–C1355.

Povlsen, G. K., Longden, T. A., Bonev, A. D., Hill-Eubanks, D. C., & Nelson, M. T. (2016). Uncoupling of neurovascular communication after transient global cerebral ischemia is caused by impaired parenchymal smooth muscle Kir channel function. *Journal of Cerebral Blood Flow and Metabolism, 36*, 1195–1201. http://dx.doi.org/10.1177/0271678X16638350.

Quayle, J. M., Bonev, A. D., Brayden, J. E., & Nelson, M. T. (1994). Calcitonin gene-related peptide activated ATP-sensitive K^+ currents in rabbit arterial smooth muscle via protein kinase A. *The Journal of Physiology, 475*(1), 9–13.

Quayle, J. M., Nelson, M. T., & Standen, N. B. (1997). ATP-sensitive and inwardly rectifying potassium channels in smooth muscle. *Physiological Reviews*, 77(4), 1165–1232.

Quinn, K. V., Cui, Y., Giblin, J. P., Clapp, L. H., & Tinker, A. (2003). Do anionic phospholipids serve as cofactors or second messengers for the regulation of activity of cloned ATP-sensitive K^+ channels? *Circulation Research, 93*(7), 646–655. http://dx.doi.org/10.1161/01.RES.0000095247.81449.8E.

Quinn, K. V., Giblin, J. P., & Tinker, A. (2004). Multisite phosphorylation mechanism for protein kinase A activation of the smooth muscle ATP-sensitive K^+ channel. *Circulation Research, 94*(10), 1359–1366. http://dx.doi.org/10.1161/01.RES.0000128513.34817.c4.

Randall, M. D. (1995). The involvement of ATP-sensitive potassium channels and adenosine in the regulation of coronary flow in the isolated perfused rat heart. *British Journal of Pharmacology, 116*(7), 3068–3074.

Renigunta, V., Schlichthorl, G., & Daut, J. (2015). Much more than a leak: Structure and function of K(2)p channels. *Pflügers Archiv, 467*(5), 867–894. http://dx.doi.org/10.1007/s00424-015-1703-7.

Richmond, K. N., Tune, J. D., Gorman, M. W., & Feigl, E. O. (1999). Role of K^+ ATP channels in local metabolic coronary vasodilation. *The American Journal of Physiology, 277*(6 Pt. 2), H2115–H2123.

Richmond, K. N., Tune, J. D., Gorman, M. W., & Feigl, E. O. (2000). Role of K(ATP)(+) channels and adenosine in the control of coronary blood flow during exercise. *Journal of Applied Physiology (Bethesda, Md.: 1985), 89*(2), 529–536.

Roberts, O. L., Kamishima, T., Barrett-Jolley, R., Quayle, J. M., & Dart, C. (2013). Exchange protein activated by cAMP (Epac) induces vascular relaxation by activating Ca^{2+}-sensitive K^+ channels in rat mesenteric artery. *The Journal of Physiology, 591*(20), 5107–5123. http://dx.doi.org/10.1113/jphysiol.2013.262006.

Robertson, B. E., Schubert, R., Hescheler, J., & Nelson, M. T. (1993). cGMP-dependent protein kinase activates Ca-activated K channels in cerebral artery smooth muscle cells. *The American Journal of Physiology, 265*(1 Pt. 1), C299–C303.

Rogers, P. A., Chilian, W. M., Bratz, I. N., Bryan, R. M., Jr., & Dick, G. M. (2007). H2O2 activates redox- and 4-aminopyridine-sensitive Kv channels in coronary vascular smooth muscle. *American Journal of Physiology. Heart and Circulatory Physiology, 292*(3), H1404–H1411. http://dx.doi.org/10.1152/ajpheart.00696.2006.

Rusch, N. J. (2009). BK channels in cardiovascular disease: A complex story of channel dysregulation. *American Journal of Physiology. Heart and Circulatory Physiology, 297*(5), H1580–H1582. http://dx.doi.org/10.1152/ajpheart.00852.2009.

Rusch, N. J., Delucena, R. G., Wooldridge, T. A., England, S. K., & Cowley, A. W. (1992). A Ca^{2+}-dependent K^+ current is enhanced in arterial membranes of hypertensive rats. *Hypertension, 19*(4), 301–307.

Rusch, N. J., & Liu, Y. (1997). Potassium channels in hypertension: Homeostatic pathways to buffer arterial contraction. *The Journal of Laboratory and Clinical Medicine, 130*(3), 245–251.

Sadoshima, J., Akaike, N., Kanaide, H., & Nakamura, M. (1988). Cyclic AMP modulates Ca-activated K channel in culture smooth muscle cells of rat aortas. *The American Journal of Physiology, 255*(4 Pt. 2), H754–H759.

Saito, Y., McKay, M., Eraslan, A., & Hester, R. L. (1996). Functional hyperemia in striated muscle is reduced following blockade of ATP-sensitive potassium channels. *American Journal of Physiology. Heart and Circulatory Physiology, 270*(5), H1649–H1654.

Samaha, F. F., Heineman, F. W., Ince, C., Fleming, J., & Balaban, R. S. (1992). ATP-sensitive potassium channel is essential to maintain basal coronary vascular tone in vivo. *The American Journal of Physiology*, *262*(5 Pt. 1), C1220–C1227.

Sampson, L. J., Davies, L. M., Barrett-Jolley, R., Standen, N. B., & Dart, C. (2007). Angiotensin II-activated protein kinase C targets caveolae to inhibit aortic ATP-sensitive potassium channels. *Cardiovascular Research*, *76*(1), 61–70. http://dx.doi.org/10.1016/j.cardiores.2007.05.020.

Satake, N., Shibata, M., & Shibata, S. (1996). The inhibitory effects of iberiotoxin and 4-aminopyridine on the relaxation induced by beta 1- and beta 2-adrenoceptor activation in rat aortic rings. *British Journal of Pharmacology*, *119*(3), 505–510.

Sawmiller, D. R., Ashtari, M., Urueta, H., Leschinsky, M., & Henning, R. J. (2006). Mechanisms of vasoactive intestinal peptide-elicited coronary vasodilation in the isolated perfused rat heart. *Neuropeptides*, *40*(5), 349–355. http://dx.doi.org/10.1016/j.npep.2006.07.004.

Schleifenbaum, J., Kohn, C., Voblova, N., Dubrovska, G., Zavarirskaya, O., Gloe, T., … Gollasch, M. (2010). Systemic peripheral artery relaxation by KCNQ channel openers and hydrogen sulfide. *Journal of Hypertension*, *28*(9), 1875–1882. http://dx.doi.org/10.1097/HJH.0b013e32833c20d5.

Schwab, A., Hanley, P., Fabian, A., & Stock, C. (2008). Potassium channels keep mobile cells on the go. *Physiology (Bethesda)*, *23*, 212–220. http://dx.doi.org/10.1152/physiol.00003.2008.

Scornik, F. S., Codina, J., Birnbaumer, L., & Toro, L. (1993). Modulation of coronary smooth muscle KCa channels by Gs alpha independent of phosphorylation by protein kinase A. *The American Journal of Physiology*, *265*(4 Pt. 2), H1460–H1465.

Scornik, F. S., & Toro, L. (1992). U46619, a thromboxane A_2 agonist, inhibits K_{Ca} channel activity from pig coronary artery. *The American Journal of Physiology*, *262*(3 Pt. 1), C708–C713.

Seo, E. Y., Kim, H. J., Zhao, Z. H., Jang, J. H., Jin, C. Z., Yoo, H. Y., … Kim, S. J. (2014). Low K(+) current in arterial myocytes with impaired K(+)-vasodilation and its recovery by exercise in hypertensive rats. *Pflügers Archiv*, *466*(11), 2101–2111. http://dx.doi.org/10.1007/s00424-014-1473-7.

Sepulveda, F. V., Pablo Cid, L., Teulon, J., & Niemeyer, M. I. (2015). Molecular aspects of structure, gating, and physiology of pH-sensitive background K2P and Kir K^+-transport channels. *Physiological Reviews*, *95*(1), 179–217. http://dx.doi.org/10.1152/physrev.00016.2014.

Sharifi-Sanjani, M., Zhou, X., Asano, S., Tilley, S., Ledent, C., Teng, B., … Mustafa, S. J. (2013). Interactions between A(2A) adenosine receptors, hydrogen peroxide, and KATP channels in coronary reactive hyperemia. *American Journal of Physiology. Heart and Circulatory Physiology*, *304*(10), H1294–H1301. http://dx.doi.org/10.1152/ajpheart.00637.2012.

Shi, Y., Chen, X., Wu, Z., Shi, W., Yang, Y., Cui, N., … Harrison, R. W. (2008). cAMP-dependent protein kinase phosphorylation produces interdomain movement in SUR2B leading to activation of the vascular KATP channel. *The Journal of Biological Chemistry*, *283*(12), 7523–7530. http://dx.doi.org/10.1074/jbc.M709941200.

Shi, Y., Wu, Z., Cui, N., Shi, W., Yang, Y., Zhang, X., … Jiang, C. (2007). PKA phosphorylation of SUR2B subunit underscores vascular KATP channel activation by beta-adrenergic receptors. *American Journal of Physiology. Regulatory, Integrative and Comparative Physiology*, *293*(3), R1205–R1214. http://dx.doi.org/10.1152/ajpregu.00337.2007.

Shimizu, S., Yokoshiki, H., Sperelakis, N., & Paul, R. J. (2000). Role of voltage-dependent and Ca(2+)-activated K(+) channels on the regulation of isometric force in porcine coronary artery. *Journal of Vascular Research*, *37*(1), 16–25. http://dx.doi.org/10.1159/000025709.

Siegl, D., Koeppen, M., Wolfle, S. E., Pohl, U., & de Wit, C. (2005). Myoendothelial coupling is not prominent in arterioles within the mouse cremaster microcirculation in vivo. *Circulation Research*, *97*(8), 781–788. http://dx.doi.org/10.1161/01.RES.0000186193.22438.6c.

Smith, P. D., Brett, S. E., Luykenaar, K. D., Sandow, S. L., Marrelli, S. P., Vigmond, E. J., & Welsh, D. G. (2008). KIR channels function as electrical amplifiers in rat vascular smooth muscle. *The Journal of Physiology*, *586*(4), 1147–1160. http://dx.doi.org/10.1113/jphysiol.2007.145474.

Sobey, C. G., & Faraci, F. M. (1999). Inhibitory effect of 4-aminopyridine on responses of the basilar artery to nitric oxide. *British Journal of Pharmacology*, *126*(6), 1437–1443. http://dx.doi.org/10.1038/sj.bjp.0702439.

Son, Y. K., Park, W. S., Ko, J. H., Han, J., Kim, N., & Earm, Y. E. (2005). Protein kinase A-dependent activation of inward rectifier potassium channels by adenosine in rabbit coronary smooth muscle cells. *Biochemical and Biophysical Research Communications*, *337*(4), 1145–1152. http://dx.doi.org/10.1016/j.bbrc.2005.09.176.

Sonkusare, S. K., Dalsgaard, T., Bonev, A. D., & Nelson, M. T. (2016). Inward rectifier potassium (Kir2.1) channels as end-stage boosters of endothelium-dependent vasodilators. *The Journal of Physiology*, *594*, 3271–3285. http://dx.doi.org/10.1113/JP271652.

Spallarossa, P., Schiavo, M., Rossettin, P., Cordone, S., Olivotti, L., Cordera, R., & Brunelli, C. (2001). Sulfonylurea treatment of type 2 diabetic patients does not reduce the vasodilator response to ischemia. *Diabetes Care*, *24*(4), 738–742.

Stepp, D. W., Kroll, K., & Feigl, E. O. (1997). K+ATP channels and adenosine are not necessary for coronary autoregulation. *The American Journal of Physiology*, *273*(3 Pt. 2), H1299–H1308.

Stott, J. B., Barrese, V., Jepps, T. A., Leighton, E. V., & Greenwood, I. A. (2015). Contribution of Kv7 channels to natriuretic peptide mediated vasodilation in normal and hypertensive rats. *Hypertension*, *65*(3), 676–682. http://dx.doi.org/10.1161/HYPERTENSIONAHA.114.04373.

Sung, D. J., Noh, H. J., Kim, J. G., Park, S. W., Kim, B., Cho, H., & Bae, Y. M. (2013). Serotonin contracts the rat mesenteric artery by inhibiting 4-aminopyridine-sensitive Kv channels via the 5-HT2A receptor and Src tyrosine kinase. *Experimental & Molecular Medicine*, *45*, e67. http://dx.doi.org/10.1038/emm.2013.116.

Suzuki, M., Li, R. A., Miki, T., Uemura, H., Sakamoto, N., Ohmoto-Sekine, Y., … Nakaya, H. (2001). Functional roles of cardiac and vascular ATP-sensitive potassium channels clarified by Kir6.2-knockout mice. *Circulation Research*, *88*(6), 570–577.

Suzuki, Y., Yamamura, H., Ohya, S., & Imaizumi, Y. (2013). Caveolin-1 facilitates the direct coupling between large conductance Ca^{2+}-activated K^+ (BKCa) and Cav1.2 Ca^{2+} channels and their clustering to regulate membrane excitability in vascular myocytes. *The Journal of Biological Chemistry*, *288*(51), 36750–36761. http://dx.doi.org/10.1074/jbc.M113.511485.

Swayze, R. D., & Braun, A. P. (2001). A catalytically inactive mutant of type I cGMP-dependent protein kinase prevents enhancement of large conductance, calcium-sensitive K^+ channels by sodium nitroprusside and cGMP. *The Journal of Biological Chemistry*, *276*(23), 19729–19737. http://dx.doi.org/10.1074/jbc.M005711200.

Taguchi, H., Heistad, D. D., Kitazono, T., & Faraci, F. M. (1994). ATP-sensitive K^+ channels mediate dilatation of cerebral arterioles during hypoxia. *Circulation Research*, *74*(5), 1005–1008.

Taguchi, H., Heistad, D. D., Kitazono, T., & Faraci, F. M. (1995). Dilatation of cerebral arterioles in response to activation of adenylate-cyclase is dependent on activation of Ca^{2+}-dependent K^+ channels. *Circulation Research*, *76*(6), 1057–1062. http://dx.doi.org/10.1161/01.res.76.6.1057.

Tajada, S., Cidad, P., Moreno-Dominguez, A., Perez-Garcia, M. T., & Lopez-Lopez, J. R. (2012). High blood pressure associates with the remodelling of inward rectifier K^+

channels in mice mesenteric vascular smooth muscle cells. *The Journal of Physiology, 590*(23), 6075–6091. http://dx.doi.org/10.1113/jphysiol.2012.236190.

Takaba, H., Nagao, T., Ibayashi, S., Kitazono, T., Fujii, K., & Fujishima, M. (1996). Altered cerebrovascular response to a potassium channel opener in hypertensive rats. *Hypertension, 28*(1), 143–146.

Tanaka, Y., Tang, G., Takizawa, K., Otsuka, K., Eghbali, M., Song, M., ... Toro, L. (2006). Kv channels contribute to nitric oxide- and atrial natriuretic peptide-induced relaxation of a rat conduit artery. *The Journal of Pharmacology and Experimental Therapeutics, 317*(1), 341–354. http://dx.doi.org/10.1124/jpet.105.096115.

Taniguchi, J., Furukawa, K. I., & Shigekawa, M. (1993). Maxi K^+ channels are stimulated by cyclic guanosine monophosphate-dependent protein kinase in canine coronary artery smooth muscle cells. *Pflügers Archiv, 423*(3–4), 167–172.

Tano, J. Y., Schleifenbaum, J., & Gollasch, M. (2014). Perivascular adipose tissue, potassium channels, and vascular dysfunction. *Arteriosclerosis, Thrombosis, and Vascular Biology, 34*(9), 1827–1830. http://dx.doi.org/10.1161/ATVBAHA.114.303032.

Tharp, D. L., & Bowles, D. K. (2009). The intermediate-conductance Ca^{2+}-activated K^+ channel (KCa3.1) in vascular disease. *Cardiovascular & Hematological Agents in Medicinal Chemistry*, 7(1), 1–11.

Tharp, D. L., Wamhoff, B. R., Turk, J. R., & Bowles, D. K. (2006). Upregulation of intermediate-conductance Ca^{2+}-activated K^+ channel (IKCa1) mediates phenotypic modulation of coronary smooth muscle. *American Journal of Physiology. Heart and Circulatory Physiology, 291*(5), H2493–H2503. http://dx.doi.org/10.1152/ajpheart.01254.2005.

Tharp, D. L., Wamhoff, B. R., Wulff, H., Raman, G., Cheong, A., & Bowles, D. K. (2008). Local delivery of the KCa3.1 blocker, TRAM-34, prevents acute angioplasty-induced coronary smooth muscle phenotypic modulation and limits stenosis. *Arteriosclerosis, Thrombosis, and Vascular Biology, 28*(6), 1084–1089. http://dx.doi.org/10.1161/ATVBAHA.107.155796.

Thengchaisri, N., & Kuo, L. (2003). Hydrogen peroxide induces endothelium-dependent and -independent coronary arteriolar dilation: Role of cyclooxygenase and potassium channels. *American Journal of Physiology. Heart and Circulatory Physiology, 285*(6), H2255–H2263. http://dx.doi.org/10.1152/ajpheart.00487.2003.

Tian, L., Coghill, L. S., McClafferty, H., MacDonald, S. H., Antoni, F. A., Ruth, P., ... Shipston, M. J. (2004). Distinct stoichiometry of BKCa channel tetramer phosphorylation specifies channel activation and inhibition by cAMP-dependent protein kinase. *Proceedings of the National Academy of Sciences of the United States of America, 101*(32), 11897–11902. http://dx.doi.org/10.1073/pnas.0402590101.

Tian, L., Duncan, R. R., Hammond, M. S., Coghill, L. S., Wen, H., Rusinova, R., ... Shipston, M. J. (2001). Alternative splicing switches potassium channel sensitivity to protein phosphorylation. *The Journal of Biological Chemistry, 276*(11), 7717–7720. http://dx.doi.org/10.1074/jbc.C000741200.

Tobin, A. A., Joseph, B. K., Al-Kindi, H. N., Albarwani, S., Madden, J. A., Nemetz, L. T., ... Rhee, S. W. (2009). Loss of cerebrovascular Shaker-type K(+) channels: A shared vasodilator defect of genetic and renal hypertensive rats. *American Journal of Physiology. Heart and Circulatory Physiology, 297*(1), H293–H303. http://dx.doi.org/10.1152/ajpheart.00991.2008.

Tomiyama, Y., Brian, J. E., Jr., & Todd, M. M. (1999). Cerebral blood flow during hemodilution and hypoxia in rats: Role of ATP-sensitive potassium channels. *Stroke, 30*(9), 1942–1947. discussion 1947–1948.

Toro, L., Amador, M., & Stefani, E. (1990). ANG II inhibits calcium-activated potassium channels from coronary smooth muscle in lipid bilayers. *The American Journal of Physiology, 258*(3 Pt. 2), H912–H915.

Toyama, K., Wulff, H., Chandy, K. G., Azam, P., Raman, G., Saito, T., … Miura, H. (2008). The intermediate-conductance calcium-activated potassium channel KCa3.1 contributes to atherogenesis in mice and humans. *The Journal of Clinical Investigation*, *118*(9), 3025–3037. http://dx.doi.org/10.1172/JCI30836.

Toyoda, K., Fujii, K., Ibayashi, S., Kitazono, T., Nagao, T., & Fujishima, M. (1997). Role of ATP-sensitive potassium channels in brain stem circulation during hypotension. *The American Journal of Physiology*, *273*(3 Pt. 2), H1342–H1346.

Trebak, M. (2012). STIM/Orai signalling complexes in vascular smooth muscle. *The Journal of Physiology*, *590*(17), 4201–4208. http://dx.doi.org/10.1113/jphysiol.2012.233353.

Troncoso Brindeiro, C. M., Fallet, R. W., Lane, P. H., & Carmines, P. K. (2008). Potassium channel contributions to afferent arteriolar tone in normal and diabetic rat kidney. *American Journal of Physiology. Renal Physiology*, *295*(1), F171–F178. http://dx.doi.org/10.1152/ajprenal.00563.2007.

Troncoso Brindeiro, C. M., Lane, P. H., & Carmines, P. K. (2012). Tempol prevents altered K(+) channel regulation of afferent arteriolar tone in diabetic rat kidney. *Hypertension*, *59*(3), 657–664. http://dx.doi.org/10.1161/HYPERTENSIONAHA.111.184218.

Tseng-Crank, J., Godinot, N., Johansen, T. E., Ahring, P. K., Strobaek, D., Mertz, R., … Reinhart, P. H. (1996). Cloning, expression, and distribution of a Ca(2+)-activated K^+ channel beta-subunit from human brain. *Proceedings of the National Academy of Sciences of the United States of America*, *93*(17), 9200–9205.

Tzeng, B. H., Chen, Y. H., Huang, C. H., Lin, S. S., Lee, K. R., & Chen, C. C. (2012). The Ca(v)3.1 T-type calcium channel is required for neointimal formation in response to vascular injury in mice. *Cardiovascular Research*, *96*(3), 533–542. http://dx.doi.org/10.1093/cvr/cvs257.

Urena, J., del Valle-Rodriguez, A., & Lopez-Barneo, J. (2007). Metabotropic Ca^{2+} channel-induced calcium release in vascular smooth muscle. *Cell Calcium*, *42*(4–5), 513–520. http://dx.doi.org/10.1016/j.ceca.2007.04.010.

Urrego, D., Tomczak, A. P., Zahed, F., Stuhmer, W., & Pardo, L. A. (2014). Potassium channels in cell cycle and cell proliferation. *Philosophical Transactions of the Royal Society of London. Series B, Biological Sciences*, *369*(1638), 20130094. http://dx.doi.org/10.1098/rstb.2013.0094.

Van de Voorde, J., Vanheel, B., & Leusen, I. (1992). Endothelium-dependent relaxation and hyperpolarization in aorta from control and renal hypertensive rats. *Circulation Research*, *70*(1), 1–8.

Vanelli, G., Chang, H. Y., Gatensby, A. G., & Hussain, S. N. (1994). Contribution of potassium channels to active hyperemia of the canine diaphragm. *Journal of Applied Physiology (Bethesda, Md.: 1985)*, *76*(3), 1098–1105.

Vanelli, G., & Hussain, S. N. (1994). Effects of potassium channel blockers on basal vascular tone and reactive hyperemia of canine diaphragm. *The American Journal of Physiology*, *266*(1 Pt. 2), H43–H51.

Vetri, F., Xu, H., Paisansathan, C., & Pelligrino, D. A. (2012). Impairment of neurovascular coupling in type 1 diabetes mellitus in rats is linked to PKC modulation of BK(Ca) and Kir channels. *American Journal of Physiology. Heart and Circulatory Physiology*, *302*(6), H1274–H1284. http://dx.doi.org/10.1152/ajpheart.01067.2011.

Vigili de Kreutzenberg, S., Kiwanuka, E., Tiengo, A., & Avogaro, A. (2003). Visceral obesity is characterized by impaired nitric oxide-independent vasodilation. *European Heart Journal*, *24*(13), 1210–1215.

von Beckerath, N., Cyrys, S., Dischner, A., & Daut, J. (1991). Hypoxic vasodilatation in isolated, perfused guinea-pig heart: An analysis of the underlying mechanisms. *The Journal of Physiology*, *442*, 297–319.

Wakatsuki, T., Nakaya, Y., & Inoue, I. (1992). Vasopressin modulates K(+)-channel activities of cultured smooth muscle cells from porcine coronary artery. *The American Journal of Physiology, 263*(2 Pt. 2), H491–H496.

Wang, R., Wang, Z., & Wu, L. (1997). Carbon monoxide-induced vasorelaxation and the underlying mechanisms. *British Journal of Pharmacology, 121*(5), 927–934. http://dx.doi.org/10.1038/sj.bjp.0701222.

Wang, R., & Wu, L. (1997). The chemical modification of KCa channels by carbon monoxide in vascular smooth muscle cells. *The Journal of Biological Chemistry, 272*(13), 8222–8226.

Wang, R., Wu, L., & Wang, Z. (1997). The direct effect of carbon monoxide on KCa channels in vascular smooth muscle cells. *Pflügers Archiv, 434*(3), 285–291.

Wang, Y., Zhang, H. T., Su, X. L., Deng, X. L., Yuan, B. X., Zhang, W., … Yang, Y. B. (2010). Experimental diabetes mellitus down-regulates large-conductance Ca^{2+}-activated K^+ channels in cerebral artery smooth muscle and alters functional conductance. *Current Neurovascular Research*, 7(2), 75–84.

Wei, E. P., & Kontos, H. A. (1999). Blockade of ATP-sensitive potassium channels in cerebral arterioles inhibits vasoconstriction from hypocapnic alkalosis in cats. *Stroke, 30*(4), 851–853. discussion 854.

Wellman, G. C., & Bevan, J. A. (1995). Barium inhibits the endothelium-dependent component of flow but not acetylcholine-induced relaxation in isolated rabbit cerebral arteries. *The Journal of Pharmacology and Experimental Therapeutics, 274*(1), 47–53.

Wellman, G. C., Nathan, D. J., Saundry, C. M., Perez, G., Bonev, A. D., Penar, P. L., … Nelson, M. T. (2002). Ca^{2+} sparks and their function in human cerebral arteries. *Stroke, 33*(3), 802–808.

Wellman, G. C., & Nelson, M. T. (2003). Signaling between SR and plasmalemma in smooth muscle: Sparks and the activation of Ca^{2+}-sensitive ion channels. *Cell Calcium, 34*(3), 211–229.

Wellman, G. C., Quayle, J. M., & Standen, N. B. (1998). ATP-sensitive K^+ channel activation by calcitonin gene-related peptide and protein kinase A in pig coronary arterial smooth muscle. *The Journal of Physiology, 507*(Pt. 1), 117–129.

Wellman, G. C., Santana, L. F., Bonev, A. D., & Nelson, M. T. (2001). Role of phospholamban in the modulation of arterial Ca(2+) sparks and Ca(2+)-activated K(+) channels by cAMP. *American Journal of Physiology. Cell Physiology, 281*(3), C1029–C1037.

Welsh, D. G., Jackson, W. F., & Segal, S. S. (1998). Oxygen induces electromechanical coupling in arteriolar smooth muscle cells: A role for L-type Ca^{2+} channels. *The American Journal of Physiology, 274*(6 Pt. 2), H2018–H2024.

Wesselman, J. P., Schubert, R., VanBavel, E. D., Nilsson, H., & Mulvany, M. J. (1997). KCa-channel blockade prevents sustained pressure-induced depolarization in rat mesenteric small arteries. *The American Journal of Physiology, 272*(5 Pt. 2), H2241–H2249.

Westcott, E. B., Goodwin, E. L., Segal, S. S., & Jackson, W. F. (2012). Function and expression of ryanodine receptors and inositol 1,4,5-trisphosphate receptors in smooth muscle cells of murine feed arteries and arterioles. *The Journal of Physiology, 590*(8), 1849–1869. http://dx.doi.org/10.1113/jphysiol.2011.222083.

Westcott, E. B., & Jackson, W. F. (2011). Heterogeneous function of ryanodine receptors, but not IP3 receptors, in hamster cremaster muscle feed arteries and arterioles. *American Journal of Physiology. Heart and Circulatory Physiology, 300*(5), H1616–H1630. http://dx.doi.org/10.1152/ajpheart.00728.2010.

Williams, D. L., Jr., Katz, G. M., Roy-Contancin, L., & Reuben, J. P. (1988). Guanosine 5′-monophosphate modulates gating of high-conductance Ca^{2+}-activated K^+ channels in vascular smooth muscle cells. *Proceedings of the National Academy of Sciences of the United States of America, 85*(23), 9360–9364.

Wilson, A. J., Jabr, R. I., & Clapp, L. H. (2000). Calcium modulation of vascular smooth muscle ATP-sensitive K(+) channels: Role of protein phosphatase-2B. *Circulation Research*, *87*(11), 1019–1025.

Winquist, R. J., Faison, E. P., Napier, M., Vandlen, R., Waldman, S. A., & Murad, F. (1985). The effects of atrial natriuretic factor on vascular smooth muscle. In J. A. Bevan, T. Godfraind, R. A. Maxwell, J. C. Stoclet, & M. Worcel (Eds.), *Vascular neuroeffector mecahnisms* (pp. 349–353). Amsterdam: Elsevier.

Winquist, R. J., Faison, E. P., & Nutt, R. F. (1984). Vasodilator profile of synthetic atrial natriuretic factor. *European Journal of Pharmacology*, *102*(1), 169–173. http://dx.doi.org/10.1016/0014-2999(84)90353-4.

Wischmeyer, E., Doring, F., & Karschin, A. (1998). Acute suppression of inwardly rectifying Kir2.1 channels by direct tyrosine kinase phosphorylation. *The Journal of Biological Chemistry*, *273*(51), 34063–34068.

Wonderlin, W. F., & Strobl, J. S. (1996). Potassium channels, proliferation and G1 progression. *The Journal of Membrane Biology*, *154*(2), 91–107.

Wu, B. N., Luykenaar, K. D., Brayden, J. E., Giles, W. R., Corteling, R. L., Wiehler, W. B., & Welsh, D. G. (2007). Hyposmotic challenge inhibits inward rectifying K^+ channels in cerebral arterial smooth muscle cells. *American Journal of Physiology. Heart and Circulatory Physiology*, *292*(2), H1085–H1094. http://dx.doi.org/10.1152/ajpheart.00926.2006.

Xi, Q., Tcheranova, D., Parfenova, H., Horowitz, B., Leffler, C. W., & Jaggar, J. H. (2004). Carbon monoxide activates KCa channels in newborn arteriole smooth muscle cells by increasing apparent Ca^{2+} sensitivity of alpha-subunits. *American Journal of Physiology. Heart and Circulatory Physiology*, *286*(2), H610–H618. http://dx.doi.org/10.1152/ajpheart.00782.2003.

Xi, Q., Umstot, E., Zhao, G., Narayanan, D., Leffler, C. W., & Jaggar, J. H. (2010). Glutamate regulates Ca^{2+} signals in smooth muscle cells of newborn piglet brain slice arterioles through astrocyte- and heme oxygenase-dependent mechanisms. *American Journal of Physiology. Heart and Circulatory Physiology*, *298*(2), H562–H569. http://dx.doi.org/10.1152/ajpheart.00823.2009.

Yamada, M., Isomoto, S., Matsumoto, S., Kondo, C., Shindo, T., Horio, Y., & Kurachi, Y. (1997). Sulphonylurea receptor 2B and Kir6.1 form a sulphonylurea-sensitive but ATP-insensitive K^+ channel. *The Journal of Physiology*, *499*(Pt. 3), 715–720.

Yamagishi, T., Yanagisawa, T., & Taira, N. (1992). K^+ channel openers, cromakalim and Ki4032, inhibit agonist-induced Ca^{2+} release in canine coronary artery. *Naunyn-Schmiedeberg's Archives of Pharmacology*, *346*, 691–700.

Yamaguchi, H., Kajita, J., & Madison, J. M. (1995). Isoproterenol increases peripheral $[Ca^{2+}]_i$ and decreases inner $[Ca^{2+}]_i$ in single airway smooth muscle cells. *The American Journal of Physiology*, *268*(3 Pt. 1), C771–C779.

Yamamura, H., Ohya, S., Muraki, K., & Imaizumi, Y. (2012). Involvement of inositol 1,4,5-trisphosphate formation in the voltage-dependent regulation of the Ca(2+) concentration in porcine coronary arterial smooth muscle cells. *The Journal of Pharmacology and Experimental Therapeutics*, *342*(2), 486–496. http://dx.doi.org/10.1124/jpet.112.194233.

Yanagisawa, T., Yamagishi, T., & Okada, Y. (1993). Hyperpolarization induced by K^+ channel openers inhibits Ca^{2+} influx and Ca^{2+} release in coronary artery. *Cardiovascular Drugs and Therapy*, 7(Suppl. 3), 565–574.

Yang, Y., Jones, A. W., Thomas, T. R., & Rubin, L. J. (2007). Influence of sex, high-fat diet, and exercise training on potassium currents of swine coronary smooth muscle. *American Journal of Physiology. Heart and Circulatory Physiology*, *293*(3), H1553–H1563. http://dx.doi.org/10.1152/ajpheart.00151.2007.

Yang, Y., Li, P. Y., Cheng, J., Mao, L., Wen, J., Tan, X. Q., … Zeng, X. R. (2013). Function of BKCa channels is reduced in human vascular smooth muscle cells from

Han Chinese patients with hypertension. *Hypertension*, *61*(2), 519–525. http://dx.doi.org/10.1161/HYPERTENSIONAHA.111.00211.

Yang, Y., Murphy, T. V., Ella, S. R., Grayson, T. H., Haddock, R., Hwang, Y. T., … Hill, M. A. (2009). Heterogeneity in function of small artery smooth muscle BKCa: Involvement of the beta1-subunit. *The Journal of Physiology*, *587*(Pt. 12), 3025–3044. http://dx.doi.org/10.1113/jphysiol.2009.169920.

Yang, Y., Shi, Y., Guo, S., Zhang, S., Cui, N., Shi, W., … Jiang, C. (2008). PKA-dependent activation of the vascular smooth muscle isoform of KATP channels by vasoactive intestinal polypeptide and its effect on relaxation of the mesenteric resistance artery. *Biochimica et Biophysica Acta*, *1778*(1), 88–96. http://dx.doi.org/10.1016/j.bbamem.2007.08.030.

Yang, Y., Sohma, Y., Nourian, Z., Ella, S. R., Li, M., Stupica, A., … Hill, M. A. (2013). Mechanisms underlying regional differences in the Ca^{2+} sensitivity of BK(Ca) current in arteriolar smooth muscle. *The Journal of Physiology*, *591*(5), 1277–1293. http://dx.doi.org/10.1113/jphysiol.2012.241562.

Yi, F., Wang, H., Chai, Q., Wang, X., Shen, W. K., Willis, M. S., … Lu, T. (2014). Regulation of large conductance Ca^{2+}-activated K^+ (BK) channel beta1 subunit expression by muscle RING finger protein 1 in diabetic vessels. *The Journal of Biological Chemistry*, *289*(15), 10853–10864. http://dx.doi.org/10.1074/jbc.M113.520940.

Yu, M., Sun, C. W., Maier, K. G., Harder, D. R., & Roman, R. J. (2002). Mechanism of cGMP contribution to the vasodilator response to NO in rat middle cerebral arteries. *American Journal of Physiology. Heart and Circulatory Physiology*, *282*(5), H1724–H1731. http://dx.doi.org/10.1152/ajpheart.00699.2001.

Yuill, K. H., McNeish, A. J., Kansui, Y., Garland, C. J., & Dora, K. A. (2010). Nitric oxide suppresses cerebral vasomotion by sGC-independent effects on ryanodine receptors and voltage-gated calcium channels. *Journal of Vascular Research*, *47*(2), 93–107. http://dx.doi.org/10.1159/000235964.

Zavaritskaya, O., Zhuravleva, N., Schleifenbaum, J., Gloe, T., Devermann, L., Kluge, R., … Schubert, R. (2013). Role of KCNQ channels in skeletal muscle arteries and periadventitial vascular dysfunction. *Hypertension*, *61*(1), 151–159. http://dx.doi.org/10.1161/HYPERTENSIONAHA.112.197566.

Zhang, Y., Gao, Y. J., Zuo, J., Lee, R. M., & Janssen, L. J. (2005). Alteration of arterial smooth muscle potassium channel composition and BKCa current modulation in hypertension. *European Journal of Pharmacology*, *514*(2–3), 111–119. http://dx.doi.org/10.1016/j.ejphar.2005.03.032.

Zhang, Y., Tazzeo, T., Chu, V., & Janssen, L. J. (2006). Membrane potassium currents in human radial artery and their regulation by nitric oxide donor. *Cardiovascular Research*, *71*(2), 383–392. http://dx.doi.org/10.1016/j.cardiores.2006.04.002.

Zhao, W., Zhang, J., Lu, Y., & Wang, R. (2001). The vasorelaxant effect of H(2)S as a novel endogenous gaseous K(ATP) channel opener. *The EMBO Journal*, *20*(21), 6008–6016. http://dx.doi.org/10.1093/emboj/20.21.6008.

Zhong, X. Z., Harhun, M. I., Olesen, S. P., Ohya, S., Moffatt, J. D., Cole, W. C., & Greenwood, I. A. (2010). Participation of KCNQ (Kv7) potassium channels in myogenic control of cerebral arterial diameter. *The Journal of Physiology*, *588*(Pt. 17), 3277–3293. http://dx.doi.org/10.1113/jphysiol.2010.192823.

Zhou, X. B., Feng, Y. X., Sun, Q., Lukowski, R., Qiu, Y., Spiger, K., … Wieland, T. (2015). Nucleoside diphosphate kinase B-activated intermediate conductance potassium channels are critical for neointima formation in mouse carotid arteries. *Arteriosclerosis, Thrombosis, and Vascular Biology*, *35*(8), 1852–1861. http://dx.doi.org/10.1161/ATVBAHA.115.305881.

Zhou, X., Teng, B., Tilley, S., Ledent, C., & Mustafa, S. J. (2014). Metabolic hyperemia requires ATP-sensitive K^+ channels and H2O2 but not adenosine in isolated mouse hearts. *American Journal of Physiology. Heart and Circulatory Physiology*, *307*(7), H1046–H1055. http://dx.doi.org/10.1152/ajpheart.00421.2014.

Zhou, W., Wang, X. L., Lamping, K. G., & Lee, H. C. (2006). Inhibition of protein kinase Cbeta protects against diabetes-induced impairment in arachidonic acid dilation of small coronary arteries. *The Journal of Pharmacology and Experimental Therapeutics*, *319*(1), 199–207. http://dx.doi.org/10.1124/jpet.106.106666.

Zhu, P., Beny, J. L., Flammer, J., Luscher, T. F., & Haefliger, I. O. (1997). Relaxation by bradykinin in porcine ciliary artery. Role of nitric oxide and K(+)-channels. *Investigative Ophthalmology & Visual Science*, *38*(9), 1761–1767.

Zitron, E., Gunth, M., Scherer, D., Kiesecker, C., Kulzer, M., Bloehs, R., … Karle, C. A. (2008). Kir2.x inward rectifier potassium channels are differentially regulated by adrenergic alpha1A receptors. *Journal of Molecular and Cellular Cardiology*, *44*(1), 84–94. http://dx.doi.org/10.1016/j.yjmcc.2007.10.008.

Zitron, E., Kiesecker, C., Luck, S., Kathofer, S., Thomas, D., Kreye, V. A., … Karle, C. A. (2004). Human cardiac inwardly rectifying current IKir2.2 is upregulated by activation of protein kinase A. *Cardiovascular Research*, *63*(3), 520–527. http://dx.doi.org/10.1016/j.cardiores.2004.02.015.

CHAPTER FOUR

Sodium–Calcium Exchanger in Pig Coronary Artery

A.K. Grover[1]
McMaster University, Hamilton, ON, Canada
[1]Corresponding author: e-mail address: groverak@mcmaster.ca

Contents

Abstract

This review focuses on the sodium–calcium exchangers (NCX) in the left anterior descending coronary artery smooth muscle. Bathing tissues in Na^+-substituted solutions caused them to contract. In cultured smooth muscle cells, it increased the cytosolic Ca^{2+} concentration and extracellular entry of $^{45}Ca^{2+}$. All three activities were attributed to NCX since they were inhibited by NCX inhibitors. The tissues also expressed the sarco/endoplasmic reticulum (SER) Ca^{2+} pump SERCA2b whose activity was much greater than that of NCX. Inhibiting SERCA2b with thapsigargin decreased the NCX-mediated $^{45}Ca^{2+}$ accumulation by the cells. The decrease was not observed in cells loaded with the Ca^{2+}-chelator BAPTA. The results are consistent with a limited diffusional space model with a proximity between NCX and SERCA2b. NCX molecules appear to be colocalized with the subsarcolemmal SERCA2b based on studies on membrane flotation experiments and microscopic fluorescence imaging of antibody-labeled cells. Thapsigargin inhibition of SERCA2b moved NCX even closer to SER. This provides a model for the NCX-mediated Ca^{2+} refilling of SER in the arterial smooth muscle. The model for the NCX-mediated refilling of the depleted SER proposed for smooth muscle did not apply to endothelium in which NCX levels were greater and SERCA levels were

Advances in Pharmacology, Volume 78
ISSN 1054-3589
http://dx.doi.org/10.1016/bs.apha.2016.06.001

lower than in smooth muscle. The effect of thapsigargin on the NCX-mediated Ca^{2+} accumulation which was observed in smooth muscle was absent in the endothelium. We propose that the coupling between NCX and smooth muscle may be tissue dependent.

ABBREVIATIONS

EC endothelial cells
NCX Na^{+}–Ca^{2+} exchangers
PM plasma membrane
SER sarco/endoplasmic reticulum
SERCA SERCa^{2+} pump
SMC smooth muscle cells

1. INTRODUCTION

This reviews focuses on the sodium–calcium exchanger (NCX) in smooth muscle and endothelium of the coronary arteries. A brief background on NCX is given first.

NCX is a plasma membrane (PM) resident protein which exchanges Na^{+} from one side of the membrane with Ca^{2+} on the other side. It is bidirectional—it can carry out Ca^{2+} efflux from cells in its forward mode and Ca^{2+} entry in its reverse mode. The exchanger is electrogenic with a likely stoichiometry of 3 Na^{+} to 1 Ca^{2+} (Barzilai & Rahamimoff, 1987; Blaustein & Lederer, 1999; Philipson & Nishimoto, 1980; Rasgado-Flores & Blaustein, 1987; Reeves & Sutko, 1979; Van Breemen, Aaronson, & Loutzenhiser, 1978).

Typically free cytosolic calcium (Ca_i^{2+}) in resting cells is 0.1 μM Ca^{2+} or less and it increases to 0.5–1 μM during excitation. Extracellular Ca^{2+} remains somewhat constant at 1.5 mM. Intracellular Na^{+} (5–15 mM) in most cells is much lower than the Na_{ex}^{+} (100–140 mM), while the K^{+} concentration is greater inside the cells (100–130 mM) than outside (5 mM). In resting cells, the membrane potential is internally negative with values of −60 to −90 mV in different cell types and moves toward neutrality during excitation. Thus, conditions may exit for the exchanger to expel Ca^{2+} from the cell or to allow Ca^{2+} entry. Several inhibitors have been used to examine the role of NCX (Enyedi & Penniston, 1993; Kleiboeker, Milanick, & Hale, 1992; Lytton, 2007; Molinaro et al., 2015; Shigekawa & Iwamoto, 2001). Commonly used and somewhat selective inhibitors are KB-R7943 and

SEA 0400. There are also several peptide inhibitors, such as FACACF, FMAF, callipeltin A, and XIP, of these only XIP is relatively specific for NCX, but it also has its own limitations.

The molecular biology and structure of NCX have been reviewed extensively (Khananshvili, 2013, 2014; Nicoll, Ottolia, Goldhaber, & Philipson, 2013; Ottolia & Philipson, 2013; Sharma & O'Halloran, 2014). The mammalian NCX proteins are encoded by three genes: NCX1, 2, and 3 (Lytton, 2007). NCX1 is ubiquitously expressed. However, a higher level of expression is observed in cardiac muscle, kidney, and brain, and a lower expression is observed in most other tissues. NCX2 and 3 expression is limited to tissues such as brain (Lytton, 2007; Nicoll et al., 2013). NCX1 undergoes tissue-specific splicing involving cassette exons (Khananshvili, 2014; Quednau, Nicoll, & Philipson, 1997). There are several good reviews on the regulation of NCX (Khananshvili, 2013, 2014; Lytton, 2007; Matsuda, Takuma, & Baba, 1997; Nicoll et al., 2013; Ottolia & Philipson, 2013; Ottolia, Torres, Bridge, Philipson, & Goldhaber, 2013; Quednau et al., 1997; Sharma & O'Halloran, 2014; Shigekawa & Iwamoto, 2001). NCX activity is regulated by concentrations of Na_i^+ and Ca_i^{2+}, intracellular pH, ATP, PIP_2, phosphoarginine, various kinases and phosphatases, phospholemman, and redox agents (Blaustein & Lederer, 1999; Khananshvili, 2013; Kuster et al., 2010; Lytton, 2007; Ottolia et al., 2013; Sharma & O'Halloran, 2014). Phospholemman (a 15-kDa protein) is an important regulator of NCX (Iwamoto et al., 2001; Lytton, 2007; Matsuda et al., 1997). The phosphorylated phospholemman acts as an endogenous inhibitor of NCX activity. The pathophysiology of NCX is not clearly understood, although it has been examined extensively (Brini, Cali, Ottolini, & Carafoli, 2014; Chen & Li, 2012; Goldhaber & Philipson, 2013; Iwamoto, 2005; Iwamoto, Kita, & Katsuragi, 2005; Li, Jiang, & Stys, 2000; Pignataro et al., 2013; Rose & Karus, 2013).

2. NCX IN CORONARY ARTERY SMOOTH MUSCLE

Coronary arteries supply blood to the heart, and this blood supply may be regulated by vasoconstriction and vasodilatation of these arteries. The force for the contraction is generated by shortening of the arterial smooth muscle cells (SMC). Although other factors also play a role, a simplified picture is that the vasoconstriction accompanies actomyosin activation upon a rise in Ca_i^{2+} and that the vasodilatation accompanies lowering of Ca_i^{2+}.

Major mechanisms which can alter Ca_i^{2+} in coronary artery cells are voltage-operated Ca^{2+} channels and receptor-operated Ca^{2+} channels, plasma membrane Ca^{2+} pumps (PMCA), sarco/endoplasmic reticulum Ca^{2+} pumps (SERCA), NCX, and mitochondrial channels. The roles of these pathways differ in electromechanical and pharmacomechanical coupling, and also between large and small arteries. The interactions between the various pathways are unclear. For example, one can place smooth muscle tissues or cells in nominally Ca^{2+}-free solutions and examine the role of the Ca^{2+} released from the sarco/endoplasmic reticulum (SER) in response to agents such as norepinephrine or ryanodine. Similarly, one can decrease the extracellular Na^+ to very low levels and examine the role of NCX-mediated Ca^{2+} entry in its reverse mode. However, neither of the two experiments contributes to the roles of these pathways during contractility since they are so far removed from the conditions that may normally occur. They give no clues about any interactions between the two pathways.

Our lab has used pig coronary artery as a model to understand the role of NCX. The goals were to determine the relative abundance of NCX in various coronary artery tissues and to explore if there are any interactions between NCX and SERCA. The lab used deendothelialized rings from the left anterior descending (LAD) artery for contractility studies, microsomal membranes for Western blot studies, and mRNA isolated from them for RT-PCR. SMC and endothelial cells (EC) cultured from LAD were used in $^{45}Ca^{2+}$ flux and other studies. The observations with SMC will be discussed first as the level and types of NCX and Ca^{2+} pumps present, functional/spatial coupling between NCX and SERCA2, and changes in this coupling upon SER depletion of Ca^{2+}. A comparison with EC will be made subsequently.

3. NCX IN SMC

The NCX-mediated contractions due to smooth muscle were determined by Na^+ loading the deendothelialized artery rings and then placing them in the Na^+-free (NMG^+-containing) solution (Qayyum, Al-Bondokji, Kuszczak, Samson, & Grover, 2009). The difference in the force of contraction between the two is attributed to the NCX-mediated Ca^{2+} entry. This difference is abolished by the NCX inhibitors KB-R7943 and SEA 0400 (Fig. 1A). The NCX-mediated contraction in these experiments depends on the extracellular concentration of $CaCl_2$

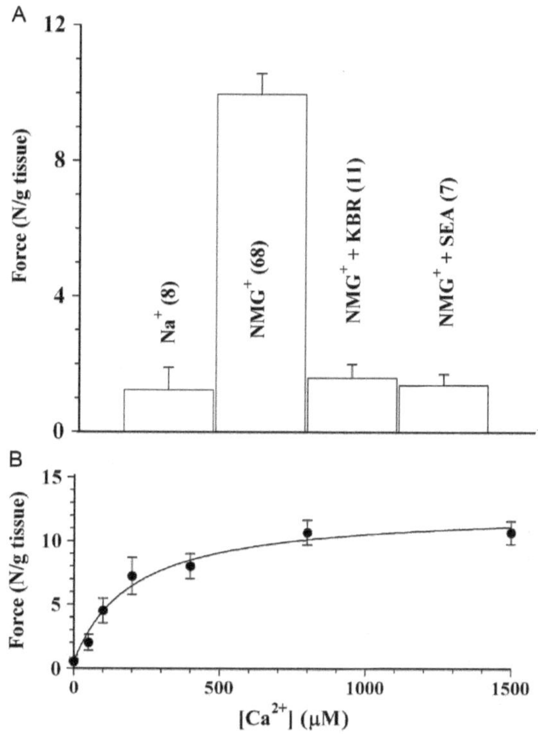

Fig. 1 NCX-mediated contraction in deendothelialized coronary arteries. (A) Characterization of NCX-mediated contraction. Force produced by Na^+-loaded tissues in Na^+-containing or Na^+-substituted solutions (NMG^+ containing) in the presence or absence of the NCX inhibitors KB-R7943 (KBR) and SEA 0400 (SEA). The values given are mean ± S.E.M. of the number of tissues as indicated for each group in parenthesis. The maximum value of the NCX-mediated force of contraction was 55 ± 5% of the maximum force obtained with membrane depolarization with 60 mM KCl. (B) External Ca^{2+} concentration dependence of the NCX-mediated contraction.

(Fig. 1B). Thus, the contractility experiments were consistent with the presence of NCX in the LAD smooth muscle.

Cultured SMC were used to determine if a similar experiment on cultured SMC would show an increase in Ca_i^{2+} and in $^{45}Ca^{2+}$ influx mediated by NCX in reverse mode. The increase in Ca_i^{2+} was monitored using the Ca^{2+}-sensitive dye Fluo 4. The Na^+-loaded cells, when placed in the Na^+-free (NMG^+-containing) solution, showed a faster increase in the fluorescence than those in the Na^+-containing solution (Fig. 2A). The effect of the Na^+-free solution was inhibited by SEA 0400 and KB-R7943 (Fig. 2B). For monitoring NCX-mediated $^{45}Ca^{2+}$ influx, SMC were loaded with Na^+

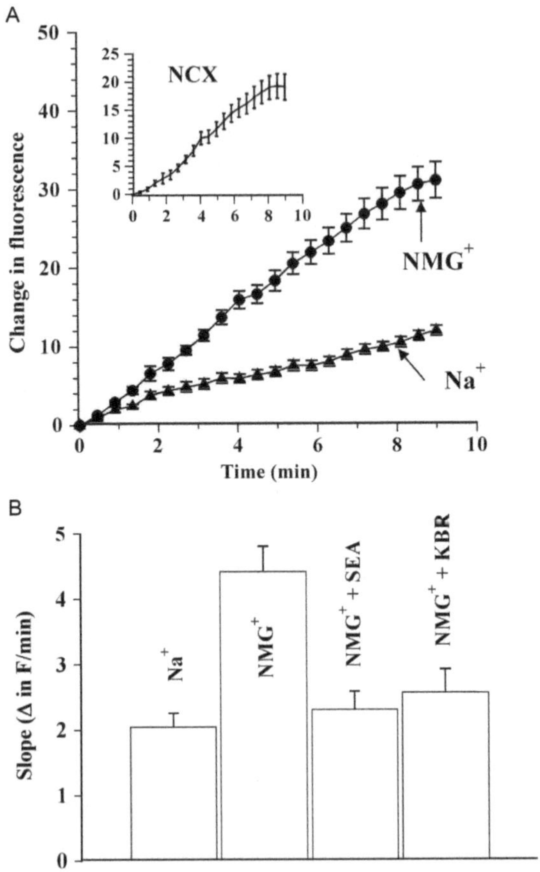

Fig. 2 NCX-mediated increase in Ca_i^{2+} as monitored using fluorescence of the Ca^{2+}-sensitive probe Fluo 4. (A) Time course of increase in Ca_i^{2+}. The cells were loaded with the dye and then Na^+ loaded, washed, and placed in a buffer containing (in mM) 20 MOPS, 20 $MgCl_2 \cdot 6H_2O$ (pH 7.4 at 37°C with Tris), 0.01 nitrendipine, 0.1 EGTA with 140 mM Na^+ or NMG^+. Basal fluorescence was monitored. $CaCl_2$ was added to a concentration of 0.3 mM and the monitoring was continued. NCX-mediated increase is the value of the difference between the two curves. (B) Inhibition of NCX-mediated rate of increase in the fluorescence (Δ in F/min) by SEA 0400 (3 μM) and KB-R7943 (10 μM) on the NCX-mediated rate of increase in fluorescence monitored for first 5 min.

and then placed in Na^+-containing or Na^+-free (NMG^+-containing) solutions and $^{45}Ca^{2+}$ to examine the accumulation of $^{45}Ca^{2+}$. SMC showed a larger the accumulation of $^{45}Ca^{2+}$ in the Na^+-free (NMG^+-containing) solutions than in the Na^+ containing (Fig. 3A). Initial experiments showed that the accumulation due to NCX was greater in the cells preloaded with the Ca^{2+}-chelator BAPTA that would maintain a very low Ca_i^{2+} concentration (indicating its dependence on a free Ca^{2+} gradient) and hence this is the

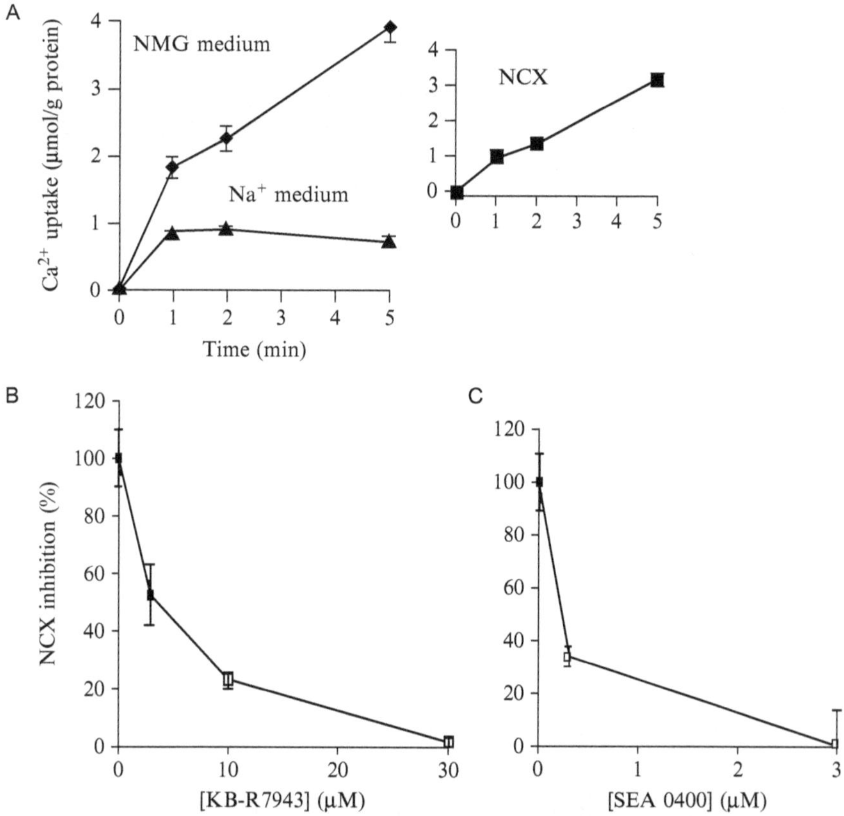

Fig. 3 ^{45}Ca uptake via NCX in cultured SMC. (A) The cells were loaded with 20 μM BAPTA–AM for 2 h in 37°C in the culture medium. The cells were Na^+ loaded in 20 mM MOPS–Tris buffer (pH 7.4 at 37°C) with 1 mM ouabain, 25 μM nystatin, and 10 μM nitrendipine. The cells were then quickly washed twice in 2 mL of the MOPS–Tris buffer with only nitrendipine. They were then placed in $^{45}Ca^{2+}$ solutions with 20 mM MOPS–Tris, 50 μM $CaCl_2$ with trace amounts of $^{45}Ca^{2+}$, 10 μM nitrendipine, and either 140 mM Na^+ (Na^+-containing) or 140 mM NMG^+ (Na^+-free) solutions, and then $^{45}Ca^{2+}$ uptake was examined. NCX activity was determined as the difference in the uptake between the Na^+-containing and the Na^+-free solutions. (B and C) Inhibition of ^{45}Ca uptake via NCX by different concentrations of KB-R7943 and SEA 0400. *Modified from Davis, K. A., Samson, S. E., Hammel, K. E., Kiss, L., Fulop, F., & Grover, A. K. (2009). Functional linkage of Na+-Ca^{2+}-exchanger to sarco/endoplasmic reticulum Ca^{2+} pump in coronary artery: Comparison of smooth muscle and endothelial cells.* Journal of Cellular and Molecular Medicine, 13, *1775–1783.*

condition used for the experiments in Fig. 3. The difference between the two was taken as the NCX activity. This activity was inhibited by KB-R7943 and SEA 0400 in a concentration-dependent manner (Fig. 3B and C) but not by the Na^+–H^+-exchange inhibitor cariporide (data not shown).

It was also abolished by the Na^+-ionophore monensin (indicating its dependence on an Na^+ gradient) (Davis et al., 2009). Thus, NCX-mediated Ca^{2+} entry caused a contraction in the smooth muscle rings and in cultured SMC it increased accumulation of $^{45}Ca^{2+}$ and produced an increase in Ca_i^{2+}. All three were inhibited by KB-R7943 and SEA 0400, and the NCX-mediated accumulation of $^{45}Ca^{2+}$ was shown to be dependent on concentration gradients of free Ca^{2+} and Na^+.

The above assays point to the presence of NCX activity in SMC. Yet, alternative explanations such as the involvement of other cation channels cannot be completely ruled out. Therefore, it was also determined if SMC contained mRNA and protein for NCX1. RT-PCR of the isolated mRNA showed that SMC showed the presence of bands for NCX1 (Szewczyk et al., 2007). SMC expressed mainly the splice NCX1.3 and smaller amounts of NCX1.7. The mRNA for the NCX endogenous inhibitor phospholemman was also present in SMC (Szewczyk et al., 2007). All the RT-PCR bands were purified and their identity confirmed by sequencing. SMC also showed the presence of NCX1 in Western blots (Szewczyk et al., 2007). All these results are consistent with the presence of NCX in SMC.

In addition to NCX1, the coronary artery SMC also express SERCA2 and PMCA4 (Table 1). SERCA activity (1–2 μmol/g protein/min) is much higher than the NCX activity (0.2–0.3 μmol/g protein/min). PMCA activity is also higher than that of NCX activity in SMC (Table 1). The three transporters also differ in their kinetic parameters for Ca^{2+}.

4. FUNCTIONAL COUPLING OF NCX AND SER IN SMC

Pharmacomechanical coupling in SMC depends on the release of the Ca^{2+} stored in the SER, and it has also been shown that not all of the released store is sequestered by SERCA into the SER (Van Breemen et al., 1978; Van Breemen, Fameli, & Evans, 2013). At least in part the SER is refilled with the extracellular Ca^{2+} and NCX has been suggested to contribute to this refilling. Further, it has been shown that the SERCA inhibitor thapsigargin may deplete the Ca^{2+} stored in the SER and move it closer to PM to allow for refilling via store-operated Ca^{2+} channels (Elmoselhi & Grover, 1999; Pani et al., 2008). We hypothesized that Ca^{2+} which enters the cells via an NCX1-mediated pathway is sequestered into the SER by the SERCA2 pump. If so, inhibiting the SERCA2 pump would locally increase the Ca^{2+} concentration between NCX1 and SER and thus inhibit the

Table 1 Properties of Smooth Muscle (SMC) of LAD

NCX	
Activity (μmol/g protein/min)	0.2–0.3
NCX/SERCA activity ratio	0.14–0.2
Isoforms expressed	NCX1.3≫NCX1.7
K_m for Ca^{2+} (μM)	2.6–7
K_m for Na^+ (mM)	25–45
Phospholemman mRNA	Detectable
SERCA	
Activity (μmol/g protein/min)	1–2
Isoforms expressed	SERCA2b
Hill coefficient for Ca^{2+}	2
K_m for Ca^{2+} (μM)	0.27 ± 0.03
pH optimum	6.8
Phospholamban inhibition	Yes
Calmodulin-dependent kinase II activation	Yes
ROS susceptibility	Very high
Catalase	Low
PMCA	
Activity (μmol/g protein/min)	1.1
Isoforms expressed	PMCA4≫PMCA1
Hill coefficient for Ca^{2+}	1
K_m for Ca^{2+} (μM)	<0.5

NCX1-mediated Ca^{2+} uptake by the cells. Toward this hypothesis, we examined the effects of thapsigargin on the NCX1-mediated Ca^{2+} uptake by cultured SMC.

The NCX-mediated $^{45}Ca^{2+}$ entry decreased in the presence of thapsigargin ($p < 0.05$, Fig. 4A) (Davis et al., 2009). However, when the cells were loaded with BAPTA which would maintain an extremely low concentration of the free cytoplasmic (and also subsarcolemmal) Ca^{2+}, the inhibition of SERCA by thapsigargin would have no effect. Indeed,

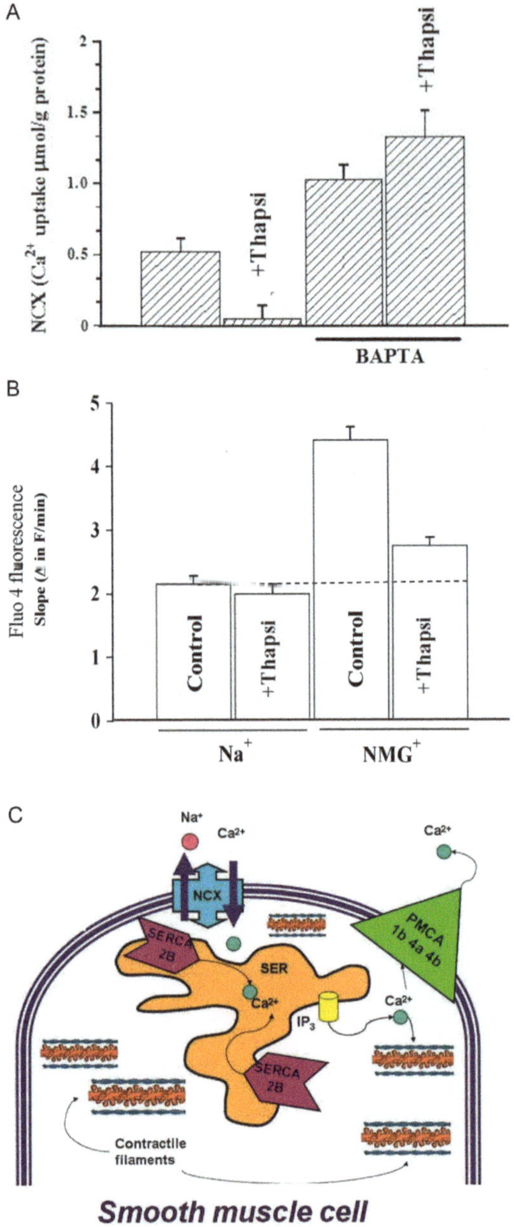

Fig. 4 The effect of thapsigargin on the NCX-mediated Ca^{2+} entry in cultured SMC. (A) NCX-mediated $^{45}Ca^{2+}$ entry. The NCX-mediated uptake was determined with or without BAPTA loading as for Fig. 2. BAPTA loading the cells increased the uptake. Thapsigargin inhibited the NCX-mediated uptake in control cells but not in the cells loaded with BAPTA. (B) Thapsigargin inhibited the NCX-mediated rate of increase in the fluorescence in cells in the NMG^+-containing buffer but not in cells in the Na^+-containing buffer. (C) Model to explain the results in (A) and (B). *Panel (A) modified from Davis, K. A., Samson, S. E., Hammel, K. E., Kiss, L., Fulop, F., & Grover, A. K. (2009). Functional linkage of Na^+-Ca^{2+}-exchanger to sarco/endoplasmic reticulum Ca^{2+} pump in coronary artery: Comparison of smooth muscle and endothelial cells.* Journal of Cellular and Molecular Medicine, 13, *1775–1783.*

thapsigargin did not affect the Ca^{2+} uptake in the BAPTA-loaded cells ($p<0.05$, Fig. 4A). The effect of thapsigargin was also examined on the NCX-mediated increase in Ca_i^{2+} monitored using fluorescence of the Ca^{2+}-sensitive probe Fluo 4. The Na^+-loaded cells, when placed in the Na^+-free (NMG^+-containing) solution, showed faster increase in the fluorescence than those in the Na^+-containing solution, but this effect was retarded by thapsigargin (Fig. 4B). Thus, thapsigargin inhibited the reverse mode NCX-mediated Ca^{2+} entry and the increase in Ca_i^{2+}. These results are consistent with a model in which SER acts as a sink for the NCX1-mediated Ca^{2+} entry. However, the need for SER to act as a sink is eliminated upon lowering $[Ca^{2+}]_i$ by BAPTA chelation and hence NCX1 is driven in full force even in the presence of thapsigargin. Fig. 4C presents a model to explain this functional linkage. In this model, NCX is in proximity of the subsurface SER.

5. COLOCALIZATION OF NCX1 AND SERCA2 IN SMC

The model in Fig. 4 suggested a colocalization of NCX1 with SERCA2. We first tested the idea that this close apposition of SERCA2 and NCX1 involves lipid rafts, which are specialized membrane microdomain platforms with several colocalized proteins (Pani et al., 2008; Simons & Ikonen, 1997). Lipid rafts are not readily solubilized with nonionic detergents. The microsomes isolated from deendothelialized LAD were treated with the nonionic detergent Tween 20 and fractionated by sucrose density gradient centrifugal flotation. Fig. 5A shows Western blots of the membrane fractions of different densities for the distribution of NCX1, SERCA2, lipid raft markers (caveolin-1, flotillin-2, prion protein, GM1 gangliosides), nonlipid raft PM marker transferrin receptor protein, and cytoskeletal markers (clathrin, actin, and myosin). The distribution of cholesterol and protein is shown in Fig. 5B. Fig. 5C–G shows the codistribution of NCX1 with SERCA2 and the various markers. NCX1 migrated very closely with SERCA2 and to some extent also with caveolin-1 and flotillin-2 as compared to the other markers. NCX1 codistributed only somewhat with gangliosides, prion protein, transferrin receptors, and total cholesterol and not at all with actin, myosin, or clathrin. Similar results were obtained when the experiments were repeated with a nondetergent method of treating microsomes with sonication at high pH for lipid raft isolation (not shown). These observations are consistent with

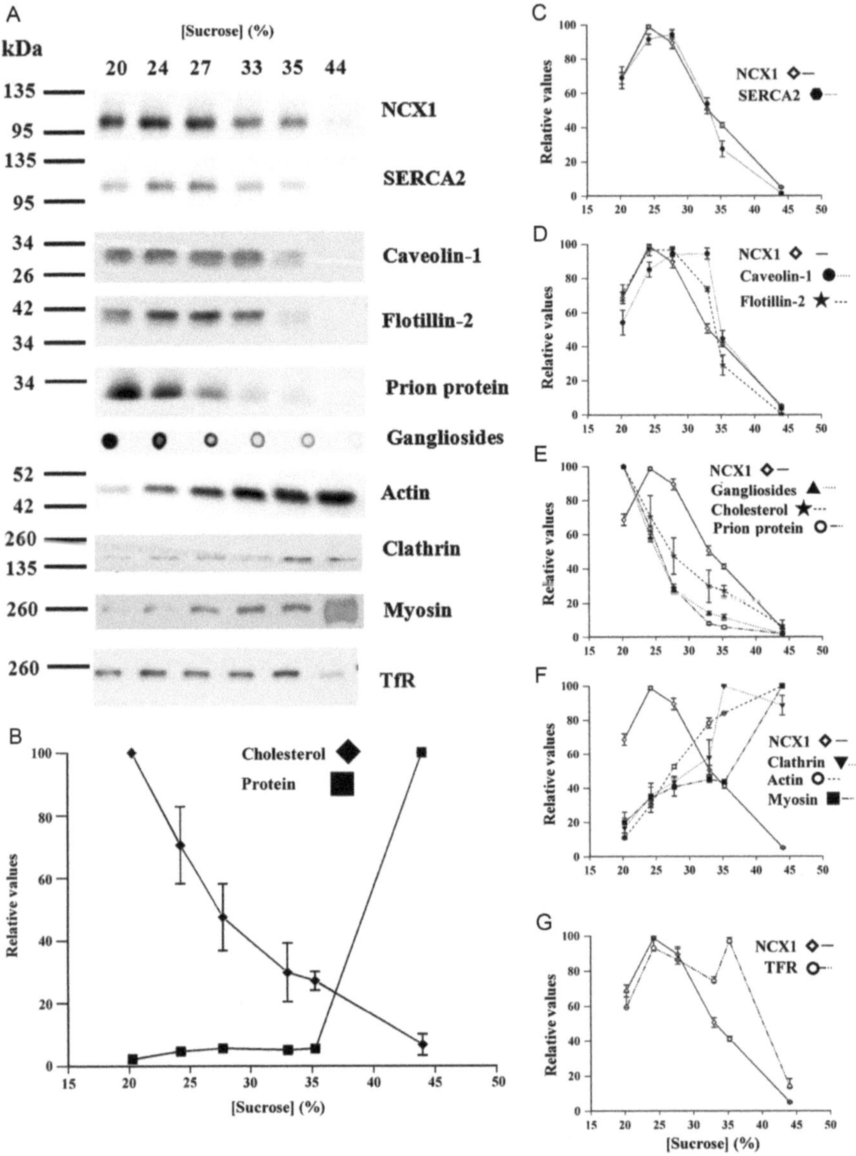

Fig. 5 Immunoblot analysis of protein distribution of detergent-treated microsomes upon centrifugal flotation. (A) For Western blots, the amounts of protein loaded per well were 2 μg for NCX1, clathrin, myosin, and transferrin receptor (TfR); 1 μg for SERCA2 and caveolin-1; 0.5 for flotillin-2 and prion protein; and 0.2 μg for actin. Dot blots with 0.2 μg protein/spot were used for GM1-ganglioside detection. (B) Total protein and cholesterol distribution. The values in (B) and the band intensities of all fractions are expressed relative to the fractions with the highest values taken as 100. The values are mean ± S.E.M. of 3–5 measurements using one membrane preparation. (B–G) Western blot quantification based on analysis of 3–5 membrane preparations similar to (A). *Modified from Kuszczak, I., Samson, S. E., Pande, J., Shen, D. Q., & Grover, A. K. (2011). Sodium-calcium exchanger and lipid rafts in pig coronary artery smooth muscle.* Biochimica et Biophysica Acta, 1808, *589–596.*

a colocalization of NCX1 and SERCA2 in lipid raft domains, even though NCX1 is localized in PM and SERCA2 in the SER.

Immunofluorescence microscopy was used next to determine if NCX1 is proximal to SERCA2 or caveolin-1 (Kuszczak, Kuner, Samson, & Grover, 2010; Pani et al., 2008). A problem in analyzing images for colocalization of the low abundance NCX1 protein with the higher abundance SERCA2 protein is that the localization of NCX1 with SERCA2 (if present) may be asymmetric. NCX1 may be in proximity to some SERCA2 molecules but many of the SERCA2 molecules may not be in proximity to NCX1. Manders' overlap coefficients have been used to measure such an asymmetric fidelity. Pearson's correlation coefficient and Manders' overlap coefficients are the two most commonly used methods for examining colocalization of particles in fluorescence image microscopy. We have described the rationale and a detailed protocol for the analysis of this asymmetrical fidelity of localization in the appendix to one of our papers (Akolkar, Pande, Samson, & Grover, 2012). Briefly, if the fluorescence intensity of the pixels with the anti-NCX1 antibody is R_i and for the anti-SERCA2 pixels it is G_i, the Pearson's correlation coefficient would be $\Sigma_i (R_i - R_{mean})(G_i - G_{mean})/(\Sigma_i (R_i - R_{mean})^2 \Sigma_i (G_i - G_{mean})^2)^{0.5}$. This value would be very sensitive to noise and become useful only for the extremely high correlation values. Here, this cross-correlation coefficient would then indicate a poor colocalization because the more abundant SERCA2 may be present in a much larger number of pixels than NCX1. Manders' colocalization coefficients with Costes' correction may be used here for two reasons. First because they take the intensity of only the colocalized pixels into account, and second because they would separately give $M1$ as a value of overlapping of NCX1 with SERCA2 and $M2$ for SERCA2 overlapping with NCX1. It uses the following equations:

$$M1 = \Sigma_i(R_{i,col})/\Sigma_i R_i. \text{ Here } R_{i,col} = R_i \text{ when } G_i > G_{it}, \text{ and } R_{i,col} = 0 \text{ when } G_i < G_{it}$$

$$M2 = \Sigma_i(G_{i,col})/\Sigma_i G_i. \text{ Here } G_{i,col} = G_i \text{ when } R_i > R_{it}, \text{ and } G_{i,col} = 0 \text{ when } R_i < R_{it}$$

Here, $R_{i,col}$ are the NCX1 pixels colocalized with SERCA2, $G_{i,col}$ are the SERCA pixels colocalized with NCX1 and R_{it} and G_{it} are the corresponding threshold values.

Stacks of images of cell surface domains were analyzed by this method. Mander's coefficients were determined for the one-way

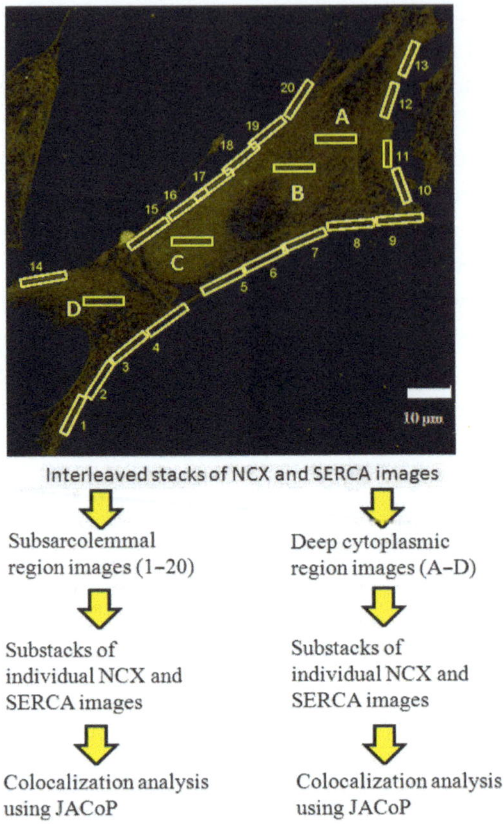

Fig. 6 Protocol for NCX1 colocalization with SERCA2 in SMC by immunofluorescence microscopy based on Akolkar et al. (2012). Confocal images were obtained by dual excitation immunofluorescence for anti-NCX1 and anti-SERCA2 antibodies. Four contiguous slices with the brightest staining for NCX1 were selected from the middle of the stack. The stacks were interleaved with the corresponding images for SERCA2. Several regions of 10–15 × 1–1.5 μm in the subsarcolemmal domain (numbered 1–15 in the figure) and cytoplasmic domains (labeled A, B, C, D) were chosen from stacks for further analysis. For analysis of each region, the stack was rotated such that the long axis of the image was horizontal. The stack of images was then cropped to obtain the desired region and deinterleaved to obtain substacks for NCX1 and SERCA2. The substacks were analyzed for colocalization using Just Another Colocalization Plugin (JACoP) (Bolte & Cordelieres, 2006). Stacks obtained from the cytoplasmic domains showed extremely poor signal for NCX1, and much lower values for *M*1, *M*2, and Pearson's coefficients than the subsarcolemmal domains.

localization of antibodies that bound to NCX1 as shown in the protocol in Fig. 6. Anti-NCX1 was detected at 568 nm using antirabbit Alexa 568 and 4G5, an anti-SERCA2 was detected at 488 nM with goat antimouse Alexa 488. As a positive control, the proximity was determined of the signal obtained for one protein to another signal for the same protein. For example, the Mander's coefficient for colocalization was close to 1 for the antibody IID8 (detected at 568 nm using Zenon Alexa Fluor 568 mouse labeling kit) that reacts with SERCA2 and with the antibody 4G5 (detected at 488 nm) which also reacts with the same protein. As a negative control, the pixels in one image stack were randomized and then the Mander's coefficients were determined. The randomization decreased the Mander's coefficients considerably. The values obtained for the different Mander's coefficients are shown in Table 2. Mander's coefficient $M1$ for NCX1–SERCA2 (0.8841 ± 0.0262, $n=25$) was significantly higher than the $M2$ for SERCA2–NCX1 (0.6887 ± 0.0482, $n=25$) indicating that NCX1 was in proximity of SERCA2, but only some of the SERCA2 molecules were in proximity of NCX1. This is also consistent with the thinking that NCX1 is closely associated with SER, but SER may also be localized to other parts of the cell than where NCX1 is found. As expected, the stacks in the cytoplasmic domains (A, B, C, D in Fig. 6) gave extremely low signal for NCX1, were very noisy, and gave very low values for $M1$ and $M2$. Detailed analysis of the cytoplasmic domains is provided in appendix 1 of a previous publication (Akolkar et al., 2012; Davis et al., 2006). Another image analysis method not described here was image shifting. It confirmed the result that NCX1 is colocalized with the subsarcolemmal SERCA2 (Kuszczak et al., 2010; Pani et al., 2008). Thus, the immunofluorescence microscopy experiments showed a one-way association between NCX1 and the subsurface SERCA2 as suggested by the model in Fig. 4. Note that deeper SER is not included in this analysis which is based only on subsarcolemmal domains.

Table 2 Mander's Colocalization Coefficients from Dual Wavelength Immunomicroscopy

Antibody Pair	Mander's Coefficient (Mean ± S.E.M.)	Number of Stacks
NCX1–SERCA2	0.8841 ± 0.0262	25
SERCA2–NCX1	0.6887 ± 0.0482	25
NCX1–CAV1	0.5698 ± 0.0456	25
CAV1–NCX1	0.6959 ± 0.0395	25
SERCA2–CAV1	0.4306 ± 0.0756	19
CAV1–SERCA2	0.7319 ± 0.0532	19

6. EFFECT OF THAPSIGARGIN ON COLOCALIZATION OF NCX1 AND SERCA2 IN SMC

The possibility was considered that emptying the SER limits the space between NCX1 and SERCA2 by increasing the proximity between NCX1 and SER. The decreased space would lead to an increased Ca^{2+} concentration when SERCA2 is inhibited. This concept was tested using dual wavelength immunomicroscopy (Akolkar et al., 2012). Steps involved in the colocalization analysis are shown in Fig. 6. The effect of thapsigargin was examined on the overlap between NCX1 and SERCA2 in the subsarcolemmal domains of cells in NMG^{+}-containing medium. The results from 500–600 regions are summarized for NCX1–SERCA2 cross-colocalization, the fraction of NCX1 overlapping with SERCA2, and the fraction of SERCA2 overlapping with NCX1 were determined. The thapsigargin treatment of the Na^{+}-loaded cells bathing in the NMG^{+}-containing solution increased the values for $M1$, $M2$, and R_{original} (Table 3). A similar experiment showed that thapsigargin treatment of the Na^{+}-loaded cells placed in Na^{+}-containing solution also increased the proximity between NCX1 and SERCA2. Thus, the presence of a Na^{+} gradient is not required for the thapsigargin-induced increase in the proximity of the two proteins. As a control, the cross-correlation coefficient values were determined in the pixels in the images that had been randomized. The values of $M1$, $M2$, and R_{random} coefficient were very low as expected and they remained low even with the thapsigargin treatment.

The proximity of NCX1 to SERCA2 was also analyzed as the frequency distribution of the R, $M1$, and $M2$ values (Fig. 7). Thapsigargin treatment shifted the frequency distributions to higher values for R_{original} (Chi square $= 47.631$, $p < 0.0001$), $M1$ (Chi square $= 54.699$, $p < 0.0001$), and $M2$ (Chi square $= 45.250$, $p < 0.0001$). Thus, the mean value and frequency distribution analyses showed that thapsigargin treatment increased the cross-colocalization between NCX1 and SERCA2 (R_{original}) and the fraction of NCX1 codistributed with SERCA2 ($M1$). It also increased the SERCA2 codistributed with NCX1 ($M2$) but to a smaller extent.

A model for the NCX-mediated SER refilling is shown in Fig. 8. Ca^{2+} depletion from SER would lead to a decrease in the subsarcolemmal space between NCX1 and SERCA2. This would allow Ca^{2+} that entered via NCX1 to be immediately sequestered into the SER and the Ca^{2+} gradient to remain low enough for NCX1-mediated Ca^{2+} entry to continue. Others

Table 3 The Effect of Thapsigargin Treatment on Colocalization Parameters in Na^+-Loaded Cells in NMG^+ Solution

	NMG (Mean ± S.E.M.) (*n*)	NMG–Thapsi (Mean ± S.E.M.) (*n*)	*t*-Value	*p*-Value
$M1$	0.617 ± 0.005 (514)	0.663 ± 0.004 (617)	7.254	<0.001
$M2$	0.595 ± 0.005 (514)	0.616 ± 0.004 (617)	3.402	0.001
$M1-M2$	0.022 ± 0.004 (514)	0.047 ± 0.003 (617)	5.159	<0.001
$M1/M2$	1.051 ± 0.007 (514)	1.085 ± 0.006 (617)	3.881	<0.001
$R_{original}$	0.356 ± 0.007 (514)	0.415 ± 0.006 (617)	6.704	<0.001
R_{random}	0.088 ± 0.004 (514)	0.087 ± 0.004 (617)	0.150	0.881

Note: $M1$ and $M2$ are Mander's colocalization for NCX1–SERCA2 and SERCA2–NCX1. $R_{original}$ is the Pearson's cross-correlation coefficient, and R_{random} is the Pearson's cross-correlation coefficient after using Costes randomization.

have proposed alternative models. In one of these models, localized cytosolic subplasmalemmal [Na^+] elevations termed Lnats act as functioning units to allow for such entry (Poburko, Fameli, Kuo, & Van Breemen, 2008; Van Breemen et al., 2013). In another, Na^+ gradient is continuously modulated for NCX to continue operating in this mode (Poburko et al., 2008; Van Breemen et al., 2013). Similar conclusions have been obtained in different cell types for the effects of the store-operated Ca^{2+} channel proteins and the SER (Pani et al., 2008). In one study depleting Ca^{2+} from the SER leads to an increase in proximity between the sarcolemmal protein Orai and the SER protein STIM to form the store-operated Ca^{2+} channels. In the sub-sarcolemmal SER, SERCA2 is also colocalized with STIM. Thapsigargin also increased the association between the lipid raft resident TRPC1 protein with STIM. Since NCX1 is also a lipid raft resident protein, one cannot rule out the possibility that the observed thapsigargin-induced increase in the proximity between NCX1 and SERCA2 results from the spatial reorganization of several proteins.

7. COMPARISON OF NCX IN CORONARY ARTERY SMC AND EC

Although EC do not generate the force of contraction, they play a major role in regulating this force and in pathophysiology of the arteries (Van der et al., 2013). EC can be activated by an increase in Ca_i^{2+} to activate eNOS thereby generating large amounts of NO which can cause

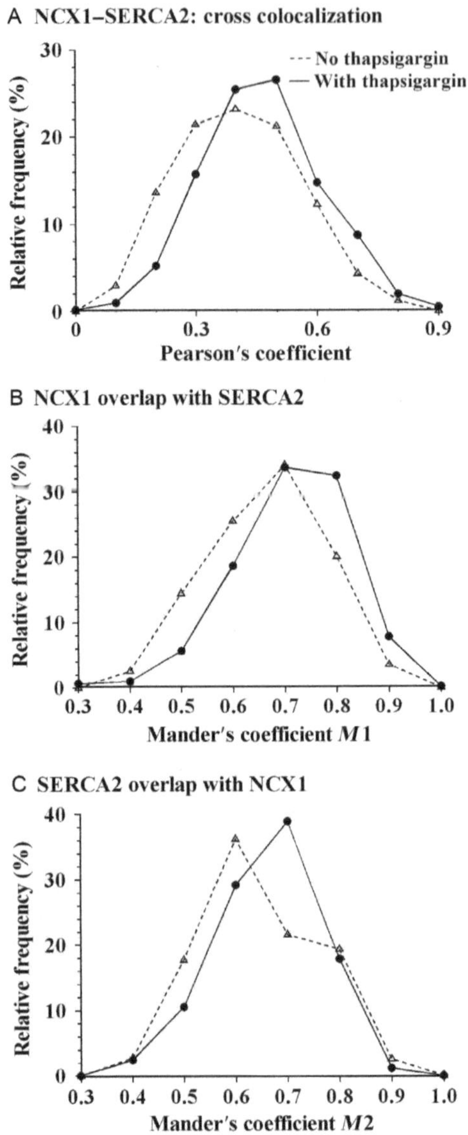

Fig. 7 Effect of thapsigargin on overlap between NCX1 and SERCA2 in the subsarcolemmal domains of cells in NMG^+-containing medium. The results from 500–600 subsarcolemmal region stacks are summarized for NCX1–SERCA2 cross-colocalization, fraction of NCX1 overlapping with SERCA2, and fraction of SERCA2 overlapping with NCX1. Relative frequencies are plotted. Thapsigargin did not affect *M*1, *M*2, and Pearson's coefficient values in the stacks obtained from the cytoplasmic domains. *Modified from Akolkar, G., Pande, J., Samson, S. E., & Grover, A. K. (2012). Thapsigargin decreases the Na(+)-Ca(2+) exchanger mediated Ca(2+) entry in pig coronary artery smooth muscle.* Biochimica et Biophysica Acta, 1818, *730–737.*

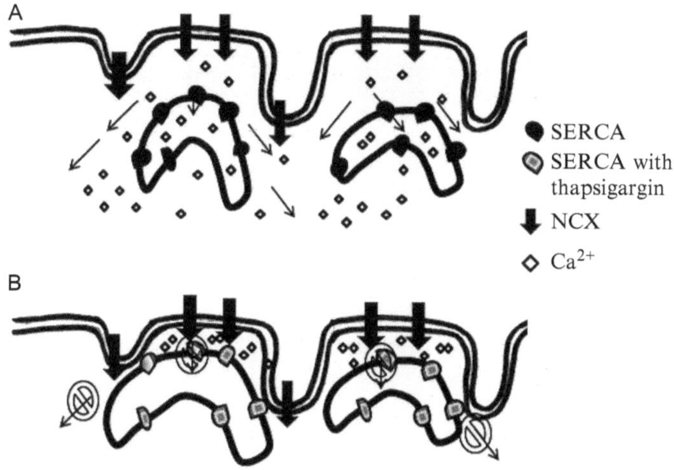

Fig. 8 Schematic model for interactions of NCX1 and SERCA2 in SMC. (A) Cytosolic Ca^{2+} resulting from the NCX entry is sequestered into the SER by SERCA2. (B) On inhibition of SERCA2 by thapsigargin, Ca^{2+} cannot be sequestered into SER, thereby increasing the cytosolic Ca^{2+} near PM. SERCA is observed to move closer to NCX, and NCX activity is diminished. Note that only the subsarcolemmal SER is depicted here and the deep SER is not shown.

vasodilatation, and NCX may play a role in it (Schneider et al., 2002; Teubl, Groschner, Kohlwein, Mayer, & Schmidt, 1999). EC can also produce peptides like endothelins, which can modulate the force generated by SMC (Anggard, Botting, & Vane, 1990). During oxidative stress, EC may also release stored ascorbate upon an increase in Ca_i^{2+} (Davis et al., 2006). Thus, a change in Ca_i^{2+} in EC may influence the coronary physiology by a variety of different mechanisms but may differ from SMC in how it comes about.

Cultured EC from LAD were used to determine the level of NCX activity by examining the accumulation of $^{45}Ca^{2+}$ (Davis et al., 2009; Qayyum et al., 2009; Shigekawa & Iwamoto, 2001; Szewczyk et al., 2007). As discussed for SMC in Fig. 2, EC were loaded with Na^+ and then placed in Na^+-containing or Na^+-free (NMG^+-containing) solutions and $^{45}Ca^{2+}$ to examine the accumulation of $^{45}Ca^{2+}$. The NCX activity in EC was also abolished by the Na^+-ionophore monensin, inhibited by the NCX inhibitors KB-R7943 and SEA 0400 but not by the Na^+–H^+-exchange inhibitor cariporide (Davis et al., 2009). Preloading EC with the Ca^{2+}-chelator BAPTA also increased the accumulation due to NCX. A comparison of the NCX-mediated $^{45}Ca^{2+}$ entry activities in SMC and EC is shown in Fig. 9. The level of NCX activity was much greater in EC than in SMC.

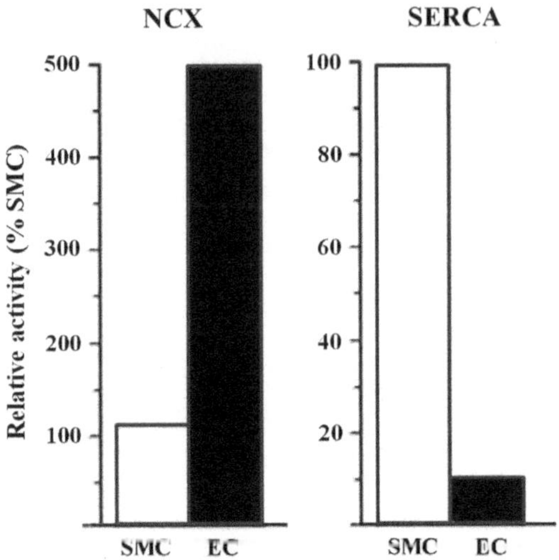

Fig. 9 Comparison of activities of NCX and SERCA in SMC and EC. ^{45}Ca uptake via NCX in cultured SMC and EC with the protocol described in Fig. 2. In 22 experiments for SMC and 14 for EC, the NCX-mediated $^{45}Ca^{2+}$ uptake in 5 min was significantly greater in EC (Davis et al., 2009). Conversely the SERCA-mediated uptake was significantly greater in SMC than in EC.

The RT-PCR experiments showed that EC also expressed mainly the splice NCX1.3 and smaller amounts of NCX1.7 but in greater amounts than SMC. The abundance of NCX1 protein determined in the Western blots was also greater in EC than in SMC (Szewczyk et al., 2007). The mRNA for the NCX endogenous inhibitor phospholemman was not detected in EC (Szewczyk et al., 2007). Thus, the higher level of the NCX activity in EC than in SMC correlates with a greater abundance of the NCX protein.

EC and SMC also differed in the ion transporters PMCA and SERCA (Szewczyk et al., 2007). SMC expressed higher levels of SERCA protein of the isoform SERCA2b, while EC expressed low levels of SERCA3 and extremely low levels of SERCA2b. The EC contained very little SERCA activity as compared to SMC but a much greater level of NCX activity (Fig. 9). EC expressed mostly PMCA1 but SMC expressed mostly PMCA4. To our knowledge, the ramifications of the relative activities of NCX, SERCA, and PMCA in SMC vs EC have not been determined. They may impact the interactions between different pathways, responses to the antihypertensive agents like dihydropyridines, ability of the cells to adhere to other cell types, and effects of ROS and hypoxia.

Because the activity of NCX1 in EC is greater than in SMC, while the SERCA activity in EC is extremely low (Fig. 9), one would expect that in EC only a very small proportion of the NCX1 molecules to be coupled to SERCA2. The effect of thapsigargin on the NCX-mediated $^{45}Ca^{2+}$ uptake was examined EC with and without BAPTA loading with the protocols as described for SMC in Fig. 4. However, thapsigargin did not inhibit the $^{45}Ca^{2+}$ uptake in control EC or in the cells loaded with BAPTA ($p > 0.05$) (Fig. 10A). Therefore, it is hypothesized that in EC, NCX1 is not linked to SERCA2 or it is not linked as strongly as it is in SMC. Fig. 10B is a model for the relationship between NCX1 and SERCA2.

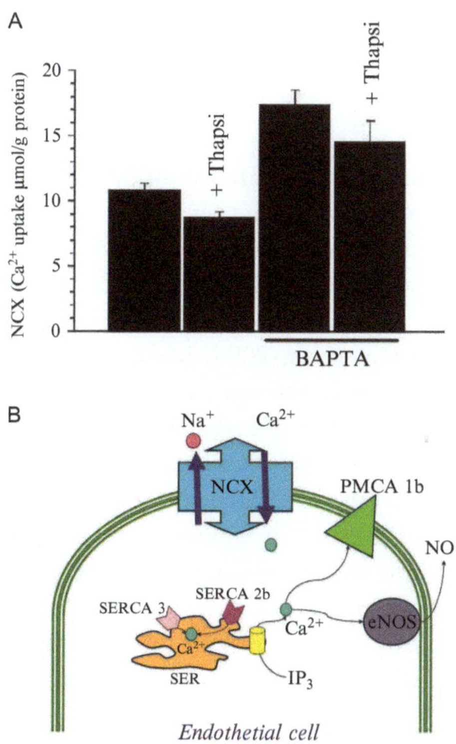

Fig. 10 The effect of thapsigargin on the NCX-mediated Ca^{2+} entry in cultured EC. (A) NCX-mediated $^{45}Ca^{2+}$ entry. The NCX-mediated uptake was determined as for Fig. 2. BAPTA loading the cells increased the uptake. Thapsigargin did not inhibit the NCX-mediated uptake in control cells or in the cells loaded with BAPTA. (B) Model to explain the results in (A). *Modified from Davis, K. A., Samson, S. E., Hammel, K. E., Kiss, L., Fulop, F., & Grover, A. K. (2009). Functional linkage of Na^+-Ca^{2+}-exchanger to sarco/endoplasmic reticulum Ca^{2+} pump in coronary artery: Comparison of smooth muscle and endothelial cells.* Journal of Cellular and Molecular Medicine, 13, *1775–1783.*

In EC, NCX1 and SERCA2 are not in proximity. This model is consistent with a greater abundance of NCX1 in EC and a very low activity of SERCA2 in these cells.

8. CONCLUSION

The review focused on the properties and roles of NCX in the LAD coronary artery SMC and their comparison with EC. Both the cell types express mainly NCX1.3. However, its expression is much lower in SMC than in EC, and in SMC it is also regulated by phospholemman which is virtually absent in EC. The interactions between NCX and SERCA2 were examined in SMC. In SMC, the thapsigargin inhibition of SERCA2 could lead to an increase in the Ca^{2+} concentration in the subsarcolemmal space between NCX and SER. This may be the reason why thapsigargin decreased the NCX-mediated entry of extracellular Ca^{2+}. Consistent with this role are the findings of colocalization of NCX with SERCA2 and the increase in this colocalization in the presence of thapsigargin. This increase was examined by immunofluorescence microscopy. However, the more recently developed method of total internal reflection fluorescence microscopy would be more appropriate to determine if SER moves closer to the PM upon Ca^{2+} depletion (Fish, 2009; Masi, Cicchi, Carloni, Pavone, & Arcangeli, 2010). Such experiments should also include other proteins suggested to interact with NCX1 (Pande et al., 2012; Poburko et al., 2008; Van Breemen et al., 2013). Contractility experiments suggest their presence in the SMC, but they need to be characterized and their role established (Pande et al., 2012). It is possible that the store-operated Ca^{2+} channels also play a role in this refilling. There is very little known about the role of the coronary artery NCX in heart failure, and this would be a relevant area to explore. It is interesting to note that thapsigargin did not inhibit the NCX-mediated entry of extracellular Ca^{2+} in EC which contain higher NCX activity and much lower SERCA activity. This leads to a cautionary note that the mechanism proposed for SMC may not be universally applicable. SMC from large and small coronary arteries are known to have differences in the Ca^{2+} pathways they utilize for contractility (Qayyum et al., 2009). Relative activities for NCX and SERCA in different endothelial beds remain to be determined, but it is known that they do not express similar proportions of different SERCA isoforms (Khan, Sandhu, Misquitta, & Grover, 2000).

NCX is present in greater abundance in the EC. The role of NCX in activation of EC to produce NO may be important to examine. Phosphorylated eNOS in EC may form a complex with NCX1 and inhibit it, this interaction may also be important in the disease process (Kiang, McClain, Warke, Krishnan, & Tsokos, 2003). The effects of oxidative stress on NCX1 during ER refilling in SMC may present a challenge since SERCA2 is inhibited by the reactive oxygen species peroxide, superoxide, and peroxynitrite (Grover & Samson, 1988; Grover, Samson, & Fomin, 1992; Grover, Samson, Robinson, & Kwan, 2003).

CONFLICT OF INTEREST

The author is a Professor Emeritus at the Department of Medicine, McMaster University and has no conflicts of interest to declare.

ACKNOWLEDGMENTS

This review contains a summary of the work on NCX1 on coronary artery conducted in our lab. The contribution of the following workers is acknowledged: S.E. Samson, K.A. Davis, G. Akolkar, I. Kuszczak, M. Szewczyk, F. Simpson, J. Pande, D.Q. Shen, R. Kuner, G. Dimmers, L. Skelley, K.E. Hammel, L. Kiss, F. Fulop, S. Robinson, C.Y. Kwan, F. Qayuum, and I. Al-Bondokji. No new funds were used in the writing of this review.

REFERENCES

Akolkar, G., Pande, J., Samson, S. E., & Grover, A. K. (2012). Thapsigargin decreases the Na(+)-Ca(2+) exchanger mediated Ca(2+) entry in pig coronary artery smooth muscle. *Biochimica et Biophysica Acta*, *1818*, 730–737.

Anggard, E. E., Botting, R. M., & Vane, J. R. (1990). Endothelins. *Blood Vessels*, *27*, 269–281.

Barzilai, A., & Rahamimoff, H. (1987). Stoichiometry of the sodium-calcium exchanger in nerve terminals. *Biochemistry*, *26*, 6113–6118.

Blaustein, M. P., & Lederer, W. J. (1999). Sodium/calcium exchange: Its physiological implications. *Physiological Reviews*, *79*, 763–854.

Bolte, S., & Cordelieres, F. P. (2006). A guided tour into subcellular colocalization analysis in light microscopy. *Journal of Microscopy*, *224*, 213–232.

Brini, M., Cali, T., Ottolini, D., & Carafoli, E. (2014). Neuronal calcium signaling: Function and dysfunction. *Cellular and Molecular Life Sciences*, *71*, 2787–2814.

Chen, S., & Li, S. (2012). The Na+/Ca(2)+ exchanger in cardiac ischemia/reperfusion injury. *Medical Science Monitor*, *18*, RA161–RA165.

Davis, K. A., Samson, S. E., Best, K., Mallhi, K. K., Szewczyk, M., Wilson, J. X., et al. (2006). Ca^{2+}-mediated ascorbate release from coronary artery endothelial cells. *British Journal of Pharmacology*, *147*, 131–139.

Davis, K. A., Samson, S. E., Hammel, K. E., Kiss, L., Fulop, F., & Grover, A. K. (2009). Functional linkage of Na^{+}-Ca^{2+}-exchanger to sarco/endoplasmic reticulum Ca^{2+} pump in coronary artery: Comparison of smooth muscle and endothelial cells. *Journal of Cellular and Molecular Medicine*, *13*, 1775–1783.

Elmoselhi, A. B., & Grover, A. K. (1999). ET(B)-mediated contraction differs between left descending coronary artery and its next branch. *Molecular and Cellular Biochemistry, 201*, 99–103.

Enyedi, A., & Penniston, J. T. (1993). Autoinhibitory domains of various Ca^{2+} transporters cross-react. *The Journal of Biological Chemistry, 268*, 17120–17125.

Fish, K. N. (2009). Total internal reflection fluorescence (TIRF) microscopy. *Current Protocols in Cytometry*. Chapter 12, Unit 12.

Goldhaber, J. I., & Philipson, K. D. (2013). Cardiac sodium-calcium exchange and efficient excitation-contraction coupling: Implications for heart disease. *Advances in Experimental Medicine and Biology, 961*, 355–364.

Grover, A. K., & Samson, S. E. (1988). Effect of superoxide radical on Ca^{2+} pumps of coronary artery. *The American Journal of Physiology, 255*, C297–C303.

Grover, A. K., Samson, S. E., & Fomin, V. P. (1992). Peroxide inactivates calcium pumps in pig coronary artery. *The American Journal of Physiology, 263*, H537–H543.

Grover, A. K., Samson, S. E., Robinson, S., & Kwan, C. Y. (2003). Effects of peroxynitrite on sarcoplasmic reticulum Ca^{2+} pump in pig coronary artery smooth muscle. *American Journal of Physiology. Cell Physiology, 284*, C294–C301.

Iwamoto, T. (2005). Sodium-calcium exchange inhibitors: Therapeutic potential in cardiovascular diseases. *Future Cardiology, 1*, 519–529.

Iwamoto, T., Kita, S., & Katsuragi, T. (2005). Salt-sensitive hypertension, Na^{+}/Ca^{2+} exchanger, and vascular smooth muscle. *Trends in Cardiovascular Medicine, 15*, 273–277.

Iwamoto, T., Kita, S., Uehara, A., Inoue, Y., Taniguchi, Y., Imanaga, I., et al. (2001). Structural domains influencing sensitivity to isothiourea derivative inhibitor KB-R7943 in cardiac Na(+)/Ca(2+) exchanger. *Molecular Pharmacology, 59*, 524–531.

Khan, I., Sandhu, V., Misquitta, C. M., & Grover, A. K. (2000). SERCA pump isoform expression in endothelium of veins and arteries: Every endothelium is not the same. *Molecular and Cellular Biochemistry, 203*, 11–15.

Khananshvili, D. (2013). The SLC8 gene family of sodium-calcium exchangers (NCX)—Structure, function, and regulation in health and disease. *Molecular Aspects of Medicine, 34*, 220–235.

Khananshvili, D. (2014). Sodium-calcium exchangers (NCX): Molecular hallmarks underlying the tissue-specific and systemic functions. *Pflügers Archiv: European Journal of Physiology, 466*, 43–60.

Kiang, J. G., McClain, D. E., Warke, V. G., Krishnan, S., & Tsokos, G. C. (2003). Constitutive NO synthase regulates the Na^{+}/Ca^{2+} exchanger in human T cells: Role of [Ca2+]i and tyrosine phosphorylation. *Journal of Cellular Biochemistry, 89*, 1030–1043.

Kleiboeker, S. B., Milanick, M. A., & Hale, C. C. (1992). Interactions of the exchange inhibitory peptide with Na-Ca exchange in bovine cardiac sarcolemmal vesicles and ferret red cells. *The Journal of Biological Chemistry, 267*, 17836–17841.

Kuster, G. M., Lancel, S., Zhang, J., Communal, C., Trucillo, M. P., Lim, C. C., et al. (2010). Redox-mediated reciprocal regulation of SERCA and Na^{+}-Ca^{2+} exchanger contributes to sarcoplasmic reticulum Ca^{2+} depletion in cardiac myocytes. *Free Radical Biology and Medicine, 48*, 1182–1187.

Kuszczak, I., Kuner, R., Samson, S. E., & Grover, A. K. (2010). Proximity of Na+-Ca2+-exchanger and sarco/endoplasmic reticulum Ca2+ pump in pig coronary artery smooth muscle: Fluorescence microscopy. *Molecular and Cellular Biochemistry, 339*, 293–300.

Li, S., Jiang, Q., & Stys, P. K. (2000). Important role of reverse Na(+)-Ca(2+) exchange in spinal cord white matter injury at physiological temperature. *Journal of Neurophysiology, 84*, 1116–1119.

Lytton, J. (2007). Na+/Ca2+ exchangers: Three mammalian gene families control Ca2+ transport. *The Biochemical Journal, 406*, 365–382.

Masi, A., Cicchi, R., Carloni, A., Pavone, F. S., & Arcangeli, A. (2010). Optical methods in the study of protein-protein interactions. *Advances in Experimental Medicine and Biology, 674*, 33–42.

Matsuda, T., Takuma, K., & Baba, A. (1997). Na(+)-Ca^{2+} exchanger: Physiology and pharmacology. *Japanese Journal of Pharmacology, 74*, 1–20.

Molinaro, P., Pannaccione, A., Sisalli, M. J., Secondo, A., Cuomo, O., Sirabella, R., et al. (2015). A new cell-penetrating peptide that blocks the autoinhibitory XIP domain of NCX1 and enhances antiporter activity. *Molecular Therapy, 23*, 465–476.

Nicoll, D. A., Ottolia, M., Goldhaber, J. I., & Philipson, K. D. (2013). 20 Years from NCX purification and cloning: Milestones. *Advances in Experimental Medicine and Biology, 961*, 17–23.

Ottolia, M., & Philipson, K. D. (2013). NCX1: Mechanism of transport. *Advances in Experimental Medicine and Biology, 961*, 49–54.

Ottolia, M., Torres, N., Bridge, J. H., Philipson, K. D., & Goldhaber, J. I. (2013). Na/Ca exchange and contraction of the heart. *Journal of Molecular and Cellular Cardiology, 61*, 28–33.

Pande, J., Dimmers, G., Akolkar, G., Skelley, L., Samson, S. E., & Grover, A. K. (2012). Store operated Ca^{2+} entry dependent contraction of coronary artery smooth muscle: Inhibition by peroxide pretreatment. *Cell Calcium, 51*, 149–154.

Pani, B., Ong, H. L., Liu, X., Rauser, K., Ambudkar, I. S., & Singh, B. B. (2008). Lipid rafts determine clustering of STIM1 in endoplasmic reticulum-plasma membrane junctions and regulation of store-operated Ca^{2+} entry (SOCE). *The Journal of Biological Chemistry, 283*, 17333–17340.

Philipson, K. D., & Nishimoto, A. Y. (1980). Na^{+}-Ca^{2+} exchange is affected by membrane potential in cardiac sarcolemmal vesicles. *The Journal of Biological Chemistry, 255*, 6880–6882.

Pignataro, G., Cuomo, O., Vinciguerra, A., Sirabella, R., Esposito, E., Boscia, F., et al. (2013). NCX as a key player in the neuroprotection exerted by ischemic preconditioning and postconditioning. *Advances in Experimental Medicine and Biology, 961*, 223–240.

Poburko, D., Fameli, N., Kuo, K. H., & Van Breemen, C. (2008). Ca^{2+} signaling in smooth muscle: TRPC6, NCX and LNats in nanodomains. *Channels (Austin, Tex.), 2*, 10–12.

Qayyum, F., Al-Bondokji, I., Kuszczak, I., Samson, S. E., & Grover, A. K. (2009). Sodium-calcium exchange mediated contraction in left anterior descending and left ventricular branch arteries. *Journal of Cellular and Molecular Medicine, 13*, 3742–3752.

Quednau, B. D., Nicoll, D. A., & Philipson, K. D. (1997). Tissue specificity and alternative splicing of the Na^{+}/Ca^{2+} exchanger isoforms NCX1, NCX2, and NCX3 in rat. *The American Journal of Physiology, 272*, C1250–C1261.

Rasgado-Flores, H., & Blaustein, M. P. (1987). Na/Ca exchange in barnacle muscle cells has a stoichiometry of 3 Na^{+}/1 Ca^{2+}. *The American Journal of Physiology, 252*, C499–C504.

Reeves, J. P., & Sutko, J. L. (1979). Sodium-calcium ion exchange in cardiac membrane vesicles. *Proceedings of the National Academy of Sciences of the United States of America, 76*, 590–594.

Rose, C. R., & Karus, C. (2013). Two sides of the same coin: Sodium homeostasis and signaling in astrocytes under physiological and pathophysiological conditions. *Glia, 61*, 1191–1205.

Schneider, J. C., El Kebir, D., Chereau, C., Mercier, J. C., Dall'Ava-Santucci, J., & Dinh-Xuan, A. T. (2002). Involvement of Na(+)/Ca(2+) exchanger in endothelial NO production and endothelium-dependent relaxation. *American Journal of Physiology. Heart and Circulatory Physiology, 283*, H837–H844.

Sharma, V., & O'Halloran, D. M. (2014). Recent structural and functional insights into the family of sodium calcium exchangers. *Genesis, 52*, 93–109.

Shigekawa, M., & Iwamoto, T. (2001). Cardiac Na(+)-Ca(2+) exchange: Molecular and pharmacological aspects. *Circulation Research*, *88*, 864–876.
Simons, K., & Ikonen, E. (1997). Functional rafts in cell membranes. *Nature*, *387*, 569–572.
Szewczyk, M. M., Davis, K. A., Samson, S. E., Simpson, F., Rangachari, P. K., & Grover, A. K. (2007). Ca^{2+}-pumps and Na^{2+}-Ca^{2+}-exchangers in coronary artery endothelium versus smooth muscle. *Journal of Cellular and Molecular Medicine*, *11*, 129–138.
Teubl, M., Groschner, K., Kohlwein, S. D., Mayer, B., & Schmidt, K. (1999). Na(+)/Ca (2+) exchange facilitates Ca(2+)-dependent activation of endothelial nitric-oxide synthase. *The Journal of Biological Chemistry*, *274*, 29529–29535.
Van Breemen, C., Aaronson, P., & Loutzenhiser, R. (1978). Sodium-calcium interactions in mammalian smooth muscle. *Pharmacological Reviews*, *30*, 167–208.
Van Breemen, C., Fameli, N., & Evans, A. M. (2013). Pan-junctional sarcoplasmic reticulum in vascular smooth muscle: Nanospace Ca^{2+} transport for site- and function-specific Ca^{2+} signalling. *The Journal of Physiology*, *591*, 2043–2054.
Van der, H. K., Gijsen, F. J., Narracott, A., Hsiao, S., Halliday, I., Gunn, J., et al. (2013). The effects of stenting on shear stress: Relevance to endothelial injury and repair. *Cardiovascular Research*, *99*, 269–275.

CHAPTER FIVE

Ca^{2+}/Calmodulin-Dependent Protein Kinase II in Vascular Smooth Muscle

F.Z. Saddouk, R. Ginnan, H.A. Singer[1]

Department of Molecular and Cellular Physiology, Albany Medical College, Albany, NY, United States
[1]Corresponding author: e-mail address: singerh@mail.amc.edu

Contents

Abstract

Ca^{2+}-dependent signaling pathways are central regulators of differentiated vascular smooth muscle (VSM) contractile function. In addition, Ca^{2+} signals regulate VSM gene transcription, proliferation, and migration of dedifferentiated or "synthetic" phenotype VSM cells. Synthetic phenotype VSM growth and hyperplasia are hallmarks of pervasive vascular diseases including hypertension, atherosclerosis, postangioplasty/in-stent restenosis, and vein graft failure. The serine/threonine protein kinase Ca^{2+}/calmodulin-dependent protein kinase II (CaMKII) is a ubiquitous mediator of intracellular Ca^{2+} signals. Its multifunctional nature, structural complexity, diversity of isoforms, and splice variants all characterize this protein kinase and make study of its activity and function challenging. The kinase has unique autoregulatory mechanisms, and emerging studies suggest that it can function to integrate Ca^{2+} and reactive oxygen/nitrogen species signaling. Differentiated VSM expresses primarily CaMKIIγ and -δ isoforms. CaMKIIγ isoform expression correlates closely with the differentiated phenotype, and some studies link its function to regulation of contractile activity and Ca^{2+} homeostasis. Conversely, synthetic phenotype VSM cells primarily express CaMKIIδ and substantial evidence links it to

Advances in Pharmacology, Volume 78
ISSN 1054-3589
http://dx.doi.org/10.1016/bs.apha.2016.08.003

regulation of gene transcription, proliferation, and migration of VSM in vitro, and vascular hypertrophic and hyperplastic remodeling in vivo. CaMKIIδ and -γ isoforms have opposing functions at the level of cell cycle regulation, proliferation, and VSM hyperplasia in vivo. Isoform switching following vascular injury is a key step in promoting vascular remodeling. Recent availability of genetically engineered mice with smooth muscle deletion of specific isoforms and transgenics expressing an endogenous inhibitor protein (CAMK2N) has enabled a better understanding of CaMKII function in VSM and should facilitate future studies.

ABBREVIATIONS

CaM calmodulin
CaMKII Ca^{2+}/calmodulin-dependent protein kinase II
CREB cAMP-responsive element-binding protein
EGFR epidermal growth factor receptor
ER endoplasmic reticulum
FRET fluorescence resonance energy transfer
MLCK myosin light-chain kinase
Myh11 smooth muscle myosin heavy chain
CDKN1A (p21) cyclin-dependent kinase inhibitor 1A
p53 tumor promoter p53
PDGF platelet-derived growth factor
SRF serum response factor
VSM vascular smooth muscle

1. INTRODUCTION

Calcium/calmodulin-dependent protein kinase II (CaMKII) is a ubiquitously expressed multifunctional serine/threonine protein kinase activated by a common second messenger, namely free intracellular Ca^{2+}. CaMKII was first identified in and purified from brain and liver and then characterized with respect to enzymatic activity and regulation. These studies revealed a complex holoenzyme structure, unique autoregulatory properties, and multiple genes encoding CaMKII isoforms (Hudmon & Schulman, 2002). Structural complexity and diversity of isoform expression have presented ongoing challenges in the development of reagents capable of specifically identifying and manipulating expression and activity. Yet with careful application of combined pharmacological, molecular, and more recently genetic approaches, CaMKII has emerged as a key regulator of diverse cellular functions and pathologies. To date, CaMKII function has been most extensively studied in brain where it regulates neurotransmission, synaptic plasticity, and memory (Wayman, Lee, Tokumitsu, Silva, & Soderling, 2008), and in heart where it has a prominent role in regulation of intracellular Ca^{2+} homeostasis, arrhythmias, and progression of heart failure (Gray & Brown, 2014; Mollova, Katus, & Backs, 2015).

Vascular smooth muscle (VSM) is a key structural component of blood vessels and by contracting and relaxing functions to regulate arterial blood flow and pressure as well as venous capacitance (Brozovich et al., 2016). VSM also plays key roles in vascular remodeling in response to chronic changes in hemodynamics, injury, and disease (Owens, Kumar, & Wamhoff, 2004). This review will focus on CaMKII structure and function in VSM. It summarizes substantial progress that has been made in defining CaMKII isozyme expression and activation profiles, signaling networks, and function in the contexts of contractility, cell proliferation, migration, and vascular remodeling. Our commentary attempts to highlight the complexity of CaMKII structure and efforts made toward understanding the functional consequences. Where possible we will discuss progress resulting from, and the potential offered by, molecular approaches and genetic mouse models for dissecting CaMKII isozyme function in VSM.

2. CaMKII STRUCTURE AND EXPRESSION IN VSM

CaMKII is structurally complex and is expressed as a large holoenzyme composed of 12–14 individual protein kinase subunits (Bhattacharyya et al., 2016). Electron microscopy and X-ray diffraction studies indicate that subunit monomers in dodecamers are organized in two stacked hexameric rings with a central core region composed of association domains and with radiating catalytic domains (Gaertner et al., 2004). In mammals, four CaMKII genes denoted *Camk2a*, *Camk2b*, *Camk2d*, and *Camk2g*, respectively, encode 50–65 kDa CaMKIIα, -β, -δ, and -γ isoforms that are differentially expressed in essentially all cells. CaMKIIα and -β isoforms are highly enriched and abundant in neural tissues (Hudmon & Schulman, 2002), while -δ and -γ isoforms are more ubiquitously expressed, including in VSM as described later. The multifunctional nature of CaMKII results from a "loose" consensus phosphorylation sequence (R/K-X-X-S/T) that is common in many proteins (Soderling & Stull, 2001). The exon structure of all four genes in mammals is highly conserved and the expressed isoforms share greater than 90% amino acid identity (Tombes, Faison, & Turbeville,

2003). As might be expected from this high degree of sequence conservation, basic enzymatic properties between recombinant highly purified isoforms are similar with only modest differences in ATP, CaM, and peptide substrate affinities (Gaertner et al., 2004; House, Zachar, Ginnan, Van Riper, & Singer, 2008). Primary sequence identity has also made the design of isoform-specific immunochemical reagents challenging and in many systems has resulted in only superficial understanding of CaMKII isoform expression and function.

CaMKII genes, as distinguished from other CaM kinases by the presence of an association domain and specific threonine autophosphorylations in the autoregulatory domain (described later), are also highly conserved across species (Tombes et al., 2003). Single genes encoding forms of the kinase meeting the above criteria have been identified in species as primitive as marine sponges, *Caenorhabditis elegans* and *Drosophila*, but not yeast or plants. Phylogenetic analyses indicate that the four mammalian genes are derived from gene duplications with the ancestral gene most likely represented by CaMKIIδ (Tombes et al., 2003). Conservation of the genes suggests important evolutionarily conserved functions, including regulation of cell division.

The basic structure of a CaMKII monomer, as represented by CaMKIIδ and -γ isoforms, is depicted in Fig. 1. Each kinase monomer contains an N-terminal catalytic/regulatory domain and C-terminal association domain,

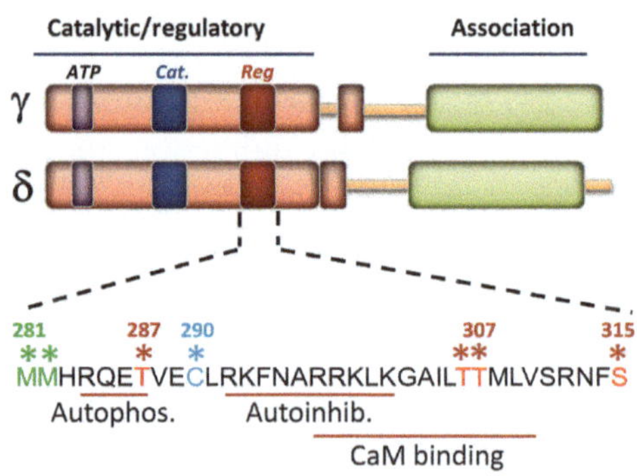

Fig. 1 CaMKIIδ and -γ monomer structures. CaMKII monomers consist of an N-terminal catalytic/regulatory domain and a C-terminal association domain which are conserved (*broad bars*) between all isoforms. A variable region (*narrow bars*) with alternatively spliced exons links these domains. CaMKIIδ is also alternatively spliced at the C-terminus. The regulatory domain contains autoinhibitory and Ca^{2+}/CaM-binding domains as well as the site of autophosphorylation (Thr^{287}) that generates autonomous activity. Additional phosphorylation ($Thr^{306,307}$), methionine oxidation ($Met^{281,282}$), and cysteine nitrosylation (Cys^{290}) sites are indicated.

the latter responsible for isoform interaction and holoenzyme structure. CaMKII monomers are generally not detectable in cells indicating rapid self-assembly of holoenzymes posttranslationally. Truncated constructs expressing only the catalytic regulatory domain are monomeric but retain full enzymatic activity regulated by Ca^{2+}/CaM (Hagiwara, Ohsako, & Yamauchi, 1991). Extensive alternative exon splicing occurs in all isoforms resulting in a variable region separating the two major domains, and in the case of CaMKIIδ, there are additional alternatively spliced exons at the C-terminus. Based on coimmunoprecipitation and fluorescence resonance energy transfer (FRET) analyses (Lantsman & Tombes, 2005), holoenzymes can be heteromultimeric, providing enormous potential for structural heterogeneity in cells expressing multiple alternatively spliced gene products.

CaMKII has been purified from arterial smooth muscle, and early studies of potential substrates and function have been reviewed (Singer, Abraham, & Schworer, 1996). Studies of CaMKII expression in differentiated arterial VSM and dedifferentiated cultured cells using conventional cDNA cloning and reverse transcriptase-polymerase chain reaction (RT-PCR) approaches resulted in the identification of novel CaMKIIδ (Schworer, Rothblum, Thekkumkara, & Singer, 1993; Zhou & Ikebe, 1994) and CaMKIIγ (Gangopadhyay et al., 2003; Singer, Benscoter, & Schworer, 1997) splice variants that are also variably expressed in heart and brain. Alternative splicing patterns and spliced exon amino acid sequences are now known to be conserved across species (Tombes et al., 2003), and the CaMKIIδ and -γ variants originally identified in rat and swine VSM are also expressed in humans.

In the case of the CaMKIIδ gene, three variants were first identified by sequencing PCR products from rat aortic and skeletal muscle tissue and designated the δ_2–δ_4 isoforms (Schworer et al., 1993). The isoforms were demonstrated to be expressed in a tissue-specific manner with the originally described δ_1 form expressed in brain, δ_2 and δ_3 forms expressed in aorta, δ_3 in cardiomyocytes, and δ_4 in skeletal muscle. Cultured aortic VSM expressed primarily the δ_2 variant. Shortly following this report, the Schulman lab (Edman & Schulman, 1994) also identified the δ_3 isoform in heart (designated δ_B by their nomenclature) and demonstrated that the alternatively spliced exon encoding an 11-amino acid sequence in this variant was responsible for nuclear targeting of holoenzymes (Srinivasan, Edman, & Schulman, 1994). This exon is also alternatively spliced in some CaMKIIα and -γ variants. Nuclear targeting can be inhibited by phosphorylation of a specific serine residue in the sequence (Heist, Srinivasan, &

Schulman, 1998). Thus, the subcellular localization of CaMKII variants in cultured VSM and cardiomyocytes is distinct with a perinuclear cytosolic distribution of δ_2 in cultured VSM (Mercure, Ginnan, & Singer, 2008; Van Riper, Schworer, & Singer, 2000) and nuclear distribution of δ_3 in cardiomyocytes. In heart, activation of nuclear CaMKIIδ_B is linked to hypertrophic signaling by phosphorylating type II histone deacetylases (HDAC4/5) and derepressing genes involved in the hypertrophic response (Backs et al., 2009). Moreover, activation of the CaMKIIδ_C variant (CaMKIIδ_2), which lacks the nuclear localization signal and is largely restricted to the cytosol, selectively phosphorylated proteins affecting intracellular Ca^{2+} dynamics, indicating distinct functions for these two major CaMKIIδ gene splice variants (Zhang et al., 2007). The expression and properties of individual CaMKIIδ splice variants have not been carefully studied in differentiated VSM.

Variable domains in CaMKIIβ have been reported to target the kinase to actin filaments (Leary, Lasda, & Bayer, 2006). While CaMKIIβ isoforms are not abundantly expressed in VSM, it is untested but possible that low-level expression may be sufficient to target a fraction of CaMKII holoenzymes to actin filaments. In VSM, a novel CaMKIIγ splice variant identified in the ferret aorta appears to localize with vimentin intermediate filaments and α-actinin-containing dense bodies and couples kinase activity to ERK1/2 activation (Marganski, Gangopadhyay, Je, Gallant, & Morgan, 2005). We have provided evidence that the variable C-terminal domain in CaMKIIδ isoforms contains a proline-rich region that localizes the kinase in a signaling complex that includes the Src-family tyrosine kinase FYN (Ginnan et al., 2013). This mechanism, which is discussed in more detail in a later section, provides cross talk between Ca^{2+} signaling and tyrosine kinase signaling pathways involved in cell motility and proliferation. It can be safely assumed that there is much yet to learn regarding the properties and function of specific CaMKII isoform splice variants in VSM. Interested investigators are also encouraged to consider the functional implications of context-dependent regulation of CaMKII isoform expression (discussed later) and/or regulated alternative splicing of CaMKII transcripts.

3. CaMKII ACTIVATION IN VSM

3.1 Thr^{287} Autophosphorylation and Autonomous Activity

As depicted in Fig. 2, CaMKII activation results from Ca^{2+}/CaM binding in the CaMKII regulatory domain which causes conformational changes that

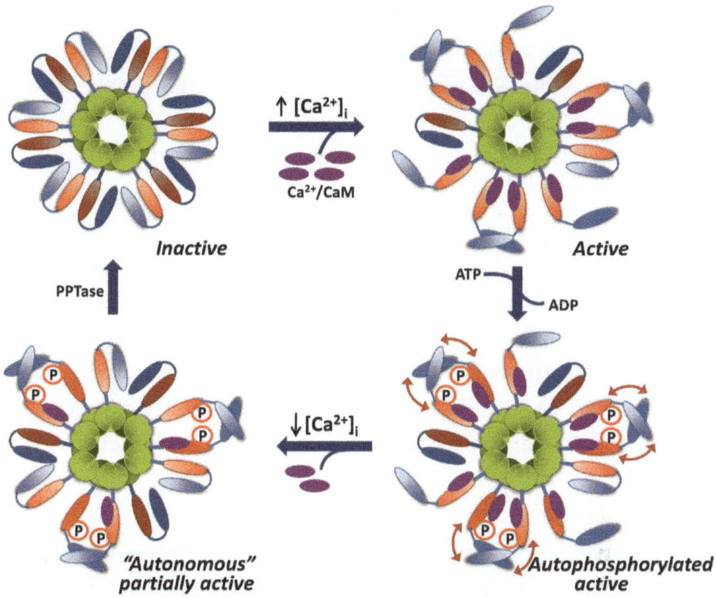

Fig. 2 CaMKII holoenzyme activation. CaMKII holoenzymes form via interaction of association domains. Upon elevation of intracellular free [Ca^{2+}], Ca^{2+}/calmodulin complexes (Ca^{2+}/CaM) bind inactive monomers (*upper left*) in the regulatory domain, resulting in a conformational change that releases the catalytic region from autoinhibition and results in full activity (*upper right*). Adjacent activated subunits autophosphorylate on Thr^{287} in a cooperative, intraholoenzyme, intersubunit reaction (*lower right*) which results in CaM-trapping and Ca^{2+}-independent activity (or autonomous) activity at low [Ca^{2+}] (*lower left*). Phosphatases (PP1, PP2A) remove Thr^{287} phosphorylation, inactivating the kinase.

release an adjacent autoinhibitory domain and expose the catalytic cleft. A characteristic feature of CaMKII is autophosphorylation on Thr^{287} (Thr^{286} in CaMKIIα) which results in "autonomous" activity that persists following dissociation of the Ca^{2+}/CaM complex (Schworer, Colbran, Keefer, & Soderlings, 1988). Interestingly, this reaction is intersubunit between activated (Ca^{2+}/CaM-bound) CaMKII monomers and is therefore highly cooperative within the holoenzyme structure (Mukherji & Soderling, 1994). Thr^{287} phosphorylation is not required for activity but increases the affinity of CaMKII for Ca^{2+}/CaM by more than 1000-fold (Meyer, Hanson, Stryer, & Schulman, 1992). These unique properties allow the kinase to retain activity beyond the time frame of a transient Ca^{2+} signal and accumulate activity in a Ca^{2+} transient frequency-dependent manner, although the extent of autonomous activity may be substrate dependent (Coultrap, Buard, Kulbe, Dell'Acqua, & Bayer, 2010). Studies in mice expressing a CaMKIIα Thr286Ala mutant have linked Thr^{286} autophosphorylation and autonomous activity to the processes of long-term potentiation and learning (Giese, Fedorov, Filipkowski, & Silva, 1998). The functional implications of CaMKII autonomous activity outside of the brain are largely unexplored. From a practical standpoint, assay of CaMKII autonomous activity using a specific peptide substrate modeled on the autoinhibitory domain (autocamtide) or direct assessment of $Thr^{286/287}$ phosphorylation using a sequence-specific antiphosphopeptide antibody provides useful indices of cellular CaMKII activation that can be applied to cell extracts and histological specimens.

After optimization of cell extraction conditions to stabilize CaMKII autophosphorylation, we were able to apply the autonomous activity assay and evaluate the sensitivity of CaMKII to activation by intracellular Ca^{2+} in cultured rat aortic VSM cells (Abraham, Benscoter, Schworer, & Singer, 1996). In cells selectively permeabilized to Ca^{2+} and depleted of sarcoplasmic reticulum Ca^{2+} by pretreatment with thapsigargin, adding Ca^{2+} back activated CaMKII with an apparent EC_{50} of 692 nM $[Ca^{2+}]_i$. Maximal autonomous activity reached 68% of total Ca^{2+}/CaM-stimulated activity. These values were similar to those estimated under optimal conditions using purified kinase in vitro and reflect a nearly complete activation of CaMKII in permeabilized VSM cells. The results also indicated that free CaM was not limiting in this setting. Despite the relatively low sensitivity of the kinase to Ca^{2+}, physiological stimuli such as angiotensin II, vasopressin, and platelet-derived growth factor (PDGF), which stimulated small transient increases in free $[Ca^{2+}]_i$ in VSM, were nevertheless able to activate CaMKII and generate autonomous activity to as much as 40% of total Ca^{2+}/CaM-stimulated activity in intact cells. From these data, we inferred that a pool of CaMKIIδ in VSM is localized to efficiently sense agonist-induced Ca^{2+} transients resulting from release of intracellular stores. In support of this concept, CaMKIIδ_2 (CaMKIIδc), the predominant expressed isoform in cultured VSM, colocalizes and concentrates with endoplasmic reticulum (ER) (Van Riper et al., 2000) and Golgi (Ginnan et al., 2013) markers. As ER is a well-characterized source of intracellular Ca^{2+} and recent studies have documented regulated release of Ca^{2+} from Golgi stores in cardiomyocytes (Yang et al., 2015), CaMKII is therefore positioned to sense Ca^{2+} release from both intracellular stores.

Kinetics of agonist-induced autonomous CaMKII activity in VSM closely track with intracellular free Ca^{2+} kinetics (Abraham et al., 1996), and phosphatase inhibitors, including okadaic acid and calyculin A, markedly enhance agonist-induced levels of autonomous activity (Mishra-gorur, Singer, & Castellot, 2002). These results indicate that under the conditions studied, endogenous phosphatase activities efficiently reverse Thr^{287} phosphorylation and suggest limited functional consequences of autonomous CaMKII activity in VSM. However, this point may need to be reconsidered and studied in the context of subcellular microdomains with variable ratios of CaMKII and regulated phosphatase activities.

Similarly, in intact carotid artery medial smooth muscle, autonomous CaMKII activity transiently reached 80% of the total Ca^{2+}/CaM kinase activity in response to histamine stimulation, while KCl depolarization manifested a slow increase in autonomous activity reaching 30% of the total activity (Rokolya & Singer, 2000). The efficiency by which the G-protein receptor agonist activated CaMKII compared to KCl depolarization with Ca^{2+} entry through plasma membrane Ca^{2+} channels is consistent with a pool of kinase in close proximity to intracellular Ca^{2+} stores, or potentially, plasma membrane store-operated channels. The subcellular localization of CaMKII has not been extensively studied in differentiated VSM and structural features resulting in physical proximity to Ca^{2+} stores are not known. It is interesting to consider that an alternative transcriptional initiation site in CaMKIIα results in a protein (αKAP) that retains the conserved association domain but lacks a catalytic/regulatory domain (Bayer, Harbers, & Schulman, 1998). The N-terminal sequence in αKAP is a membrane association domain and incorporation of αKAP subunits into CaMKII holoenzymes targets the kinase to intracellular membranes. Subcellular localization of CaMKII holoenzymes and coupling to sources of $[Ca^{2+}]_i$ entry or release may depend on isoform composition and dictate available substrates and function.

A number of other autophosphorylation sites in CaMKII have been documented that can affect activity or autoregulatory properties, but the functional importance of these events in intact cellular functions is largely unexplored. Following $Thr^{286/287}$ autophosphorylation and generation of autonomous activity in vitro, a second series of autophosphorylations on Thr^{305} and Thr^{306} in the CaMKIIα-binding domain ($Thr^{306,307}$ in other isoforms) can inhibit subsequent reactivation of the kinase by Ca^{2+}/CaM (Colbran & Soderling, 1990). In arterial smooth muscle, the Morgan lab linked Thr^{305} phosphorylation to inhibition of CaMKII reactivation and tonic contraction suggesting the functional relevance of autophosphorylation on this site, as well as on Thr^{287} which is a prerequisite event (Munevar et al., 2008). Recently, the same group demonstrated that phosphorylation of Ser^{26} in the ATP-binding domain of CaMKIIγ inhibited activity (Yilmaz, Gangopadhyay, Leavis, Grabarek, & Morgan, 2013). Moreover, the site was shown to be phosphorylated in contracting arterial preparations and selectively dephosphorylated by a specific phosphatase, suggesting functional relevance.

3.2 CaMKII Oxidation and Nitrosylation

Recently, CaMKII activation by methionine oxidation has emerged as a functionally important posttranslational modification and this sense CaMKII can be viewed as an integrator of Ca^{2+} and redox signaling (Anderson, 2015; House, Potier, Bisaillon, Singer, & Trebak, 2008). The original description of this mechanism required prior Ca^{2+}/CaM-dependent activation of the kinase to expose $Met^{281/282}$ in the CaMKII regulatory domain to oxidation (Erickson et al., 2008). Oxidation of $Met^{281/282}$ results in autonomous activity via a mechanism similar to Thr^{287} autophosphorylation. Mice with surgically induced myocardial infarction (MI) had increased cardiac CaMKII $Met^{281/282}$ oxidation and knockout of methionine sulfoxide reductase A (MsrA), a reductase that regenerates methionine from its oxidized form, enhanced MI-induced CaMKII oxidation with increased myocardial dysfunction and mortality compared to controls. Recent studies have also documented nitric oxide-stimulated S-nitrosylation of Cys^{273} and Cys^{290} in the regulatory region of CaMKIIδ (Erickson et al., 2015). In heart, Cys^{290} nitrosylation in CaMKIIδ was found to be dependent upon prior activation by Ca^{2+}/CaM and functions similar to Thr^{287} phosphorylation and $Met^{281/282}$ oxidation to provide autonomous activity (Erickson et al., 2015). However, in unactivated kinase, nitrosylation of Cys^{273} inhibited CaMKII activation. CaMKIIδ has been suggested to regulate inducible nitric oxide synthase trafficking and activity in cultured VSM (Jones, Jourd'heuil, Salerno, Smith, & Singer, 2007) a process that is enhanced in cells from diabetic rats (Di Pietro et al., 2013). This suggests a complex interplay between these two regulatory molecules that merits additional investigation.

Multiple posttranslational modifications in the autoregulatory domain could help explain the efficiency by which CaMKII is activated by physiological stimuli in VSM in situ. In support of this concept, $Met^{281/282}$ in CaMKII can be oxidized in cultured VSM in response to normal growth stimuli or addition of H_2O_2 (Zhu et al., 2014). VSM cells from genetically engineered mice expressing a CaMKII mutant with $Met^{281/282}$ replaced by Valine residues (M2V) proliferated normally, but motility and apoptosis stimulated by H_2O_2 were inhibited (Zhu et al., 2014). Surprisingly, injury-induced vascular remodeling was not inhibited in M2V mice, but CaMKII expression was upregulated, suggesting a compensatory effect. The efficiency of CaMKII methionine oxidation and cysteine nitrosylation has recently been shown to be isoform dependent (Coultrap, Zaegel, & Bayer, 2014), and additional studies are required to dissect the functional importance of these posttranslational modifications in VSM CaMKII.

4. CaMKII FUNCTION IN VSM

4.1 Differentiated VSM Contractile Function

Given its activation by physiological contractile stimuli and multifunctional nature, CaMKII might be expected to regulate VSM contractility by either directly targeting contractile proteins or indirectly targeting associated regulatory proteins and/or proteins involved in intracellular free Ca^{2+} homeostasis. In fact a number of such proteins have been identified based on in vitro studies but few have been established unambiguously as substrates for CaMKII in intact arterial systems. Perhaps the best-studied candidate substrate for CaMKII in this context is myosin light-chain kinase (MLCK), a dedicated Ca^{2+}/CaM-activated protein kinase which plays an essential role in activation of smooth muscle myosin ATPase activity and contraction (Kamm & Stull, 2001). Specifically, phosphorylation of a threonine residue in MLCK's CaM-binding domain by PKA or CaMKII in vitro desensitizes MLCK to activation by Ca^{2+}/CaM. Early studies in intact contractile tracheal (Tansey et al., 1992) and arterial preparations (Van Riper, Weaver, Stull, & Rembold, 1995) established that phosphorylation and desensitization of MLCK, presumably mediated by CaMKII, occur during an agonist-stimulated contractile response. Based on this mechanism in isolation, CaMKII would be predicted to negatively regulate arterial VSM contractility. However, application of selective pharmacological inhibitors (KN-62, KN-93) in isolated arterial preparations results in strong inhibition of sustained contractile activity indicating the opposite, a net positive function for CaMKII in regulating VSM contractility (Kim et al., 2000; Rokolya & Singer, 2000). This suggests that although MLCK is a bona fide substrate for CaMKII in intact VSM, the functional consequence of negative regulation of MLCK by CaMKII must be obscured by other CaMKII-regulated processes, including Ca^{2+} dynamics. A diverse array of ion channels, Ca^{2+} release receptors, and membrane ATPases have been directly shown or implicated as CaMKII substrates, including L-type Ca^{2+} channels, TRPC channels, IP_3 and ryanodine receptors, phospholamban, and SERCA. Because the primary literature on this topic is extensive, the reader is referred to reviews that address this in the context of neuronal (Hudmon & Schulman, 2002), cardiac (Colbran, 2004), and smooth muscle (Perrino, 2011) systems.

Recent studies using a transgenic mouse that overexpresses an endogenous CaMKII inhibitor (CAMK2N) in smooth muscle demonstrated that CaMKII inhibition did not alter myogenic tone or vasoconstriction of isolated mesenteric arteries but paradoxically did inhibit agonist-induced increases in free $[Ca^{2+}]_i$ homeostasis and phosphorylation of MLCK (Prasad et al., 2013). Baseline blood pressures were also not different between control and transgenic mice expressing CAMK2N. Similarly, global knockout of CaMKIIδ (Li et al., 2011) or conditional smooth muscle-specific knockout of CaMKIIγ (Saddouk et al., 2016) has no reported effects on baseline blood pressure or VSM contractility. Interpretation of the CaMKII "loss of function" studies discussed in the preceding paragraphs is complicated by potential off-target effects of the pharmacological inhibitors used (Ledoux, Chartier, & Leblanc, 1999) and genetic adaptations secondary to chronic inhibition of CaMKII developmentally and postnatally. Moreover, while these studies provide some insights into "net" effects of CaMKII inhibition in regulating VSM contractility and blood pressure, the approaches provide little information regarding CaMKII isoform-specific substrates and mechanisms that are regulated in vivo.

To our knowledge, few studies have taken an isoform-specific approach to dissect the function of CaMKII in regulating VSM contractility. The Morgan laboratory reported that antisense oligonucleotides suppressing expression of CaMKIIγ isoforms inhibited ferret aorta contractions ex vivo, indicating a net positive function for the kinase in regulating contractile activity (Kim et al., 2000). Subsequent studies identified the intermediate protein vimentin and dense body protein α-actinin as potential cytoskeletal targets for a novel CaMKIIγ splice variant and suggested a positive function in regulating ERK1/2 activation, myosin phosphorylation, and contractile force (Gangopadhyay et al., 2003). More detailed information regarding CaMKII isoform expression and localization in differentiated smooth muscle preparations and inducible approaches to target expression or activity of specific isoforms in VSM postnatally are required to gain a picture of specific in vivo substrates and kinase function in the context of VSM contractility. Recent availability of genetically engineered mice with floxed CaMKIIδ (Backs et al., 2009) and -γ (Backs et al., 2010) alleles and transgenics with Cre-recombinase under control of a tamoxifen-inducible smooth muscle myosin heavy chain (*Myh11*) promoter (Wirth et al., 2008) provides opportunities to specifically dissect CaMKII isoform function in differentiated VSM.

4.2 Synthetic Phenotype Function

4.2.1 VSM Phenotype Switching

In the mature vasculature, differentiated VSM cells maintain a low proliferative index and express a distinct set of genes that determine the specialized contractile phenotype. In contrast to cardiac and skeletal muscle that are terminally differentiated, smooth muscle cells can undergo reversible changes in phenotype between the differentiated state and a dedifferentiated "synthetic" state in response to a variety of environmental cues including growth factors, mechanical stimulation, and inflammation. The synthetic phenotype is characterized by a high proliferative index, motility, and production of proteins involved in extracellular matrix remodeling. Synthetic phenotype VSM cells contribute to a number of pervasive vascular diseases and conditions including atherosclerotic plaque development and stabilization, aneurysms, transplant vasculopathy, and restenosis following angioplasty (Owens et al., 2004). In addition, synthetic phenotype venous smooth muscle promotes both maturation of arterialized vein grafts and arteriovenous fistulas (used for dialysis access), as well as a failure due to neointimal hyperplasia and occlusion (Lu et al., 2014). Differentiated VSM undergoes phenotypic switching following vascular injury in vivo or primary cell culture, and as discussed in detail later, both systems have been widely used as model systems to dissect regulatory pathways involved.

Smooth muscle phenotype switching has been studied extensively with regard to changes in expression of contractile proteins and transcriptional and epigenetic mechanisms (Owens et al., 2004). Less appreciated are changes in intracellular Ca^{2+} handling and expression of numerous components of Ca^{2+} signaling pathways. Loss of the contractile phenotype is accompanied by downregulation of L-type voltage-gated Ca^{2+} channels, the sarco–endoplasmic reticulum Ca^{2+} ATPase type 2a (SERCA2a), and ryanodine receptors (Barlow, Rose, Pulver-Kaste, & Lounsbury, 2006; House, Potier, et al., 2008; House, Zachar, et al., 2008). Switching to the synthetic phenotype manifests a shift from the voltage-dependent Ca^{2+} influx to store-operated Ca^{2+} entry that results from upregulation of stromal interaction molecule 1 (STIM1), a sarco–endoplasmic reticulum Ca^{2+} sensor (Bisaillon et al., 2010) and ORAI1, a Ca^{2+} release-activated Ca^{2+} channel (Zhang et al., 2011). Ca^{2+} signals from these channels promote synthetic phenotype VSM proliferation, migration, and matrix remodeling functions.

A significant challenge is to understand how Ca^{2+} signals are transduced to regulate VSM phenotype switching and/or synthetic phenotype function, and whether specific Ca^{2+} signaling pathways can be targeted to inhibit vascular remodeling associated with injury and disease. As a key mediator of Ca^{2+} signals, CaMKII is implicated in transcriptional regulation as well as direct regulation of cell cycle progression and motility. In differentiated VSM cells, mRNA transcripts for both CaMKIIγ and CaMKIIδ variants are expressed (Schworer et al., 1993; Singer et al., 1997). However, CaMKIIδ_2 is the predominant isoform expressed in cultured VSM (House, Ginnan, Armstrong, & Singer, 2007). The predominance of CaMKIIδ isoforms in synthetic phenotype VSM has also been documented in vivo, in arterial smooth muscle following injury (House & Singer, 2008). Thus, like numerous proteins involved in generation of Ca^{2+} signals, the Ca^{2+} signal effector CaMKII is also dynamically regulated during phenotype switching and, as discussed later, promotes VSM synthetic phenotype functions and plays key roles in vascular remodeling in response to injury.

4.2.2 CaMKII and Gene Transcription

In addition to acutely regulating cell function, Ca^{2+} signals play a central role in the regulation of gene expression, a process termed excitation–transcription (E–T) coupling in excitable cells, including VSM (Barlow et al., 2006; Owens et al., 2004). E–T coupling in differentiated VSM appears important in modulating contractile protein gene expression and therefore provides long-term regulation of VSM function. Potential Ca^{2+}-regulated transcription factors include SRF (serum response factor), NFAT, CREB (cAMP-responsive element-binding protein), and MEF2 (myocyte-enhancing factor 2), and Ca^{2+}-dependent effectors include Rho/ROCK, calcineurin, and Ca^{2+}/CaM-activated kinases, respectively. Given the possibilities of nonspecific effects using pharmacological approaches, and until recently, the lack of genetic mouse models with knockout or overexpressed CaMKII isoforms, the mechanisms by which CaMKII integrates into the E–T coupling paradigm in differentiated VSM remain poorly understood. However, some progress has been made in this regard in dedifferentiated cultured VSM where CaMKIIδ isoform-dependent regulation of MEF2C and CREB activities has been documented.

MEF2, an MADS-box family of DNA transcription factors, has important functions in cardiovascular development and muscle differentiation (Black & Olson, 1998). Paradoxically, the MEF2C isoform is also upregulated in the neointima of rat carotid arteries after balloon catheter injury and has been associated with the VSM synthetic phenotype (Firulli et al., 1996). CaMKII was first shown in heart to regulate MEF2 activity by phosphorylating type II histone deacetylases (HDAC4/HDAC5), stimulating nucleocytoplasmic shuttling and subsequent derepression of MEF2-regulated genes (Zhang et al., 2007). In VSM, angiotensin II-induced hypertrophy was reported to be dependent on CaMKII phosphorylation of HDAC4 and subsequent activation of MEF2 (Li et al., 2010). Although CaMKII-dependent regulation of HDAC4/5 in cardiomyocytes was reported to be dependent on the nuclear-localized CaMKIIδ_B (CaMKIIδ_3) isoform (Backs et al., 2009), our lab used molecular approaches to demonstrate that activation of the cytosolic CaMKIIδ_2 (CaMKIIδ_C) isoform also resulted in phosphorylation of HDAC4 and HDAC5 in cultured VSM in response to angiotensin II or PDGF stimulation (Ginnan, Sun, Schwarz, & Singer, 2012). This resulted in Mef2 activation and induced expression of Mef2-regulated marker genes, including monocyte chemoattractant protein-1 (MCP-1). A recent study used KN-93 to inhibit CaMKII in cultured VSM and confirmed inhibition of HDAC4 phosphorylation (Usui, Morita, Okada, & Yamawaki, 2014). CaMKII was also reported to form a complex with the scaffolding protein GIT1 (G-protein-coupled receptor kinase-interacting protein-1), phospholipase Cγ, and HDAC5 (Pang et al., 2008). This "calcium signaling complex" facilitates the phosphorylation of HDAC5 by CaMKII and subsequent MEF2 activation in response to angiotensin II stimulation.

Multifunctional CaMKII and CaMKIV mediate Ca^{2+}-dependent regulation of CREB transcriptional activity. Expression of CREB is downregulated in several vascular diseases, including hypertension, dyslipidemia, and atherosclerosis, as well as in response to vascular injury, suggesting a vasculoprotective effect of CREB activity (Schauer et al., 2010). Phosphorylation of Ser^{133} in CREB is required for activation, and based on in vitro studies, this site is a substrate for protein kinase A, CaMKIV and CaMKII, while phosphorylation of Ser^{142} by CaMKII has been reported to inhibit CREB activity by interfering with CREB dimerization and protein interactions to form an active promoter complex (Sun, Enslen, Myung, & Maurer, 1994). In cultured VSM cells and intact cerebral arteries, Ca^{2+} entry through voltage-dependent Ca^{2+} channels and release from store-operated Ca^{2+} channels result in increased phosphorylation of CREB and CREB transcriptional activity (Pulver, Rose-Curtis, Roe, Wellman, & Lounsbury, 2004).

However, our recent studies in cultured VSM using siRNA approaches to silence CaMKIIδ expression or overexpression of constitutively active and kinase-negative CaMKIIδ$_2$ mutants to manipulate activity identified CREB Ser^{142} as the substrate for CaMKIIδ$_2$ in situ (Liu, Sun, Singer, Ginnan, & Singer, 2013). Signaling by CaMKIIδ$_2$ resulted in nuclear export of CREB, negative regulation of CREB promoter activity, decreased binding of CREB to promoters of target genes *Sik1* and *Rgs2*, and attenuated expression of these genes induced by thrombin (Liu et al., 2013). Conversely, CaMKII activation in neural systems has been repeatedly associated with CREB activation (Ma, Groth, Wheeler, Barrett, & Tsien, 2011). In this context, a recent report provided evidence that in neural cells, membrane-associated CaMKIIγ, specifically compared to CaMKIIβ, indirectly activates CREB by translocating CaM into the nucleus (Ma & Tsien, 2014). These studies raise the possibility that positive and negative regulation of CREB activity in response to Ca^{2+}-dependent stimuli is CaMKII isoform dependent. In VSM, additional studies are in order to clearly define the role of different CaMKII isoforms in the regulation of CREB.

4.2.3 CaMKII Cross Talk with ERK1/2 and Tyrosine Kinases in VSM

In addition to direct regulation of gene transcription by phosphorylating repressors and activators, CaMKII may indirectly modulate these processes through cross talk with other signaling pathways that impinge on gene regulation. ERK1/2, members of the mitogen-activated family of protein kinases (MAPK), and downstream signaling events regulate multiple transcription factors in VSM including NF-kB (Nakano et al., 1998), c-JUN (Iyoda et al., 2012), and EGR-1 (Thiel, Mayer, Müller, Stefano, & Rössler, 2010). CaMKIIδ was first shown to mediate Ca^{2+} signal-dependent activation of ERK1/2 in cultured VSM in response to both Ca^{2+} ionophores and G-protein-coupled receptor stimuli, including angiotensin II, ATP, and thrombin (Abraham, Benscoter, Schworer, & Singer, 1997; Muthalif, Benter, Uddin, & Malik, 1996). Subsequent studies indicated that "transactivation" of EGF receptor (EGF) and nonreceptor tyrosine kinases were required for CaMKII-dependent activation of ERK1/2 (Ginnan & Singer, 2002). Specifically, the nonreceptor tyrosine kinases PYK2, a homolog of focal adhesion kinase (FAK), and a Src-family kinase (SFK) were implicated as intermediates in CaMKII-dependent activation of ERK. The precise nature of PYK2 involvement in mediating the CaMKII-dependent epidermal growth factor receptor (EGFR) transactivation has not been elucidated, but another study in VSM cells showed PYK2 interaction with the EGFR in response to proteinase receptor 1 (PAR1) activation (Schauwienold, Sastre, Genzel, Schaefer, & Reusch, 2008). The scaffolding protein GPCR kinase 2-interacting protein (GIT1) may facilitate CaMKII-dependent activation of ERK1/2 in VSM cells. GIT1 scaffolds both CaMKIIδ and MEK1, an upstream activator of ERK1/2, and is required for angiotensin II-dependent activation of ERK1/2 through mechanisms involving Src-PLCγ signaling and EGFR transactivation (Pang et al., 2008). One study indicated that adhesion-dependent activation of ERK1/2 involves a physical interaction between CaMKII and Raf1 that results in the direct activation of Raf1 by CaMKII with a consequent increase in the ERK activity (Illario et al., 2003). Since the initial reports in VSM, CaMKII has been implicated in ERK1/2 activation in other systems, including heart where it promotes cardiac hypertrophy (Cipolletta et al., 2015).

4.2.4 CaMKII Regulation of VSM Cell Motility

VSM cell migration is a property of "synthetic phenotype" cells that have not acquired, or have lost, differentiated contractile protein markers and function. Migration of VSM cells occurs during development, wound healing, and angiogenesis and contributes to the progression of vascular disease. The overall function of CaMKII in VSM cell migration has been well studied in vitro and was first reported over two decades ago (Bilato et al., 1995). Adhesion and spreading of VSM cells (Lu, Armstrong, Ginnan, & Singer, 2005), stimulation with the growth and chemotactic factor PDGF (Abraham et al., 1996), or "scratch wounding" a monolayer of VSM cells grown in serum (Mercure et al., 2008) results in activation of CaMKII. Both pharmacological inhibitors and siRNA-mediated knockdown of CaMKIIδ isoforms in VSM cells have been shown to inhibit PDGF- and serum-stimulated motility with most studies, indicating a net positive role for CaMKIIδ in regulating VSM motility (Bilato et al., 1995; Mercure et al., 2008). Mechanisms underlying CaMKII activation and downstream protein substrates in migrating cells are still not well understood.

Although Ca^{2+} signals have been implicated as regulators of motility (Wei, Wang, Zheng, & Cheng, 2012), this knowledge has not been well integrated into models of cell migration, in part due to lack of knowledge of specific mediators. In one study, spontaneous localized Ca^{2+} transients were imaged in the leading lamella of migrating fibroblasts and found to promote directional cell migration and turning to a chemoattractant, consistent with a positive role in leading edge remodeling and cell polarization (Wei et al., 2009). In fibroblasts, CaMKII was localized near cell membranes in migrating cells using total internal reflection fluorescence microscopy (Easley, Brown, Horwitz, & Tombes, 2008). Both CaMKII inhibition and overexpression of a constitutively active mutant disrupted focal adhesion dynamics and cell motility, suggesting a key role for the kinase in regulating focal adhesion turnover. Using indirect immunofluorescence and an anti-phospho-Thr^{287}, we localized a pool of activated CaMKII in the lamella of migrating VSM (Mercure et al., 2008). Suppression of CaMKIIδ expression inhibited RAC activation and VSM cell polarization as assessed by Golgi polarization relative to the leading edge. Thus, CaMKII is activated during VSM migration and, as a multifunctional serine/threonine kinase, is positioned to coordinate complex cellular events that promote cellular polarization and motility (Fig. 3).

CaMKII substrates involved in cell migration remain unknown; however, results from work in differentiated muscle discussed earlier suggest that cytoskeletal proteins are possible targets. Another general mechanism by

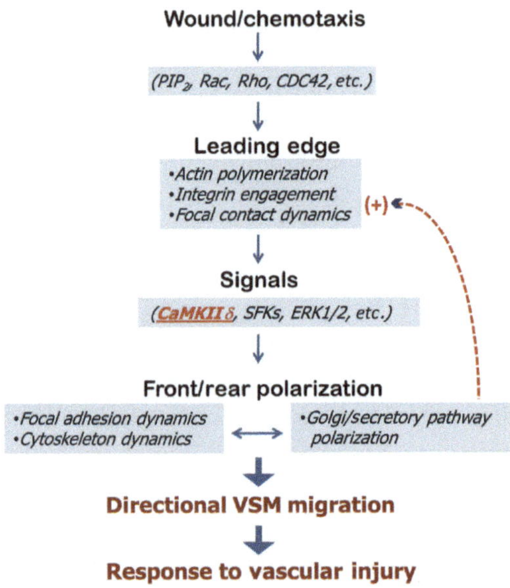

Fig. 3 Model of CaMKIIδ function in VSM cell migration. Front/rear polarization of cells in response to injury or chemotaxis is required for directional VSM cell migration, which contributes to vascular remodeling. Chemotactic stimuli result in generation of membrane signals (PIP_2) and cellular gradients of activated small GTPases (Rac, Rho, CDC42) which promote leading edge formation and advancement. CaMKIIδ and other intracellular signaling mediators including Src-family kinases (SFKs) and ERK1/2 regulate focal adhesion and cytoskeleton dynamics required for cell migration and reinforcement of front/rear polarization.

which CaMKII could exert diverse cellular effects during a complex process like cell migration is through interaction with other intracellular signaling pathways, including ERK1/2 and tyrosine kinases as described earlier. Adhesion-dependent activation of CaMKII mediates activation of ERK1/2 dependent on FAK activation (Lu et al., 2005). Major targets of FAK, PYK2, and SFKs and ERK1/2 signaling pathways in adhering or migrating cells are focal contacts and adhesions (Parsons, Horwitz, & Schwartz, 2010).

Of the SFKs, FYN in particular has been implicated in mechanosensing and regulation of focal adhesion dynamics (Kostic & Sheetz, 2006; Zeng et al., 2003) and cell migration (Lamalice, Houle, & Huot, 2006). Activation of FYN in response to α_v/β_3 integrin engagement (Zeng et al., 2003) requires activation of an integral membrane receptor protein tyrosine phosphatase (RPTPα) and dephosphorylation of an inhibitory C-terminal tyrosine residue (Tyr^{531}) in FYN (Bodrikov et al., 2005; Vacaresse, Moller, Danielsen, Okada, & Sap, 2008). RPTPα has also been implicated in the regulation of focal adhesion dynamics, and RPTPα-null cells display focal adhesion and migration deficits very similar to *Rac1*- or SFK-null cells (Vacaresse et al., 2008). FYN, which is myristoylated and dually palmitoylated, RPTPα, and CaMKIIα colocalize in lipid rafts forming a signaling complex. RPTPα may be a relevant CaMKII substrate in this complex as an activation of CaMKII with neural cell adhesion molecule results in serine phosphorylation and activation of RPTPα, FYN activation, and neurite outgrowth (Bodrikov, Sytnyk, Leshchyns'ka, den Hertog, & Schachner, 2008).

We recently reported a similar signaling complex involving CaMKIIδ_2, FYN, and RPTPα in rat VSM (Ginnan et al., 2013). Activation of cells with Ca^{2+}-dependent stimuli decreased protein interactions and correlated with FYN activation, while inhibition of CaMKII activity with KN-93 stabilized the complex. Confocal immunofluorescence and FRET microscopy confirmed colocalization of CaMKIIδ_2 and FYN in a perinuclear compartment. SiRNA-mediated suppression of CaMKIIδ or FYN inhibited focal adhesion turnover kinetics as assessed by FRAP (fluorescence recovery after photobleaching) and VSM motility. Interestingly, the CaMKIIδ_2/FYN interaction appears to be mediated through a proline-rich sequence in the alternatively spliced CaMKIIδ_2 C-terminus. Ectopic expression of CaMKIIδ_6, which lacks this sequence, inhibited complex formation and cell motility. Thus, FYN and RPTPα represent potential targets of CaMKII and mediators of CaMKII-dependent effects on focal adhesion dynamics and leading edge polarization.

A proposed general model for how CaMKIIδ promotes VSM cell polarization and directional migration is depicted in Fig. 3.

In vivo, regulated expression of extracellular matrix components and degrading enzymes are key factors in smooth muscle cell migration (Gerthoffer, 2007). Using VSM cells from *Camk2d*-null mice, Scott et al. demonstrated that CaMKII promotes VSM migration and remodeling by increasing the expression of *Mmp9* (Scott et al., 2012). MMP9 protein expression and VSM migration in response to serum, PDGF-BB and TNFα, were inhibited in *Camk2d*-null cells and rescued with MMP9 overexpression. The mechanism mediating CaMKIIδ-dependent regulation of MMP9 expression was identified to be posttranscriptional regulation of *Mmp9* mRNA stability.

4.2.5 CaMKII Regulation of VSM Cell Proliferation

Although there is accumulating evidence for CaMKII-dependent regulation of cell proliferation in other systems, the mechanisms coupling specific CaMKII isoforms to regulation of cell proliferation are not completely understood (Skelding, Rostas, & Verrills, 2011). In VSM cultures, inhibition of CaMKII activity by overexpressing a kinase-negative CaMKIIδ_2 mutant or silencing CaMKIIδ expression with siRNA slows serum-stimulated cell proliferation with increased numbers of multinucleated cells and cells accumulating at the G2/M checkpoint as assessed by flow cytometry (House et al., 2007). Acute modulation of CaMKII isoform expression from γ to δ_2 isoforms occurs within 30 h of dispersing VSM cells and correlates with VSM phenotype switching, DNA synthesis, and cell proliferation. Preventing CaMKIIδ_2 upregulation in primary dispersed cells by transducing with short hairpin RNAs targeting CaMKIIδ inhibits proliferation without interfering with the phenotype switch as assessed by downregulation of *Myh11* and α-smooth muscle actin (*Acta2*). Studies in global *Camk2d* knockout mice and in aortic VSM cells cultured from these mice confirmed a role for CaMKIIδ in promoting VSM proliferation (Li et al., 2011). Decreased proliferation in the knockouts was associated with activation of AKT with increased expression of p21 and its transcriptional activator, p53. These results were interpreted to indicate that CaMKIIδ-dependent enhancement of VSM cell proliferation was mediated by suppression of an endogenously active p53- and p21-dependent cell cycle inhibitory pathway. Although cell cycle analysis was not performed, we note that p53 and p21 can regulate both the G1/S and G2/M checkpoints.

Because CaMKIIδ is the predominant isoform expressed in cultured VSM and CaMKIIγ expression is strongly downregulated, it is not clear from the loss-of-function experiments described earlier if inhibition of VSM proliferation is due specifically to loss of CaMKIIδ, or more generally to a loss of total CaMKII activity. This was recently addressed indirectly by comparing effects of overexpressed CaMKIIδ_2 to overexpressed CaMKIIγ_C in cultured VSM in gain-of-function experiments (Saddouk et al., 2016). Surprisingly, CaMKIIγ overexpression inhibited proliferation, while comparable levels of CaMKIIδ overexpression had no effect. The results were confirmed in cells from *Camk2d* knockouts and in primary cultures of VSM following initial cell seeding, during a period when endogenous CaMKIIγ is undergoing active downregulation. Inhibition of VSM cell proliferation by CaMKIIγ overexpression was associated with increased expression of p53, and CaMKIIγ overexpression had no effect on proliferation of VSM isolated from p53-null animals. The p53 target gene CDKN1A (p21) was also upregulated following CaMKIIγ overexpression, while CDC2 (*Cdk1*), *Cdc20* and *Cdc25*, genes reported to be transcriptionally repressed by p53, were downregulated (Saddouk et al., 2016). This pattern of gene expression is consistent with cell cycle inhibition at both G1/S and G2/M checkpoints (Stark & Taylor, 2006). Thus, CaMKII has isoform-specific effects in VSM with signaling through CaMKIIδ promoting and CaMKIIγ inhibiting proliferation (Fig. 4). The mechanistic basis for opposing effects on cell proliferation and on regulation of p53/p21 cell cycle

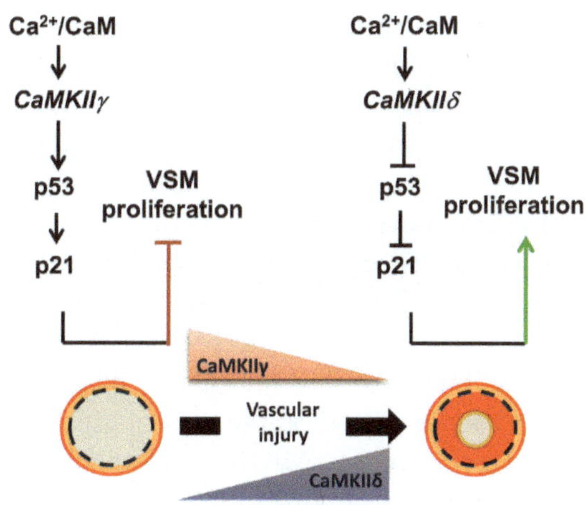

Fig. 4 Reciprocal regulation of VSM proliferation by CaMKIIγ and -δ isoforms. CaMKIIγ inhibits VSM proliferation by promoting expression of cell cycle inhibitors, p53 and p21. Vascular injury with VSM phenotype switching results in decreased CaMKIIγ and increased CaMKIIδ, protein expression. CaMKIIδ inhibits expression of p53 and p21, promoting VSM proliferation and vascular wall thickening due to neointimal hyperplasia.

inhibitors is not known. Possibilities include differences in intrinsic substrate specificities or subcellular localization and coupling to substrates.

4.3 CaMKII Isozymes in Vascular Remodeling In Vivo

CaMKIIδ and -γ isoform functions in regulating VSM gene transcription, cell proliferation, and cell motility predict that the kinase would regulate vascular remodeling in response to hemodynamic stresses, injury, or disease. In support of this concept, administration of the CaMKII inhibitor KN-93 to rats made hypertensive by chronic infusion of angiotensin II, decreased renal CaMKII activity, reduced blood pressure, and inhibited renal intravascular injury (Muthalif et al., 2002). Subsequent studies using KN-93 confirmed decreased angiotensin II-induced hypertension and aortic medial wall hypertrophy in mice (Li et al., 2010). More recently, sm22α-cre-driven overexpression of CAMK2N, the endogenous CaMKII inhibitor protein, was found to inhibit angiotensin II-induced hypertension in mice (Prasad et al., 2015). CAMK2N expression decreased baroreceptor activity and decreased angiotensin II-induced genes related to extracellular matrix composition and remodeling. Based on in vitro studies, the angiotensin II/CaMKII hypertrophic pathway was dependent on phosphorylation of HDAC4 and transcriptional activation of MEF2. In a mouse model of high-flow-dependent outward remodeling, global knockout of CaMKIIδ or knock-in of CaMKIIδ with $Met^{280/281}$ to Valine mutations in mice inhibited remodeling in the right common carotid artery following complete ligation of the left carotid (Scott et al., 2013). Although these studies could not pinpoint the effects specifically to VSM CaMKII, they support the concept that CaMKII coordinates chronic Ca^{2+} and reactive oxygen species signals to regulate vascular remodeling.

Studies from this laboratory demonstrated that transition of VSM from differentiated to synthetic phenotype following injury (House & Singer, 2008) or primary cell culture (House et al., 2007) is marked by increased expression of the CaMKIIδ$_2$ isoform and reciprocal decreased expression of CaMKIIγ isoforms. Intraluminal administration of adenovirus expressing a shRNA targeting *Camk2d* mRNA at the time of injury resulted in attenuated injury-induced increases in CaMKIIδ$_2$ in medial VSM with decreased proliferation, migration, and neointima formation (House & Singer, 2008), suggesting a role for CaMKIIδ in promoting vasculoproliferative remodeling. This basic finding was confirmed in a mouse carotid ligation model in which global deletion of *Camk2d* markedly decreased ligation-induced medial wall VSM proliferation and nearly prevented injury-induced neointimal hyperplasia (Li et al., 2011). Decreased proliferation was associated with increased medial wall p53 and p21 expression consistent with in vitro studies in cultured VSM. However, a limitation of these global loss-of-function approaches is that they do not unambiguously distinguish CaMKIIδ function in VSM from other cells, including resident or circulating vascular progenitors, endothelial cells, and immune cells, all of which can contribute to vascular remodeling (Tang et al., 2012).

Recently, we started to address this limitation using conditional approaches to knockout CaMKII isoforms specifically in smooth muscle and related mesenchymal cells using mice with floxed *Camk2* alleles crossed with transgenics expressing Cre-recombinase under control of an sm22α (transgelin) promoter. These approaches confirmed that smooth muscle knockout of *Camk2d* inhibited ligation-induced VSM proliferation and neointimal hyperplasia (unpublished studies). However, parallel studies using conditional *Camk2g* knockouts resulted in the opposite effect, i.e., increased VSM cell proliferation and decreased expression of p21 in ligated arteries (Saddouk et al., 2016). Moreover, overexpression of CaMKIIγ following balloon catheter injury in the rat carotid model rescued CaMKIIγ expression to preinjury levels and strongly attenuated neointimal hyperplasia. Thus, CaMKIIδ and -γ have opposing isoform-specific effects on vascular proliferation in vitro, as discussed earlier, and in vivo in response to vascular injury. CaMKIIγ appears to serve as a brake on CaMKIIδ-dependent function, and downregulation of CaMKIIγ expression following vascular injury may be permissive for CaMKIIδ to promote synthetic VSM phenotype functions. The mechanistic basis for this finding at the level of CaMKII holoenzyme structure, localization, and substrate coupling remains to be determined.

Carotid ligation and balloon injury have been reported to stimulate extensive medial wall VSM death and regulated apoptosis or necrosis could contribute to the vascular remodeling responses (Tang et al., 2012). CaMKII has been reported to be a key intermediate in ER-stress responses and apoptosis in a number of systems including cardiomyocytes (Zhu et al., 2007) and lung epithelial cells (Winters et al., 2016). CaMKII has also recently been reported to be a substrate for RIP3 (receptor-interacting protein kinase 3) and mediate programmed cell necrosis (necroptosis) in cardiomyocytes (Zhang et al., 2016). Although there is little reported evidence for CaMKII-dependent increases in apoptotic rates in cultured VSM from either *Camk2d* or *Camk2g* knockouts, or in vivo in these models following carotid ligation (Saddouk et al., 2016), this issue may need to be examined in more detail focusing on specific stimuli in vitro and time points following vascular injury in vivo.

The experiments discussed earlier indicate that dynamic regulation of CaMKII isoform expression in VSM has important functional consequences. A recent study suggests that at least one endogenous regulator that could account for CaMKIIδ protein upregulation in synthetic phenotype VSM is the microRNA-30 family of small noncoding RNAs (Liu et al., 2016). MiR-30 is strongly downregulated in VSM following vascular injury and in cultured cells relative to differentiated arterial tissue. Reexpression in vitro, or in vivo following rat carotid balloon injury, suppresses CaMKIIδ expression, VSM cell proliferation, and neointimal hyperplasia in response to vascular injury. CaMKIIδ has also been shown to be regulated by mir143/145 in VSM and is downregulated in Dicer-null mice which cannot process microRNA (Turczyńska et al., 2012). Mechanisms controlling rapid decreases in expression of CaMKIIγ mRNA and protein following vascular injury are not known. However, CaMKIIγ expression correlates closely with differentiated VSM markers and can be enhanced by overexpression of myocardin, albeit through an SRF-independent mechanism (Saddouk et al., 2016). Future studies of *Camk2g* promoter activation and/or epigenetic modifications in the *Camk2g* gene may be required to understand its regulation during VSM phenotype modulation.

5. CONCLUSION

As a multifunctional protein kinase, CaMKII can be expected to mediate or modulate diverse cellular processes and serve an integrative function. In the context of VSM, significant progress has been made toward our understanding of CaMKII function in the contexts of both differentiated VSM and synthetic phenotype VSM functions, as summarized in Fig. 5. Yet, there is much left to learn, particularly with regard to identification of specific substrates and mechanisms underlying CaMKII function. Structural complexity and the possibilities of isoform- and splice variant-specific subcellular localization, protein interactions, and substrate specificities present significant challenges in the design and interpretation of studies aimed at defining CaMKII function in all systems, including VSM. Future progress will depend on careful application of combined pharmacological and molecular/genetic approaches and design of isoform-specific reagents and approaches.

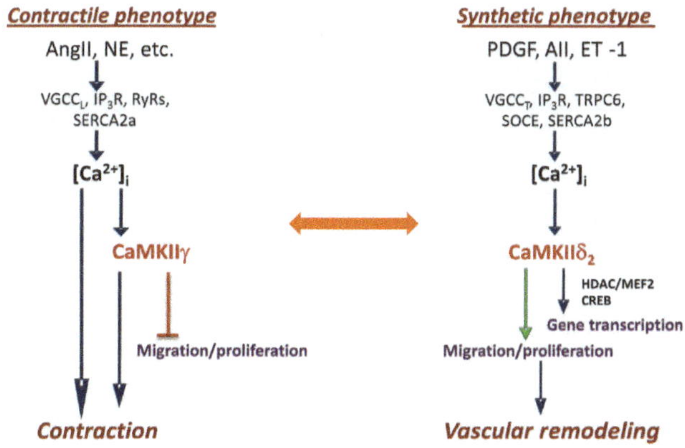

Fig. 5 Model of CaMKII isoform function in vascular smooth muscle. G-protein receptor-coupled ligands such as angiotensin II (AngII) and norepinephrine (NE) stimulate Ca^{2+} entry and release that result in contraction of differentiated VSM cells (*left*). Upon injury or cell culture VSM cells transition to a synthetic phenotype (*right*), a process that includes changes in expression of Ca^{2+} signaling proteins, including L-type voltage-gated Ca^{2+} channels ($VGCC_L$), IP_3 receptors (IP_3R), ryanodine receptors (RyRs), STIM1/Orai1 store-operated Ca^{2+} entry (SOCE), TRPC6 channels, and SERCA pump isoforms. CaMKIIγ isoforms are downregulated and CaMKIIδ upregulated in the switch to the synthetic phenotype. CaMKIIδ promotes synthetic phenotype functions (gene transcription, migration, proliferation) and ultimately vascular remodeling. CaMKIIγ promotes contraction and inhibits synthetic VSM phenotype function.

CONFLICT OF INTEREST

The authors have no conflicts of interest to disclose.

REFERENCES

Abraham, S. T., Benscoter, H., Schworer, C. M., & Singer, H. A. (1996). In situ Ca^{2+} dependence for activation of Ca^{2+}/calmodulin-dependent protein kinase II in vascular smooth muscle cells. *The Journal of Biological Chemistry*, *271*(5), 2506–2513.

Abraham, S. T., Benscoter, H. A., Schworer, C. M., & Singer, H. A. (1997). A role for Ca^{2+}/calmodulin-dependent protein kinase II in the mitogen-activated protein kinase signaling cascade of cultured rat aortic vascular smooth muscle. *Circulation Research*, *81*(4), 575–584.

Anderson, M. E. (2015). Oxidant stress promotes disease by activating CaMKII. *Journal of Molecular and Cellular Cardiology*, *89*, 160–167.

Backs, J., Backs, T., Neef, S., Kreusser, M. M., Lehmann, L. H., Patrick, D. M., … Olson, E. N. (2009). The delta isoform of CaM kinase II is required for pathological cardiac hypertrophy and remodeling after pressure overload. *Proceedings of the National Academy of Sciences of the United States of America*, *106*(7), 2342–2347.

Backs, J., Stein, P., Backs, T., Duncan, F. E., Grueter, C. E., McAnally, J., … Olson, E. N. (2010). The gamma isoform of CaM kinase II controls mouse egg activation by regulating cell cycle resumption. *Proceedings of the National Academy of Sciences of the United States of America*, *107*(1), 81–86.

Barlow, C. A., Rose, P., Pulver-Kaste, R. A., & Lounsbury, K. M. (2006). Excitation-transcription coupling in smooth muscle. *The Journal of Physiology*, *570*(Pt. 1), 59–64.

Bayer, K. U., Harbers, K., & Schulman, H. (1998). αKAP is an anchoring protein for a novel CaM kinase II isoform in skeletal muscle. *The EMBO Journal*, *17*(19), 5598–5605.

Bhattacharyya, M., Stratton, M. M., Going, C. C., McSpadden, E. D., Huang, Y., Susa, A. C., … Kuriyan, J. (2016). Molecular mechanism of activation-triggered subunit exchange in Ca^{2+}/calmodulin-dependent protein kinase II. *eLife*, *5*, 1–32.

Bilato, C., Pauly, R. R., Melillo, G., Monticone, R., Gorelick-Feldman, D., Gluzband, Y. A., … Crow, M. T. (1995). Intracellular signaling pathways required for rat vascular smooth muscle cell migration. Interactions between basic fibroblast growth factor and platelet-derived growth factor. *The Journal of Clinical Investigation*, *96*(4), 1905–1915.

Bisaillon, J. M., Motiani, R. K., Gonzalez-Cobos, J. C., Potier, M., Halligan, K. E., Alzawahra, W. F., … Trebak, M. (2010). Essential role for STIM1/Orai1-mediated calcium influx in PDGF-induced smooth muscle migration. *American Journal of Physiology. Cell Physiology*, *298*(5), C993–C1005.

Black, B. L., & Olson, E. N. (1998). Transcriptional control of muscle development by myocyte enhancer factor-2 (MEF2) proteins. *Annual Review of Cell and Developmental Biology*, *14*(1), 167–196.

Bodrikov, V., Leshchyns'ka, I., Sytnyk, V., Overvoorde, J., Den Hertog, J., & Schachner, M. (2005). RPTPα is essential for NCAM-mediated p59 FYN activation and neurite elongation. *Journal of Cell Biology*, *168*(1), 127–139.

Bodrikov, V., Sytnyk, V., Leshchyns'ka, I., den Hertog, J., & Schachner, M. (2008). NCAM induces CaMKIIalpha-mediated RPTPalpha phosphorylation to enhance its catalytic activity and neurite outgrowth. *The Journal of Cell Biology*, *182*(6), 1185–1200.

Brozovich, F. V., Nicholson, C. J., Degen, C. V., Gao, Y. Z., Aggarwal, M., & Morgan, K. G. (2016). Mechanisms of vascular smooth muscle contraction and the basis for pharmacologic treatment of smooth muscle disorders. *Pharmacological Reviews*, *68*(2), 476–532.

Cipolletta, E., Rusciano, M. R., Maione, A. S., Santulli, G., Sorriento, D., Del Giudice, C., … Illario, M. (2015). Targeting the CaMKII/ERK interaction in the heart prevents cardiac hypertrophy. *PLoS One*, *10*(6), 1–23.

Colbran, R. J. (2004). Targeting of calcium/calmodulin-dependent protein kinase II. *The Biochemical Journal*, *378*(Pt. 1), 1–16.

Colbran, R. J., & Soderling, R. (1990). Calcium/calmodulin-independent autophosphorylation sites of calcium/calmodulin-dependent protein kinase II. Studies on the effect of phosphorylation of threonine 305/306 and serine 314 on calmodulin binding using synthetic peptides. *The Journal of Biological Chemistry*, *265*(19), 11213–11219.

Coultrap, S. J., Buard, I., Kulbe, J. R., Dell'Acqua, M. L., & Bayer, K. U. (2010). CaMKII autonomy is substrate-dependent and further stimulated by Ca^{2+}/calmodulin. *The Journal of Biological Chemistry*, *285*(23), 17930–17937.

Coultrap, S. J., Zaegel, V., & Bayer, K. U. (2014). CaMKII isoforms differ in their specific requirements for regulation by nitric oxide. *FEBS Letters*, *588*(24), 4672–4676.

Di Pietro, N., Di Tomo, P., Di Silvestre, S., Giardinelli, A., Pipino, C., Morabito, C., … Pandolfi, A. (2013). Increased iNOS activity in vascular smooth muscle cells from diabetic rats: Potential role of Ca(2+)/calmodulin-dependent protein kinase II delta 2 (CaMKIIδ(2)). *Atherosclerosis*, *226*(1), 88–94.

Easley, C. A., Brown, C. M., Horwitz, A. F., & Tombes, R. M. (2008). CaMK-II promotes focal adhesion turnover and cell motility by inducing tyrosine dephosphorylation of FAK and paxillin. *Cell Motility and the Cytoskeleton*, *65*(8), 662–674.

Edman, C. F., & Schulman, H. (1994). Identification and characterization of delta B-CaM kinase and delta C-CaM kinase from rat heart, two new multifunctional Ca^{2+}/calmodulin-dependent protein kinase isoforms. *Biochimica et Biophysica Acta, 1221*, 89–101.

Erickson, J. R., Joiner, M. A., Guan, X., Kutschke, W., Yang, J., Oddis, C. V., … Anderson, M. E. (2008). A dynamic pathway for calcium-independent activation of CaMKII by methionine oxidation. *Cell, 133*(3), 462–474.

Erickson, J. R., Nichols, C. B., Uchinoumi, H., Stein, M. L., Bossuyt, J., & Bers, D. M. (2015). S-nitrosylation induces both autonomous activation and inhibition of calcium/calmodulin-dependent protein kinase II δ. *Journal of Biological Chemistry, 290*(42), 25646–25656.

Firulli, A. B., Miano, J. M., Bi, W., Johnson, A. D., Casscells, W., Olson, E. N., & Schwarz, J. J. (1996). Myocyte enhancer binding factor-2 expression and activity in vascular smooth muscle cells. Association with the activated phenotype. *Circulation Research, 78*(2), 196–204.

Gaertner, T. R., Kolodziej, S. J., Wang, D., Kobayashi, R., Koomen, J. M., Stoops, J. K., & Waxham, M. N. (2004). Comparative analyses of the three-dimensional structures and enzymatic properties of alpha, beta, gamma and delta isoforms of Ca^{2+}-calmodulin-dependent protein kinase II. *The Journal of Biological Chemistry, 279*(13), 12484–12494.

Gangopadhyay, S. S., Barber, A. L., Gallant, C., Grabarek, Z., Smith, J. L., & Morgan, K. G. (2003). Differential functional properties of calmodulin-dependent protein kinase IIgamma variants isolated from smooth muscle. *The Biochemical Journal, 372*(Pt. 2), 347–357.

Gerthoffer, W. T. (2007). Mechanisms of vascular smooth muscle cell migration. *Circulation Research, 100*(5), 607–621.

Giese, K. P., Fedorov, N. B., Filipkowski, R. K., & Silva, A. J. (1998). Autophosphorylation at Thr286 of the alpha calcium-calmodulin kinase II in LTP and learning. *Science (New York, NY), 279*(5352), 870–873.

Ginnan, R., & Singer, H. A. (2002). CaM kinase II-dependent activation of tyrosine kinases and ERK1/2 in vascular smooth muscle. *American Journal of Physiology. Cell Physiology, 282*(4), C754–C761.

Ginnan, R., Sun, L. Y., Schwarz, J. J., & Singer, H. A. (2012). MEF2 is regulated by CaMKIIδ2 and a HDAC4-HDAC5 heterodimer in vascular smooth muscle cells. *The Biochemical Journal, 444*(1), 105–114.

Ginnan, R., Zou, X., Pfleiderer, P. J., Mercure, M. Z., Barroso, M., & Singer, H. A. (2013). Vascular smooth muscle cell motility is mediated by a physical and functional interaction of Ca^{2+}/calmodulin-dependent protein kinase IIδ2 and FYN. *The Journal of Biological Chemistry, 288*(41), 29703–29712.

Gray, C. B. B., & Brown, J. H. (2014). CaMKIIdelta subtypes: Localization and function. *Frontiers in Pharmacology, 5*, 1–8.

Hagiwara, T., Ohsako, S., & Yamauchi, T. (1991). Studies on the regulatory domain of Ca^{2+}/calmodulin-dependent protein kinase I1 by expression of mutated cDNAs in Escherichia coli. *The Journal of Biological Chemistry, 266*(25), 16401–16408.

Heist, E. K., Srinivasan, M., & Schulman, H. (1998). Phosphorylation at the nuclear localization signal of Ca^{2+}/calmodulin-dependent protein kinase II blocks its nuclear targeting. *The Journal of Biological Chemistry, 273*(31), 19763–19771.

House, S. J., Ginnan, R. G., Armstrong, S. E., & Singer, H. A. (2007). Calcium/calmodulin-dependent protein kinase II-delta isoform regulation of vascular smooth muscle cell proliferation. *American Journal of Physiology. Cell Physiology. 292*(6), C2276–C2287.

House, S. J., Potier, M., Bisaillon, J., Singer, H. A., & Trebak, M. (2008). The non-excitable smooth muscle: Calcium signaling and phenotypic switching during vascular disease. *Pflügers Archiv/European Journal of Physiology, 456*(5), 769–785.

House, S. J., & Singer, H. A. (2008). CaMKII-delta isoform regulation of neointima formation after vascular injury. *Arteriosclerosis, Thrombosis, and Vascular Biology, 28*(3), 441–447.
House, S. J., Zachar, M. M., Ginnan, R. G., Van Riper, D., & Singer, H. A. (2008). Ca^{2+}/calmodulin-dependent protein kinase II signaling in vascular smooth muscle. In A. K. -Srivastava & M. B. Anand-Srivastava (Eds.), *Advances in biochemistry in health and disease. Signal transduction in the cardiovascular system in health and disease* (pp. 339–355). New York: Springer.
Hudmon, A., & Schulman, H. (2002). Neuronal Ca^{2+}/calmodulin-dependent protein kinase II: The role of structure and autoregulation in cellular function. *Annual Review of Biochemistry, 71*, 473–510.
Illario, M., Cavallo, A. L., Bayer, K. U., Di Matola, T., Fenzi, G., Rossi, G., & Vitale, M. (2003). Calcium/calmodulin-dependent protein kinase II binds to Raf-1 and modulates integrin-stimulated ERK activation. *Journal of Biological Chemistry, 278*(46), 45101–45108.
Iyoda, T., Zhang, F., Sun, L., Hao, F., Schmitz-Peiffer, C., Xu, X., & Cui, M.-Z. (2012). Lysophosphatidic acid induces early growth response-1 (Egr-1) protein expression via protein kinase Cδ-regulated extracellular signal-regulated kinase (ERK) and c-Jun N-terminal kinase (JNK) activation in vascular smooth muscle cells. *The Journal of Biological Chemistry, 287*(27), 22635–22642.
Jones, R. J., Jourd'heuil, D., Salerno, J. C., Smith, S. M. E., & Singer, H. A. (2007). iNOS regulation by calcium/calmodulin-dependent protein kinase II in vascular smooth muscle. *American Journal of Physiology. Heart and Circulatory Physiology, 292*(6), H2634–H2642.
Kamm, K. E., & Stull, J. T. (2001). Dedicated myosin light chain kinases with diverse cellular functions. *Journal of Biological Chemistry, 276*(7), 4527–4530.
Kim, I., Je, H. D., Gallant, C., Zhan, Q., Riper, D. V., Badwey, J. A., … Morgan, K. G. (2000). Ca^{2+}-calmodulin-dependent protein kinase II-dependent activation of contractility in ferret aorta. *The Journal of Physiology, 526*(Pt. 2), 367–374.
Kostic, A., & Sheetz, M. P. (2006). Fibronectin rigidity response through FYN and p130Cas recruitment to the leading edge. *Molecular Biology of the Cell, 17*(2), 2684–2695.
Lamalice, L., Houle, F., & Huot, J. (2006). Phosphorylation of Tyr1214 within VEGFR-2 triggers the recruitment of Nck and activation of FYN leading to SAPK2/p38 activation and endothelial cell migration in response to VEGF. *Journal of Biological Chemistry, 281*(45), 34009–34020.
Lantsman, K., & Tombes, R. M. (2005). CaMK-II oligomerization potential determined using CFP/YFP FRET. *Biochimica et Biophysica Acta—Molecular Cell Research, 1746*(1), 45–54.
Leary, H. O., Lasda, E., & Bayer, K. U. (2006). CaMKIIbeta association with the actin cytoskeleton is regulated by alternative splicing. *Molecular Biology of the Cell, 17*(11), 4656–4665.
Ledoux, J., Chartier, D., & Leblanc, N. (1999). Inhibitors of calmodulin-dependent protein kinase are nonspecific blockers of voltage-dependent K^+ channels in vascular myocytes. *The Journal of Pharmacology and Experimental Therapeutics, 290*(3), 1165–1174.
Li, H., Li, W., Gupta, A. K., Mohler, P. J., Anderson, M. E., & Grumbach, I. M. (2010). Calmodulin kinase II is required for angiotensin II-mediated vascular smooth muscle hypertrophy. *American Journal of Physiology Heart and Circulatory Physiology, 298*(2), H688–H698.
Li, W., Li, H., Sanders, P. N., Mohler, P. J., Backs, J., Olson, E. N., … Anderson, E. (2011). The multifunctional Ca^{2+}/calmodulin-dependent kinase II delta (CaMKIIdelta) controls neointima formation after carotid ligation and vascular smooth muscle cell proliferation through cell cycle regulation by p21. *The Journal of Biological Chemistry, 286*(10), 7990–7999.

Liu, Y. F., Spinelli, A., Sun, L. Y., Jiang, M., Singer, D. V., Ginnan, R., … Singer, H. A. (2016). MicroRNA-30 inhibits neointimal hyperplasia by targeting Ca(2+)/calmodulin-dependent protein kinase IIdelta (CaMKIIdelta). *Scientific Reports*, *6*, 26166.

Liu, Y., Sun, L.-Y., Singer, D. V., Ginnan, R., & Singer, H. A. (2013). CaMKIIδ-dependent inhibition of cAMP-response element-binding protein activity in vascular smooth muscle. *The Journal of Biological Chemistry*, *288*(47), 33519–33529.

Lu, K. K., Armstrong, S. E., Ginnan, R., & Singer, H. A. (2005). Adhesion-dependent activation of CaMKII and regulation of ERK activation in vascular smooth muscle. *American Journal of Physiology. Cell Physiology*, *289*(5), C1343–C1350.

Lu, D. Y., Chen, E. Y., Wong, D. J., Yamamoto, K., Protack, C. D., Williams, W. T., … Dardik, A. (2014). Vein graft adaptation and fistula maturation in the arterial environment. *Journal of Surgical Research*, *188*(1), 162–173.

Ma, H., Groth, R. D., Wheeler, D. G., Barrett, C. F., & Tsien, R. W. (2011). Excitation-transcription coupling in sympathetic neurons and the molecular mechanism of its initiation. *Neuroscience Research*, *70*(1), 2–8.

Ma, H., & Tsien, R. W. (2014). γCaMKII shuttles Ca^{2+}/CaM to the nucleus to trigger CREB phosphorylation and gene expression. *Cell*, *159*, 281–294.

Marganski, W. A., Gangopadhyay, S. S., Je, H.-D., Gallant, C., & Morgan, K. G. (2005). Targeting of a novel Ca^{2+}/calmodulin-dependent protein kinase II is essential for extracellular signal-regulated kinase-mediated signaling in differentiated smooth muscle cells. *Circulation Research*, *97*(6), 541–549.

Mercure, M. Z., Ginnan, R., & Singer, H. A. (2008). CaM kinase II delta2-dependent regulation of vascular smooth muscle cell polarization and migration. *American Journal of Physiology. Cell Physiology*, *294*(6), C1465–C1475.

Meyer, T., Hanson, P. I., Stryer, L., & Schulman, H. (1992). Calmodulin trapping by calcium-calmodulin-dependent protein kinase. *Science*, *256*(5060), 1199–1202.

Mishra-gorur, K., Singer, H. A., & Castellot, J. J. (2002). Heparin inhibits phosphorylation and autonomous activity of Ca^{2+}/calmodulin-dependent protein kinase II in vascular smooth muscle cells. *The American Journal of Pathology*, *161*(5), 1893–1901.

Mollova, M. Y., Katus, H. A., & Backs, J. (2015). Regulation of CaMKII signaling in cardiovascular disease. *Frontiers in Pharmacology*, *6*, 1–8.

Mukherji, S., & Soderling, T. R. (1994). Regulation of Ca^{2+}/calmodulin-dependent protein-kinase II by inter- and intrasubunit-catalyzed autophosphorylations. *Journal of Biological Chemistry*, *269*(19), 13744–13747.

Munevar, S., Gangopadhyay, S. S., Gallant, C., Colombo, B., Sellke, F. W., & Morgan, K. G. (2008). CaMKIIT287 and T305 regulate history-dependent increases in alpha agonist-induced vascular tone. *Journal of Cellular and Molecular Medicine*, *12*(1), 219–226.

Muthalif, M. M., Benter, I. F., Uddin, M. R., & Malik, K. U. (1996). Calcium/calmodulin-dependent protein kinase II α mediates activation of mitogen-activated protein kinase and cytosolic phospholipase A2 in norepinephrine-induced arachidonic acid release in rabbit aortic smooth muscle cells. *The Journal of Biological Chemistry*, *47*(271), 30149–30157.

Muthalif, M. M., Karzoun, N. A., Benter, I. F., Gaber, L., Ljuca, F., Uddin, M. R., … Malik, K. U. (2002). Functional significance of activation of calcium/calmodulin-dependent protein kinase II in angiotensin II-induced vascular hyperplasia and hypertension. *Hypertension*, *39*(2 Pt. 2), 704–709.

Nakano, H., Shindo, M., Sakon, S., Nishinaka, S., Mihara, M., Yagita, H., & Okumura, K. (1998). Differential regulation of IkappaB kinase alpha and beta by two upstream kinases, NF-kappaB-inducing kinase and mitogen-activated protein kinase/ERK kinase kinase-1. *Proceedings of the National Academy of Sciences of the United States of America*, *95*(7), 3537–3542.

Owens, G. K., Kumar, M. S., & Wamhoff, B. R. (2004). Molecular regulation of vascular smooth muscle cell differentiation in development and disease. *Physiological Reviews*, *84*, 767–801.

Pang, J., Yan, C., Natarajan, K., Cavet, M. E., Massett, M. P., Yin, G., & Berk, B. C. (2008). GIT1 mediates HDAC5 activation by angiotensin II in vascular smooth muscle cells. *Arteriosclerosis, Thrombosis, and Vascular Biology*, *28*(5), 892–898.

Parsons, J. T., Horwitz, A. R., & Schwartz, M. A. (2010). Cell adhesion: Integrating cytoskeletal dynamics and cellular tension. *Nature Reviews. Molecular Cell Biology*, *11*(9), 633–643.

Perrino, B. A. (2011). Regulation of gastrointestinal motility by Ca^{2+}/calmodulin-stimulated protein kinase II. *Archives of Biochemistry and Biophysics*, *510*(2), 174–181.

Prasad, A. M., Morgan, D. A., Nuno, D. W., Ketsawatsomkron, P., Bair, T. B., Venema, A. N., … Grumbach, I. M. (2015). Calcium/calmodulin-dependent kinase II inhibition in smooth muscle reduces angiotensin II-induced hypertension by controlling aortic remodeling and baroreceptor function. *Journal of the American Heart Association*, *4*(6), e001949.

Prasad, A. M., Nuno, D. W., Koval, O. M., Ketsawatsomkron, P., Li, W., Li, H., … Grumbach, I. M. (2013). Differential control of calcium homeostasis and vascular reactivity by Ca^{2+}/calmodulin dependent kinase II. *Hypertension*, *62*(2), 434–441.

Pulver, R. A., Rose-Curtis, P., Roe, M. W., Wellman, G. C., & Lounsbury, K. M. (2004). Store-operated Ca^{2+} entry activates the CREB transcription factor in vascular smooth muscle. *Circulation Research*, *94*(10), 1351–1358.

Rokolya, A., & Singer, H. A. (2000). Inhibition of CaM kinase II activation and force maintenance by KN-93 in arterial smooth muscle. *American Journal of Physiology. Cell Physiology*, *278*(3), C537–C545.

Saddouk, F. Z., Sun, L., Liu, Y. F., Jiang, M., Singer, D. V., Backs, J., … Singer, H. A. (2016). Ca^{2+}/calmodulin-dependent protein kinase II-g (CaMKII g) negatively regulates vascular smooth muscle cell proliferation and vascular remodeling. *The FASEB Journal*, *30*, 1051–1064.

Schauer, I. E., Knaub, L. A., Lloyd, M., Watson, P. A., Gliwa, C., Lewis, K. E., … Reusch, J. E. B. (2010). CREB downregulation in vascular disease: A common response to cardiovascular risk. *Arteriosclerosis, Thrombosis, and Vascular Biology*, *30*(4), 733–741.

Schauwienold, D., Sastre, A. P., Genzel, N., Schaefer, M., & Reusch, H. P. (2008). The transactivated epidermal growth factor receptor recruits Pyk2 to regulate Src kinase activity. *Journal of Biological Chemistry*, *283*(41), 27748–27756.

Schworer, C. M., Colbran, R. J., Keefer, J. R., & Soderlings, T. R. (1988). Ca^{2+}/calmodulin-dependent protein kinase II. *The Journal of Biological Chemistry*, *263*(27), 13486–13489.

Schworer, C. M., Rothblum, I., Thekkumkara, T. J., & Singer, H. A. (1993). Identification of novel isoforms of the δ-subunit of Ca^{2+}/calmodulin-dependent protein kinase II. Differential expression in rat brain and aorta. *Journal of Biological Chemistry*, *268*(19), 14443–14449.

Scott, J. A., Klutho, P. J., El Accaoui, R., Nguyen, E., Venema, A. N., Xie, L., … Grumbach, I. M. (2013). The multifunctional Ca^{2+}/calmodulin-dependent kinase IIδ (CaMKIIδ) regulates arteriogenesis in a mouse model of flow-mediated remodeling. *PLoS One*, *8*(8), e71550.

Scott, J. A., Xie, L., Li, H., Li, W., He, J. B., Sanders, P. N., … Grumbach, I. M. (2012). The multifunctional Ca^{2+}/calmodulin-dependent kinase II regulates vascular smooth muscle migration through matrix metalloproteinase 9. *American Journal of Physiology. Heart and Circulatory Physiology*, *302*(10), H1953–H1964.

Singer, H. A., Abraham, S. T., & Schworer, C. M. (1996). Calcium/calmodulin-dependent protein kinase II. In M. Barany (Ed.), *Biochemistry of smooth muscle contraction* (pp. 143–153). New York: Academic Press.

Singer, H. A., Benscoter, H. A., & Schworer, C. M. (1997). Novel Ca^{2+}/calmodulin-dependent protein kinase II gamma-subunit variants expressed in vascular smooth muscle, brain, and cardiomyocytes. *Journal of Biological Chemistry, 272*(14), 9393–9400.

Skelding, K. A., Rostas, J. A. P., & Verrills, N. M. (2011). Controlling the cell cycle: The role of calcium/calmodulin-stimulated protein kinases I and II. *Cell Cycle, 10*(4), 631–639.

Soderling, T., & Stull, J. (2001). Structure and regulation of calcium/calmodulin-dependent protein kinases. *Chemical Reviews, 101*(8), 2341–2352.

Srinivasan, M., Edman, C. F., & Schulman, H. (1994). Alternative splicing introduces a nuclear localization signal that targets multifunctional CaM kinase to the nucleus. *The Journal of Cell Biology, 126*(4), 839–852.

Stark, G. R., & Taylor, W. R. (2006). Control of the G2/M transition. *Molecular Biotechnology, 32*(3), 227–248.

Sun, P., Enslen, H., Myung, P. S., & Maurer, R. A. (1994). Differential activation of CREB by Ca^{2+}/calmodulin-dependent protein kinases type II and type IV involves phosphorylation of a site that negatively regulates activity. *Genes & Development, 8*(21), 2527–2539.

Tang, Z., Wang, A., Yuan, F., Yan, Z., Liu, B., Chu, J. S., ... Li, S. (2012). Differentiation of multipotent vascular stem cells contributes to vascular diseases. *Nature Communications, 3*, 875.

Tansey, M. G., Word, R. A., Hidaka, H., Singer, H. A., Schworer, C. M., Kamm, K. E., & Stull, J. T. (1992). Phosphorylation of myosin light chain kinase by the multifunctional calmodulin-dependent protein kinase II in smooth muscle cells. *The Journal of Biological Chemistry, 267*(18), 12511–12516.

Thiel, G., Mayer, S. I., Müller, I., Stefano, L., & Rössler, O. G. (2010). Egr-1-A Ca(2+)-regulated transcription factor. *Cell Calcium, 47*(5), 397–403.

Tombes, R. M., Faison, M. O., & Turbeville, J. (2003). Organization and evolution of multifunctional Ca^{2+}/CaM-dependent protein kinase genes. *Gene, 322*(2003), 17–31.

Turczyńska, K. M., Sadegh, M. K., Hellstrand, P., Swärd, K., Albinsson, S., Turczynska, K. M., ... Albinsson, S. (2012). MicroRNAs are essential for stretch-induced vascular smooth muscle contractile differentiation via microRNA (miR)-145-dependent expression of L-type calcium channels. *Journal of Biological Chemistry, 287*(23), 19199–19206.

Usui, T., Morita, T., Okada, M., & Yamawaki, H. (2014). Histone deacetylase 4 controls neointimal hyperplasia via stimulating proliferation and migration of vascular smooth muscle cells. *Hypertension, 63*(2), 397–403.

Vacaresse, N., Moller, B., Danielsen, E. M., Okada, M., & Sap, J. (2008). Activation of c-Src and FYN kinases by protein-tyrosine phosphatase RPTPalpha is substrate-specific and compatible with lipid raft localization. *Journal of Biological Chemistry, 283*(51), 35815–35824.

Van Riper, D., Schworer, C. M., & Singer, H. A. (2000). Ca^{2+}-induced redistribution of Ca^{2+}/calmodulin-dependent protein kinase II associated with an endoplasmic reticulum stress response in vascular smooth muscle. *Molecular and Cellular Biochemistry, 213*(1-2), 83–92.

Van Riper, D. A., Weaver, B. A., Stull, J. T., & Rembold, C. M. (1995). Myosin light chain kinase phosphorylation in swine carotid artery contraction and relaxation. *The American Journal of Physiology, 268*(6 Pt. 2), H2466–H2475.

Wayman, G. A., Lee, Y.-S., Tokumitsu, H., Silva, A. J., & Soderling, T. R. (2008). Calmodulin-kinases: Modulators of neuronal development and plasticity. *Neuron, 59*(6), 914–931.

Wei, C., Wang, X., Chen, M., Ouyang, K., Song, L.-S., & Cheng, H. (2009). Calcium flickers steer cell migration. *Nature, 457*(7231), 901–905.
Wei, C., Wang, X., Zheng, M., & Cheng, H. (2012). Calcium gradients underlying cell migration. *Current Opinion in Cell Biology, 24*(2), 254–261.
Winters, C. J., Koval, O., Murthy, S., Allamargot, C., Sebag, S. C., Paschke, J. D., … Grumbach, I. M. (2016). CaMKII inhibition in type II pneumocytes protects from bleomycin-induced pulmonary fibrosis by preventing Ca^{2+}-dependent apoptosis. *American Journal of Physiology. Lung Cellular and Molecular Physiology, 310*(1), L86–L94.
Wirth, A., Benyó, Z., Lukasova, M., Leutgeb, B., Wettschureck, N., Gorbey, S., … Offermanns, S. (2008). G12-G13-LARG-mediated signaling in vascular smooth muscle is required for salt-induced hypertension. *Nature Medicine, 14*(1), 64–68.
Yang, Z., Kirton, H. M., Macdougall, D. A., Boyle, J. P., Deuchars, J., Frater, B., … Steele, D. S. (2015). The Golgi apparatus is a functionally distinct Ca^{2+} store regulated by the PKA and Epac branches of the β1-adrenergic signaling pathway. *Science Signaling, 8*(398), 1–12.
Yilmaz, M., Gangopadhyay, S. S., Leavis, P., Grabarek, Z., & Morgan, K. G. (2013). Phosphorylation at Ser26 in the ATP-binding site of Ca^{2+}/calmodulin-dependent kinase II as a mechanism for switching off the kinase activity. *Bioscience Reports, 33*(2), 259–268.
Zeng, L., Si, X., Yu, W. P., Thi Le, H., Ng, K. P., Teng, R. M. H., … Pallen, C. J. (2003). PTPα regulates integrin-stimulated FAK autophosphorylation and cytoskeletal rearrangement in cell spreading and migration. *Journal of Cell Biology, 160*(1), 137–146.
Zhang, W., Halligan, K. E., Zhang, X., Bisaillon, J. M., Gonzalez-Cobos, J. C., Motiani, R. K., … Trebak, M. (2011). Orai1-mediated ICRAC is essential for neointima formation after vascular injury. *Circulation Research, 109*(5), 534–542.
Zhang, T., Kohlhaas, M., Backs, J., Mishra, S., Phillips, W., Dybkova, N., … Brown, J. H. (2007). CaMKIIdelta isoforms differentially affect calcium handling but similarly regulate HDAC/MEF2 transcriptional responses. *The Journal of Biological Chemistry, 282*(48), 35078–35087.
Zhang, T., Zhang, Y., Cui, M., Jin, L., Wang, Y., Lv, F., … Xiao, R.-P. (2016). CaMKII is a RIP3 substrate mediating ischemia- and oxidative stress-induced myocardial necroptosis. *Nature Medicine, 22*(2), 175–182.
Zhou, Z. L., & Ikebe, M. (1994). New isoforms of Ca^{2+}/calmodulin-dependent protein kinase II in smooth muscle. *Biochemical Journal, 299*, 489–495.
Zhu, L. J., Klutho, P. J., Scott, J. A., Xie, L., Luczak, E. D., Dibbern, M. E., … Grumbach, I. M. (2014). Oxidative activation of the Ca(2+)/calmodulin-dependent protein kinase II (CaMKII) regulates vascular smooth muscle migration and apoptosis. *Vascular Pharmacology, 60*(2), 75–83.
Zhu, W., Woo, A. Y. H., Yang, D., Cheng, H., Crow, M. T., & Xiao, R. P. (2007). Activation of CaMKIIdeltaC is a common intermediate of diverse death stimuli-induced heart muscle cell apoptosis. *Journal of Biological Chemistry, 282*(14), 10833–10839.

CHAPTER SIX

Protein Kinase C as Regulator of Vascular Smooth Muscle Function and Potential Target in Vascular Disorders

H.C. Ringvold, R.A. Khalil[1]
Brigham and Women's Hospital, Harvard Medical School, Boston, MA, United States
[1]Corresponding author: e-mail address: raouf_khalil@hms.harvard.edu

Contents

Advances in Pharmacology, Volume 78
ISSN 1054-3589
http://dx.doi.org/10.1016/bs.apha.2016.06.002

Abstract

Vascular smooth muscle (VSM) plays an important role in maintaining vascular tone. In addition to Ca^{2+}-dependent myosin light chain (MLC) phosphorylation, protein kinase C (PKC) is a major regulator of VSM function. PKC is a family of conventional Ca^{2+}-dependent α, β, and γ, novel Ca^{2+}-independent δ, ε, θ, and η, and atypical ξ, and ι/λ isoforms. Inactive PKC is mainly cytosolic, and upon activation it undergoes phosphorylation, maturation, and translocation to the surface membrane, the nucleus, endoplasmic reticulum, and other cell organelles; a process facilitated by scaffold proteins such as RACKs. Activated PKC phosphorylates different substrates including ion channels, pumps, and nuclear proteins. PKC also phosphorylates CPI-17 leading to inhibition of MLC phosphatase, increased MLC phosphorylation, and enhanced VSM contraction. PKC could also initiate a cascade of protein kinases leading to phosphorylation of the actin-binding proteins calponin and caldesmon, increased actin–myosin interaction, and VSM contraction. Increased PKC activity has been associated with vascular disorders including ischemia–reperfusion injury, coronary artery disease, hypertension, and diabetic vasculopathy. PKC inhibitors could test the role of PKC in different systems and could reduce PKC hyperactivity in vascular disorders. First-generation PKC inhibitors such as staurosporine and chelerythrine are not very specific. Isoform-specific PKC inhibitors such as ruboxistaurin have been tested in clinical trials. Target delivery of PKC pseudosubstrate inhibitory peptides and PKC siRNA may be useful in localized vascular disease. Further studies of PKC and its role in VSM should help design isoform-specific PKC modulators that are experimentally potent and clinically safe to target PKC in vascular disease.

ABBREVIATIONS

AGE advanced glycation end products
ALDH2 aldehyde dehydrogenase 2
AngII angiotensin II
BK_{Ca} large conductance Ca^{2+}-activated K^+ channel
BP blood pressure
Ca^{2+} calcium
$[Ca^{2+}]_i$ intracellular free Ca^{2+} concentration
CaD caldesmon
CAM calmodulin
CaP calponin
cAMP cyclic adenosine monophosphate
cGMP cyclic guanosine monophosphate

CNP C-type natriuretic peptide
CPI-17 PKC-potentiated phosphatase inhibitor protein-17
DAG diacyglycerol
ECM extracellular matrix
ER endoplasmic reticulum
ERK extracellular signal-regulated kinase
ET-1 endothelin-1
HSP heat shock protein
HR1 homology region 1
ICAM-1 intercellular adhesion molecule-1
IP_3 inositol 1,4,5-trisphosphate
IRS1 insulin receptor substrate 1
K_v voltage-gated K^+ channel
MARCKS myristoylated alanine-rich C kinase substrate
MCP-1 monocyte chemoattractant protein-1
MLC myosin light chain
PDBu phorbol 12,13-dibutyrate
PDGF platelet-derived growth factor
PDK phosphoinositide-dependent kinase
PKC protein kinase C
PKG cGMP-dependent protein kinase
PMA phorbol 12-myristate 13-acetate
PLC phospholipase C
PS phosphatidylserine
RAGE AGE receptor
ROS reactive oxygen species
VCAM-1 vascular cell adhesion molecule-1
VEGF vascular endothelial growth factor
VGCC voltage-gated Ca^{2+} channel
VSM vascular smooth muscle

1. INTRODUCTION

Protein kinase C (PKC) is a ubiquitous enzyme found in almost all cell types including the endothelium, vascular smooth muscle (VSM), and fibroblasts of blood vessels. PKC phosphorylates serine and threonine residues in a large number of protein substrates and regulates many cellular processes. PKC exhibits significant and sometimes opposite effects in different tissues and is widely implicated in multiple physiological and pathological processes. The versatility of the effects of PKC is best illustrated in the observation that it could induce both vascular contraction and relaxation. For instance, PKC may mediate the release of endothelium-derived contracting factors such as endothelin-1 (ET-1) and promote vasoconstriction, but could also mediate endothelial nitric oxide (NO) synthesis and promote vasodilation (Wang et al., 2015). PKC could also affect vascular fibroblasts causing increases in transforming growth factor-β, and extracellular matrix (ECM) production, thus promoting vascular remodeling (Ding, Chai, et al., 2011; Geraldes & King, 2010). In addition to its effects on the endothelium and ECM, PKC plays a major role in the regulation of VSM function.

Several excellent reviews have discussed many of the biochemical aspects of PKC and their substrates (Kanashiro & Khalil, 1998; Khalil, 2013; Mochly-Rosen, Das, & Grimes, 2012; Mukherjee, Roy, Saha, & Mukherjee, 2016; Newton, 2010; Nishizuka, 1992). Also, the multiple effects of PKC in different cellular processes have made it an important target in many diseases. Understanding the basic biochemical properties of PKC and its effects in the vascular system should help to provide the basis for targeting PKC in different vascular disorders. The purpose of this chapter is to highlight the role of PKC as a major regulator of VSM function with emphasis on recent discoveries and their relevance to vascular disease. We used data published in PubMed and other databases, as well as data from our laboratory to first provide a brief background on PKC biochemistry, its different isoforms, tissue distribution, substrates, and different activators and inhibitors. We will discuss some findings that challenged the concept that PKC translocation is necessary for its activation, and other theories on how to modulate PKC activity by targeting different sites in its regulatory and catalytic domains. We will then discuss the potential role of PKC in vascular disorders and the potential benefits of PKC inhibitors in the management of vascular disease. While the focus of the chapter is on PKC in VSM, in the instances that there is little information available in VSM, the effects of PKC on other systems will be described.

2. PKC STRUCTURE AND ISOFORMS

In 1977, Nishizuka and colleagues discovered PKC in rat brain extract (Takai, Kishimoto, Inoue, & Nishizuka, 1977). PKC was initially defined as a kinase that is activated by proteolysis, but was soon found to be activated by diacyglycerol (DAG) (Takai, Kishimoto, Kikkawa, Mori, & Nishizuka, 1979), and later by phorbol ester, a tumor promoter (Castagna et al., 1982). The PKC molecule comprises a N-terminal regulatory domain and a C-terminal catalytic domain between which lies the V3 hinge region (Kishimoto et al., 1989) (Fig. 1). The conventional PKC isoforms α, βI, βII, and γ have four conserved regions (C1, C2, C3, and C4) and five variable

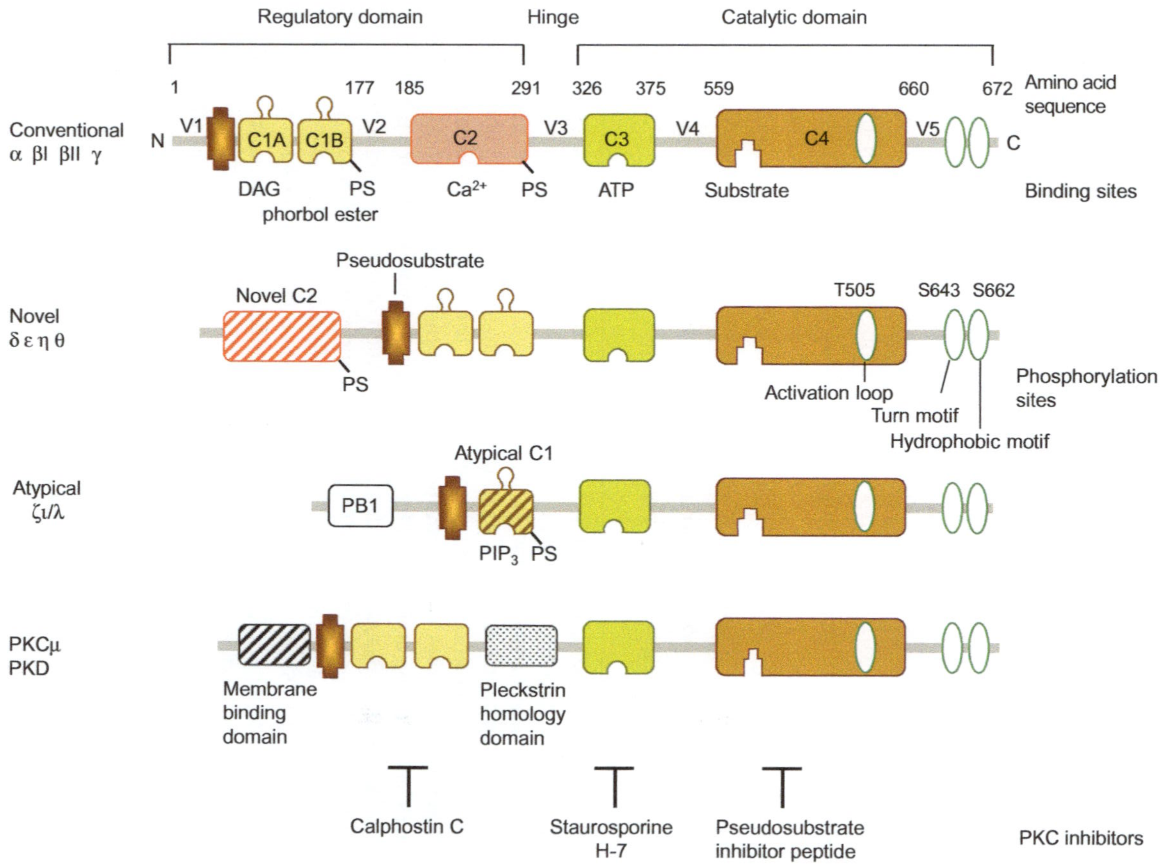

Fig. 1 See legend on next page.

regions (V1, V2, V3, V4, and V5). The regulatory domain is comprised of the two conserved C1 and C2 regions. The C1 region contains cysteine-rich zinc finger-like motifs as well as lipid-binding sites surrounded by a band of hydrophobic residues that penetrate the lipid bilayer and provide a strong driving force to anchor PKC to DAG-containing membranes. In addition to the membrane-docking interactions caused by binding of the C1 region to DAG, full-length PKC may stably associate with membranes by a second membrane-targeting interaction involving the C2 region (Steinberg, 2008). Furthermore, the C1 region binds the PKC cofactor phosphatidylserine (PS) (Kim, Gallant, & Morgan, 2013), and PS may also bind to the C2 region (Poli, Mongiorgi, Cocco, & Follo, 2014). An autoinhibitory pseudosubstrate sequence, common to all protein kinases, is embedded in the N-terminal regulatory domain immediately preceding the C1 region and has an amino acid sequence between residues 19 and 36 that resembles the PKC substrate phosphorylation site (Newton, 2001). The catalytic or kinase activity domain contains a conserved C3 region, an ATP-/Mg-binding site in a narrow hydrophobic pocket, and a binding site for the phospho-acceptor sequence in the substrate (Mochly-Rosen et al., 2012). The C4 region

Fig. 1 PKC structure. PKC contains a N-terminal "regulatory domain" and a C-terminal "catalytic domain," between which lies the V3 hinge region. The regulatory domain contains two conserved C1 and C2 regions, and the pseudosubstrate region. The catalytic or kinase activity domain contains a C3 ATP/Mg-binding site and a C4-binding site for the phospho-acceptor sequence in the substrate proteins. The catalytic domain also contains phosphorylation sites in the activation loop, turn motif, and hydrophobic motif. The figure illustrates the amino acids for PKCδ phosphorylation sites, which may vary among different PKCs. Based on the domain composition of the N-terminals, the PKC family is classified into conventional cPKCs α, βI, βII, and γ; novel nPKCs δ, ε, η, and θ; and atypical aPKCs ι/λ and ζ isoforms. cPKCs consist of four conserved (C1–C4) and five variable regions (V1–V5) and are activated by Ca^{2+}, DAG, and PS. The C1 region contains the recognition site for phosphatidylserine (PS), DAG, and phorbol esters, while the C2 region contains the binding site for Ca^{2+}. PS can also be bound by the C2 region. Both cPKCs and nPKCs have twin C1 regions (C1A and C1B) and a C2 region. The order of C1 and C2 regions is switched in nPKCs compared with cPKCs. The nPKCs have a variant form of C2 region that is insensitive to Ca^{2+} activation, but still binds lipids. The aPKCs do not have a C2 region and hence not activated by Ca^{2+} or DAG, and have a variant form of C1 that is not duplicated, but retains lipid-binding activity and are activated by DAG and PS. The aPKCs also encode the protein–protein-interacting region Phox and Bem 1 (PB1) that binds with ZIP/p62, Par6, or MEK5 through a PB1–PB1 domain interaction and hence control the localization of aPKCs. Other related kinases include PKCμ (PKD). PKC inhibitors could compete with DAG at the C1 region (calphostin) or ATP at the ATP-binding site (staurosporine) or the true PKC substrate (pseudosubstrate inhibitor peptide).

constitutes the substrate-binding part of the PKC molecule (Newton, 1995). The catalytic domain also contains three key phosphorylation and autophosphorylation sites in the C-terminal activation loop, turn motif, and hydrophobic motif.

PKC belongs to the AGC family of serine/threonine protein kinases that are related at their primary sequence and are named after the first identified "founding members" cyclic adenosine monophosphate (cAMP)-dependent protein kinase (PKA), cyclic guanosine monophosphate (cGMP)-dependent protein kinase (PKG), and PKC (Linch et al., 2014). PKC is the largest serine/threonine kinase family, comprising ~2% of the human kinome (Mellor & Parker, 1998). The PKC family is encoded by 9 different genes and consists of 10 isoforms (Hoque et al., 2014). The N-terminal regulatory domain contains a highly 60–80% homologous C1 region among different PKC isoforms (Mochly-Rosen et al., 2012). Based on the structure of the N-terminal domain, PKC isoforms are classified into conventional cPKCs α, βI, βII, and γ; novel nPKCs δ, ε, η, and θ; and atypical aPKCs ζ and ι/λ isoforms (Salamanca & Khalil, 2005) (Fig. 1). The cPKCs consist of four conserved regions (C1–C4) and five variable regions (V1–V5), and are activated by calcium (Ca^{2+}), DAG, and PS. The C1 region contains the recognition site for PS, DAG, and phorbol esters while the C2 region is rich in acidic residues and contains the binding site for Ca^{2+} (Newton, 1995). In cPKCs, the C2 region comprises ~130 residue eight-stranded antiparallel β-sandwich structures with three interstrand Ca^{2+}-binding loops responsible for Ca^{2+}-dependent anionic phospholipid binding (Steinberg, 2015). Both cPKCs and nPKCs have twin C1 regions (C1A and C1B) and a C2 region, but the ordering of the C1 and C2 regions is switched in nPKCs compared to cPKCs (Steinberg, 2015). Also, nPKCs have a variant C2 region that lacks the critical Ca^{2+}-coordinating aspartic acid residues that are highly conserved in cPKCs, making it insensitive to Ca^{2+} (Newton, 2001). As a result, the nPKC C2 region does not bind Ca^{2+}, and nPKCs are activated by DAG and PS, but not Ca^{2+}. The C1 region of nPKCs has a higher affinity for DAG than the C1 region of cPKCs, and functions as a lipid-binding membrane-targeting module in a Ca^{2+}-independent manner (Dries, Gallegos, & Newton, 2007). The C2 region of PKCδ does not bind lipids, but has a protein–protein interaction domain that binds phosphotyrosine residues flanked by the consensus sequence (Y/F)-(S/A)-(V/I)-pY-(Q/R)-X-(Y/F). PKCδ contains several tyrosine phosphorylation sites throughout its structure, including in the highly conserved regulatory and kinase domains and the intervening more flexible variable hinge region, most of which are unique to PKCδ (Benes et al., 2005). The aPKCs do not have a C2 region but have a variant form of C1 and are therefore activated by PS but not Ca^{2+} or DAG (Newton, 2001). However, aPKCs do retain lipid-binding activity, and the C1 region confers DAG binding that is not duplicated, unlike the C1A–C1B tandem repeat found in cPKCs and nPKCs (Linch et al., 2014). The aPKCs also uniquely encode the protein–protein-interacting Phox and Bem 1 (PB1) region in the N-terminus domain, which binds ZIP/p62, Par6, or MEK5 through a PB1–PB1 domain interaction that controls the localization of aPKCs (Hirano et al., 2004).

PKCμ and PKCν are often considered a fourth class of PKC isoforms or members of a different family called protein kinase D (PKD) (Poli et al., 2014; Rozengurt, 2011). Other PKC-related kinases (PRKs) include PRK1–3 and are also considered a fourth group of the PKC family (Mellor & Parker, 1998). PRKs are similar in structure to PKCs except for the C1 region, but do not bind Ca^{2+}, DAG, or phorbol esters, and have homology region 1 (HR1) motifs responsible for RhoA binding (Hage-Sleiman, Hamze, Reslan, Kobeissy, & Dbaibo, 2015).

The greatest homology between PKC isoforms is in the highly conserved catalytic domain (~70%). Also, similar to other Ser/Thr protein kinases, PKC isoforms have a highly conserved ATP-binding site. The exception to the catalytic domain homology is the variable V5 region, consisting of 60–70 different amino acids. PKC isoforms also differ in the intervening V3 hinge region and the C2 region of the regulatory domain (Mochly-Rosen et al., 2012). PKCβI and βII are generated by alternative splicing from a single gene, but differ at their C-terminal 50 (βI) or 52 (βII) residues (Newton, 1995). The amino acid for each phosphorylation site also varies in different PKCs. For example, the activation loop contains a phosphorylatable Thr497 in PKCα, T500 in PKCβII, T505 in PKCδ, and T538 in PKCθ, and the turn motif contains a T638 in PKCα, T641 in PKCβI and PKCβII, S643 in PKCδ, and S676 and S685 in PKCθ, while the hydrophobic motif contains a Ser657 in PKCα, S660 in PKCβII, S662 in PKCδ, and S695 flanked by bulky hydrophobic residues in PKCθ (Steinberg, 2008; Xu et al., 2004).

3. PKC DISTRIBUTION AND TRANSLOCATION

PKCs are found in varying amounts in different tissues and cells, including various vascular beds. PKCα, δ, and ζ are universally expressed in almost all blood vessels examined, while other PKCs show specific distribution in specific vascular beds (Kanashiro & Khalil, 1998; Khalil, 2013) (Table 1). In human VSMCs, the expression of PKCα, β, δ, and ε, but not PKCζ, is relatively high (Grange et al., 1998). In endothelial cells, however, the levels of PKCδ are lower than PKCζ, demonstrating how PKC distribution varies depending on the vascular cell type (Magid & Davies, 2005; Mattila, Majuri, Tiisala, & Renkonen, 1994). In resting cells, PKCα, β, and γ are localized mainly in the cytosolic fraction, and activated PKC undergoes translocation from the cytosolic to the particulate and membrane fraction (Kraft & Anderson, 1983) (Fig. 2); although other PKCs may show redistribution in specific cell membranes (Khalil, 2013).

The mechanisms of PKC movement from the cytosol to the membrane are not fully understood. Simple diffusion and other physicochemical forces may drive the movement of PKC inside the cell, while specific targeting mechanisms would allow its translocation to different cell membranes and tight binding to its target location. Targeting mechanisms include conformation changes, altered hydrophobicity, lipid modification, protein–protein interaction, targeting sequences, and phosphorylation (Khalil, 2013; Saito, Ito, Takakuwa, Tamura, & Kinjo, 2003).

Binding of Ca^{2+} or DAG may cause conformational changes that unfold the PKC molecule and result in exposure of the substrate region, increased PKC hydrophobicity, and binding to membrane lipids (Newton, 1995). Changes in the plasma membrane lipid domains could also influence the subcellular distribution of PKC. The VSM plasma membrane is composed of several domains of focal adhesions alternating with zones rich in caveolae, and both harbor a subset of membrane-associated proteins. Caveolae appear to be a major cell surface location for PKC. For instance, PKCα is constitutively present and exhibits binding activity in caveolae, and does not bind to noncaveolae membranes, which constitute over 90% of the plasma membrane (Mineo, Ying, Chapline, Jaken, & Anderson, 1998). Local fluctuations in $[Ca^{2+}]$ may affect the amount of a specific PKC isoforms retained in caveolae, reflecting the ion requirement for PKC binding to caveolae. For instance, caveolae contain PKCα only in the presence of Ca^{2+}, and PKCλ only in the absence of Ca^{2+} from the isolation buffer, while retention of PKCε in caveolae is not dependent on Ca^{2+} (Mineo et al., 1998). Caveolins are scaffold proteins that could help PKCα and ζ localize to the caveolar microdomains where they are subsequently activated (Oka et al., 1997). In rabbit femoral and renal arteries at rest, PKCζ is localized in punctuate plasma membrane aggregates alternating with vinculin (supporting its location in caveolae), and in a perinuclear location, and these

Table 1 Distribution of PKC Isoforms and Potential Scaffold Proteins in Major Tissues and Representative Blood Vessels

PKC	MW (kDa)	Main Tissue Distribution	Blood Vessel Examined	Location in Resting State	Location in Activated State	Potential Scaffold	Reference
cPKCs							
α	74–82	Universally expressed	Rat aorta	Cytosolic	Nuclear	RACK1	Watanabe, Hachiya, Hagiwara, and Hidaka (1989), Wetsel et al. (1992), Haller, Quass, Lindschau, Luft, and Distler (1994), Khalil, Lajoie, and Morgan (1994), Ohanian et al. (1996), Kanashiro, Altirkawi, and Khalil (2000), and Hoque et al. (2014)
			Rat mesenteric artery	Cytosolic/ Membrane	Cytosolic/ Membrane	p32	
			Rat carotid, porcine coronary artery, bovine aorta, ferret portal vein	Cytosolic	Plasma membrane	RACK1, AKAPs, HSP, p32	
β	80–82	Adipose tissue, liver, kidney, spleen, skeletal muscle, brain	Rat aorta	Cytosolic	Nuclear	RACK1, p32	Haller et al. (1994), Hoque et al. (2014), and Mehta (2014)
			Rat carotid	Cytosolic	Membrane	RACK1, AKAPs, HSP, p32, 14-3-3	
γ	70–82	Adrenal gland, brain	Rat mesenteric artery	Cytosolic	Cytosolic	RACK1, AKAPs, HSP importins, 14-3-3	Wetsel et al. (1992), Ohanian et al. (1996), and Hoque et al. (2014)

nPKCs							
δ	76–82	Universally expressed	Rat aorta	Cytoskeleton/ Organelle	Cytoskeleton/ Organelle	RACK1, p32	Liou and Morgan (1994), Ohanian et al. (1996), Zhao, Xia, and Chen (2012), and Hoque et al. (2014)
			Rat mesenteric artery	Membrane	Membrane	AKAPs, HSP, p32, 14-3-3	
ϵ	90–97	Pancreas, kidney, brain	Rat mesenteric artery, porcine coronary artery	Cytosolic/ Membrane	Cytosolic/ Membrane	AKAPs, p32	Khalil, Lajoie, Resnick, and Morgan (1992), Wetsel et al. (1992), Ohanian et al. (1996), Kanashiro, Altirkawi, et al. (2000), and Hoque et al. (2014)
			Ferret aorta	Cytosolic	Surface membrane	RACK1, RACK2 AKAPs, HSP, p32, 14-3-3	
η	80	Lung, skin, brain	NIH 3T3 fibroblasts	Cytosolic/ Membrane	Membrane		Goodnight, Mischak, Kolch, and Mushinski (1995), Suzuki et al. (2009), and Hoque et al. (2014)
θ	79	Lymphoid and hematopoetic cells, skeletal muscle				AKAPs, HSP, Importins, p32, 14-3-3	Hoque et al. (2014) and Hage-Sleiman et al. (2015)

Continued

Table 1 Distribution of PKC Isoforms and Potential Scaffold Proteins in Major Tissues and Representative Blood Vessels—cont'd

PKC	MW (kDa)	Main Tissue Distribution	Blood Vessel Examined	Location in Resting State	Location in Activated State	Potential Scaffold	Reference
aPKCs							
ζ	64–82	Universally expressed	Rat aorta, ferret aorta, and portal vein	Perinuclear	Intranuclear	AKAPs, HSP, Importins, p32, 14-3-3	Khalil et al. (1992), Wetsel et al. (1992), Liou and Morgan (1994), Ohanian et al. (1996), and Hoque et al. (2014)
			Rat mesenteric artery	Cytosolic	Cytosolic		
ι/λ	70	Testis, ovary, kidney, brain	Rabbit femoral artery and portal vein	Cytosolic	Cytosolic	AKAPs, HSP, importins, 14-3-3	Akimoto et al. (1994), Gailly, Gong, Somlyo, and Somlyo (1997), and Hoque et al. (2014)

MW, molecular weight.

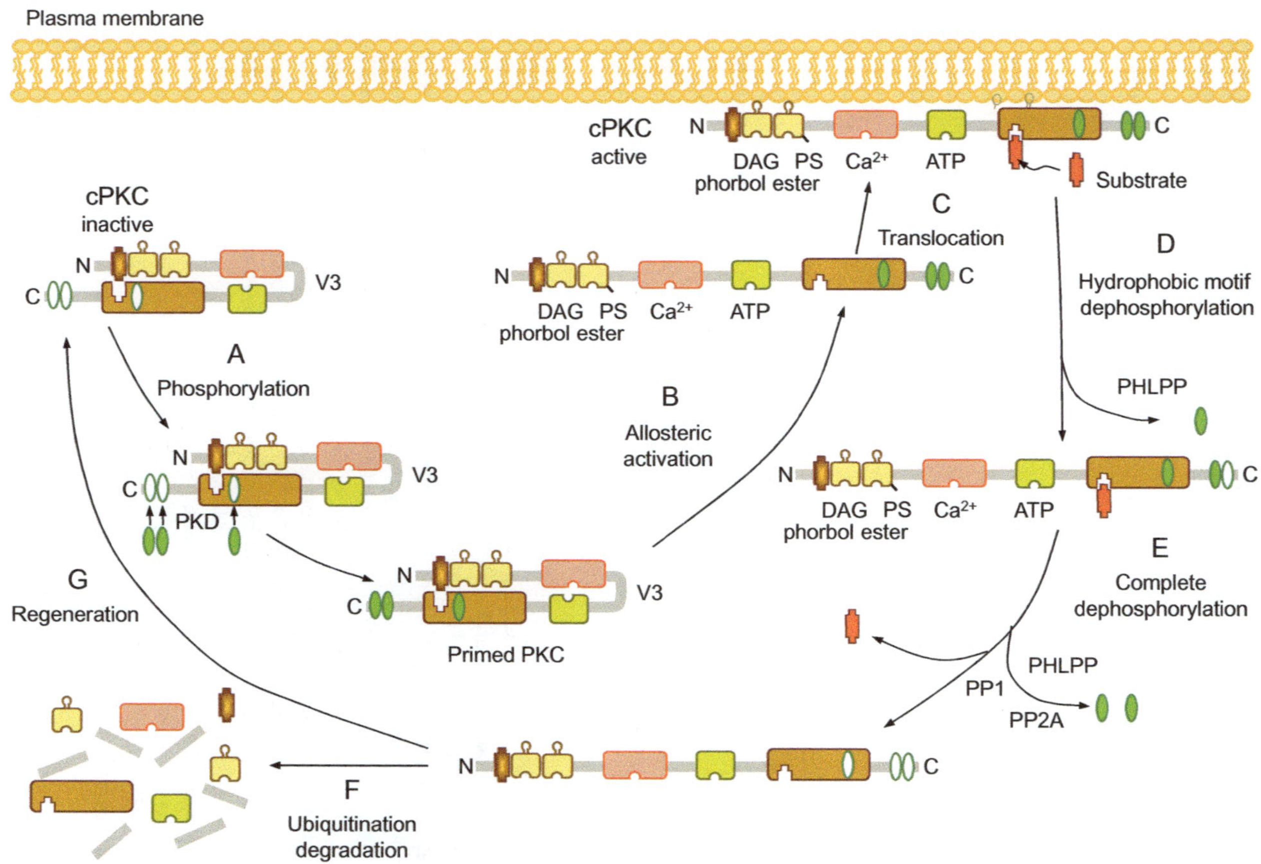

Fig. 2 See legend on next page.

locations may be conducive to regulating VSM intracellular free Ca^{2+} concentration ($[Ca^{2+}]_i$) (Ratz & Miner, 2009).

The plasma membrane lipids are also segregated into cholesterol-rich lipid rafts and glycerophospholipid-rich nonraft regions, an arrangement that is critical for preserving the membrane protein architecture and for the translocation of proteins. In VSMC membrane, lipid segregation is supported by annexins that target membrane sites of distinct lipid composition, and each annexin requires different $[Ca^{2+}]$ for its translocation to the plasma membrane, thus allowing a spatially confined graded response to external stimuli and plasmalemmal localization of PKC (Draeger, Wray, & Babiychuk, 2005). Several members of the annexin family function as PKC substrates and can promote membrane association of PKC (Dubois, Oudinet, Mira, & Russo-Marie, 1996; Xu & Rumsby, 2004). PKC isoforms interact with unique members of the annexin family, and PKCβ, ε and α interact with annexin I, II, and VI, respectively (Mochly-Rosen, Khaner, & Lopez, 1991). Also, a transient interaction between annexin V and PKCδ occurs in cells after PKCδ stimulation, but before its translocation to the particulate fraction, suggesting that PKCδ requires binding to annexin V for its translocation, and whether other PKCs require annexin binding before translocation is unclear (Kheifets, Bright, Inagaki, Schechtman, & Mochly-Rosen, 2006).

Fig. 2 Activation, translocation, substrate interaction, and deactivation of cPKCs. In the PKC cytosolic and inactive state, the pseudosubstrate binds the catalytic site in the C4 region, leading to folding of the regulatory and catalytic domain. Before it becomes catalytically competent, nascent PKC undergoes phosphorylation at three phosphorylation sites. The first and rate-limiting phosphorylation of the activation loop is catalyzed by phosphoinositide-dependent kinase (PDK). Consequently, a negative charge is introduced that properly aligns residues to form a competent catalytic domain and facilitate subsequent autophosphorylation at the turn motif and hydrophobic motif, a process that keeps PKC in a catalytically competent and protease resistant conformation. PKC activators such as PS, DAG, phorbol esters, and Ca^{2+} promote full allosteric activation and translocation of PKC to the plasma membrane. Allosteric activation also induces an open conformation state, making PKC susceptible to phosphatases and proteases, and allows PKC to either enter an autophosphorylation/dephosphorylation cycle or undergoes proteolytic degradation, PKC dephosphorylation terminates its kinase activity and is carried out by the PP2C member pleckstrin homology domain leucine-rich repeat protein phosphatase (PHLPP) at the hydrophobic motif, starting the process that consequently drives complete dephosphorylation of PKC by PP1/PP2A protein phosphatases at the turn motif. Dephosphorylation also predisposes "naked" PKC to ubiquitination and degradation, requiring new synthesis of the enzyme.

Myristoylated alanine-rich C kinase substrate (MARCKS) may play a role in PKC membrane binding. MARCKS is a major PKC substrate that is bound to F-actin and may function as a cross-bridge between cytoskeletal actin and the plasma membrane (Hartwig et al., 1992). Phosphorylation of MARKS by PKC may have an electrostatic effect that affects its protein affinity to the plasma membrane and consequently interferes with its actin cross-linking and causes its displacement from the plasma membrane. On the other hand, dephosphorylation of MARCKS causes its reassociation with the plasma membrane via its stably attached myristic acid membrane-targeting moiety (Thelen, Rosen, Nairn, & Aderem, 1991).

Protein–protein interactions are crucial in signal transduction, and binding sites for arginine-rich polypeptides have been identified in the PKC molecule distal to its catalytic site, allowing targeting of PKC to precise substrates at specific cellular locations. Scaffold proteins could participate in the compartmentalization of PKC to the membrane and include receptor for activated C kinase (RACK), substrates that interacts with C kinase (STICK), receptor for inactive C kinase (RICK), and A-kinase activating proteins (AKAPs) (Ron & Kazanietz, 1999). RACKs and STICKs bind to active PKCs, whereas RICKs and AKAPs interact with inactive PKCs. Binding of a specific activated PKC to its RACK provides access to and phosphorylation of its substrates (Qvit & Mochly-Rosen, 2010). Also, the binding of RACK increases the phosphorylation capacity of PKC several-fold independently from the substrate identity (Ron & Mochly-Rosen, 1994). RACKs may also target PKC to cytoskeletal elements (Ron & Mochly-Rosen, 1994). The interaction of PKC and RACK is isoform specific and is largely mediated by the C2 region of cPKCs (Ron, Luo, & Mochly-Rosen, 1995), and peptide fragments of this region have been developed as modulators of PKC activity (Mochly-Rosen & Kauvar, 2000). These short peptides induce activation and translocation of the corresponding PKC isoform by mimicking the action of the RACK on the isoform and, therefore, are termed "pseudo RACKs" (ΨRACK) (Churchill, Qvit, & Mochly-Rosen, 2009; Dorn et al., 1999). Disruption of the interaction between the ψεRACK and the RACK-binding site is a critical rate-limiting step in translocation of PKCε (Schechtman et al., 2004). Other scaffold proteins including 14-3-3, heat shock protein (HSP), importins, and even actin, have been suggested to tether PKC isozymes to organelles and membranes (Adwan, Ohm, Jones, Humphries, & Reyland, 2011; Lum, Balaburski, Murphy, Black, & Black, 2013; Prekeris, Mayhew, Cooper, & Terrian, 1996; Toker, Ellis, Sellers, & Aitken, 1990; Welch, Jones, & Scott, 2010). Additional unique protein–protein interactions between specific PKC isoforms and their substrates might provide further anchoring and specificity at the subcellular sites. For instance, for PKCε, a myofilament-binding site in the C2 region (Huang & Walker, 2004), an intrasarcoplasmic reticulum calsequestrin-binding site (Rodriguez, Chen, Smith, & Mochly-Rosen, 1999), a neurocytoskeletal elements-binding site (Zeidman, Lofgren, Pahlman, & Larsson, 1999), a unique actin-binding site within the C1 region of εPKC, and a Golgi-binding site (Csukai, Chen, De Matteis, & Mochly-Rosen, 1997; Prekeris et al., 1996) have been identified (Churchill et al., 2009). PKCϵ exhibits a unique association with Golgi membranes via its zinc finger domain and specifically modulates Golgi function, and the zinc finger domain might act as a specific localization signal (Lehel, Olah, Jakab, & Anderson, 1995). PKCϵ may also translocate from Golgi to the plasma membrane by two distinct mechanisms; as a rapid, vesicle-independent process was observed with holo PKCϵ (which requires the presence of the pseudosubstrate and/or hinge regions), and a slow, vesicle-dependent pathway was observed with the zinc finger fragment (Lehel, Olah, Petrovics, Jakab, & Anderson, 1996).

Different parts of the PKC molecule play varying roles in mediating PKC activation and translocation. The pseudosubstrate and hinge regions facilitate plasma membrane and cytoskeletal association (Lehel, Olah, Jakab, Szallasi, et al., 1995). Also, the V5 variable region contributes to the regulation of PKCα activity through multiple mechanisms involving stabilizing the kinase through direct interactions with its N-terminal, interacting with the pseudosubstrate in the N-terminal regulatory domain, and mediating subcellular localization through interaction with RACK (Yang & Igumenova, 2013).

Although the interaction of cPKCs at the plasma membrane has been well studied, less is known about the activity of nPKCs and aPKCs at the plasma membrane and other cell membranes including those of the nucleus and other cell organelles such as the mitochondria, endoplasmic reticulum (ER), and Golgi (Fig. 3). For instance, c-src-dependent phosphorylation of tyrosine Y256 in PKCι, through enhanced interaction with the nuclear transporter protein importin-β, results in its translocation to the nucleus (White, Seibenhener, & Wooten, 2002). Also, the ER membrane may represent the main target for PKCδ recruitment rather than Golgi or mitochondrial membranes. Other nPKCs such as PKCε also display the same translocation pattern following ATP binding. Considering the importance of the ER in protein synthesis and modification, the functions of nPKCs at the ER membrane need further investigations (Hui, Reither, Kaestner, & Lipp, 2014).

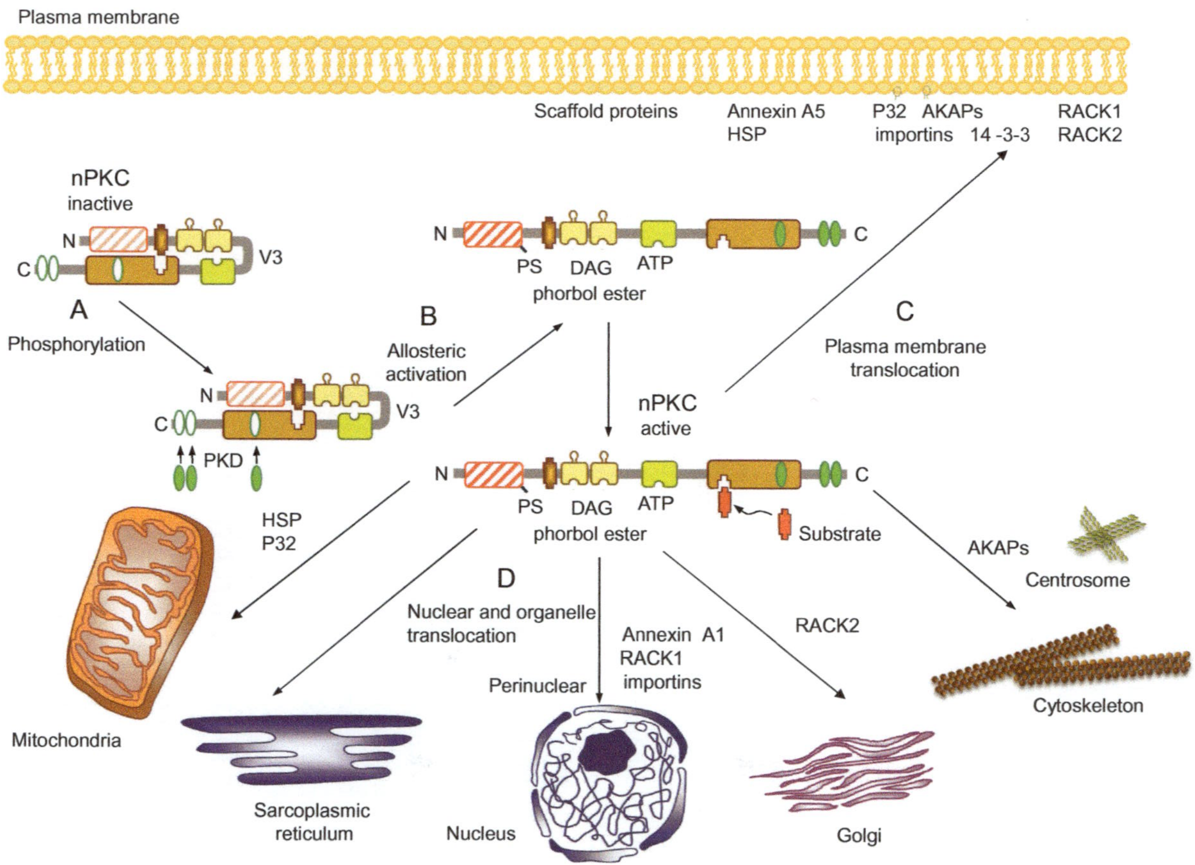

Fig. 3 Activation and translocation of novel PKCs (nPKCs). After undergoing phosphorylation (A) and allosteric activation (B), nPKCs may remain cytosolic or use scaffold proteins to translocate to the plasma membrane (C), or to the perinuclear region, the nucleus, and cell organelles (D) such as the mitochondria, sarcoplasmic reticulum, Golgi, the cytoskeleton, and centrosome.

While translocation to cell membranes was traditionally considered the hallmark of PKC activation, this allosteric model for PKC activation by lipid cofactors and the concept that membrane translocation is essential to PKC activation have been challenged. For instance, the model assumes that the cellular actions of PKC are membrane limited, as it focuses on the role of lipid cofactor in mediating translocation and delivery of the enzyme to the membranes. However, PKC may be localized in the cell periphery in both resting and stimulated tissues. Immunohistochemical studies have shown that the distribution of PKCα in the longitudinal and circular layers of the swine stomach tonic fundus and phasic antrum under resting conditions does not differ, being predominantly localized near the smooth muscle plasma membrane, and stimulation of either tissue with phorbol 12,13-dibutyrate (PDBu) or carbachol does not alter this peripheral PKCα distribution (Zhang, Hermanson, & Eddinger, 2013). Also, PKC is found in other cell compartments like the mitochondria and soluble fraction of cells subjected to oxidative stress, a known activator of PKC (Konishi et al., 2001; Steinberg, 2015). PKCs in the soluble fraction of VSM could also phosphorylate sarcomeric proteins in the contractile apparatus, located distant from the membrane lipid bilayer (Steinberg, 2015). PKC translocation may also be dependent on cytoskeletal elements and active transport along the cytoskeleton, suggesting that other forms of protein–protein interactions may be involved in the translocation process (Dykes, Fultz, Norton, & Wright, 2003; Kheifets & Mochly-Rosen, 2007; Schmalz, Kalkbrenner, Hucho, & Buchner, 1996). Another misconception of the canonical model of PKC activation is related to the concept that PKC's catalytic activity is an inherent property of the enzyme that is not altered by the activation process; a model that does not adequately explain the diverse and, in some cases, opposing actions of certain PKCs (Steinberg, 2015).

4. PKC PHOSPHORYLATION

In the inactive state of PKC, the pseudosubstrate binds the catalytic site in the C4 region, and both the regulatory and catalytic domains are folded together (House & Kemp, 1987). In the activated state, the PKC molecule is unfolded, the pseudosubstrate is dissociated from the C4 region, and PKC is ready to target its specific substrate. Before it becomes catalytically competent and able to respond to its allosteric activators, nascent PKCs undergo conformational changes at three conserved serine/threonine residue phosphorylation sites in the C-terminal domain (Newton, 1995) (see

Fig. 2). Phosphorylation could change protein conformation or electric charge and consequently affect its lipid affinity and binding to the plasma membrane. Phosphorylation of PKC itself via autophosphorylation or by a putative PKC kinase may determine its localization. The cPKCs, nPKCs, and probably aPKCs are dependent to varying degrees on phosphorylation of the sites at the activation loop, turn motifs, and hydrophobic motif (Newton, 2003; Parekh, Ziegler, & Parker, 2000). These phosphorylations are thought to keep PKC in a catalytically competent and protease resistant conformation. Full activation of PKC by allosteric activators induces an open conformation that makes the enzyme susceptible to both phosphatases and proteases, leading to either repeated autophosphorylation/dephosphorylation cycles, or proteolytic degradation of the PKC molecule and the synthesis of de novo enzyme (Newton, 2010; Parekh et al., 2000).

The first and rate-limiting phosphorylation of the activation loop at the conserved threonine, is catalyzed by phosphoinositide-dependent kinase (PDK), and is critical for activation of PKC (Le Good et al., 1998; Newton, 2001). In the absence of PDK-1, PKC is prone to rapid degradation before turning into catalytically competent enzyme (Balendran, Hare, Kieloch, Williams, & Alessi, 2000). Mutation of phosphorylatable Thr-residues in the activation loop abolishes PKC activity, supporting its essential role in PKC activation (Cazaubon, Bornancin, & Parker, 1994; Liu, Graham, Li, Fisher, & Shaw, 2002). As a result of phosphorylation of the activation loop, a negative charge is introduced that properly aligns residues to form a competent catalytic domain and facilitate the subsequent autophosphorylation of two sites in the C-terminus, one at the "turn motif," so named because it corresponds to a phosphorylation site in PKA localized at the apex of a turn, and the other at the more C-terminal hydrophobic motif (Behn-Krappa & Newton, 1999). The hydrophobic motif is an important and direct mediator of PKC stability, functioning as a docking site for PDK-1 through its repeated negatively charged aspartate sequence called PDK-1 interacting fragment (Balendran et al., 2000; Newton, 2003); an interaction that allows PDK-1 to access the activation loop (Hage-Sleiman et al., 2015).

There are differences in the phosphorylation process in different PKCs. In cPKCs, both the turn motif and the hydrophobic motif are autophosphorylated, whereas in nPKCs autophosphorylation occurs only in the turn motif, and phosphorylation in the hydrophobic motif is carried out by other kinases (Hage-Sleiman et al., 2015). For PKCδ, autophosphorylation of its turn motif contributes to its relative stability and solubility. In VSM, autophosphorylation of PKCα and ε may be regulated by α-adrenergic receptor agonists, and the actin-binding protein calponin (CaP) may be involved as α-adrenergic agonists induce translocation of CaP from the contractile filaments to the cortex of VSMCs (Kim et al., 2013). Also, aPKCs are phosphorylated at the activation loop and turn motif, but naturally contain glutamate "phosphomimetic" residues in their hydrophobic motif (Cameron et al., 2007; Newton, 2003; Parekh et al., 2000), while the hydrophobic motif of nPKCs contains an aspartate residue (Cameron et al., 2007).

PKC phosphorylation may occur only during maturation of the newly synthesized enzyme, as has been shown with PKCα, or is dynamically regulated, as has been shown with nPKCs (Cenni et al., 2002; Rybin et al., 2004, 2003). For example, phosphorylation of multiple sites may be required for activation of mature PKCs as has been shown during H_2O_2-induced tyrosine phosphorylation of PKCδ (Konishi et al., 1997). Also, in cardiomyocytes, PKCδ, and ε appear to undergo phosphorylation of the activation loop and the hydrophobic motif even in the absence of allosteric regulators (Rybin et al., 2003), supporting that the regulatory pathways of PKC are isoform and cell specific.

There has been some discussion whether phosphorylation of the hydrophobic motif of cPKCs and nPKCs occur via autophosphorylation or through trans-phosphorylation by upstream kinases (Cameron et al., 2007; Ziegler et al., 1999). PDK1 and mTOR are potential upstream kinases that may be key to these phosphorylations (Dutil, Toker, & Newton, 1998; Jacinto & Lorberg, 2008; Le Good et al., 1998). For instance, phosphorylation of the turn motif by the mTORC2 complex may trigger autophosphorylation of the hydrophobic motif (Ikenoue, Inoki, Yang, Zhou, & Guan, 2008; Sarbassov et al., 2004).

The scaffold protein 14-3-3 has been identified as a partner of phosphorylated PKCε in mammalian cells. Phosphorylation of PKCε on Ser346 and Ser368 is required for binding to 14-3-3, and in turn locks the enzyme in an open, active, and lipid-independent conformation (Linch et al., 2014; Saurin et al., 2008). On the other hand, direct interaction between PKCθ and 14-3-3 tau has been observed in T cells, and 14-3-3 overexpression inhibits PKCθ translocation and function (Meller et al., 1996).

Other phosphorylation patterns may be specific to certain PKC isoforms. For instance, tyrosine phosphorylation has been implicated as a mechanism to regulate PKCδ catalytic activity, and the tyrosine-phosphorylated enzyme is constitutively active and no longer requires DAG as a cofactor

(Konishi et al., 2001). PKCδ has tyrosine phosphorylation sites, and a pathway involving tyrosine phosphorylation may underlie redox control of PKCδ activity. A Src family kinase (Lck)-driven phosphorylation at Tyr311 in rodent PKCδ (Tyr313 in human PKCδ) is believed to mediate H_2O_2-dependent increase in PKCδ activity (Konishi et al., 2001; Steinberg, 2015), and has been implicated in different PKCδ-dependent cellular responses (Lu et al., 2007; Nakashima et al., 2008; Steinberg, 2015).

PKC phosphorylation is often used as a marker of its activation and in testing whether certain effects are mediated by PKC. For instance, a PKCα autophosphorylation site has been used as a marker of its activity (Durgan, Michael, Totty, & Parker, 2007; Ng et al., 1999). However, the use of PKC phosphorylation as a marker of PKC activation may not be definitive. For example, while phospho-S299 could be a useful marker of activated PKCδ, PKCδ is phosphorylated at other sites and undergoes in vitro autophosphorylation at three sites within its V3 region (S299, S302, S304), each of which is evolutionarily conserved and unique to PKCδ. S299-phosphorylated PKCδ is localized at both the plasma and nuclear membranes, making it the best marker of the activated enzyme (Durgan et al., 2007). While S643 is also an important PKCδ autophosphorylation site (Li, Zhang, Bottaro, & Pierce, 1997), it is not ideally used as a marker of activation because it is relatively resistant to dephosphorylation and remains phosphorylated even when PKCδ releases its activator DAG and adopts a "closed" conformation (Durgan et al., 2007; Parekh et al., 2000).

The kinase activity of PKCs is terminated by dephosphorylation, and this usually occurs when PKC is in an "open" conformation, unbound by the pseudosubstrate or constitutively active (Dutil, Keranen, DePaoli-Roach, & Newton, 1994; Gao, Brognard, & Newton, 2008; Lee, Smith, Pettit, & Bingham Smith, 1996). For cPKCs and nPKCs, dephosphorylation is carried out by the PP2C member pleckstrin homology domain leucine-rich repeat protein phosphatase (PHLPP) at the hydrophobic motif, which starts the process that drives PKC to be totally dephosphorylated and degraded by PP1/PP2A protein phosphatases at the turn motif (Ahn et al., 2007; Gao et al., 2008; Newton, 2010; Sontag, Sontag, & Garcia, 1997). In some instances, phosphatases may have an indirect effect on PKC, for example dephosphorylation of the PKCθ downstream molecules CARMA1 by PP2A leads to PKCθ deactivation (Eitelhuber et al., 2011). Dephosphorylation predisposes "naked" protein kinases to ubiquitination and degradation (Katzmann, Odorizzi, & Emr, 2002). Partial inhibition of phosphorylation is caused by binding with HSP70, thus promoting rephosphorylation of PKCs and their subsequent reactivation (Gao & Newton, 2006; Poli et al., 2014).

PKC-priming phosphorylation may also be influenced by the inferred allosteric behavior caused by MgATP binding. Nucleotide pocket occupation confers on PKC a conformation that is both conducive to the action of upstream kinases and protective from the action of antagonistic phosphatases. Studies of kinase-inactive mutant forms of PKC isotypes that typically comprise mutation of the highly conserved lysine residue responsible for coordination of the α-β phosphates of ATP, have shown that when the kinase domain is compromised through mutation of the highly conserved lysine it fails to be primed, but can be fully primed upon binding an ATP-competitive high-affinity inhibitor. In one study, investigators expressed an inactive green fluorescent protein (GFP) tagged PKCε K437M mutant (where K437 refers to the conserved lysine that contacts the ATP α-β phosphates) in HEK cells using a tetracycline-inducible system. This led to the accumulation in the cells of inactive PKCε lacking appreciable phosphorylation of the priming sites, while wild-type PKCε expressed under the same conditions was constitutively phosphorylated at all priming sites. Importantly, treatment with the PKC inhibitor bisindolylmaleimide I induced rapid phosphorylation of the priming sites of PKCε K437M-expressing cells, but did not increase phosphorylation of wild-type PKCε, supporting that the conformation induced by occupation of the nucleotide pocket of PKCε K437M with an inhibitor might be sufficient to promote priming. This may be relevant to the natural occupation of the nucleotide pocket with MgATP, as the PKCε M468A gatekeeper mutant confers sensitivity to the PKD inhibitor 1-naphthyl-PP1 (NaPP1) and becomes fully primed at steady state. However, activation with phorbol esters, which relieves the regulatory domain inhibition, permits catalytic action, and turnover of MgATP/ADP, leading to rapid dephosphorylation. This is likely due to reduced steady-state occupancy of the nucleotide-binding pocket, as the mutant PKCε dephosphorylation can be blocked by NaPP1. Also, following complete phorbol ester-induced dephosphorylation of the PKCε M468A mutant, phosphorylation can be reestablished by treatment with NaPP1 (Cameron, Escribano, Saurin, Kostelecky, & Parker, 2009; Linch et al., 2014).

5. PKC ACTIVATORS

PKCs are activated by a variety of hormones such as adrenaline and angiotensin II (AngII), growth factors including epidermal growth factor and insulin, and neurotransmitters like dopamine and endorphin (Mochly-Rosen et al., 2012). These stimulants generally interact with their plasma membrane receptors leading to activation of phospholipase C (PLC) and hydrolysis of phosphatidylinositol 4,5-bisphosphate (PIP_2) into inositol 1,4,5-trisphosphate (IP_3) and DAG. IP_3 stimulates Ca^{2+} release from the intracellular stores in the endoplasmic reticulum, while DAG activates PKC. Activation of PKC can also occur in the absence of receptor activation, as high levels of cytosolic Ca^{2+} can directly activate PLC and lead to activation of PKC (Mochly-Rosen et al., 2012). PKC isoforms respond differently to Ca^{2+}, PS, DAG, and other phospholipid degradation products. cPKCs bind Ca^{2+} in a phospholipid-dependent manner, and Ca^{2+} may form a "bridge" holding the protein and phospholipid complex together at the membrane (Bazzi & Nelsestuen, 1990). PS is indispensable for activation of PKC. Phosphatidylinositol and phosphatidic acid activate PKC at high Ca^{2+} concentrations. DAG activates PKC by reducing its Ca^{2+} requirement and enhancing its membrane association (Nishizuka, 1992). PKC activators also include lipids derived from sources other than glycerolipid hydrolysis such as *cis*-unsaturated free fatty acids and lysophosphatidylcholine, ceramide (a sphingomyelinase product), phosphatidylinositol 3,4,5-trisphosphate, and cholesterol sulfate (Nishizuka, 1995).

Phorbol esters such as PDBu, phorbol 12-myristate 13-acetate (PMA), and 12-O-tetradecanoylphorbol-13-acetate (TPA) can substitute for DAG in PKC activation. Phorbol esters stabilize PKC–membrane association by reducing its apparent K_m for Ca^{2+} (Kanashiro & Khalil, 1998). PMA binds to PKC 1000-fold more strongly than DAG (Dries & Newton, 2008; Sanchez-Bautista, Corbalan-Garcia, Perez-Lara, & Gomez-Fernandez, 2009). Interestingly, the binding of PMA to the C1B domain of PKCs alone may not induce a significant conformational change in the protein or release the pseudosubstrate domain from the catalytic core, but may generate a hydrophobic cap covering polar groups and thus helping PKCs to insert into membrane lipid bilayer (Zhang, Kazanietz, Blumberg, & Hurley, 1995).

DAG analogs and phorbol esters are not specific for a particular PKC isoform, and may have other effects unrelated to PKC. For example, PMA, a widely used PKC activator, recruits members of both cPKCs and nPKCs to the plasma membrane (Hui et al., 2014). Also, 1,2-dioctanoyl-sn-glycerol (DiC8), a DAG analog used commonly to activate PKC, blocks K_v, BK_{Ca}, and K_{ATP} channels of mesenteric artery VSM in a PKC-independent manner. 1-Oleoyl-2-acetyl-sn-glycerol (OAG) is a related compound that activates PKC but without blocking K^+ channels, and is therefore a preferred pharmacological tool over DiC8 (Rainbow, Parker, & Davies, 2011).

Posttranslational modifications could activate certain PKC isoforms. For example, proteolysis between the regulatory and the catalytic domains activates PKCδ (Persaud, Hoang, Huang, & Basu, 2005). Other posttranslational modifications including oxidation, acetylation, and nitration and phosphorylation could also activate PKC (Steinberg, 2008). Oxidants such as H_2O_2 can directly activate PKC, and both the regulatory and catalytic domains of PKC are susceptible to oxidative modification (Gopalakrishna & Anderson, 1989). PKC contains unique structural features that are especially susceptible to oxidative modification, like the zinc-binding cysteine-rich motifs of the N-terminal regulatory domain (Gopalakrishna & Jaken, 2000). Also, hydroquinone, catechol, and whole cigarette smoke condensate have been shown to activate PKC in Lewis lung carcinoma cells (Gopalakrishna, Chen, & Gundimeda, 1994).

6. PKC SUBSTRATES

When PKC is not catalytically active, the basic autoinhibitory pseudosubstrate is protected from proteolysis by an acidic patch in the substrate-binding site. When PKC is activated, it phosphorylates arginine-rich protein substrates, which neutralize the acidic patch and displace the pseudosubstrate from its binding site in the kinase core (House & Kemp, 1987; Newton, 1995). The amino acid sequence near the substrate phosphorylation site may assist in PKC substrate recognition. Several PKC substrates have been identified (Table 2), and PKC isotypes may show some substrate specificity. For instance, PKCα, β, and γ are potent histone IIIS kinases, while PKCδ, ϵ, and η have a poor capacity to phosphorylate histone (Kanashiro & Khalil, 1998). However, PKC isoforms may show overlapping specificities for substrates derived from modification of their pseudosubstrate regions. For example, the PKC targeting protein AKAP79 binds the catalytic core of all PKCs through a pseudosubstrate-like mechanism (Bogard & Tavalin, 2015; Faux et al., 1999).

PKC substrates include the anchoring proteins STICKs such as MARCKs, MacMARCKs, α-, β-, and γ-adducin, clone 72 (SseCKs), GTP-binding proteins, and cytoskeletal proteins (Hage-Sleiman et al., 2015; Ron & Kazanietz, 1999; Ron & Mochly-Rosen, 1994). PKC causes phosphorylation of the inhibitory GTP-binding protein Gi, facilitating the

Table 2 Representative PKC Substrates and the Effect of Their Phosphorylation

Substrate	Effect of Substrate Phosphorylation	Reference
Histones		
H3T45	DNA fragmentation, apoptosis	Hurd et al. (2009)
H3T6	Prevents LSD1 from demethylating H3K4 during androgen receptor-dependent gene activation. Promotes cell proliferation	Metzger et al. (2010)
Membrane-Bound Proteins		
MARCKS (myristoylated, alanine-rich C kinase substrate)	MARCKS is bound to F-actin. Functions as cross-bridge between cytoskeletal actin and plasma membrane	Hartwig et al. (1992)
The inhibitory GTP-binding protein Gi	Facilitates the dissociation of the αi subunit from adenylyl cyclase and thereby relieves it from inhibition	Kanashiro and Khalil (1998)
Ion channels		
BK_{Ca} channels	Inhibition, leading to membrane depolarization, activation of L-type voltage-gated Ca^{2+} channels, and increased $[Ca^{2+}]_i$ and vascular tone, e.g., in pulmonary artery and porcine coronary artery	Minami, Fukuzawa, and Nakaya (1993), Lange, Gebremedhin, Narayanan, and Harder (1997), Taguchi, Kaneko, and Kubo (2000), Barman, Zhu, and White (2004), Crozatier (2006), Ledoux, Werner, Brayden, and Nelson (2006), and Zhu, Xiao, and Zhang (2013)
Voltage-gated K^+ channel	Inhibition. Increases vascular tone	Cogolludo, Moreno, Bosca, Tamargo, and Perez-Vizcaino (2003), Novokhatska et al. (2013), Zhu et al. (2013), and Brueggemann et al. (2014)
K_{ATP} channels	Inhibition. Alters the channel properties by modifying kinetics and/or the number of channels at the cell membrane, e.g., in mesenteric artery	Levitan (1994), Bonev and Nelson (1996), Light (1996), and Zhu et al. (2013)

Continued

Table 2 Representative PKC Substrates and the Effect of Their Phosphorylation—cont'd

Substrate	Effect of Substrate Phosphorylation	Reference
Store-operated Ca^{2+} channel	HEK293 cells. Inhibition	Shi et al. (2004)
Ion pumps and exchangers		
Ca^{2+}-ATPase activation	Activation. Promotes Ca^{2+} extrusion. Explains transient nature of agonist-induced increase in VSM $[Ca^{2+}]_i$	Salamanca and Khalil (2005)
α1 subunit of Na^+/K^+-ATPase	Inhibition. Alters membrane potential and intracellular concentrations of Na^+ and K^+	Bertorello, Aperia, Walaas, Nairn, and Greengard (1991)
Na^+/H^+ antiport exchanger	Activation. Increases cytoplasmic pH, which increases contraction	Aviv (1994), Austin and Wray (2000), and Wray and Smith (2004)
Cytoskeletal and regulatory proteins		
Vinculin	Controls cell shape and adhesion	Perez-Moreno, Avila, Islas, Sanchez, and Gonzalez-Mariscal (1998)
Vimentin	Recycles β1-integrins to plasma membrane	Ivaska et al. (2005)
CPI-17	Enhances myofilament force sensitivity to Ca^{2+}. Inhibits MLC phosphatase, increases MLC phosphorylation and enhances VSM contraction, e.g., in rabbit femoral artery	Woodsome, Eto, Everett, Brautigan, and Kitazawa (2001)
Calponin	Allows actin–myosin interaction and enhances VSM contraction	Parker, Takahashi, Tao, and Morgan (1994)
Raf	Initiates a cascade involving MAPK kinase (MEK) and MAPK, and phosphorylation of the actin-binding protein caldesmon (CaD) which reverses its inhibition of MgATPase activity and thus increases actin–myosin interaction and VSM contraction	Khalil, Menice, Wang, and Morgan (1995) and Kim, Appel, Vetterkind, Gangopadhyay, and Morgan (2008)

Table 2 Representative PKC Substrates and the Effect of Their Phosphorylation—cont'd

Substrate	**Effect of Substrate Phosphorylation**	**Reference**
20-kDa MLC and MLCK	Counteracts Ca^{2+}-induced actin–myosin interaction and force development, e.g., in rabbit mesenteric artery	Inagaki et al. (1987)
Ribosomal protein kinases		
S6KβII	Nucleocytoplasmic shuttling of S6KβII. Regulates protein synthesis and the G1/S transition in the cell cycle	Valovka et al. (2003)
Other		
Arginine-rich protein substrates	Neutralizes the acidic patch in the substrate-binding site. Displaces PKC pseudosubstrate from the kinase core	House and Kemp (1987) and Newton (1995)

dissociation of its αi subunit from adenylyl cyclase and thereby relieves it from inhibition (Kanashiro & Khalil, 1998). PKC could also phosphorylate and activate cell migration-related molecules such as focal adhesion kinase, paxillin, and vinculin (Ding, Tsao, Chai, Mochly-Rosen, & Zhou, 2011; Kappert et al., 2010; Lewis, Cheresh, & Schwartz, 1996; Li, Wernig, Leitges, Hu, & Xu, 2003). PKC phosphorylation of vinculin, a cytoskeletal protein localized at adhesion plaques, could control cell shape and adhesion (Perez-Moreno et al., 1998). PKC could also phosphorylate substrates involved in protein trafficking. Efficient recycling of β1-integrins to the plasma membrane requires PKCε-regulated phosphorylation of vimentin, an intermediate filament protein upregulated upon epithelial cell transformation. Inhibition of PKC and vimentin phosphorylation causes integrins to become trapped in vesicles and attenuates directional cell motility. In vitro reconstitution assays showed that PKCε dissociates from integrin-containing endocytic vesicles in a selectively phosphorylated vimentin-containing complex. Mutations of PKC-regulated sites on vimentin lead to the accumulation of intracellular PKCε/integrin positive vesicles, while introduction of wild-type vimentin promotes cell motility in a PKCε-dependent manner, supporting that PKC-mediated phosphorylation of vimentin is a key process in integrin trafficking and cell motility (Ivaska et al., 2005). PKC also plays a role in phosphorylation and nucleocytoplasmic shuttling of S6KβII, one of the forms of the ribosomal protein S6 kinase (S6K) involved in the regulation of protein synthesis and the G1/S transition in the cell cycle, and this PKC-mediated phosphorylation is induced by mitogens such as PMA, EGF, IGF-1, and platelet-derived growth factor (PDGF) (Valovka et al., 2003).

The list of PKC substrates is growing and many of these substrates could be playing a role in VSM contraction and growth (Table 2).

7. PKC INHIBITORS

PKC inhibitors include compounds that could interact with the PKC molecule, interfere with PKC binding to its substrates, decrease PKC synthesis, or counteract the effects of PKC. Several PKC inhibitors interact directly with PKC at different sites of the PKC molecule (Table 3). The first-generation PKC inhibitors such as H7 and staurosporine are nonspecific pan-PKC inhibitors that block all PKC isoforms and are toxic for clinical use (Clarke & Dodson, 2007). H7 and staurosporine are ATP-competitive small molecule inhibitors that bind to and compete with ATP at the ATP site of the catalytic domain, and therefore display severe side effects in vivo (Mochly-Rosen et al., 2012). The poor selectivity of ATP-binding drugs is also due to their interaction with ATP-binding kinases other than PKC, since the hydrophobic pocket is conserved throughout the kinome (Roffey et al., 2009). Some PKC inhibitors targeting the ATP-binding site such as indolcarbazole and bisindoylmaleimide have shown selectivity to specific PKC isoforms. Ruboxistaurin is a class of bisindoylmaleimide and a relatively selective PKCβ inhibitor (Geraldes & King, 2010; Koya et al., 1997). PKC inhibitors competing at the DAG/phorbol ester or the PS-binding site may be more specific. Calphostin C binds to the C1 domain, mimicking DAG binding (Mochly-Rosen et al., 2012). Interestingly, extended exposure to phorbol esters can specifically downregulate PKCα, β, and γ (Kanashiro, Altirkawi, et al., 2000), but the tumor-promoting properties of phorbol esters limit their clinical use.

Peptides that interfere with the intramolecular interactions within PKC have been developed (Churchill et al., 2009). For instance, myr-ΨPKC is a myristoylated peptide based on the substrate motif of PKCα and β that inhibits TPA-induced PKC activation and phosphorylation of MARCKS (Eichholtz et al., 1993). Other peptides disrupt protein/protein interactions between the PKC regulatory domain and RACK

Table 3 PKC Inhibitors

Class/Inhibitor	Chemistry	Site of Action	Isoform Selectivity	K_d or IC_{50}	Reference
Isoquinolines H-7	1-(5-Isoquinolinesulfonyl)-2-methylpiperazines	ATP-binding site		PKCβI 3.5 μM PKCζ 6 μM	Howcroft and Lindquist (1991)
Benzophenones Chelerythrine	1,2-Dimethoxy-12-methyl[1,3]benzodioxolo[5,6-c]phenanthridin-12-ium	ATP-binding site	Pan-PKCs	0.66 μm	Ding, Tsao, et al. (2011)
Balanol	2-{[2,6-Dihydroxy-4-({[(3S,4R)-3-[(4-hydroxybenzene)amido]azepan-4-yl]oxy}carbonyl)phenyl]carbonyl}-3-hydroxybenzoic acid	ATP-binding site	Pan-PKCs PKCβII > βI > η > δ > α > ε	4-9 μm	Pande, Ramos, and Gago (2008) and Mochly-Rosen et al. (2012)
Indolocarbazoles Gö6976	5,6,7,13-Tetrahydro-13-methyl-5-oxo-12H-indolo[2,3-a]pyrrolo[3,4-c]carbazole-12-propanenitrile	Catalytic domain	PKCα, β1	PKCα 2.3, βI 6.2 nM	Martiny-Baron et al. (1993) and Grandage, Everington, Linch, and Khwaja (2006)
Gö6983	1H-Pyrrole-2,5-dione, 3-[1-[3-(dimethylamino)propyl]-5-methoxy-1H-indol-3-yl]-4-(1H-indol-3-yl)-	ATP-binding site Suppresses PKCμ autophosphorylation	Pan-PKC inhibitor potent: PKCα, β, γ, δ Less potent: PKCζ	PKCα 7, β 7, γ 6, δ 10, ζ 60 nM	Gschwendt et al. (1996) and Peterman, Taormina, Harvey, and Young (2004)
Enzastaurin (LY317615)	3-(1-methyl-1H-indol-3-yl)-4-(1-(1-(pyridin-2-ylmethyl)piperidin-4-yl)-1H-indol-3-yl)-1H-pyrrole-2,5-dione	ATP-binding site	Potent: PKCβ Less potent: PKCα, γ, ε	PKCα 39, β 6, γ 83, ε 110 nM	Graff et al. (2005) and Rovedo, Krett, and Rosen (2011)
LY379196		ATP-binding site	PKCβ	3–6 μM	Slosberg et al. (2000)

Continued

Table 3 PKC Inhibitors—cont'd

Class/Inhibitor	Chemistry	Site of Action	Isoform Selectivity	K_d or IC_{50}	Reference
Staurosporine (CGP41251)	9,13-Epoxy-1H,9H-diindolo[1,2,3-gh:3′,2′,1′-lm]pyrrolo[3,4-j][1,7] benzodiazonin-1-one, 2,3,10,11,12,13-hexahydro-10-methoxy-9-methyl-11-(methylamino)-, [9S-(9α,10β,11β,13α)]-	ATP-binding site	Pan-PKCs Potent: PKCα, γ, η Less potent: PKCδ, ε	PKCα 2, γ 5, δ 20, η 4 nM	Tamaoki et al. (1986) and Meggio et al. (1995)
CGP53353	5,6-Bis[(4-Fluorophenyl)amino]-1H-isoindole-1,3(2H)-dione	ATP-binding site	PKCβ	PKCβI 3.8, βII 0.41 μM	Deng, Xie, Wang, Xia, and Nie (2012)
UCN-01	7-Hydroxystaurosporine	ATP-binding site	cPKCs	25–50 nM	Tamaoki (1991)
Sotrastaurin (AEB071)	3-(1H-indol-3-yl)-4-(2-(4-methylpiperazin-1-yl)quinazolin-4-yl)-1H-pyrrole-2,5-dione	ATP-binding site	Pan-PKC, especially PKCθ	PKCα 0.95, βI 0.64, δ 2.1, ε 3.2, η 1.8, θ 0.22 nM (Ki)	Evenou et al. (2009) and Naylor et al. (2011)
Staurosporine analogs Ruboxistaurin (LY333531)	(9S)-9-[[(Dimethyl-d6)amino]methyl]-6,7,10,11-tetrahydro-9H,18H-5,21:12,17-dimethenodibenzo[e,k] pyrrolo[3,4-h][1,4,13] oxadiazacyclohexadecine-18,20(19H)-dione hydrochloride	ATP-binding site	PKCβI, βII	PKCβI 4.7, βII: 5.9 nM	Aiello et al. (2011)
Midostaurin (PKC412, CGP41251)	(9S,10R,11R,13R)-2,3,10,11,12,13-Hexahydro-10-methoxy-9-methyl-11-(methylamino)-9,13-epoxy-1H,9H-diindolo[1,2,3-gh:3′,2′,1′-lm]pyrrolo [3,4-j][1,7]benzodiamzonine-1-one	ATP-binding site	Pan-PKCs	12 nM	Millward et al. (2006)

Bisindolylmaleimide (GF 109203X, Gö 6850)	3-(1-(3-(Dimethylamino)propyl)-1H-indol-3-yl)-4-(1H-indol-3-yl)-1H-pyrrole-2,5-dione	ATP-binding site	Pan-PKC, especially PKCα, βI	PKCα 8.4, βI 18, βII 16, γ 20, δ 210, ε 132, ζ 5800 nM	Toullec et al. (1991) and Gekeler et al. (1996)
Ro 31-8220	Carbamimidothioic acid, 3-[3-[2,5-dihydro-4-(1-methyl-1H-indol-3-yl)-2,5-dioxo-1H-pyrrol-3-yl]-1H-indol-1-yl]propyl ester, methanesulfonate	Catalytic domain	Pan-PKC: PKCα, βI, βII, γ, ε	PKCα 5, βI 24, βII: 14, γ 27, ε 24 nM	Wilkinson, Parker, and Nixon (1993) and Davies, Reddy, Caivano, and Cohen (2000)
SCH47112		ATP-binding site			Reynolds, McCombie, Shankar, Bishop, and Fisher (1997)
Dicationic, lipophilic compounds dequalinium Cl	Quinolinium, 1,1′-(1,10-decanediyl)bis [4-amino-2-methyl-, chloride (1:2)	Covalently modifies the C2-domain	All PKC	7–18 μM	Castle, Haylett, Morgan, and Jenkinson (1993), Manetta et al. (1993), and Roffey et al. (2009)
Flavonoid Myricitrin	4H-1-Benzopyran-4-one, 3-[(6-deoxy-α-L-mannopyranosyl)oxy]-5,7-dihydroxy-2-(3,4,5-trihydroxyphenyl)-	Prevents PKCα and PKCε activation by phorbol esters	PKCα, ε		Meotti et al. (2006)

Continued

Table 3 PKC Inhibitors—cont'd

Class/Inhibitor	Chemistry	Site of Action	Isoform Selectivity	K_d or IC_{50}	Reference
Quercetin	4H-1-Benzopyran-4-one, 2-(3,4-dihydroxyphenyl)-3,5,7-trihydroxy-		Slight PKC inhibitor		Navarro-Nunez, Lozano, Martinez, Vicente, and Rivera (2010)
Benzothiazole riluzole	6-(Trifluoromethoxy)benzothiazol-2-amine	ATP-binding site	PKCα		Noh, Hwang, Shin, and Koh (2000)
Perylenequinone Calphostin C (UCN-1028C)	1-[3,10-Dihydroxy-12-[2-(4-hydroxyphenoxy)carbonyloxypropyl]-2,6,7,11-tetramethoxy-4,9-dioxoperylen-1-yl]propan-2-yl benzoate	Regulatory domain: competes at the binding site for DAG and phorbol esters.	cPKCs, nPKCs	50 nM	Ogiwara, Negishi, Chik, and Ho (1998)
Phenolic ketone Rottlerin (Mallotoxin)	5,7-Dihydroxy-2,2-dimethyl-6-(2,4,6-trihydroxy-3-methyl-5-acetylbenzyl)-8-cinnamoyl-1,2-chromene)	ATP-binding site	PKCδ Other nPKCs	PKCδ 5 μM Other PKCs 30 μM	Gschwendt et al. (1994)
Macrolactone Bryostatin 1 (NSC 339555)	(1S,3S,5Z,7R,8E,11S,12S,13E,15S,17R,21R,23R,25S)-25-(Acetyloxy)-1,11,21-trihydroxy-17-[(1R)-1-hydroxyethyl]-5,13-bis(2-methoxy-2-oxoethylidene)-10,10,26,26-tetramethyl-19-oxo-18,27,28,29-tetraoxatetracyclo [21.3.1.1^{3,7}.1^{11,15}]nonacos-8-en-12-yl (2*E*,4*E*)-2,4-octadienoate	C1 domain of PKC: competes with phorbol ester and diacylglycerol binding	Twofold selectivity for PKCε over PKCα and PKCδ (short-term administration activates PKC, long-term inhibits)		Kraft, Smith, and Berkow (1986), Roffey et al. (2009), and Mochly-Rosen et al. (2012)

Membrane lipids Sphingosine (D-erythro-Sphingosine)	2-Amino-4-octadecene-1,3-diol; trans-4-Sphingenine	Regulatory domain: competitive inhibitor with phosphatidylserine		2.8 μM	Khan, Dobrowsky, el Touny, and Hannun (1990)
N,*N*-dimethyl-D-erythro-sphingosine	(E,2S,3R)-2-(Dimethylamino)octadec-4-ene-1,3-dio			12 μM	Kim and Im (2008)
Taxol Tamoxifen	2-[4-[(*Z*)-1,2-Diphenylbut-1-enyl] phenoxy]-*N*,*N*-dimethylethanamine	Regulatory domain	cPKCs	–	Zarate et al. (2007)
Purine nucleoside Sangivamycin	4-Amino-5-carboxamide-7-(D-ribofuranosyl)pyrrolo[2,3-d] pyrimidine	ATP-binding site	–	10 μM	Osada, Sonoda, Tsunoda, and Isono (1989)
Carbonitrile 5-Vinyl-3-pyridinecarbonitriles		Catalytic domain	PKCθ	PKCθ 4.7 nM	Tumey et al. (2009)
Pyrimidine 2,4-Diamino-5-nitropyrimidine		Catalytic domain	PKCθ	–	Cywin et al. (2007)
Sterols Spheciosterol sulfate A		Catalytic domain	PKCζ	PKCζ 1.59 μM	Whitson et al. (2009)
Spheciosterol sulfate B		Catalytic domain	PKCζ	PKCζ 0.53 μM	
Spheciosterol sulfate C		Catalytic domain	PKCζ	PKCζ 0.11 μM	

Continued

Table 3 PKC Inhibitors—cont'd

Class/Inhibitor	Chemistry	Site of Action	Isoform Selectivity	K_d or IC_{50}	Reference
Antisense oligonucleotides					
Isis3521 (CGP64128A, Aprinocarsen)	20-mer phosphorothioate oligodeoxynucleotide	Inhibits PKCα mRNA expression	PKCα	–	Lahn, Sundell, and Moore (2003)
Isis9606	19-mer phophorothioate oligodeoxynucleotide	Inhibits PKCα mRNA	PKCα	–	Levesque, Dean, Sasmor, and Crooke (1997)
Short peptides					
Myristoylated-pseudosubstrate peptide inhibitor	Peptide sequence: myr-FARKGALRQ	Substrate-binding site	cPKCs	–	Eichholtz, de Bont, de Widt, Liskamp, and Ploegh (1993)
αV5-3	Peptide sequence: QLVIAN	Site: aa 642–647	PKCα	–	Kim, Thorne, Sun, Huang, and Mochly-Rosen (2011)
βIV5-3	Peptide sequence: KLFIMN	Inhibits PKC translocation Site: aa 646–651	PKCβI	–	Ferreira et al. (2011)
βIIV5-3	Peptide sequence: QEVIRN	Inhibits PKC translocation Site: aa 645–650	PKCβII	–	Stebbins and Mochly-Rosen (2001)

βC2–4	Peptide sequence: SLNPEWNET	Site: aa 218–226	All cPKCs	–	Ron et al. (1995)
δV1–1 (KAI-9803, Delcasertib)	Peptide sequence: SFNSYELGSL	RACK-binding site Inhibits translocation Site: aa 8–17	PKCδ	–	Chen et al. (2001)
εV1–2 (KAI-1678)	Peptide sequence: EAVSLKPT	RACK-binding site Inhibits translocation Site: aa 14–21	PKCε	–	Gray, Karliner, and Mochly-Rosen (1997)
KCe-12 and KCe-16		Substrate-binding site	PKCε	–	Yonezawa, Kurata, Kimura, and Inoko (2009)
ZIP	Peptide sequence: SIYRRGARRWRKL	ζ-Pseudo substrate	PKCζ and aPKCs	–	Braun and Mochly-Rosen (2003)
γV5-3	Peptide sequence: RLVLAS	Site: aa 659–664	PKCγ	–	Sweitzer et al. (2004)
Other					
α-tocopherol, adriamycin, aminoacridine, apigenin, cercosporin, chlorpromazine, dexniguldipine, polymixin B, trifluoperazine, UCN-02					

aa, amino acid.

(Mochly-Rosen et al., 2012). The interaction of PKC and RACK is isoform selective and largely involves the C2 region of cPKC, and peptide fragments of this region may function as selective cPKCs inhibitors (Ron et al., 1995). Also, a peptide derived from the PKC-binding proteins annexin I and RACKI inhibits translocation of PKCβ (Ron & Mochly-Rosen, 1994).

Peptides derived from the pseudosubstrate region show autoinhibitory effect on PKC activity and are attractive PKC inhibitors (Bogard & Tavalin, 2015; Eichholtz et al., 1993; House & Kemp, 1987). The autoinhibitory role of the PKC pseudosubstrate has been suggested as deletion of the pseudosubstrate site abrogates the inhibitory effect of the regulatory domain of PKCα on the full-length enzyme (Parissenti, Kirwan, Kim, Colantonio, & Schimmer, 1998). Synthetic oligopeptides based on pseudosubstrate sequence are specific PKC inhibitors because they exploit its substrate specificity and do not interfere with ATP binding. The synthetic peptide (19–36) inhibits PKC autophosphorylation and protein substrate phosphorylation. Replacement of Arg-27 with alanine in the peptide [Ala-27] PKC (19–31) increases the IC_{50} for inhibition of substrate phosphorylation. A structure–function study of the PKC pseudosubstrate sequence R19FARK-GALRQKNV31 examined the role of specific residues using an alanine substitution scan. Arg-22 was the most important determinant in the inhibitor sequence, since substitution of this residue by alanine gave a 600-fold increase in the IC_{50}. Substitutions of other basic residues with Ala-19, Ala-23, and Ala-27 also increased the IC_{50} 5-, 11-, and 24-fold, respectively. The importance of basic residues in determining the potency of the pseudosubstrate peptide reflects the requirement of these residues in peptide substrate phosphorylation. Gly-24, Leu-26, and Gln-28 residues were also important for pseudosubstrate inhibitor potency. The large increase in the IC_{50} for the [A22]PKC(19–31) peptide makes it a valuable control in studies utilizing the pseudosubstrate peptide to examine functional roles of PKC (House & Kemp, 1990). Another reason pseudosubstrate inhibitors were thought to be more specific inhibitors for PKC isoforms is that the pseudosubstrate region provides a large interface for multiple points of contact (Bogard & Tavalin, 2015; Churchill et al., 2009). However, this is not always the case as a cell-penetrating myristoylated peptide derived from the pseudosubstrate domain of PKCζ, and termed PKCζ pseudosubstrate inhibitor peptide (ZIP) shows affinity for all PKC isoforms causing disruption of PKC targeting and translocation, suggesting that pseudosubstrates of PKC isoforms may possess several invariant well-conserved residues (Bogard & Tavalin, 2015). Also, mutation of the alanine in the pseudosubstrate with serine or glutamate, mimics the charge of a phosphorylated residue and in effect activates PKC (Kheifets & Mochly-Rosen, 2007; Parissenti et al., 1998; Pears, Kour, House C, Kemp, & Parker, 1990).

Compounds that counteract the effects of PKC include activators of β-adrenoceptors and antioxidants. For example, in portal vein, stimulation of β-adrenoceptors opposes the effects of PKC and causes vasodilatation and reduces the activity of store-operated channels via a cAMP-dependent protein kinase (PKA) pathway (Albert & Large, 2006; Liu, Large, & Albert, 2005). Also, antioxidants may inactivate PKC. The PKC catalytic domain contains several reactive cysteines that can be targeted by antioxidants such as selenocompounds, vitamin E, and polyphenolic agents such as curcumin (Boscoboinik, Szewczyk, Hensey, & Azzi, 1991; Gopalakrishna & Jaken, 2000; Liu, Lin, & Lin, 1993). In VSM, α-tocopherol inhibits the expression, activity, and phosphorylation of PKCα and decrease VSM proliferation, and PKC activity in VSM gradually declines as the α-tocopherol level rises. These effects are not mimicked by β-tocopherol or probucol (Engin, 2009), and, in effect, β-tocopherol may oppose the inhibitory effects of α-tocopherol (Clement, Tasinato, Boscoboinik, & Azzi, 1997). Interestingly, hyperglycemia-induced retinal vascular dysfunction in different animal models can be prevented by α-tocopherol via inhibition of the DAG-PKC pathway (Engin, 2009). Also, high doses of vitamin E may decrease hyperglycemia-induced DAG and PKC activity and reverse some of the changes in the retinal and renal vessels in diabetes (Bursell & King, 1999). On the other hand, glutathione may inhibit PKC by a nonredox mechanism (Ward, Pierce, Chung, Gravitt, & O'Brian, 1998).

Posttranslational modifications of PKC may alter its function. S-nitrosylation, a ubiquitous protein modification in redox-based signaling that forms S-nitrosothiol from nitric oxide (NO) on cysteine residues, decreases PKC activity and signaling, and impairs contraction in mouse aorta, and may represent a key mechanism in conditions associated with decreased vascular reactivity (Choi, Tostes, & Webb, 2011).

Transgenic animals, knockout mice, and antisense techniques have been useful in studying the effects of PKC downregulation in vivo (Table 4). Isoform-specific PKC knockout mice have demonstrated a critical role of PKC in several tissues including endocrine and vascular cells, and further characterization of the PKC knockout vascular phenotype should shed more light on the role of PKC in the vascular system. Also, antisense and siRNA for specific PKC isoforms are now available and can be used to study the role of PKC in various cell functions. ISSI-3521 is a phosphorothioate antisense

Table 4 PKC Knockout Mouse Models, Their Prominent Phenotype, and Major Implications

PKC Knockout	Prominent Phenotype	Implications	Reference
PKC$\alpha^{-/-}$	Increased BP in knockout mice fed a high-salt diet. Principal cells of renal cortical collecting ducts show increased number of epithelial Na channel (ENaC) per cell-attached patch clamp, increased membrane localization of α-, β-, and γ-subunits of ENaC, and increased open probability of ENaC channel	PKCα reduces ENaC membrane accumulation and open probability	Bao et al. (2014)
	In skeletal muscles and adipocytes, enhanced insulin signaling to insulin receptor substrate (IRS) 1-dependent PI_3K, PKB, and PKCλ, and downstream processes, glucose transport, and activation of ERK	PKCα serves as a tonic endogenous inhibitor of IRS-1-dependent PI_3K, PKB, and PKCλ during insulin stimulation of glucose transport and ERK	Leitges et al. (2002)
	Peripheral CD3(+)T cells show impaired CD3/CD28 Ab- and MHC alloantigen-induced T cell proliferation and IFN-γ production. PKC$\alpha^{-/-}$ mice give diminished protein Ag chicken egg albumin (OVA)-specific IgG2a and IgG2b responses following OVA immunization experiments	PKCα is necessary for T cell-dependent IFN-γ production and IgG2a/2b Ab responses	Pfeifhofer et al. (2006)
PKC$\beta^{-/-}$	ApoE$^{-/-}$ and PKC$\beta^{-/-}$/ApoE$^{-/-}$ mice rendered diabetic with streptozotocin. Diabetes accelerated atherosclerosis in the aorta, increased the level of phosphorylated ERK1/2 and Jun-N-terminus kinase MAPK and augmented vascular expression of inflammatory mediators, and monocyte/macrophage infiltration and CD11c(+) cells accumulation, and these processes were diminished by pharmacological inhibition of PKCβ and in diabetic PKC$\beta^{-/-}$/ApoE$^{-/-}$ mice	PKCβ is linked to diabetic atherosclerosis through modulation of gene transcription, cell signaling, and inflammation in the vascular wall. PKCβ could be a potential therapeutic target for prevention and treatment of diabetic atherosclerosis	Kong et al. (2013)

$PKC\gamma^{-/-}$	Exposure to hyperbaric oxygen was associated with increased thicknesses of the inner nuclear and ganglion cell layers of the retina. Destruction of the outer plexiform layer. Significant degradation of the retina Damage to the outer segments of the photoreceptor layer and ganglion cell layer	PKCγ may protect retina from damage by hyperbaric oxygen. Hyperbaric oxygen, should be used with care particularly in patients with a genetic disease such as spinocerebellar ataxia type 14 with nonfunctional PKCγ	Yevseyenkov et al. (2009)
$PKC\delta^{-/-}$	Thickening of the articular cartilage and calcified bone-cartilage interface. Increased number of hypertrophic chondrocytes in the articular cartilage Loss of demarcation between articular cartilage and bone was concomitant with irregular chondrocyte morphology and arrangement Increased intensity of calcein labeling in the interface of the growth plate and metaphysis Reduced level of glycosaminoglycan production	PKCδ plays a role in the osteochondral plasticity of the interface between articular cartilage and the osteochondral junction	Yang et al. (2015)
	Increased white blood cells and platelet counts, and bone marrow and splenic megakaryocytes Increased megakaryocyte number and DNA content Altered thrombopoietin-induced signaling and increased ERK and Akt308 phosphorylation in megakaryocytes Faster recovery and heightened rebound thrombocytosis after thrombocytopenic challenge	PKCδ is important for megakaryopoiesis by regulating thrombopoietin-induced signaling	Kostyak, Bhavanasi, Liverani, McKenzie, and Kunapuli (2014)

Continued

Table 4 PKC Knockout Mouse Models, Their Prominent Phenotype, and Major Implications—cont'd

PKC Knockout	Prominent Phenotype	Implications	Reference
	Fertility analysis has shown that mating pairs produce fewer pups per litter than wild-type pairs. Reduced number of total implantations in females. Sperms showed decreased capacity to penetrate the zona pellucida. Pregnant females exhibit a high incidence of embryonic loss postimplantation	PKCδ is important for key reproductive functions and fertility in both males and females	Ma, Baumann, and Viveiros (2015)
PKCε$^{-/-}$	Embryonic fibroblasts exhibit reduced insulin uptake which was associated with decreased insulin receptor phosphorylation. Changed localization of insulin receptor with colocalization with membrane microdomains marker flotillin-1. Reduced redistribution of insulin receptor by insulin stimulation Reduced expression of CEACAM1, a receptor substrate which modulates insulin clearance	PKCε affects insulin uptake through promotion of receptor-mediated endocytosis, and that this may be mediated by regulation of CEACAM1 expression	Pedersen, Diakanastasis, Stockli, and Schmitz-Peiffer (2013)
PKCη$^{-/-}$	Poor proliferation of T cells in response to stimulation by antigen Defective homeostatic proliferation, a function requiring recognition of self antigens Higher ratio of CD4+ to CD8+ T cells compared to that of wild-type mice	PKCη performs functions that are important for homeostasis and activation of T cells	Fu et al. (2011)
PKCθ$^{-/-}$	The thymus contains less mature single positive T cells than wild type. Thymocytes show defective activation of transcription factors AP-1, NFAT, and NFkB and impaired phosphorylation of ERK after T cell receptor stimulation in vitro	PKCθ plays a role in positive selection of thymocytes in a pathway leading to the activation of ERK, AP-1, NFAT, and NFkB	Gruber, Pfeifhofer-Obermair, and Baier (2010)

PKC$\zeta^{-/-}$	Impaired secretion of T helper 2 (Th2) cytokines, as well as the nuclear translocation and tyrosine phosphorylation of Stat6 and Jak1 activation, essential downstream targets of IL-4 signaling Dramatic inhibition of ovalbumin-induced allergic airway disease	PKCζ is critical for IL-4 signaling and Th2 differentiation. Asthma is a disease of chronic airway inflammation in which T helper (Th) 2 cells play a critical role, and PKCζ can be a therapeutic target in asthma	Martin et al. (2005)
PKC$\lambda^{-/-}$	Tissue-specific knockout in muscle shows impaired insulin-stimulated glucose transport and insulin resistance Knockout in liver shows impaired insulin-stimulated lipid synthesis and insulin-hypersensitivity Knockout in adipocytes shows diminished insulin-stimulated activity and glucose transport, ERK levels and activity Diminished adiposity and serum leptin levels	PKCλ plays a role in insulin-stimulated glucose transport and ERK signaling in muscle, liver, and adipocytes	Sajan et al. (2014)

oligonucleotide that has been targeted to the 3′-untranslated region of PKCα mRNA, and has shown a highly specific reduction of PKCα protein expression in cancer cell lines and human tumor xenograft models (Roffey et al., 2009; Song, Tang, Yuan, Zhu, & Liu, 2003).

Thus some challenges remain with the development of drugs that target specific PKC isoforms. These challenges are largely posed by the ~70% homologous structure of the catalytic domain within the PKC family. Pharmacological tools that target the C2 region could be more selective, as the C2 region is the less conserved among different PKCs (Mochly-Rosen et al., 2012). The V5 region may also be a good target for isoform-specific modulators of PKC activity. PKC isoforms interact with their substrates at sequences unique for the individual isoforms and this interaction can be selectively disrupted by peptide inhibitors that share the same substrate sequence. Also, protein–protein interactions can regulate the subcellular localization of specific PKC isoforms. Further research of PKC substrate interaction sites and PKC protein–protein interactions would shed more light on the various PKC-mediated effects in different systems and provide more specific targets for future therapy of PKC-related disorders (Mochly-Rosen et al., 2012).

8. VASCULAR EFFECTS OF PKC

PKC isoforms have diverse effects in different vascular cell types, with prominent effects on VSM. The role of each PKC isoform in certain vascular responses has been supported by measuring PKC gene expression, protein levels and PKC activity, and by determining the effects of pharmacological isoform-specific PKC inhibitors as well as knockout mice and transgenic rats (Mehta, 2014).

8.1 PKC and VSM Contraction

It is widely accepted that Ca^{2+}-dependent myosin light chain (MLC) phosphorylation is a major determinant of VSM contraction (Kamm & Stull, 1989; Rembold & Murphy, 1988) (Fig. 4). Agonist-induced activation of membrane receptors causes an increase in $[Ca^{2+}]_i$ due to initial Ca^{2+} release from the sarcoplasmic reticulum and maintained Ca^{2+} entry from the extracellular space. Ca^{2+} binds calmodulin (CAM) to form a Ca^{2+}–CAM complex, which activates MLC kinase (MLCK), causes phosphorylation of the 20-kDa MLC, and increases the activity of actin-activated Mg^{2+}-ATPase, leading to actin–myosin interaction and VSM contraction (Kamm &

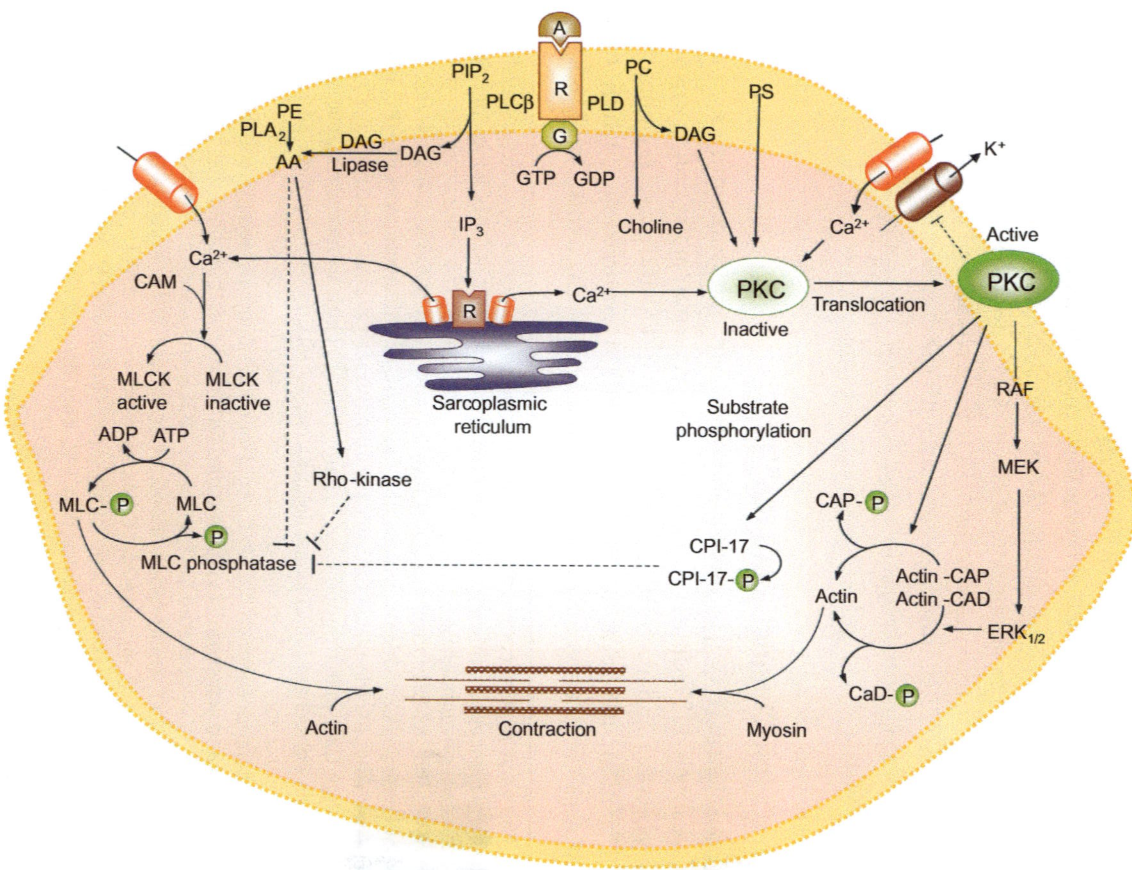

Fig. 4 See legend on next page.

Stull, 1989; Rembold & Murphy, 1988). VSM relaxation is initiated by a decrease in $[Ca^{2+}]_i$ due to Ca^{2+} uptake by the sarcoplasmic reticulum and Ca^{2+} extrusion by the plasmalemmal Ca^{2+} pump and Na^+–Ca^{2+} exchanger. The decrease in $[Ca^{2+}]_i$ causes dissociation of the Ca^{2+}–CAM complex and the phosphorylated MLC is dephosphorylated by MLC phosphatase.

PKC can affect VSM contraction by several mechanisms including regulation of ion channels and pumps and in turn $[Ca^{2+}]_i$, Ca^{2+} sensitization of the contractile proteins, or activation of Ca^{2+}-independent contraction pathways. PKC translocation to the cell surface could also trigger a cascade of protein kinases that ultimately interact with the contractile myofilaments and cause VSM contraction. In some instances, PKC may inhibit VSM contraction.

8.2 PKC, Ion Channels, and $[Ca^{2+}]_i$

PKC can change $[Ca^{2+}]_i$ by modulating the activity of plasmalemmal K^+ and Ca^{2+} channels. K^+ channels play a role in the regulation of the resting membrane potential, and inactivation of K^+ channels in VSMCs causes membrane depolarization, elevation of $[Ca^{2+}]_i$, and VSM contraction (Nelson & Quayle, 1995). Membrane depolarization activates Ca^{2+} entry via L-type

Fig. 4 Pathways of VSM contraction. The interaction of an agonist (A) with its specific α-adrenergic receptor (R) and its coupled heterotrimeric GTP-binding protein (G) activates phospholipase C (PLCβ) which stimulates the hydrolysis of phosphatidylinositol 4,5-bisphosphate (PIP_2) into IP_3 and diacylglycerol (DAG) as well as phospholipase D (PLD) which stimulates the hydrolysis of phosphatidylcholine (PC) into choline and DAG. IP_3 stimulates Ca^{2+} release from the sarcoplasmic reticulum. Agonists also stimulate Ca^{2+} influx through Ca^{2+} channels. Ca^{2+} binds calmodulin (CAM), activates myosin light chain (MLC) kinase (MLCK), causes MLC phosphorylation, and initiates VSM contraction. DAG in the presence of PS, and in case of cPKCs Ca^{2+}, cause activation and translocation of PKC. PKC could inhibit K^+ channels leading to membrane depolarization and activation of voltage-gated Ca^{2+} channels, but could also inhibit Ca^{2+} entry through store-operated Ca^{2+} channels (SOCs) and transient receptor potential channels (TRPCs). PKC could cause phosphorylation of CPI-17, which in turn inhibits MLC phosphatase and increases MLC phosphorylation and VSM contraction. PKC-induced phosphorylation of the actin-binding protein calponin (CaP) allows more actin to bind myosin and enhances contraction. PKC may also activate a protein kinase cascade involving Raf, MAPK kinase (MEK), and MAPK ($ERK_{1/2}$) leading to phosphorylation of the actin-binding protein caldesmon (CaD) and enhanced contraction. DAG is transformed by DAG lipase into arachidonic acid (AA). Also, activation of phospholipase A2 (PLA2) increases the hydrolysis of phosphatidylethanolamine (PE) into AA. AA could activate RhoA/Rho-kinase, which in turn inhibits MLC phosphatase and further enhances the Ca^{2+} sensitivity of contractile proteins. *Dashed line* indicates inhibition.

voltage-gated Ca^{2+} channels (L-VGCC) and may also cause Ca^{2+} release from IP_3- and ryanodine-sensitive intracellular Ca^{2+} stores leading to increase in $[Ca^{2+}]_i$ (Kizub, Klymenko, & Soloviev, 2014; Nauli, Williams, Akopov, Zhang, & Pearce, 2001). Large conductance Ca^{2+}-activated K^+ channels (BK_{Ca}) are the predominant K^+ channels in VSMCs (Ghatta, Nimmagadda, Xu, & O'Rourke, 2006; Nelson & Quayle, 1995). PKC activators such as PDBu inhibit BK_{Ca} leading to increases in vascular tone in both physiological and pathophysiological conditions (Barman et al., 2004; Kizub, Pavlova, Ivanova, & Soloviev, 2010; Novokhatska et al., 2013; Taguchi et al., 2000), and in various vascular beds including pulmonary (Barman et al., 2004), coronary (Minami et al., 1993), cerebral (Lange et al., 1997), and uterine vessels (Hu et al., 2011). PKC activators inhibit BK_{Ca} by phosphorylation of the channel protein and decreasing its sensitivity to activation by cGMP-dependent protein kinase (Crozatier, 2006; Ledoux et al., 2006).

Voltage-gated K^+ channels (K_v) also play a role in the regulation of VSM function, and can be modulated by vasoconstrictors such as arginine vasopressin, ET-1 and AngII via a mechanism involving PKC. In rat mesenteric artery VSMCs, vasopressin regulates Kv7.4 and Kv7.5 subunits of Kv7 channels via activation of PKC. PKCα-dependent phosphorylation of the K^+ channel proteins on serine residues is sufficient to reduce Kv7 channel activity, and the extent of PKC-mediated Kv7.4 and Kv7.5 phosphorylation and K^+ current suppression depends on the subunit composition of the channel proteins (Brueggemann et al., 2014). Also, thromboxane A2 may induce pulmonary vasoconstriction by a mechanism involving PKCζ and inhibition of K_v (Cogolludo et al., 2003). PKC isoforms may contribute differently to the vasoconstrictor-induced effects on different K^+ channels. In rabbit coronary arterial VSMCs, ET-1 and AngII inhibit K_v currents by activating PKCε, and inhibit K_{IR} channel activity by activating PKCα (Park et al., 2005, 2006).

PKC may also regulate K_{ATP} channels, and vasoconstrictor agonists may inhibit K_{ATP} through PKC signaling (Nelson & Quayle, 1995; Quayle, Nelson, & Standen, 1997). Phorbol esters inhibit K_{ATP} currents in mesenteric arteries (Bonev & Nelson, 1996). Although the mechanism via which PKC regulates K_{ATP} is not well defined, in human embryonic kidney cells (HEK293) PKC-mediated AngII- and PDBu-induced inhibition of K_{ATP} channel may involve channel complexes composed of four Kir6.1 and their associated SUR2B subunits (Thorneloe et al., 2002). Also, trafficking studies have shown that PKC may initiate internalization of the channel complex leading to decreased K_{ATP} channel activity (Manna et al., 2010).

PKC-mediated phosphorylation of K_{ATP} may also alter the channel properties, kinetics, and/or number at the cell membrane (Levitan, 1994; Light, 1996).

8.3 PKC, Ion Pumps and Cotransporters, and $[Ca^{2+}]_i$

Plasmalemmal Ca^{2+}-ATPase (PMCA) and sarcoplasmic reticulum Ca^{2+}-ATPase (SERCA) are important Ca^{2+} homeostasis mechanisms in VSM. PKC may activate PMCA or SERCA, an action that promotes Ca^{2+} extrusion and reuptake and lead to a decrease in VSM $[Ca^{2+}]_i$. In isolated cardiac sarcoplasmic reticulum preparations, PKC activates the Ca^{2+}-transport ATPase (Limas, 1980). Also, the α1 subunit of Na^+/K^+-ATPase may serve as a PKC substrate, and PKC-mediated inhibition of Na^+/K^+ pump causes changes in the membrane potential and the intracellular concentrations of Na^+ and K^+ (Bertorello et al., 1991). PKC activation by phorbol esters and permeable DAG analogs may also phosphorylate and activate the Na^+/H^+ antiport exchanger and thereby increase the cytoplasmic pH leading to alkalinization, which generally increases vascular contraction (Austin & Wray, 2000; Aviv, 1994; Rosoff, Stein, & Cantley, 1984; Wray & Smith, 2004).

8.4 PKC and Ca^{2+}-Sensitization of Contractile Proteins

Activation of PKC could increase the myofilament force sensitivity to $[Ca^{2+}]_i$, thereby maintaining VSM contraction with smaller increases in $[Ca^{2+}]_i$. The nPKC isoforms play an important role in mediating VSM contraction through a Ca^{2+} sensitizing pathway, and inhibition of nPKCs attenuates norepinephrine-induced VSM contraction (Wang et al., 2015). PKC-induced Ca^{2+} sensitization could involve phosphorylation of regulatory proteins in the VSM contractile myofilaments and the cytoskeleton. PKC phosphorylates CPI-17, which in turn inhibits MLC phosphatase, increases MLC phosphorylation, and thereby enhances VSM contraction (Woodsome et al., 2001). PKC could also inhibit MLC phosphatase via the phosphorylation of the myosin targeting subunit of myosin phosphatase (MYPT1) (El-Yazbi, Abd-Elrahman, & Moreno-Dominguez, 2015). Activation of PKCα could cause phosphorylation of CaP, a VSM differentiation marker, and an actin-associated regulatory protein, and thereby reverses its inhibition of actin-activated myosin ATPase, allowing more actin to bind myosin, and enhancing VSM contraction. Interestingly, CaP may activate PKC in vitro in the absence of lipid cofactors, and knockdown of CaP inhibits PKC-dependent contraction in ferret arterial VSM (Je, Gangopadhyay, Ashworth, & Morgan, 2001; Kim et al., 2008).

PKC may contribute to VSM force production in a MLC phosphorylation-independent manner. In rat middle cerebral artery, PKC activation by PDBu is associated with sustained force generation and vasoconstriction that is much larger than that expected with the same level of MLC phosphorylation achieved by 5-HT (El-Yazbi et al., 2015).

PKC may also be involved in mechanical stretch-induced vascular myogenic response. In rat cerebral artery VSMCs, PKC activators increase stretch-activated channel activity and induce depolarization, and these effects are blocked by PKC inhibitors (Slish, Welsh, & Brayden, 2002). It is possible that increases in tension on vascular myocytes lead to stimulation of PLC, hydrolysis of phosphoinositides and production of DAG, which activates PKC and stimulate the myogenic response (Albert & Large, 2006). PKC may primarily affect the maintained phase of stretch-induced contraction by changing the Ca^{2+} sensitivity of the contractile elements (Nakayama & Tanaka, 1993). Some studies suggest that PKCθ and μ may participate in stretch-induced VSM mechanotransduction, as cyclic stretch of VSM specifically activates these PKC (Yang et al., 2014). However, other studies have shown activation of PKCδ by cyclic stretch in VSM (Li et al., 2003) and PKCε activation by mechanical stretch in cardiomyocytes (Bullard, Hastings, Davis, Borg, & Price, 2007; Klein et al., 2005).

The nature and extent of the PKC-activated pathway could vary depending on the vasoconstrictor tested, the vascular bed examined and in arteries vs veins. Vasoconstrictors such as AngII, ET-1, serotonin, norepinephrine, and neuropeptide Y activate PKC-dependent pathways, causing VSMC membrane depolarization and contraction (Cole, Malcolm, Walsh, & Light, 2000; Quayle et al., 1997; Zhu et al., 2013). Of note, AngII activates multiple PKC isoforms in VSM (Griendling, Ushio-Fukai, Lassegue, & Alexander, 1997) and PKC may increase VSM contraction via other pathways involving downregulation of atrial natriuretic peptide (ANP) receptor and the binding of ANP to VSM, and thereby preventing ANP-induced inhibition of contraction (Jiao & Yang, 2015). Although the role of venous function in blood pressure (BP) control has been underappreciated, its contribution is significant in the deoxycorticosterone salt rat model of HTN where ET-1 was found to elevate venomotor tone and contribute to HTN (Tykocki, Wu, Jackson, & Watts, 2014). The PKC inhibitor chelerythrine attenuates ET-1-induced contraction in both the aorta and vena cava, suggesting that ET-1 acts via PKC to mediate VSM contraction of both arteries and veins. However, in the aorta, ET-1-induced contraction is largely dependent on PLC activation and IP_3-mediated

Ca^{2+} release, while in the vena cava ET-1-induced contraction is unaffected by the IP_3 receptor antagonist 2-APB. Also, only the vena cava contracts in response to the DAG analog OAG, highlighting the differences in the venous and arterial pathways of contraction (Tykocki et al., 2014). It should be noted that endothelium-derived NO regulates VSM tone by activating guanylate cyclase, increasing cGMP and producing vasodilation, and PKC could inhibit NO-mediated vasodilation by inhibiting guanylate cyclase, leading to decreases in intracellular cGMP and increased vasoconstriction (Johnson & Barman, 2004).

8.5 PKC and Cytoskeletal Proteins

Studies in cerebral resistance arteries have shown that PKC could mediate myogenic constriction through dynamic reorganization of the cytoskeleton and increased actin polymerization (Moreno-Dominguez et al., 2014). Also, both in the presence and absence of Ca^{2+}, PKC may promote cerebral vasoconstriction by increasing the phosphorylation of paxillin and HSP27, reducing G-actin content, and promoting actin cytoskeleton reorganization (El-Yazbi et al., 2015). The relative contribution of PKC to cytoskeletal modification vs other mechanisms of VSM contraction appears to be more significant in the cerebral circulation. In rat middle cerebral arteries, PDBu-induced PKC constriction is more sensitive to disruption of actin cytoskeleton compared to inhibition of cross-bridge cycling, providing evidence for the pivotal contribution of PKC-mediated cytoskeletal actin polymerization to force generation in cerebral resistance arteries (El-Yazbi et al., 2015).

PKC may also modulate certain genes that code for structural proteins such as fibronectin and type IV collagen, by changing the binding of nuclear transcription factors to the promoter regions on responsive genes (Clarke & Dodson, 2007). PKC also affects the gene expression of the regulator of G-protein signaling 2 (RGS2), which may affect vascular tone. In cultured VSMCs, adrenotensin increases RGS2 expression, while the PKC inhibitor chelerythrine reduces RGS2 expression, suggesting that adrenotensin increases gene expression via a PKC-dependent pathway (Mao et al., 2013).

8.6 PKC-Dependent Signaling Cascades

The interaction of PKC with its substrate may trigger a cascade of protein kinases that ultimately stimulate VSM contraction (Fig. 4). PKC may affect Akt signaling PKC (Ding, Tsao, et al., 2011; Radhakrishnan, Maile, Ling, Graves, & Clemmons, 2008). Also, mitogen-activated protein kinases

(MAPK) such as extracellular signal-regulated kinase (ERK), p38 and JNK, are common downstream effectors of PKC (Ginnan & Singer, 2005; Yamaguchi et al., 2004). PKC, MAPK, and c-Raf-1 have been implicated in VSM growth. MAPK is a Ser/Thr kinase that is activated by its dual phosphorylation at Thr and Tyr residues. In quiescent undifferentiated cultured VSMCs, MAPK is mainly cytosolic, but translocates to the nucleus during activation by mitogens (Mii, Khalil, Morgan, Ware, & Kent, 1996). Tyrosine kinase and MAPK activities have also been identified in differentiated contractile VSM. MAPK transiently translocates to the surface membrane during early activation of VSM, but undergoes redistribution to the cytoskeleton during maintained VSM activation (Khalil et al., 1995). It has been suggested that during VSM activation, DAG promotes translocation of cytosolic PKCε to the surface membrane, where it is fully activated. Activated PKCε stimulates the translocation of cytosolic MAPK kinase (MEK) and MAPK to the plasmalemma, where they form a surface membrane kinase complex. PKC causes phosphorylation and activation of MEK, which in turn phosphorylates MAPK at both Thr and Tyr residues. Tyrosine phosphorylation targets MAPK to the cytoskeleton, where it phosphorylates the actin-binding protein caldesmon (CaD) and reverses its inhibition of MgATPase activity and thus increases actin–myosin interaction and VSM contraction (Khalil et al., 1995; Kim et al., 2008). This is supported by the observations that in aortic VSM, phenylephrine activates a pathway involving CaP-dependent PKC autophosphorylation and activation followed by a much delayed ERK activation, CaD phosphorylation, and VSM contraction (Kim et al., 2013). These PKC-dependent pathways occur in parallel with the previously described transient spike in $[Ca^{2+}]_i$ and MLC phosphorylation in VSM (Kim et al., 2013). Interestingly, biochemical studies have shown that in either the presence or absence of extracellular Ca^{2+}, PKC activation by PDBu does not directly change the phospho-content of the thin filament proteins CaP or CaD (El-Yazbi et al., 2015), supporting that other kinases downstream of PKC may be needed to cause phosphorylation of CaP or CaD in vivo, and further demonstrating the complexity of signaling at the whole cell level (Kim et al., 2013).

8.7 PKC and Vasodilation

PKC may directly affect the permeability of VSM Ca^{2+} channels. In VSMC, agonists of G-protein-coupled receptors could activate receptor-operated Ca^{2+} channels (ROCs) including store-operated Ca^{2+} channels (SOCs) and transient receptor potential channels (TRPCs), and PKC may modulate these channels. Studies have suggested that low levels of DAG could activate TRPC6 via a PKC-independent mechanism, while high levels of DAG inhibit TRPC6 SOCs activity. In mesenteric and ear artery VSMC, DAG exerts an inhibitory action on TRPCs through a PKC-dependent pathway, and such mechanism may limit ROC activity at high agonist concentrations (Large, Saleh, & Albert, 2009). PKC inhibits the TRPC6 SOCs activity in a Ca^{2+}-dependent manner (Albert & Large, 2006; Shi et al., 2004).

The 20-kDa MLC and MLCK also serve as substrates for PKC, and their phosphorylation could counteract the Ca^{2+}-induced actin–myosin interaction and force development (Inagaki et al., 1987). In human VSMCs, PKC activation stimulates secretion of C-type natriuretic peptide (CNP) (Mendonca, Doi, Glerum, & Sellitti, 2006), which could function as an endogenous vasodilator (Ahluwalia, MacAllister, & Hobbs, 2004). PKCα and δ mediate most of the increase in CNP mRNA induced by the PKC activator PMA, and PDGF increases CNP in SMCs via a PKCδ-dependent pathway (Mendonca, Koles, & Sellitti, 2012).

9. PHYSIOLOGICAL CHANGES IN PKC

PKC levels may vary with certain physiological changes such as age, exercise, gender, sex hormone status, and pregnancy.

9.1 Age-Related Changes in PKC

Studies have shown age-dependent decrease in PKC activity and its translocation has in postmortem human brains (Wang, Pisano, & Friedman, 1994). Also, in platelets, PKC activity in both the cytosolic and membrane fractions and its redistribution in response to stimulation of cell surface receptors are reduced in elderly men. Interestingly, age-related decrease in PKC activity is mitigated in older men who maintain moderately high levels of aerobic fitness as they age (Wang, Bashore, & Friedman, 1995). Also, in rats, PKCε expression decreases gradually with age particularly among male rats (Li et al., 2014). Thus, PKC activity and its translocation may serve as biological markers of aging, and physical exercise may slow the changes in PKC during the aging process (Wang, Bashore, Tran, & Friedman, 2000).

9.2 Sex Differences in PKC

Sex hormone status has emerged as an important modulator of vascular physiology and cardiovascular risk, and PKC expression/activity may be different in males vs females. Low testosterone levels in men may be associated with a higher risk of cardiovascular disease (Weidemann & Hanke, 2002), and testosterone reduces neointimal plaque development in male rabbit aortas (Hanke, Lenz, Hess, Spindler, & Weidemann, 2001). PKCδ plays a role in mediating testosterone-induced apoptosis and inhibition of VSMC proliferation (Bowles, Maddali, Dhulipala, & Korzick, 2007). Overexpression of PKCδ in rat aortic VSMCs inhibits growth and proliferation, decreases thymidine incorporation, induces G0/G1 arrest, reduces cyclin D1 and E, and increases p27kip1, and PKCδ knockdown with siRNA diminishes the downregulation of cyclin D1 and E, and the upregulation of p21cip1 (Bowles et al., 2007; Fukumoto et al., 1997). Cleavage of PKCδ by caspase 3 and nuclear accumulation of catalytic PKCδ could be an important component of the apoptotic response induced by testosterone. It is believed that testosterone-induced increase in full-length PKCδ could cause an increase in caspase 3-mediated production of the 40-kDa catalytic fragment of PKCδ and lead to VSMC apoptosis (Bowles et al., 2007). PKCδ null mice exhibit decreased VSMC apoptosis and exacerbated vein graft arteriosclerosis (Leitges et al., 2001). The testosterone-induced PKCδ-dependent G1/S cell cycle arrest and stimulation of apoptosis may explain some of its beneficial effects on coronary vasculoproliferative disease, restenosis, and atherosclerosis (Bowles et al., 2007). While local conversion of testosterone to estrogen via aromatase could mediate some of the beneficial effects of testosterone (Yamada, Kimura, Harada, & Nakayama, 1990), it may not be involved in testosterone-induced PKCδ-mediated inhibition of coronary VSMC proliferation (Bowles et al., 2007).

PKC may also mediate some of the vascular effects of female sex hormones. For instance, females tolerate shock and sepsis better than males likely through a protective GPR30-estrogen receptor-mediated vascular response involving PKC (Angele, Frantz, & Chaudry, 2006; Li et al., 2014). Also, in mesenteric arteries from normal and shocked rats, estrogen increases the expression/activity of PKCε, and PKCε psuedosubstrate inhibitory peptide antagonizes the effect of estrogen on vascular reactivity in shocked rats (Li et al., 2014).

Sex differences in the expression/activity of PKC isoforms have also been observed in aortic VSM of male and female Wistar–Kyoto (WKY) and spontaneously hypertensive rats (SHR). VSM contraction and the expression/activity of PKCα, δ, and ζ are less in intact female compared with male WKY, and the sex-related differences are greater in VSM from SHR compared with WKY rats (Kanashiro & Khalil, 2001). PDBu-induced contraction and PKC activity are greater in ovariectomized (OVX) females than in intact female rats, and treatment of OVX females with 17β-estradiol subcutaneous implants reduces PDBu contraction and PKC activity to a greater extent in SHR than WKY rats. These observations have suggested sex-related reduction in VSM contraction and the expression/activity of PKCα, δ, and ζ in females compared with males, and that these differences are possibly mediated by estrogen, and are enhanced in hypertension (Kanashiro & Khalil, 2001).

9.3 Pregnancy-Related Changes in PKC

Normal pregnancy is associated with physiological changes in uterine blood flow caused by changes in uterine arterial Ca^{2+}-dependent phasic contraction and maintained DAG/PKC-mediated tonic contraction (Ford, 1995). PKC inhibitors decrease thromboxane A2-induced contraction in uterine and mesenteric arteries of nonpregnant rats and in mesenteric of pregnant rats, supporting a role of PKC in mediating VSM contraction during pregnancy (Goulopoulou, Hannan, Matsumoto, & Webb, 2012). PKC activity changes during the course of pregnancy, and PKC activity and vascular contraction are reduced in uterine artery of late pregnant ewes and gilts and aorta of late pregnant rats (Farley & Ford, 1992; Kanashiro, Cockrell, Alexander, Granger, & Khalil, 2000; Magness, Rosenfeld, & Carr, 1991). Also, the expression and subcellular redistribution of PKCα, δ, and ζ are reduced in aortic VSM of late pregnant rats (Kanashiro, Alexander, Granger, & Khalil, 1999; Kanashiro, Cockrell, et al., 2000). A decrease in PKC signaling is chiefly responsible for the decreased contractions in pregnant uterine arteries in order to maintain low basal uterine vascular tone and to accommodate the increased uterine blood flow during pregnancy (Xiao, Buchholz, & Zhang, 2006). The pregnancy-associated decrease in uterine vascular tone and increase in uterine blood flow may be caused by increased steroid hormones and their receptors. The sex steroids estrogen and progesterone have been shown to attenuate PKC-mediated signaling in uterine arterial VSMCs and uterine artery contraction and myogenic tone, partly through upregulation of K^+ channel expression/activity (Xiao et al., 2006; Zhu et al., 2013).

10. PKC IN VASCULAR INJURY AND DISEASE

In addition to its effects on vascular contraction/relaxation mechanisms, PKC has been implicated in multiple pathological processes involving VSM growth/proliferation, angiogenesis/vasculogenesis, apoptosis, vascular inflammation, restenosis, oxidative stress, and ischemia–reperfusion injury. Pathological changes in PKC expression/activity could cause vascular hyperreactivity and vascular remodeling leading to vascular disorders such as systemic and pulmonary HTN, preeclampsia, diabetic vasculopathy, atherosclerosis, and coronary artery disease (Fig. 5).

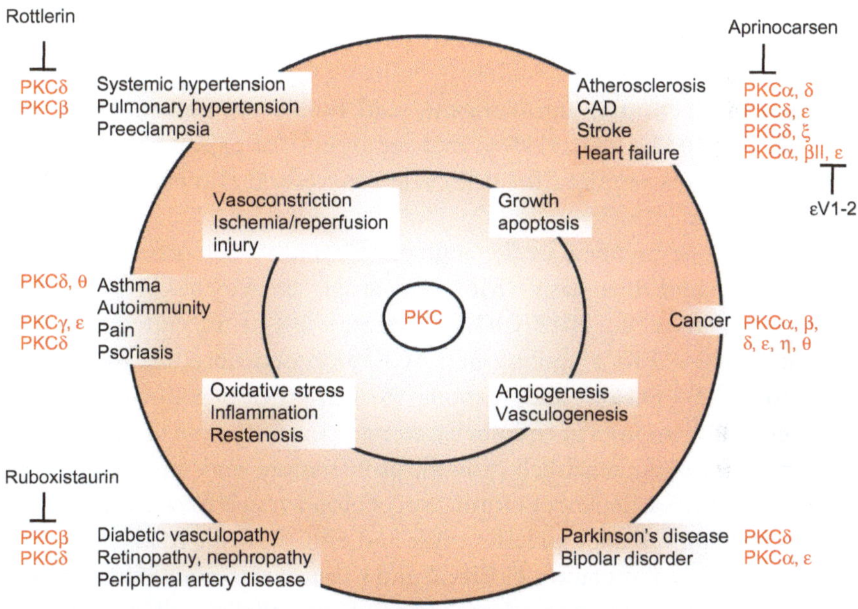

Fig. 5 Implications for PKC in pathological processes and diseases. PKC mediates many pathological processes in the vascular system and other tissue and organs, which could contribute to vascular diseases such as hypertension and atherosclerosis, and other diseases such as cancer, asthma, and autoimmune disease. The pathological changes in vascular PKC could also participate in diseases of other tissues and organs. For example, PKC through promoting angiogenesis could play a role in tumor growth. Also, PKC through promoting inflammation could play a role in asthma. PKC isoforms show varying contributions to different diseases, and isoform-specific PKC inhibitors such as the PKCα inhibitor aprinocarsen may be the key to combating their pathological effects while minimizing side effects. *CAD*, coronary artery disease.

10.1 PKC, VSM Growth, and Angiogenesis/Vasculogenesis

Studies have shown PKC translocation and localization to the nucleus in different cell types including VSM, suggesting interaction with nuclear factors and genes, and a role in the regulation of VSM growth and proliferation (Salamanca & Khalil, 2005). PKC isoforms exert different effects leading to either stimulation or suppression of cell growth (Clarke & Dodson, 2007). PKC regulates vascular endothelial growth factor (VEGF) at the gene transcription level (Carracedo, Sacher, Brandes, Braun, & Leitges, 2014; Monti et al., 2013). Also, PKCε is a powerful oncogene promoting cell growth and proliferation (Nishizuka, 1995) and has been used as a tumor biomarker (Duquesnes, Lezoualc'h, & Crozatier, 2011; Gorin & Pan, 2009). PKCβII is also an upstream regulator of early growth response-1 (Egr-1), a master switch that orchestrates the expression of diverse gene families that elicit a pathological response to hypoxia, ischemia/reperfusion, and vascular stress (Yan, Harja, Andrassy, Fujita, & Schmidt, 2006). In contrast, PKCδ is proapoptotic, antioncogene, and tumor suppressor (Duquesnes et al., 2011; Reddig et al., 1999), that suppresses the expression of positive regulatory factors required for cell cycle progression (Bowles et al., 2007; Fukumoto et al., 1997).

PKC may be involved in the angiogenesis and vasculogenesis associated with cancer and metastasis (Kim, Koyanagi, & Mochly-Rosen, 2011; Mochly-Rosen et al., 2012). VEGF activates PKCε in endothelial cells, and the selective PKCε agonist ψεRACK promotes fibroblast growth factor-2 (FGF-2) release and export to cell membrane, and induces proangiogenic responses in endothelial cells and the formation of capillary-like structures and endothelial cell growth, proliferation, and sprouting (Monti et al., 2013), PKCε-dependent formation of blood vessels may involve downstream signaling cascades including Akt and eNOS (Rask-Madsen & King, 2008). Double null mutation of PKCδ and ε causes embryonic lethality with defective blood vessel formation, impaired endothelial cell organization, dilated vessels, reduced endothelial-specific adherent junctions, decreased contact of endothelial cells with mural cells, deficient angiogenesis related transcripts, and almost undetectable α-smooth muscle actin, a classical marker for VSMC (Carracedo et al., 2014). On the other hand, PKCδ-deficient mice show increased number of SMCs and macrophages, accelerated neointimal lesions and intimal hyperplasia, and delayed reendothelialization in mouse wire-injured femoral artery. PKCδ knockdown using small hairpin RNA (shRNA) in cultured endothelial cells is also associated with reduced cell migration and accumulation of the antiangiogenesis protein vasohibin-1, and downregulation of vasohibin-1 restores the migration rate in PKCδ-deficient cells (Bai et al., 2010).

10.2 PKC and VSM Apoptosis

Apoptosis has been observed in cardiovascular diseases such as myocardial infarction, aneurysm, and ischemia/reperfusion injury. Whether apoptosis is beneficial or detrimental in vascular disease has been debated, but the finding of marked endothelial cell apoptosis in patients with peripheral vascular disease suggest that it may induce cell and tissue damage in certain conditions (Gardner et al., 2014).

PKCδ plays a role in apoptosis, and overexpression of the catalytic fragment of PKCδ alone is sufficient to induce apoptosis (Zhao et al., 2012). PKCδ is activated by a variety of proapoptotic stimuli including DNA damaging agents, ultraviolet (UV) radiation, the phorbol ester PMA, and reactive oxygen species (ROS). VSMCs from PKCδ null mice are resistant to apoptosis induced by UV, TNFα, or H_2O_2, and show defective caspase-3 activation in response to oxidative stress (Zhao et al., 2012). PKCδ is an early regulator of apoptosis, and may function upstream of the mitochondria as an integrator for various death signals in multiple cell types and under various stimuli. Cytosolic PKCδ may function at the initial stages of apoptosis. Tyrosine phosphorylation of PKCδ also occurs at the beginning of apoptosis and may be responsible for its translocation to the cell membrane, mitochondria, ER, and lysosomes (Zhao et al., 2012). A positive loop between mitochondrial-mediated caspase activation and PKCδ cleavage and activation supports the proapoptotic role of PKCδ in the cytoplasm. The nucleus may also act as a major target for PKCδ to amplify the apoptosis signal. Translocation of PKCδ to the nucleus may be essential for inducing apoptosis, and the proteolytically cleaved constitutively active catalytic fragment of PKCδ accumulates in the nucleus (DeVries, Neville, & Reyland, 2002; Zhao et al., 2012). PKCδ knockout mice are resistant to apoptosis in models of abdominal aortic aneurysm, and adenovirus-mediated delivery of PKCδ locally to the arterial wall is sufficient to restore aneurysm development in PKCδ knockout mice (Morgan et al., 2012).

In contrast with the proapoptotic properties of PKCδ, PKCδ may have antiapoptotic effects, as demonstrated in the response to the cytokine TNFα (Lu, Liu, Yamaguchi, Miki, & Yoshida, 2009; Ren et al., 2014). Silencing PKCδ expression by siRNA inhibits TNFα-mediated ERK1/2 activation

(Kilpatrick et al., 2006; Ren et al., 2014). PKCδ also interacts with the mitochondrial protein Smac, and exposure to apoptotic stimuli such as paclitaxel, disrupts the PKCδ–Smac interaction resulting in the release of Smac into the cytosol, activation of caspases in the cytochrome c/Apaf-1/caspase-9 pathway, and promotion of apoptosis. On the other hand, activation of PKCδ rescues the PKCδ–Smac interaction and suppresses paclitaxel-induced cell death (Masoumi, Cornmark, Lonne, Hellman, & Larsson, 2012; Ren et al., 2014). The factors that determine whether PKCδ exerts a pro- or anti-apoptotic role in a given cell remain to be examined.

10.3 PKC and Vascular Inflammation

Vascular inflammation is observed in cardiovascular diseases such as atherosclerosis and myocardial infarction (Ross, 1999). Proinflammatory chemokines released by VSMCs play a role in vascular inflammation and recruit inflammatory cells to the vascular wall (Brasier, 2010; Ren et al., 2014). PKCδ is upregulated in VSMCs of injured arteries such as in aneurysmal aortic tissues and in restenotic lesions, and could be involved in vascular inflammation (Morgan et al., 2012; Ren et al., 2014; Si et al., 2012). PKCδ knockout mice show diminished expression of VSM proinflammatory factors and inflammatory cell infiltration (Morgan et al., 2012). PKCδ may promote chemokine expression at the transcription level by activating NF-kB through an IkB-independent cytosolic interaction, which subsequently leads to enhanced p65 phosphorylation and DNA-binding affinity (Ren et al., 2014). Delivery of PKCδ to the aortic wall of PKCδ$^{-/-}$ mice restores aneurysm, whereas overexpression of a dominant negative PKCδ mutant in the aorta of wild-type mice attenuates aneurysm. Monocyte chemoattractant protein-1 (MCP-1) is one of several inflammatory chemokines in VSMCs induced by PKCδ-regulated genes, and could be involved in the PKCδ role in aneurysm formation (Ren et al., 2014). This is supported by reports that PKCδ gene deficiency reduces the production of MCP-1 and other cytokines by aortic VSMCs, and the ectopic administration of MCP-1 to the aortic wall of PKCδ knockout mice restores aneurysm development (Morgan et al., 2012).

PKCε is also likely involved in inflammation, as PKCε inhibition both by knockout in mice and peptide modulators suppress the acute and chronic inflammatory pain response (Hucho, Dina, & Levine, 2005; Koyanagi et al., 2007). Also, selective inhibition of PKCε with εV1–2 prolongs graft survival and improves functional recovery of the heart in cardiac transplantation models. PKCε inhibition attenuates the inflammatory response, decreases infiltration of macrophages, and T cells and the attachment of mononuclear inflammatory cells to the arterial wall, and reduces luminal narrowing and parenchymal fibrosis, thereby preserving cardiac tissue architecture after transplantation (Koyanagi et al., 2007).

PKC may play a role in the inflammation caused by prolonged Mg^{2+} deficiency (Altura et al., 2012). PKC is activated in rat VSMC exposed to short-term Mg^{2+} deficiency. In Mg^{2+} deficient animals, there may be cross-talk between PKC and the ceramide, sphingosine, NF-kB, and cytokine pathways in vascular cells. PKCζ, in particular, plays a role in de novo formation of ceramide, through a sphingolipid salvage pathway. Mg^{2+} supplements in drinking water prevents the upregulation of PKC isoforms in VSMCs when exposed to low $[Mg^{2+}]_o$ providing an effective solution to prevent inflammation induced by Mg^{2+} deficiency (Altura et al., 2014).

10.4 PKC and Vascular Restenosis

Long-term success of vascular bypass and angioplasty procedures is limited by restenosis particularly in obese and diabetic patients and PKC may contribute to vascular restenosis through the initial thrombosis and inflammation and the subsequent VSMC migration and proliferation (Ding, Chai, et al., 2011).

Thrombosis is involved in the early stages of vascular restenosis. PKCα, β, δ, and θ are expressed in platelets, and cPKCs may promote while nPKCs inhibit platelet aggregation and thrombus formation (Gilio et al., 2010). This is illustrated by reports that knocking out PKCδ or PKCθ potentiates murine platelet aggregation, and that the PKCδ inhibitor rottlerin potentiates human platelet aggregation (Ding, Chai, et al., 2011; Pula et al., 2006).

PKCα, β, and ζ also potentiate the expression of intercellular adhesion molecule-1 (ICAM-1) and vascular cell adhesion molecule-1 (VCAM-1), a key step in leucocyte recruitment, that leads to VSMC migration and proliferation and vascular stenosis (Abdala-Valencia & Cook-Mills, 2006; Javaid et al., 2003; Kouroedov et al., 2004). PKCα, β, and δ further affect VSMC migration by promoting actin polymerization and enhancing cell adhesion (Campbell & Trimble, 2005; Liu et al., 2007; Okazaki, Mawatari, Liu, & Kent, 2000). PKCε promotes VSMC migration by upregulating matrix metalloproteinases (MMPs), particularly MMP-2 and -9 (Ding, Chai, et al., 2011; Rodriguez-Pla et al., 2005; Thomas & Newby, 2010).

PKC also contributes to VSMC proliferation, the final step of vascular restenosis. PKCβ mediate synergistic proliferative effect of PDGF and high glucose on human coronary VSMCs (Ling et al., 2002), and the selective PKCβ inhibitor LY-379196 attenuates DNA synthesis and cell growth (Ding, Tsao, et al., 2011). PKCε seems to be involved in the development of neointimal hyperplasia. In rat models of aortic balloon injury, the PKCε activator ψεRACK promotes neointimal development, while the PKCε inhibitor εV1–2 reduces luminal narrowing, neointimal proliferation and VSMC ERK phosphorylation in vivo, and PDGF-induced VSMC proliferation/migration in vitro (Deuse et al., 2010).

10.5 PKC and Oxidative Stress

Oxidative agents such as H_2O_2 and superoxide activate PKC independent of classical PKC cofactors such as DAG. H_2O_2-induced activation of PKC may cause stimulation of arterial VSM L-type Ca^{2+} channels, and these effects are abolished by PKC inhibition. Also, hypoxia and vasoconstrictors such as AngII increase mitochondrial ROS production via PKC-mediated activation of NADPH oxidase (Nox) in pulmonary artery VSMCs (Doughan, Harrison, & Dikalov, 2008; Perez-Vizcaino, Cogolludo, & Moreno, 2010; Rathore et al., 2008).

PKC and ROS appear to be tightly coupled in causing vascular dysfunction, as ROS can activate PKC and vice versa (Novokhatska et al., 2013). Also, a positive feedback mechanism may amplify the production of ROS and PKC. For instance, excess ROS production may occur through PKC activation, and subsequent phosphorylation of p47 phox subunit and activation of NADPH oxidase. ROS then creates a positive feedback loop through activation of c-Src, which then amplifies NADPH oxidase activity to produce more ROS (Lyle & Griendling, 2006; Novokhatska et al., 2013).

ROS activates different PKC isoforms. In isolated pulmonary artery, H_2O_2-induced Ca^{2+} sensitization and constriction is associated with PKCα activation and abolished by PKC inhibitors (Perez-Vizcaino et al., 2010; Pourmahram et al., 2008). Also, exogenous H_2O_2, mimics hypoxia, and increases PKCε activity (Perez-Vizcaino et al., 2010; Rathore et al., 2006). In pulmonary artery VSMCs, mitochondrial-derived ROS may activate PKCε, which subsequently activates Nox-dependent ROS generation, further illustrating the positive feedback mechanism involved in hypoxia-induced increase in ROS (Perez-Vizcaino et al., 2010; Rathore et al., 2008).

PKC isoforms affect ROS production via different pathways. For instance, PKCε siRNA knockdown blocks ROS production by sphingosylphosphorylcholine (Shaifta et al., 2015). On the other hand, PKCζ appears to increase ROS through an insulin growth factor (IGF-I)-stimulated pathway, as high glucose induces NADPH oxidase 4 (Nox4) upregulation in a PKCζ/NF-kB-dependent manner in VSMCs and diabetic mice (Xi et al., 2012).

Oxidative stress may have other PKC-mediated vascular effects. In pulmonary artery, H_2O_2 may inhibit K_v channel by activating PKCα and ε. Also, in mesenteric artery, PKCα and ε may mediate the K_v channel inhibitory effect of ET-1 and AngII, respectively (Perez-Vizcaino et al., 2010; Rainbow et al., 2009). PKCζ also plays a role in the inhibition of K_v channels by U46619 and hypoxia in rat pulmonary artery VSMCs (Cogolludo et al., 2003, 2009). Although PKC is involved in hypoxia-induced ROS generation, PKCζ does not mediate K_v channel inhibition through ROS (Perez-Vizcaino et al., 2010).

10.6 PKC and Ischemia/Reperfusion Injury

PKC activity has been observed in ischemic injury in multiple tissues including the heart (Speechly-Dick, Mocanu, & Yellon, 1994), liver (Piccoletti, Bendinelli, Arienti, & Bernelli-Zazzera, 1992), and kidney (Padanilam, 2001). The regulation of cellular viability during an ischemic event may be influenced by the ratio of PKCδ and ε, as they display detrimental and protective effects, respectively (Churchill & Mochly-Rosen, 2007; Duquesnes et al., 2011).

Prolonged ischemia and reperfusion activate PKCδ more than PKCε, leading to translocation of PKCδ into the mitochondria and phosphorylation of pyruvate dehydrogenase kinase, which in turn phosphorylates pyruvate dehydrogenase, leading to a reduction in the tricarboxylic acid (TCA) cycle and ATP regeneration (Churchill, Murriel, Chen, Mochly-Rosen, & Szweda, 2005; Inagaki et al., 2003; Mochly-Rosen et al., 2012). Mitochondrial dysfunction causes increases in ROS production and lipid peroxidation leading to accumulation of ROS and toxic aldehydes, such as 4-hydroxynonenal (4HNE), that interact and inactivate macromolecules including proteins, DNA, and lipids. Mitochondrial dysfunction and increase in ROS leads to apoptosis, necrosis, and severe cardiac dysfunction (Armstrong & Whiteman, 2007; Mochly-Rosen et al., 2012).

Short bouts of ischemia and reperfusion prior to the prolonged ischemic event (ischemic preconditioning) provide cardioprotection by preferentially activating PKCε (Inagaki, Churchill, & Mochly-Rosen, 2006), which translocates into the mitochondria and prevents mitochondrial dysfunction induced by prolonged ischemia and reperfusion (Budas, Churchill, Disatnik, Sun, & Mochly-Rosen, 2010; Mochly-Rosen et al., 2012). PKCε-mediated protection occurs, in part, by PKCε-induced phosphorylation and activation of aldehyde dehydrogenase 2 (ALDH2) (Chen et al., 2008), which metabolizes aldehydes such as 4HNE, thus reducing the aldehyde load and the mitochondrial and cellular damage (Mochly-Rosen et al., 2012). In addition, mitochondrial function is preserved in preconditioned hearts through the inhibition of the mitochondrial permeation pore and K_{ATP} channel opening (Costa et al., 2006; Duquesnes et al., 2011). The reduced 4HNE levels also prevent direct inactivation of peroxisome and thus enable fast removal of aggregated proteins. Furthermore, ischemic preconditioning prevents I/R injury at reperfusion by protecting ATP-dependent 26S proteasomal function. The active proteasome also selectively degrades activated PKCδ, thus decreasing the accumulation of the proapoptotic PKCδ at cardiac mitochondria and increasing the balance in favor of the cardiac protective and prosurvival PKCε (Budas, Churchill, & Mochly-Rosen, 2007; Churchill, Ferreira, Brum, Szweda, & Mochly-Rosen, 2010; Mochly-Rosen et al., 2012). PKC-induced closure of connexons may also participate in ischemic preconditioning by an unclear mechanism (Duquesnes et al., 2011; Naitoh et al., 2009).

Ischemic stroke represents a major cause of death and disability among elderly, and the presence or absence of reperfusion is an important variable affecting outcome (Aronowski & Labiche, 2003). PKCβI and βII are increased in infarcted tissue of an ischemic stroke, whereas PKCγ increases 2–24-fold in the ischemic penumbra, but not in the infarcted tissues (Krupinski, Slevin, Kumar, Gaffney, & Kaluza, 1998; Young, Balin, & Weis, 2005). PKCγ may play a contrasting role in regulating the vulnerability of tissue to I/R-induced damage, as it functions first as a deleterious factor during evolution of intraischemic neuronal damage, then as a neuroprotective factor during postischemic reperfusion (Aronowski & Labiche, 2003). PKCγ may carry out its neuroprotective role in reversible focal ischemia by protein phosphorylation, as impaired protein phosphorylation in PKCγ knockout mice influences the overall infarct volume. Studies have shown larger infarct volumes in PKCγ knockout compared with wild type, and inhibitors of the protein phosphatase calcineurin reduced infarct volume in the PKCγ knockout mice (Aronowski, Grotta, Strong, & Waxham, 2000; Young et al., 2005).

In the cerebral circulation, PKCδ is believed to have a deleterious role in cerebral reperfusion. A model of transient middle cerebral artery occlusion demonstrated that PKCδ-null mice showed a 70% reduction in stroke size compared with wild-type mice (Chou et al., 2004; Young et al., 2005). PKCδ may mediate its detrimental effects in cerebral reperfusion by affecting neutrophil migration into ischemic tissue. PKCδ null mice show impaired neutrophil function and decreased neutrophil migration into ischemic tissues, and transplantation of bone marrow from the PKCδ-null mice into the wild-type mice reduces infarct size while bone marrow transplantation from wild-type donors increased infarction size and worsened neurological scores in PKCδ-null mice (Chou et al., 2004; Young et al., 2005). Inhibition of PKCδ improves microvascular pathology and function in transient focal ischemia in normotensive animals and chronic hypertension, and reduces ischemic damage following an ischemic event. Thus PKCδ could be an important therapeutic target for the preservation of microcerebrovascular function following stroke, and its inhibition may reduce stroke risk and damage in hypertensive patients (Bright, Steinberg, & Mochly-Rosen, 2007). PKCζ also appears to be a downstream component of NMDA-induced excitotoxic neuronal cell death, as inhibiting PKCζ and its translocation, prevents NMDA-induced cell death. PKCζ mRNA is also induced in the cerebral cortex after focal brain ischemia (Koponen et al., 2003; Young et al., 2005). In contrast, PKCε is activated during cerebral ischemia in vivo and may play a role in mediating the early cellular response to ischemic stress, possibly mediating ischemic tolerance. Systemic delivery of PKCε-selective peptide activator ψεRACK confers neuroprotection against a subsequent cerebral ischemic event when delivered immediately prior to stroke. In addition, activation of PKCε by ψεRACK decreases vascular tone and microvascular cerebral blood flow, which may contribute to the conferred protection (Bright, Sun, Yenari, Steinberg, & Mochly-Rosen, 2008).

10.7 PKC and Coronary Artery Disease

PKCδ may contribute to coronary artery disease, through increased ROS formation, decreased ATP generation, and increased apoptosis and necrosis (Churchill & Mochly-Rosen, 2007; Inagaki et al., 2003; Mochly-Rosen et al., 2012). PKCε, on the other hand, is protective, as it protects mitochondrial functions and proteasomal activity, activates ALDH2 and reduces aldehyde load (Budas et al., 2007; Chen et al., 2008; Mochly-Rosen et al., 2012; Mochly-Rosen & Kauvar, 2000). A combination of a PKCδ inhibitor and a PKCε activator could be useful for organ preservation and in prevention of ischemia–reperfusion injury and graft coronary artery disease in cardiac transplantation (Tanaka et al., 2004). In a case–control study, ischemia–reperfusion injury was the strongest alloantigen-independent factor for the subsequent development of graft coronary artery disease (Gaudin et al., 1994), and PKC modulators could modulate this pathological process.

10.8 PKC and Hypertension

Increased PKC activity could play a role in HTN, and mutations of PKC may influence the individual susceptibility to vascular hyperreactivity. For instance, a consistent association is found between the single nuclear polymorphism rs9922316 in PKCβ gene (*PRKCB*) and interindividual variation in the constriction responses of dorsal hand vein to the selective α2 adrenergic receptor agonist dexmedetomidine (Posti et al., 2013). Also, PKCδ mRNA expression and protein levels are increased in VSM from SHR rats. PKC could increase VSM contraction in HTN by altering BK_{Ca} channel conductance, as the PKC inhibitor chelerythrine restores K^+ channel activity in SHR (Novokhatska et al., 2013).

Sleep apnea could cause systemic and pulmonary HTN (Campen, Shimoda, & O'Donnell, 2005; Snow, Gonzalez Bosc, Kanagy, Walker, & Resta, 2011). Rat models of sleep apnea produced by exposure to eucapnic intermittent hypoxia display increased circulating ET-1 levels and ET-1-dependent systemic HTN, that is likely mediated by PKCδ-dependent VSM Ca^{2+} sensitization in systemic arteries (Allahdadi, Duling, Walker, & Kanagy, 2008; Snow et al., 2011). On the other hand, in the pulmonary circulation, intermittent hypoxia appears to mediate a PKCβ-dependent increase in reactivity to different receptor-mediated vasoconstrictor agonists including ET-1 (Snow et al., 2008).

PKC isoforms have been implicated in hypoxia-associated pulmonary HTN by affecting both Ca^{2+} influx and Ca^{2+} sensitization in pulmonary artery VSM. In normal small pulmonary arteries, PKC inhibitors attenuate ET-1-induced constriction and $[Ca^{2+}]_i$ as well as vasoconstrictor responses associated with store-operated Ca^{2+} entry, suggesting that PKC contributes to both Ca^{2+} sensitization and Ca^{2+} influx (Jernigan & Resta, 2014). Also, in fawnhooded rat model of pulmonary HTN, PKC inhibits BK_{Ca}, resulting in indirect activation of VGCC and pulmonary vasoconstriction (Zhu,

White, & Barman, 2008). PKC-mediated Ca^{2+} sensitization is also demonstrated by an increase in total and phosphorylated active CPI-17 levels in pulmonary arteries from newborn swine exposed to hypoxia (Dakshinamurti, Mellow, & Stephens, 2005).

PKCε may have divergent effects in pulmonary HTN. PKCε null mice show decreased acute hypoxic pulmonary vasoconstriction, increased Kv3.1b channel expression and membrane hyperpolarization (Littler et al., 2003). On the other hand, PKCε null mice show a greater increase in pulmonary arterial pressure compared to wild-type mice following chronic hypoxia exposure. The increase in pressure is reversed by inhaled NO suggesting that PKCε may be an important signaling intermediate in the hypoxic regulation of NO synthase (Littler et al., 2005).

Hypertension in pregnancy and preeclampsia are major complications of pregnancy and placental ischemia/hypoxia could be an initiating event. Chronic hypoxia enhances uterine vascular tone in pregnant sheep and is associated with an increase in PKC activity (Chang, Xiao, Huang, Longo, & Zhang, 2009). Hypoxia during pregnancy may attenuate the effects of sex steroid hormones/receptors, leading to enhanced PKC activation in pregnant uterine arteries. Increased BK_{Ca} channel activity inhibits PKC-mediated contraction in ovine uterine arteries during pregnancy, and gestational hypoxia may upregulate PKC and inhibit BK_{Ca} (Xiao, Zhu, & Zhang, 2014). Hypoxia may also inhibit K_{IR} channels via PKC-dependent mechanism, and may contribute to the maladaptation of uterine vascular hemodynamics in preeclampsia and the fetal intrauterine growth restriction in response to hypoxia (Zhu et al., 2013). Also, in cultured rat cardiomyocytes, treatment with IgG obtained from preeclamptic women enhances AT_1R-mediated response, which is ameliorated with the PKC inhibitor calphostin C, further supporting a role of PKC in preeclampsia (Wallukat et al., 1999).

10.9 PKC and Diabetic Vasculopathy

Diabetes mellitus is a complex syndrome of multiple disorders including vascular dysfunction. PKC could play a role in diabetes-related vascular pathology through multiple mechanisms including cell growth and proliferation, cell permeability, oxidative stress, increased vascular reactivity, inhibition of K^+ channels and Na^+-K^+-ATPase, activation of cytosolic phospholipase A2, vascular remodeling and increased ECM, and vascular inflammation and increased proinflammatory cytokines (Koya & King, 1998; Meier &

King, 2000; Nishizuka, 1992). In diabetes, PKC is activated by advanced glycation end (AGE) products and polyol pathway flux (Geraldes & King, 2010; Kizub et al., 2014; Thallas-Bonke et al., 2008). Also, chronic hyperglycemia stimulates synthesis of DAG and activates DAG-dependent cPKCs and nPKCs in cultured bovine aortic endothelial cells and VSM (Inoguchi et al., 1992). Fatty acids, especially the unesterified forms and their coenzyme A esters, work synergistically with DAG to activate PKC (Clarke & Dodson, 2007). PKC is also activated by ROS generated by different oxidases and the mitochondrial electron transport chain, and following AGE: RAGE (AGE receptor) interactions (Liu & Heckman, 1998).

High glucose-induced activation of PKC could cause vascular dysfunction by altering the expressions of growth factors such as VEGF, PDGF, and transforming growth factor-β (Lizotte et al., 2013; Yokota et al., 2003), which in turn affect the expression of ECM proteins (Jakus & Rietbrock, 2004). Also, in diabetes, activated PKC increases endothelial cell permeability and decreases blood flow and the production of and responsiveness to angiogenic factors, and this may contribute to the loss of capillary pericytes, retinal permeability, ischemia, and neovascularization (Aiello et al., 1994; Huang & Yuan, 1997; Lizotte et al., 2013; Pomero et al., 2003; Williams, Gallacher, Patel, & Orme, 1997). PKC also activates NADPH oxidases and increases ROS (Gao & Mann, 2009; Inoguchi et al., 2000; Kizub et al., 2014). In hyperglycemia, VSMCs show increased DNA synthesis and contraction to PMA and reduced apoptosis, and these effects are blocked by the PKC inhibitor calphostin C (Geraldes & King, 2010; Hall, Matter, Wang, & Gibbons, 2000). In diabetes, PKC may enhance vascular reactivity by inhibition of K^+ channels and promoting Ca^{2+} sensitization in VSM myofilaments (Kizub et al., 2014; Nelson & Quayle, 1995). High glucose via PKC activation and oxidative stress also reduces arterial SMC K_v current resulting in VSM depolarization and vasoconstriction (Liu, Terata, Rusch, & Gutterman, 2001; Rainbow, Hardy, Standen, & Davies, 2006; Straub et al., 2009). Diabetic patients also often have reduced nocturnal BP dip and increased vascular complications regardless of the average BP, partly due to lack of diurnal PKC inhibition (Nakano et al., 1991; Palmas et al., 2008).

PKC may contribute to diabetic nephropathy by increasing endothelial permeability to albumin and other macromolecules. PKC also induces ECM protein synthesis by mesangial cells and promotes sclerosis (Heilig et al., 2013; Henry, Busik, Brosius, & Heilig, 1999; Rovin, Yoshiumura, & Tan, 1992). In cultured mesangial cells, cGMP suppresses PKC-mediated actions including matrix protein production, and impaired NO-mediated cGMP generation in mesangial cells could amplify the PKC signal and increase matrix protein synthesis in diabetes (Craven, Studer, & DeRubertis, 1994; Derubertis & Craven, 1994; Williamson et al., 1993). Podocyte injury or loss is a hallmark of diabetic nephropathy and PKC contributes to the progression of glomerular injury (Teng, Duong, Tossidou, Yu, & Schiffer, 2014). PKC also mediates diabetic glomerulosclerosis partially through its interaction with GLUT1, which facilitates the movement of glucose into the cell (Heilig et al., 2013; Koya et al., 1997). PKC stimulates TGF-β-mediated effects including activation of the highly profibrotic cytokine CTGF, and engagement of the TGF-β receptor, triggering more GLUT1 synthesis via MAPK, and ECM protein synthesis via Smads (Heilig et al., 2013; Qi et al., 2005; Twigg et al., 2001). Also, AngII interaction with AT1 receptor stimulates DAG/PKC causing additional GLUT1 synthesis, and AngII and GLUT1 activation of PKC could promote glomerulosclerosis through both TGF-β-dependent and -independent pathways (Heilig et al., 2013; Henry et al., 1999; Koya et al., 1997).

PKC isoforms play different roles in diabetic vasculopathy. In VSMCs, high glucose activates PKCα, β, δ, and ε but not PKCζ (Haller et al., 1995; Igarashi et al., 1999; Lizotte et al., 2013). PKCβ and δ appear to be the dominant PKCs involved in diabetes. In rat VSMCs, high glucose increases the membrane fraction expression of PKCβ and PKCδ, p38 MAPK phosphorylation, and arachidonic acid release (Geraldes & King, 2010; Igarashi et al., 1999). PKCβ is implicated in insulin resistance. Transgenic mice overexpressing PKCβII exhibit decreased Akt activation in vascular cells after insulin stimulation (Geraldes & King, 2010; Naruse et al., 2006). PKC could prevent insulin actions on the PI_3K pathway at the insulin receptor substrate (IRS) level (Sampson & Cooper, 2006), but could accentuate insulin actions on the ERK1/2 pathway (Bakker et al., 2008). Thus, PKC could mediate selective insulin resistance by enhancing insulin's proatherosclerotic mechanisms via ERK1/2 signaling or inhibiting its antiatherosclerotic mechanisms by inhibiting the PI_3K/Akt pathway (Geraldes & King, 2010). PKCβ activation by hyperglycemia may play a role in mediating the microvascular disease complications of retinopathy, nephropathy, and neuropathy. Hyperglycemia leads to chronic activation of PKCβ, causing aberrant signaling and other pathologies including cytokine activation and inhibition, vascular alterations, cell cycle and transcriptional factor dysregulation, and abnormal angiogenesis (Geraldes & King, 2010; Mochly-Rosen et al., 2012). PKCβ is chiefly responsible in causing diabetic retinopathy by affecting VEGF expression through the mRNA-stabilizing human embryonic lethal abnormal vision protein, HuR, in the retina (Amadio et al., 2010; Gogula, Divakar, Satyanarayana, Kumar, & Lavanaya, 2013). PKCβ may mediate diabetes-induced increase in vascular contraction by inhibiting BK_{Ca} channel, and PKCβ inhibition restores BK_{Ca}-mediated vasodilation in diabetic mice. Also, reduced expression of the $BK_{\beta I}$ channel subunit in arteries of STZ-induced diabetic mice and in human coronary artery VSMCs cultured with high glucose has been related to increased PKCβ expression (Kizub et al., 2014; Lu et al., 2012).

The nPKCs could also contribute to insulin resistance by serine phosphorylation and inhibition of IRS1 (Ritter, Jelenik, & Roden, 2015; Yu et al., 2002). PKCδ plays a role in islet cell function and insulin response, and changes in PKCδ expression/activity among mice strains correlate with insulin resistance and glucose intolerance. Also, mice with global or liver-specific downregulation of the PKCδ gene (*PRKCD*) display increased hepatic insulin signaling and improved glucose tolerance with aging. Conversely, mice with liver-specific overexpression of PKCδ develop hepatic insulin resistance and decreased insulin signaling (Bezy et al., 2011). Diabetes-induced PKCδ activation also decreases responsiveness to PDGF leading to pericyte apoptosis, acellular capillaries, and diabetic retinopathy (Geraldes et al., 2009). PKCδ is also likely involved in poor collateral vessel formation in diabetes, as the ischemic adductor muscles of diabetic *PRKCD* knockout mice show increased blood flow and capillary density compared with diabetic $PRKCD^{+/+}$ mice. The poor angiogenesis response in ischemic diabetic muscles could be caused by PKCδ-induced expression of Src homology-2 domain-containing phosphatase-1 (SHP-1), contributing to VEGF and PDGF unresponsiveness (Lizotte et al., 2013). PKCδ may also mediate inhibition of K^+ current in aortic SMCs, and PKCδ gene silencing by siRNAs restores VSMCs K^+ current and endothelium-dependent vasodilatation in aorta of streptozotocin-induced diabetic rats (Kizub et al., 2014; Klymenko et al., 2014). Endothelium-independent vasoconstriction mediated by EP1-/EP3-receptors activation is also enhanced in mesenteric arteries of diabetic rats and highly sensitive to PKCδ inhibition (Ishida, Matsumoto, Taguchi, Kamata, & Kobayashi, 2012; Kizub et al., 2014).

Ruboxistaurin (LY333531) is an oral PKCβII inhibitor commonly used in cellular, animal, and human studies (Geraldes & King, 2010). Ruboxistaurin has been tested in diabetic retinopathy, nephropathy, and neuropathy, and is well tolerated (Aiello et al., 2011; Kizub et al., 2010; Mehta et al., 2009). Ruboxistaurin decreases vessel permeability and the onset of diabetic macular edema, improves retinal condition in diabetic patients, and prevents reduction of visual acuity (Geraldes & King, 2010; Gogula et al., 2013). While ruboxistaurin preserves visual acuity by decreasing capillary permeability or targeting the neural retina it may not delay the progression of diabetic retinopathy, and inhibiting PKCβ alone may not be sufficient to stop the early metabolic changes that drive the progression of preproliferative diabetic retinopathy (Geraldes & King, 2010). Indolylmaleimide and its derivatives are nonselective PKC inhibitors that reduced diabetic complications such as nephropathy, cardiomyopathy, and neuropathy in clinical trials (Kizub et al., 2014; Sobhia et al., 2013), but the lack of selectivity on PKC isoforms raises concerns regarding safety. Interestingly, some of the drugs already in clinical use for vascular disease may mediate some of their effects through inhibition of PKC. For instance, metformin and liraglutide (a glucagon like peptide-1) could prevent diabetic cardiovascular complications and atherosclerosis (Batchuluun et al., 2013). In cultured human endothelial cells, both metformin and liraglutide prevent high glucose-induced oxidative stress through inhibition of PKC-NADPH oxidase pathway, and these effects are enhanced when the drugs are combined. In cells treated with metformin and liraglutide, hyperglycemia fails to induce PKCβII translocation and phosphorylation of endogenous PKC. Also, both drugs inhibit p47phox translocation and NADPH oxidase activation, and prevent high glucose-induced changes in intracellular DAG level and phosphorylation of AMP-activated protein kinase (AMPK) (Batchuluun et al., 2013).

10.10 PKC and Atherosclerosis

Atherosclerosis results from deposition of lipid and chronic inflammation in the arterial wall, and PKC has been linked to many of the pathways involved in atherosclerosis. PKC expression is higher in plaques from atherosclerotic patients compared with control subjects. Also, PKC expression is higher in atherosclerotic rabbit aorta VSMC than in the control group and in unstable vs stable plaques (Sirikci, Ozer, & Azzi, 1996). PKC is also positively correlated with the increase in adipose differentiation-related protein, a protein present in higher amounts in unstable atherosclerotic vs stable plaques. PKC may further potentiate plaque-formation through endothelial dysfunction, foam cell formation, and VSMC proliferation. PKC may also contribute to the increased thickness of the intima and media of the vessel wall in atherosclerosis due to an imbalance between proliferation and apoptosis (Xu, Zhao, Wang, Sun, & Qin, 2015).

PKC isoforms play a varying role in the atherosclerotic process. Both PKCβ and δ are potential therapeutic targets as PKCβ potentiates atherosclerotic formation, and PKCδ appears to have an opposite effect (Fan, Fernandez-Hernando, & Lai, 2014). Depletion of PKCβ gene or treatment with LY333531 in apolipoprotein E-deficient mice decreases atherosclerosis by inhibiting the early growth response (Egr)-1 protein, which regulates VCAM expression and matrix metalloproteinase-2 activity (Geraldes & King, 2010; Harja et al., 2009). On the other hand, PKCδ deletion promotes arteriosclerosis, partly due to the lack of PKCδ-mediated VSMC apoptosis (Geraldes & King, 2010; Leitges et al., 2001). Also, PKCδ mediates collagen I secretion in VSMCs, and tight regulation of collagen is critical to the stability of atherosclerotic plaque. PKCδ knockout mice show marked reduction of collagen I in the arterial wall. PKCδ may also regulate the trafficking of collagen by controlling its exit from the trans-Golgi network through a mechanism involving cell division cycle 42 (Cdc42) protein (Lengfeld et al., 2012).

Vascular calcification contributes to atherosclerosis, as it reduces elasticity of blood vessels, and PKC may coordinate between cytoskeletal changes and hyperphosphatemia-induced vascular calcification. Expression and phosphorylation of both PKCα and δ decrease during inorganic phosphate (Pi)-induced VSMC calcification. Knockdown of PKCα and δ accelerates Pi-induced calcification in VSMCs and aorta in culture through upregulation of osteogenic signaling. Inhibition of PKCα and δ may also induce disassembly of microtubule and actin, respectively (Lee, Kim, & Jeong, 2014).

11. CONCLUSION

PKC is a major regulator of vascular function and a potential target in several pathological processes. Although significant information is currently available on PKC, it is important to further our knowledge of the role of PKC in vascular disease and the mechanisms behind its contribution. Research efforts have been limited by the existence of several PKC isoforms, the nonuniform expression and distribution of PKC throughout the vascular tree, and the poor specificity of chemical inhibitors (Schubert, Lidington, & Bolz, 2008). Continued research should further define the specific characteristics of the different PKC isoforms that determine their subcellular localization, phosphorylation pattern, and potential substrates. The precise knowledge of the structural aspects of PKC isoforms should allow the development of new tools to evaluate PKC function and potential new therapies. The development of FRET-based reporters of PKC activity (Braun, Garfield, & Blumberg, 2005; Violin, Zhang, Tsien, & Newton, 2003) and new peptides directed toward other domains than those presently utilized in the V1 region is a step in the right direction (Churchill et al., 2009; Duquesnes et al., 2011). Also, while PKC could play a role in vascular disease, this should not minimize its role in other pathological processes and diseases (Fig. 5). It is important to continue research of the role of PKC isoforms in various diseases, as there are still many uncertainties about their exact mechanism of action. For instance, while proapoptotic PKCδ likely acts as a tumor suppressor (Hampson et al., 2005; Zhao et al., 2012), in some cases it enhances tumorigenesis, and PKCδ-deficient mice are protected from urethane-induced lung tumor (Symonds et al., 2011; Zhao et al., 2012).

In order to enhance selectivity, it is important to determine the precise cellular location of various PKC isoforms, both in the resting and in the active states. The subcellular location of PKC may determine the state of VSM activity, and may be useful in the diagnosis/prognosis of HTN (Salamanca & Khalil, 2005). Since some PKC-mediated pathways involve PKC translocation to the nucleus, such as in apoptosis-induction by PKCδ, new nuclear targets of PKCδ such as the recently identified C/EBPa and hnRNPK could limit apoptosis (Gao et al., 2009; Zhao et al., 2012). Similarly, the translocation activator peptides ψδRACK attenuates Ccl2 production, providing a way to specifically block PKCδ-regulated proinflammatory chemokines (Ren et al., 2014). Genetic differences in PKC may also alter its effects, and studies have suggested a new role for PKC in inhibiting store-operated Ca^{2+} entry in the hypertensive pulmonary circulation of Sprague–Dawley, but not Wistar rats. The precise genetic differences responsible for this discrepancy in VSM Ca^{2+} regulation, as well as in other PKC-mediated effects, should be further explored (Snow, Kanagy, Walker, & Resta, 2009).

Modulation of PKC activity presents an attractive target for drug development in vascular disease and other related conditions. Despite the promise of PKC modulators, results in clinical trials have been mixed and often negative (Mochly-Rosen et al., 2012). Isoform-specific PKC inhibitors have shown some promise in clinical trials (Table 5). Partial prevention of the progress of malignancies was found in early phases of clinical trials of the PKC inhibitors UCN-01 and CGP41251 (Shen, 2003). PKC inhibitors could also be useful in treatment of Ca^{2+} antagonist-resistant forms of

Table 5 PKC Isoforms Involved in Specific Vascular Diseases and PKC Modulators Used as Potential Therapeutic Tools in Clinical Trials

PKC	Vascular Disease	Role of PKC	Drug	Effect on PKC	Outcome	Comments	Reference
PKCβ	Diabetic Retinopathy	Detrimental. Cytokine activation and inhibition, vascular alterations, cell cycle and transcriptional factor dysregulation, abnormal angiogenesis, increased matrix protein synthesis	Ruboxistaurin	Inhibit	Reduced sustained moderate vision loss in large studies	Under review by FDA for diabetic retinopathy	Tuttle et al. (2005), Vinik et al. (2005), Geraldes and King (2010), Aiello et al. (2011), and Mochly-Rosen et al. (2012)
			Ruboxistaurin	Inhibit	Failed to improve kidney outcomes	Studied as secondary outcome in large retinopathy trials	
			Ruboxistaurin	Inhibit	Mild decrease in symptoms	Requires validation in larger study	
PKCδ	Ischemic Heart Disease	Detrimental. Increases ROS production, decreases ATP generation, increases apoptosis and necrosis	Delcasertib for acute myocardial infarction (MI)	Inhibit	No benefit when given intravenously	Positive biomarker trend when given to patients with TIMI 0/1 flow	Inagaki et al. (2003), Churchill and Mochly-Rosen (2007), Bates et al. (2008), and Mochly-Rosen et al. (2012)
			KAI-9803: Phase I clinical trial, intracoronary injection during primary percutaneous coronary intervention	Selective PKCδ RACK antagonist	Signs of potential drug activity (not dose dependent)	Acceptable safety and tolerability	

PKCε	Ischemic Heart Disease	Protective. Protection of mitochondrial functions and proteasomal activity, activation of ALDH2 and reduction of aldehydic load	Adenosine for acute MI	↑PKCε	Reduced infarct size from 27% to 11% when given at 70 mcg/kg/min	No reduction in composite endpoint of death and CHF	Mochly-Rosen and Kauvar (2000), Ross, Gibbons, Stone, Kloner, and Alexander (2005), Budas et al. (2007), Chen et al. (2008), and Mochly-Rosen et al. (2012)
			Adenosine for coronary bypass grafting	↑PKCε	Reduction in composite AMI, mortality, need for pressors postoperatively	Requires validation in larger study	
			Acadesine for coronary bypass grafting	↑PKCε	Reduced 2-year mortality in the small group of patients who had a postoperative acute MI	No reduction in death, acute MI, or stroke	

HTN where the Ca^{2+}-independent PKC isoforms could be targeted (Salamanca & Khalil, 2005). Inhibitors of PKCβ and δ may reduce fat accumulation, improve glucose tolerance, decrease hepatosteatosis, and suppress foam cell formation in obesity and hyperlipidemia-induced atherosclerosis (Bezy et al., 2011; Fan et al., 2014; Huang, Bansode, Bal, Mehta, & Mehta, 2012). Also, a PKCβ inhibitor or PKCε activator may reduce damage secondary to endothelial dysfunction and VSMC proliferation in patients with atherosclerosis caused by long-term smoking, HTN, or diabetes (Fan et al., 2014; Harja et al., 2009; Huang et al., 2010; Monti, Donnini, Giachetti, Mochly-Rosen, & Ziche, 2010). Activators of PKCε may also be useful in coronary artery disease, and the PKCε activator acadesine reduced the 2-year mortality in patients with postoperative acute myocardial infarction after coronary bypass grafting (Mochly-Rosen et al., 2012).

Since PKC modulates many physiological functions, unwanted effects may occur when nonselective PKC inhibitors are administered systemically. Sustained delivery of peptide inhibitors of PKC for 2 months is safe in animals (Ding, Tsao, et al., 2011; Inagaki, Koyanagi, Berry, Sun, & Mochly-Rosen, 2008; Tanaka et al., 2004). Nevertheless, local delivery of PKC inhibitors may be a better approach. For example, to prevent restenosis, PKC inhibitors could be coated onto stents or balloons to be directly released into the injured area at effective concentrations. PKCβII and δ inhibitors coated stents or balloons showed efficacy and safety in animal trials (Ding, Tsao, et al., 2011).

PKC siRNA may hold the promise to target specific PKC isoforms in vascular disease. PKCδ gene silencing with the short-hairpin RNAs (shRNAs)-plasmid delivery system administered intravenously restores the vasodilator potential and normalize vascular function and high BP in SHR (Novokhatska et al., 2013). Also, PKCδ siRNA attenuates the proinflammatory effect of human CRP in diabetic rats (Jialal, Machha, & Devaraj, 2013). Further research should help design more specific and effective remedies of PKC-mediated vascular disease and other PKC-related conditions.

CONFLICT OF INTEREST

None.

ACKNOWLEDGMENTS

This work was supported by grants from National Heart, Lung, and Blood Institute (HL-65998, HL-98724, HL-111775).

REFERENCES

Abdala-Valencia, H., & Cook-Mills, J. M. (2006). VCAM-1 signals activate endothelial cell protein kinase Calpha via oxidation. *The Journal of Immunology*, *177*, 6379–6387.

Adwan, T. S., Ohm, A. M., Jones, D. N., Humphries, M. J., & Reyland, M. E. (2011). Regulated binding of importin-alpha to protein kinase Cdelta in response to apoptotic signals facilitates nuclear import. *The Journal of Biological Chemistry*, *286*, 35716–35724.

Ahluwalia, A., MacAllister, R. J., & Hobbs, A. J. (2004). Vascular actions of natriuretic peptides. Cyclic GMP-dependent and -independent mechanisms. *Basic Research in Cardiology*, *99*, 83–89.

Ahn, J. H., McAvoy, T., Rakhilin, S. V., Nishi, A., Greengard, P., & Nairn, A. C. (2007). Protein kinase A activates protein phosphatase 2A by phosphorylation of the B56delta subunit. *Proceedings of the National Academy of Sciences of the United States of America*, *104*, 2979–2984.

Aiello, L. P., Avery, R. L., Arrigg, P. G., Keyt, B. A., Jampel, H. D., Shah, S. T., et al. (1994). Vascular endothelial growth factor in ocular fluid of patients with diabetic retinopathy and other retinal disorders. *The New England Journal of Medicine*, *331*, 1480–1487.

Aiello, L. P., Vignati, L., Sheetz, M. J., Zhi, X., Girach, A., Davis, M. D., et al. (2011). Oral protein kinase c beta inhibition using ruboxistaurin: Efficacy, safety, and causes of vision loss among 813 patients (1,392 eyes) with diabetic retinopathy in the Protein Kinase C beta Inhibitor-Diabetic Retinopathy Study and the Protein Kinase C beta Inhibitor-Diabetic Retinopathy Study 2. *Retina*, *31*, 2084–2094.

Akimoto, K., Mizuno, K., Osada, S., Hirai, S., Tanuma, S., Suzuki, K., et al. (1994). A new member of the third class in the protein kinase C family, PKC lambda, expressed dominantly in an undifferentiated mouse embryonal carcinoma cell line and also in many tissues and cells. *The Journal of Biological Chemistry*, *269*, 12677–12683.

Albert, A. P., & Large, W. A. (2006). Signal transduction pathways and gating mechanisms of native TRP-like cation channels in vascular myocytes. *The Journal of Physiology*, *570*, 45–51.

Allahdadi, K. J., Duling, L. C., Walker, B. R., & Kanagy, N. L. (2008). Eucapnic intermittent hypoxia augments endothelin-1 vasoconstriction in rats: Role of PKCdelta. *American Journal of Physiology. Heart and Circulatory Physiology*, *294*, H920–H927.

Altura, B. M., Shah, N. C., Shah, G., Zhang, A., Li, W., Zheng, T., et al. (2012). Short-term magnesium deficiency upregulates ceramide synthase in cardiovascular tissues and cells: Cross-talk among cytokines, Mg2+, NF-kappaB, and de novo ceramide. *American Journal of Physiology. Heart and Circulatory Physiology*, *302*, H319–H332.

Altura, B. M., Shah, N. C., Shah, G. J., Zhang, A., Li, W., Zheng, T., et al. (2014). Short-term Mg deficiency upregulates protein kinase C isoforms in cardiovascular tissues and cells; relation to NF-kB, cytokines, ceramide salvage sphingolipid pathway and PKC-zeta: Hypothesis and review. *International Journal of Clinical & Experiment Medicine*, *7*, 1–21.

Amadio, M., Bucolo, C., Leggio, G. M., Drago, F., Govoni, S., & Pascale, A. (2010). The PKCbeta/HuR/VEGF pathway in diabetic retinopathy. *Biochemical Pharmacology*, *80*, 1230–1237.

Angele, M. K., Frantz, M. C., & Chaudry, I. H. (2006). Gender and sex hormones influence the response to trauma and sepsis: Potential therapeutic approaches. *Clinics (São Paulo, Brazil)*, *61*, 479–488.

Armstrong, J. S., & Whiteman, M. (2007). Measurement of reactive oxygen species in cells and mitochondria. *Methods in Cell Biology*, *80*, 355–377.

Aronowski, J., Grotta, J. C., Strong, R., & Waxham, M. N. (2000). Interplay between the gamma isoform of PKC and calcineurin in regulation of vulnerability to focal cerebral ischemia. *Journal of Cerebral Blood Flow and Metabolism*, *20*, 343–349.

Aronowski, J., & Labiche, L. A. (2003). Perspectives on reperfusion-induced damage in rodent models of experimental focal ischemia and role of gamma-protein kinase C. *ILAR Journal, 44*, 105–109.
Austin, C., & Wray, S. (2000). Interactions between Ca(2+) and H(+) and functional consequences in vascular smooth muscle. *Circulation Research, 86*, 355–363.
Aviv, A. (1994). Cytosolic Ca2+, Na+/H+ antiport, protein kinase C trio in essential hypertension. *American Journal of Hypertension, 7*, 205–212.
Bai, X., Margariti, A., Hu, Y., Sato, Y., Zeng, L., Ivetic, A., et al. (2010). Protein kinase C {delta} deficiency accelerates neointimal lesions of mouse injured artery involving delayed reendothelialization and vasohibin-1 accumulation. *Arteriosclerosis, Thrombosis, and Vascular Biology, 30*, 2467–2474.
Bakker, W., Sipkema, P., Stehouwer, C. D., Serne, E. H., Smulders, Y. M., van Hinsbergh, V. W., et al. (2008). Protein kinase C theta activation induces insulin-mediated constriction of muscle resistance arteries. *Diabetes, 57*, 706–713.
Balendran, A., Hare, G. R., Kieloch, A., Williams, M. R., & Alessi, D. R. (2000). Further evidence that 3-phosphoinositide-dependent protein kinase-1 (PDK1) is required for the stability and phosphorylation of protein kinase C (PKC) isoforms. *FEBS Letters, 484*, 217–223.
Bao, H. F., Thai, T. L., Yue, Q., Ma, H. P., Eaton, A. F., Cai, H., et al. (2014). ENaC activity is increased in isolated, split-open cortical collecting ducts from protein kinase Calpha knockout mice. *American Journal of Physiology. Renal Physiology, 306*, F309–F320.
Barman, S. A., Zhu, S., & White, R. E. (2004). Protein kinase C inhibits BKCa channel activity in pulmonary arterial smooth muscle. *American Journal of Physiology. Lung Cellular and Molecular Physiology, 286*, L149–L155.
Batchuluun, B., Inoguchi, T., Sonoda, N., Sasaki, S., Inoue, T., Fujimura, Y., et al. (2013). Metformin and liraglutide ameliorate high glucose-induced oxidative stress via inhibition of PKC-NAD(P)H oxidase pathway in human aortic endothelial cells. *Atherosclerosis, 232*, 156–164.
Bates, E., Bode, C., Costa, M., Gibson, C. M., Granger, C., Green, C., et al. (2008). Intracoronary KAI-9803 as an adjunct to primary percutaneous coronary intervention for acute ST-segment elevation myocardial infarction. *Circulation, 117*, 886–896.
Bazzi, M. D., & Nelsestuen, G. L. (1990). Protein kinase C interaction with calcium: A phospholipid-dependent process. *Biochemistry, 29*, 7624–7630.
Behn-Krappa, A., & Newton, A. C. (1999). The hydrophobic phosphorylation motif of conventional protein kinase C is regulated by autophosphorylation. *Current Biology, 9*, 728–737.
Benes, C. H., Wu, N., Elia, A. E., Dharia, T., Cantley, L. C., & Soltoff, S. P. (2005). The C2 domain of PKCdelta is a phosphotyrosine binding domain. *Cell, 121*, 271–280.
Bertorello, A. M., Aperia, A., Walaas, S. I., Nairn, A. C., & Greengard, P. (1991). Phosphorylation of the catalytic subunit of Na+, K(+)-ATPase inhibits the activity of the enzyme. *Proceedings of the National Academy of Sciences of the United States of America, 88*, 11359–11362.
Bezy, O., Tran, T. T., Pihlajamaki, J., Suzuki, R., Emanuelli, B., Winnay, J., et al. (2011). PKCdelta regulates hepatic insulin sensitivity and hepatosteatosis in mice and humans. *The Journal of Clinical Investigation, 121*, 2504–2517.
Bogard, A. S., & Tavalin, S. J. (2015). Protein kinase C (PKC)zeta pseudosubstrate inhibitor peptide promiscuously binds PKC family isoforms and disrupts conventional PKC targeting and translocation. *Molecular Pharmacology, 88*, 728–735.
Bonev, A. D., & Nelson, M. T. (1996). Vasoconstrictors inhibit ATP-sensitive K+ channels in arterial smooth muscle through protein kinase C. *The Journal of General Physiology, 108*, 315–323.

Boscoboinik, D., Szewczyk, A., Hensey, C., & Azzi, A. (1991). Inhibition of cell proliferation by alpha-tocopherol. Role of protein kinase C. *The Journal of Biological Chemistry*, *266*, 6188–6194.

Bowles, D. K., Maddali, K. K., Dhulipala, V. C., & Korzick, D. H. (2007). PKCdelta mediates anti-proliferative, pro-apoptic effects of testosterone on coronary smooth muscle. *American Journal of Physiology. Cell Physiology*, *293*, C805–C813.

Brasier, A. R. (2010). The nuclear factor-kappaB-interleukin-6 signalling pathway mediating vascular inflammation. *Cardiovascular Research*, *86*, 211–218.

Braun, D. C., Garfield, S. H., & Blumberg, P. M. (2005). Analysis by fluorescence resonance energy transfer of the interaction between ligands and protein kinase Cdelta in the intact cell. *The Journal of Biological Chemistry*, *280*, 8164–8171.

Braun, M. U., & Mochly-Rosen, D. (2003). Opposing effects of delta- and zeta-protein kinase C isozymes on cardiac fibroblast proliferation: Use of isozyme-selective inhibitors. *Journal of Molecular and Cellular Cardiology*, *35*, 895–903.

Bright, R., Steinberg, G. K., & Mochly-Rosen, D. (2007). DeltaPKC mediates microcerebrovascular dysfunction in acute ischemia and in chronic hypertensive stress in vivo. *Brain Research*, *1144*, 146–155.

Bright, R., Sun, G. H., Yenari, M. A., Steinberg, G. K., & Mochly-Rosen, D. (2008). epsilonPKC confers acute tolerance to cerebral ischemic reperfusion injury. *Neuroscience Letters*, *441*, 120–124.

Brueggemann, L. I., Mackie, A. R., Cribbs, L. L., Freda, J., Tripathi, A., Majetschak, M., et al. (2014). Differential protein kinase C-dependent modulation of Kv7.4 and Kv7.5 subunits of vascular Kv7 channels. *The Journal of Biological Chemistry*, *289*, 2099–2111.

Budas, G. R., Churchill, E. N., Disatnik, M. H., Sun, L., & Mochly-Rosen, D. (2010). Mitochondrial import of PKCepsilon is mediated by HSP90: A role in cardioprotection from ischaemia and reperfusion injury. *Cardiovascular Research*, *88*, 83–92.

Budas, G. R., Churchill, E. N., & Mochly-Rosen, D. (2007). Cardioprotective mechanisms of PKC isozyme-selective activators and inhibitors in the treatment of ischemia-reperfusion injury. *Pharmacological Research*, *55*, 523–536.

Bullard, T. A., Hastings, J. L., Davis, J. M., Borg, T. K., & Price, R. L. (2007). Altered PKC expression and phosphorylation in response to the nature, direction, and magnitude of mechanical stretch. *Canadian Journal of Physiology and Pharmacology*, *85*, 243–250.

Bursell, S. E., & King, G. L. (1999). Can protein kinase C inhibition and vitamin E prevent the development of diabetic vascular complications? *Diabetes Research and Clinical Practice*, *45*, 169–182.

Cameron, A. J., De Rycker, M., Calleja, V., Alcor, D., Kjaer, S., Kostelecky, B., et al. (2007). Protein kinases, from B to C. *Biochemical Society Transactions*, *35*, 1013–1017.

Cameron, A. J., Escribano, C., Saurin, A. T., Kostelecky, B., & Parker, P. J. (2009). PKC maturation is promoted by nucleotide pocket occupation independently of intrinsic kinase activity. *Nature Structural & Molecular Biology*, *16*, 624–630.

Campbell, M., & Trimble, E. R. (2005). Modification of PI3K- and MAPK-dependent chemotaxis in aortic vascular smooth muscle cells by protein kinase CbetaII. *Circulation Research*, *96*, 197–206.

Campen, M. J., Shimoda, L. A., & O'Donnell, C. P. (2005). Acute and chronic cardiovascular effects of intermittent hypoxia in C57BL/6J mice. *Journal of Applied Physiology (1985)*, *99*, 2028–2035.

Carracedo, S., Sacher, F., Brandes, G., Braun, U., & Leitges, M. (2014). Redundant role of protein kinase C delta and epsilon during mouse embryonic development. *PLoS One*, *9*, e103686.

Castagna, M., Takai, Y., Kaibuchi, K., Sano, K., Kikkawa, U., & Nishizuka, Y. (1982). Direct activation of calcium-activated, phospholipid-dependent protein kinase by tumor-promoting phorbol esters. *The Journal of Biological Chemistry*, *257*, 7847–7851.

Castle, N. A., Haylett, D. G., Morgan, J. M., & Jenkinson, D. H. (1993). Dequalinium: A potent inhibitor of apamin-sensitive K+ channels in hepatocytes and of nicotinic responses in skeletal muscle. *European Journal of Pharmacology, 236*, 201–207.

Cazaubon, S., Bornancin, F., & Parker, P. J. (1994). Threonine-497 is a critical site for permissive activation of protein kinase C alpha. *The Biochemical Journal, 301*(Pt. 2), 443–448.

Cenni, V., Doppler, H., Sonnenburg, E. D., Maraldi, N., Newton, A. C., & Toker, A. (2002). Regulation of novel protein kinase C epsilon by phosphorylation. *The Biochemical Journal, 363*, 537–545.

Chang, K., Xiao, D., Huang, X., Longo, L. D., & Zhang, L. (2009). Chronic hypoxia increases pressure-dependent myogenic tone of the uterine artery in pregnant sheep: Role of ERK/PKC pathway. *American Journal of Physiology. Heart and Circulatory Physiology, 296*, H1840–H1849.

Chen, C. H., Budas, G. R., Churchill, E. N., Disatnik, M. H., Hurley, T. D., & Mochly-Rosen, D. (2008). Activation of aldehyde dehydrogenase-2 reduces ischemic damage to the heart. *Science, 321*, 1493–1495.

Chen, L., Hahn, H., Wu, G., Chen, C. H., Liron, T., Schechtman, D., et al. (2001). Opposing cardioprotective actions and parallel hypertrophic effects of delta PKC and epsilon PKC. *Proceedings of the National Academy of Sciences of the United States of America, 98*, 11114–11119.

Choi, H., Tostes, R. C., & Webb, R. C. (2011). S-nitrosylation inhibits protein kinase C-mediated contraction in mouse aorta. *Journal of Cardiovascular Pharmacology, 57*, 65–71.

Chou, W. H., Choi, D. S., Zhang, H., Mu, D., McMahon, T., Kharazia, V. N., et al. (2004). Neutrophil protein kinase Cdelta as a mediator of stroke-reperfusion injury. *The Journal of Clinical Investigation, 114*, 49–56.

Churchill, E. N., Ferreira, J. C., Brum, P. C., Szweda, L. I., & Mochly-Rosen, D. (2010). Ischaemic preconditioning improves proteasomal activity and increases the degradation of deltaPKC during reperfusion. *Cardiovascular Research, 85*, 385–394.

Churchill, E. N., & Mochly-Rosen, D. (2007). The roles of PKCdelta and epsilon isoenzymes in the regulation of myocardial ischaemia/reperfusion injury. *Biochemical Society Transactions, 35*, 1040–1042.

Churchill, E. N., Murriel, C. L., Chen, C. H., Mochly-Rosen, D., & Szweda, L. I. (2005). Reperfusion-induced translocation of deltaPKC to cardiac mitochondria prevents pyruvate dehydrogenase reactivation. *Circulation Research, 97*, 78–85.

Churchill, E. N., Qvit, N., & Mochly-Rosen, D. (2009). Rationally designed peptide regulators of protein kinase C. *Trends in Endocrinology and Metabolism, 20*, 25–33.

Clarke, M., & Dodson, P. M. (2007). PKC inhibition and diabetic microvascular complications. *Best Practice & Research. Clinical Endocrinology & Metabolism, 21*, 573–586.

Clement, S., Tasinato, A., Boscoboinik, D., & Azzi, A. (1997). The effect of alpha-tocopherol on the synthesis, phosphorylation and activity of protein kinase C in smooth muscle cells after phorbol 12-myristate 13-acetate down-regulation. *European Journal of Biochemistry, 246*, 745–749.

Cogolludo, A., Moreno, L., Bosca, L., Tamargo, J., & Perez-Vizcaino, F. (2003). Thromboxane A2-induced inhibition of voltage-gated K+ channels and pulmonary vasoconstriction: Role of protein kinase Czeta. *Circulation Research, 93*, 656–663.

Cogolludo, A., Moreno, L., Frazziano, G., Moral-Sanz, J., Menendez, C., Castaneda, J., et al. (2009). Activation of neutral sphingomyelinase is involved in acute hypoxic pulmonary vasoconstriction. *Cardiovascular Research, 82*, 296–302.

Cole, W. C., Malcolm, T., Walsh, M. P., & Light, P. E. (2000). Inhibition by protein kinase C of the K(NDP) subtype of vascular smooth muscle ATP-sensitive potassium channel. *Circulation Research, 87*, 112–117.

Costa, A. D., Jakob, R., Costa, C. L., Andrukhiv, K., West, I. C., & Garlid, K. D. (2006). The mechanism by which the mitochondrial ATP-sensitive K+ channel opening and

H2O2 inhibit the mitochondrial permeability transition. *The Journal of Biological Chemistry, 281*, 20801–20808.
Craven, P. A., Studer, R. K., & DeRubertis, F. R. (1994). Impaired nitric oxide-dependent cyclic guanosine monophosphate generation in glomeruli from diabetic rats. Evidence for protein kinase C-mediated suppression of the cholinergic response. *The Journal of Clinical Investigation, 93*, 311–320.
Crozatier, B. (2006). Central role of PKCs in vascular smooth muscle cell ion channel regulation. *Journal of Molecular and Cellular Cardiology, 41*, 952–955.
Csukai, M., Chen, C. H., De Matteis, M. A., & Mochly-Rosen, D. (1997). The coatomer protein beta'-COP, a selective binding protein (RACK) for protein kinase Cepsilon. *The Journal of Biological Chemistry, 272*, 29200–29206.
Cywin, C. L., Dahmann, G., Prokopowicz, A. S., 3rd., Young, E. R., Magolda, R. L., Cardozo, M. G., et al. (2007). Discovery of potent and selective PKC-theta inhibitors. *Bioorganic & Medicinal Chemistry Letters, 17*, 225–230.
Dakshinamurti, S., Mellow, L., & Stephens, N. L. (2005). Regulation of pulmonary arterial myosin phosphatase activity in neonatal circulatory transition and in hypoxic pulmonary hypertension: A role for CPI-17. *Pediatric Pulmonology, 40*, 398–407.
Davies, S. P., Reddy, H., Caivano, M., & Cohen, P. (2000). Specificity and mechanism of action of some commonly used protein kinase inhibitors. *The Biochemical Journal, 351*, 95–105.
Deng, B., Xie, S., Wang, J., Xia, Z., & Nie, R. (2012). Inhibition of protein kinase C beta(2) prevents tumor necrosis factor-alpha-induced apoptosis and oxidative stress in endothelial cells: The role of NADPH oxidase subunits. *Journal of Vascular Research, 49*, 144–159.
Derubertis, F. R., & Craven, P. A. (1994). Activation of protein kinase C in glomerular cells in diabetes. Mechanisms and potential links to the pathogenesis of diabetic glomerulopathy. *Diabetes, 43*, 1–8.
Deuse, T., Koyanagi, T., Erben, R. G., Hua, X., Velden, J., Ikeno, F., et al. (2010). Sustained inhibition of epsilon protein kinase C inhibits vascular restenosis after balloon injury and stenting. *Circulation, 122*, S170–S178.
DeVries, T. A., Neville, M. C., & Reyland, M. E. (2002). Nuclear import of PKCdelta is required for apoptosis: Identification of a novel nuclear import sequence. *The EMBO Journal, 21*, 6050–6060.
Ding, Q., Chai, H., Mahmood, N., Tsao, J., Mochly-Rosen, D., & Zhou, W. (2011). Matrix metalloproteinases modulated by protein kinase Cepsilon mediate resistin-induced migration of human coronary artery smooth muscle cells. *Journal of Vascular Surgery, 53*, 1044–1051.
Ding, R. Q., Tsao, J., Chai, H., Mochly-Rosen, D., & Zhou, W. (2011). Therapeutic potential for protein kinase C inhibitor in vascular restenosis. *Journal of Cardiovascular Pharmacology and Therapeutics, 16*, 160–167.
Dorn, G. W., 2nd., Souroujon, M. C., Liron, T., Chen, C. H., Gray, M. O., Zhou, H. Z., et al. (1999). Sustained in vivo cardiac protection by a rationally designed peptide that causes epsilon protein kinase C translocation. *Proceedings of the National Academy of Sciences of the United States of America, 96*, 12798–12803.
Doughan, A. K., Harrison, D. G., & Dikalov, S. I. (2008). Molecular mechanisms of angiotensin II-mediated mitochondrial dysfunction: Linking mitochondrial oxidative damage and vascular endothelial dysfunction. *Circulation Research, 102*, 488–496.
Draeger, A., Wray, S., & Babiychuk, E. B. (2005). Domain architecture of the smooth-muscle plasma membrane: Regulation by annexins. *The Biochemical Journal, 387*, 309–314.
Dries, D. R., Gallegos, L. L., & Newton, A. C. (2007). A single residue in the C1 domain sensitizes novel protein kinase C isoforms to cellular diacylglycerol production. *The Journal of Biological Chemistry, 282*, 826–830.

Dries, D. R., & Newton, A. C. (2008). Kinetic analysis of the interaction of the C1 domain of protein kinase C with lipid membranes by stopped-flow spectroscopy. *The Journal of Biological Chemistry, 283*, 7885–7893.
Dubois, T., Oudinet, J. P., Mira, J. P., & Russo-Marie, F. (1996). Annexins and protein kinases C. *Biochimica et Biophysica Acta, 1313*, 290–294.
Duquesnes, N., Lezoualc'h, F., & Crozatier, B. (2011). PKC-delta and PKC-epsilon: Foes of the same family or strangers? *Journal of Molecular and Cellular Cardiology, 51*, 665–673.
Durgan, J., Michael, N., Totty, N., & Parker, P. J. (2007). Novel phosphorylation site markers of protein kinase C delta activation. *FEBS Letters, 581*, 3377–3381.
Dutil, E. M., Keranen, L. M., DePaoli-Roach, A. A., & Newton, A. C. (1994). In vivo regulation of protein kinase C by trans-phosphorylation followed by autophosphorylation. *The Journal of Biological Chemistry, 269*, 29359–29362.
Dutil, E. M., Toker, A., & Newton, A. C. (1998). Regulation of conventional protein kinase C isozymes by phosphoinositide-dependent kinase 1 (PDK-1). *Current Biology, 8*, 1366–1375.
Dykes, A. C., Fultz, M. E., Norton, M. L., & Wright, G. L. (2003). Microtubule-dependent PKC-alpha localization in A7r5 smooth muscle cells. *American Journal of Physiology. Cell Physiology, 285*, C76–C87.
Eichholtz, T., de Bont, D. B., de Widt, J., Liskamp, R. M., & Ploegh, H. L. (1993). A myristoylated pseudosubstrate peptide, a novel protein kinase C inhibitor. *The Journal of Biological Chemistry, 268*, 1982–1986.
Eitelhuber, A. C., Warth, S., Schimmack, G., Duwel, M., Hadian, K., Demski, K., et al. (2011). Dephosphorylation of Carma1 by PP2A negatively regulates T-cell activation. *The EMBO Journal, 30*, 594–605.
El-Yazbi, A. F., Abd-Elrahman, K. S., & Moreno-Dominguez, A. (2015). PKC-mediated cerebral vasoconstriction: Role of myosin light chain phosphorylation versus actin cytoskeleton reorganization. *Biochemical Pharmacology, 95*, 263–278.
Engin, K. N. (2009). Alpha-tocopherol: Looking beyond an antioxidant. *Molecular Vision, 15*, 855–860.
Evenou, J. P., Wagner, J., Zenke, G., Brinkmann, V., Wagner, K., Kovarik, J., et al. (2009). The potent protein kinase C-selective inhibitor AEB071 (sotrastaurin) represents a new class of immunosuppressive agents affecting early T-cell activation. *The Journal of Pharmacology and Experimental Therapeutics, 330*, 792–801.
Fan, H. C., Fernandez-Hernando, C., & Lai, J. H. (2014). Protein kinase C isoforms in atherosclerosis: Pro- or anti-inflammatory? *Biochemical Pharmacology, 88*, 139–149.
Farley, D. B., & Ford, S. P. (1992). Evidence for declining extracellular calcium uptake and protein kinase C activity in uterine arterial smooth muscle during gestation in gilts. *Biology of Reproduction, 46*, 315–321.
Faux, M. C., Rollins, E. N., Edwards, A. S., Langeberg, L. K., Newton, A. C., & Scott, J. D. (1999). Mechanism of A-kinase-anchoring protein 79 (AKAP79) and protein kinase C interaction. *The Biochemical Journal, 343*(Pt. 2), 443–452.
Ferreira, J. C., Koyanagi, T., Palaniyandi, S. S., Fajardo, G., Churchill, E. N., Budas, G., et al. (2011). Pharmacological inhibition of betaIIPKC is cardioprotective in late-stage hypertrophy. *Journal of Molecular and Cellular Cardiology, 51*, 980–987.
Ford, S. P. (1995). Control of blood flow to the gravid uterus of domestic livestock species. *Journal of Animal Science, 73*, 1852–1860.
Fu, G., Hu, J., Niederberger-Magnenat, N., Rybakin, V., Casas, J., Yachi, P. P., et al. (2011). Protein kinase C eta is required for T cell activation and homeostatic proliferation. *Science Signaling, 4*, ra84.
Fukumoto, S., Nishizawa, Y., Hosoi, M., Koyama, H., Yamakawa, K., Ohno, S., et al. (1997). Protein kinase C delta inhibits the proliferation of vascular smooth muscle cells

by suppressing G1 cyclin expression. *The Journal of Biological Chemistry*, *272*, 13816–13822.
Gailly, P., Gong, M. C., Somlyo, A. V., & Somlyo, A. P. (1997). Possible role of atypical protein kinase C activated by arachidonic acid in Ca2+ sensitization of rabbit smooth muscle. *The Journal of Physiology*, *500*(Pt. 1), 95–109.
Gao, F. H., Wu, Y. L., Zhao, M., Liu, C. X., Wang, L. S., & Chen, G. Q. (2009). Protein kinase C-delta mediates down-regulation of heterogeneous nuclear ribonucleoprotein K protein: Involvement in apoptosis induction. *Experimental Cell Research*, *315*, 3250–3258.
Gao, L., & Mann, G. E. (2009). Vascular NAD(P)H oxidase activation in diabetes: A double-edged sword in redox signalling. *Cardiovascular Research*, *82*, 9–20.
Gao, T., Brognard, J., & Newton, A. C. (2008). The phosphatase PHLPP controls the cellular levels of protein kinase C. *The Journal of Biological Chemistry*, *283*, 6300–6311.
Gao, T., & Newton, A. C. (2006). Invariant Leu preceding turn motif phosphorylation site controls the interaction of protein kinase C with Hsp70. *The Journal of Biological Chemistry*, *281*, 32461–32468.
Gardner, A. W., Parker, D. E., Montgomery, P. S., Sosnowska, D., Casanegra, A. I., Ungvari, Z., et al. (2014). Greater endothelial apoptosis and oxidative stress in patients with peripheral artery disease. *International Journal of Vascular Medicine*, *2014*, 160534.
Gaudin, P. B., Rayburn, B. K., Hutchins, G. M., Kasper, E. K., Baughman, K. L., Goodman, S. N., et al. (1994). Peritransplant injury to the myocardium associated with the development of accelerated arteriosclerosis in heart transplant recipients. *The American Journal of Surgical Pathology*, *18*, 338–346.
Gekeler, V., Boer, R., Uberall, F., Ise, W., Schubert, C., Utz, I., et al. (1996). Effects of the selective bisindolylmaleimide protein kinase C inhibitor GF 109203X on P-glycoprotein-mediated multidrug resistance. *British Journal of Cancer*, *74*, 897–905.
Geraldes, P., Hiraoka-Yamamoto, J., Matsumoto, M., Clermont, A., Leitges, M., Marette, A., et al. (2009). Activation of PKC-delta and SHP-1 by hyperglycemia causes vascular cell apoptosis and diabetic retinopathy. *Nature Medicine*, *15*, 1298–1306.
Geraldes, P., & King, G. L. (2010). Activation of protein kinase C isoforms and its impact on diabetic complications. *Circulation Research*, *106*, 1319–1331.
Ghatta, S., Nimmagadda, D., Xu, X., & O'Rourke, S. T. (2006). Large-conductance, calcium-activated potassium channels: Structural and functional implications. *Pharmacology & Therapeutics*, *110*, 103–116.
Gilio, K., Harper, M. T., Cosemans, J. M., Konopatskaya, O., Munnix, I. C., Prinzen, L., et al. (2010). Functional divergence of platelet protein kinase C (PKC) isoforms in thrombus formation on collagen. *The Journal of Biological Chemistry*, *285*, 23410–23419.
Ginnan, R., & Singer, H. A. (2005). PKC-delta-dependent pathways contribute to PDGF-stimulated ERK1/2 activation in vascular smooth muscle. *American Journal of Physiology. Cell Physiology*, *288*, C1193–C1201.
Gogula, S. V., Divakar, C., Satyanarayana, C., Kumar, Y. P., & Lavanaya, V. S. (2013). Computational investigation of pkcbeta inhibitors for the treatment of diabetic retinopathy. *Bioinformation*, *9*, 1040–1043.
Goodnight, J. A., Mischak, H., Kolch, W., & Mushinski, J. F. (1995). Immunocytochemical localization of eight protein kinase C isozymes overexpressed in NIH 3T3 fibroblasts. Isoform-specific association with microfilaments, Golgi, endoplasmic reticulum, and nuclear and cell membranes. *The Journal of Biological Chemistry*, *270*, 9991–10001.
Gopalakrishna, R., & Anderson, W. B. (1989). Ca2+- and phospholipid-independent activation of protein kinase C by selective oxidative modification of the regulatory domain. *Proceedings of the National Academy of Sciences of the United States of America*, *86*, 6758–6762.
Gopalakrishna, R., Chen, Z. H., & Gundimeda, U. (1994). Tobacco smoke tumor promoters, catechol and hydroquinone, induce oxidative regulation of protein kinase

C and influence invasion and metastasis of lung carcinoma cells. *Proceedings of the National Academy of Sciences of the United States of America*, *91*, 12233–12237.

Gopalakrishna, R., & Jaken, S. (2000). Protein kinase C signaling and oxidative stress. *Free Radical Biology & Medicine*, *28*, 1349–1361.

Gorin, M. A., & Pan, Q. (2009). Protein kinase C epsilon: An oncogene and emerging tumor biomarker. *Molecular Cancer*, *8*, 9.

Goulopoulou, S., Hannan, J. L., Matsumoto, T., & Webb, R. C. (2012). Pregnancy reduces RhoA/Rho kinase and protein kinase C signaling pathways downstream of thromboxane receptor activation in the rat uterine artery. *American Journal of Physiology. Heart and Circulatory Physiology*, *302*, H2477–H2488.

Graff, J. R., McNulty, A. M., Hanna, K. R., Konicek, B. W., Lynch, R. L., Bailey, S. N., et al. (2005). The protein kinase Cbeta-selective inhibitor, Enzastaurin (LY317615. HCl), suppresses signaling through the AKT pathway, induces apoptosis, and suppresses growth of human colon cancer and glioblastoma xenografts. *Cancer Research*, *65*, 7462–7469.

Grandage, V. L., Everington, T., Linch, D. C., & Khwaja, A. (2006). Go6976 is a potent inhibitor of the JAK 2 and FLT3 tyrosine kinases with significant activity in primary acute myeloid leukaemia cells. *British Journal of Haematology*, *135*, 303–316.

Grange, J. J., Baca-Regen, L. M., Nollendorfs, A. J., Persidsky, Y., Sudan, D. L., & Baxter, B. T. (1998). Protein kinase C isoforms in human aortic smooth muscle cells. *Journal of Vascular Surgery*, *27*, 919–926. discussion 926–927.

Gray, M. O., Karliner, J. S., & Mochly-Rosen, D. (1997). A selective epsilon-protein kinase C antagonist inhibits protection of cardiac myocytes from hypoxia-induced cell death. *The Journal of Biological Chemistry*, *272*, 30945–30951.

Griendling, K. K., Ushio-Fukai, M., Lassegue, B., & Alexander, R. W. (1997). Angiotensin II signaling in vascular smooth muscle. New concepts. *Hypertension*, *29*, 366–373.

Gruber, T., Pfeifhofer-Obermair, C., & Baier, G. (2010). PKCtheta is necessary for efficient activation of NFkappaB, NFAT, and AP-1 during positive selection of thymocytes. *Immunology Letters*, *132*, 6–11.

Gschwendt, M., Dieterich, S., Rennecke, J., Kittstein, W., Mueller, H. J., & Johannes, F. J. (1996). Inhibition of protein kinase C mu by various inhibitors. Differentiation from protein kinase c isoenzymes. *FEBS Letters*, *392*, 77–80.

Gschwendt, M., Muller, H. J., Kielbassa, K., Zang, R., Kittstein, W., Rincke, G., et al. (1994). Rottlerin, a novel protein kinase inhibitor. *Biochemical and Biophysical Research Communications*, *199*, 93–98.

Hage-Sleiman, R., Hamze, A. B., Reslan, L., Kobeissy, H., & Dbaibo, G. (2015). The Novel PKCtheta from Benchtop to Clinic. *Journal of Immunology Research*, *2015*, 348798.

Hall, J. L., Matter, C. M., Wang, X., & Gibbons, G. H. (2000). Hyperglycemia inhibits vascular smooth muscle cell apoptosis through a protein kinase C-dependent pathway. *Circulation Research*, *87*, 574–580.

Haller, H., Baur, E., Quass, P., Behrend, M., Lindschau, C., Distler, A., et al. (1995). High glucose concentrations and protein kinase C isoforms in vascular smooth muscle cells. *Kidney International*, *47*, 1057–1067.

Haller, H., Quass, P., Lindschau, C., Luft, F. C., & Distler, A. (1994). Platelet-derived growth factor and angiotensin II induce different spatial distribution of protein kinase C-alpha and -beta in vascular smooth muscle cells. *Hypertension*, *23*, 848–852.

Hampson, P., Chahal, H., Khanim, F., Hayden, R., Mulder, A., Assi, L. K., et al. (2005). PEP005, a selective small-molecule activator of protein kinase C, has potent antileukemic activity mediated via the delta isoform of PKC. *Blood*, *106*, 1362–1368.

Hanke, H., Lenz, C., Hess, B., Spindler, K. D., & Weidemann, W. (2001). Effect of testosterone on plaque development and androgen receptor expression in the arterial vessel wall. *Circulation*, *103*, 1382–1385.

Harja, E., Chang, J. S., Lu, Y., Leitges, M., Zou, Y. S., Schmidt, A. M., et al. (2009). Mice deficient in PKCbeta and apolipoprotein E display decreased atherosclerosis. *The FASEB Journal, 23*, 1081–1091.

Hartwig, J. H., Thelen, M., Rosen, A., Janmey, P. A., Nairn, A. C., & Aderem, A. (1992). MARCKS is an actin filament crosslinking protein regulated by protein kinase C and calcium-calmodulin. *Nature, 356*, 618–622.

Heilig, C. W., Deb, D. K., Abdul, A., Riaz, H., James, L. R., Salameh, J., et al. (2013). GLUT1 regulation of the pro-sclerotic mediators of diabetic nephropathy. *American Journal of Nephrology, 38*, 39–49.

Henry, D. N., Busik, J. V., Brosius, F. C., 3rd., & Heilig, C. W. (1999). Glucose transporters control gene expression of aldose reductase, PKCalpha, and GLUT1 in mesangial cells in vitro. *The American Journal of Physiology, 277*, F97–F104.

Hirano, Y., Yoshinaga, S., Ogura, K., Yokochi, M., Noda, Y., Sumimoto, H., et al. (2004). Solution structure of atypical protein kinase C PB1 domain and its mode of interaction with ZIP/p62 and MEK5. *The Journal of Biological Chemistry, 279*, 31883–31890.

Hoque, M., Rentero, C., Cairns, R., Tebar, F., Enrich, C., & Grewal, T. (2014). Annexins—Scaffolds modulating PKC localization and signaling. *Cellular Signalling, 26*, 1213–1225.

House, C., & Kemp, B. E. (1987). Protein kinase C contains a pseudosubstrate prototope in its regulatory domain. *Science, 238*, 1726–1728.

House, C., & Kemp, B. E. (1990). Protein kinase C pseudosubstrate prototope: Structure–function relationships. *Cellular Signalling, 2*, 187–190.

Howcroft, T. K., & Lindquist, R. R. (1991). The protein kinase C inhibitor 1-(5-isoquinolinesulfonyl)-2-methylpiperazine dihydrochloride (H-7) inhibits PMA-induced promiscuous cytolytic activity but not specific cytolytic activity by a cloned cytolytic T lymphocyte. *Biochemical and Biophysical Research Communications, 179*, 720–725.

Hu, X. Q., Xiao, D., Zhu, R., Huang, X., Yang, S., Wilson, S., et al. (2011). Pregnancy upregulates large-conductance Ca(2+)-activated K(+) channel activity and attenuates myogenic tone in uterine arteries. *Hypertension, 58*, 1132–1139.

Huang, C., Chang, J. S., Xu, Y., Li, Q., Zou, Y. S., & Yan, S. F. (2010). Reduction of PKCbetaII activity in smooth muscle cells attenuates acute arterial injury. *Atherosclerosis, 212*, 123–130.

Huang, Q., & Yuan, Y. (1997). Interaction of PKC and NOS in signal transduction of microvascular hyperpermeability. *The American Journal of Physiology, 273*, H2442–H2451.

Huang, W., Bansode, R. R., Bal, N. C., Mehta, M., & Mehta, K. D. (2012). Protein kinase Cbeta deficiency attenuates obesity syndrome of ob/ob mice by promoting white adipose tissue remodeling. *Journal of Lipid Research, 53*, 368–378.

Huang, X., & Walker, J. W. (2004). Myofilament anchoring of protein kinase C-epsilon in cardiac myocytes. *Journal of Cell Science, 117*, 1971–1978.

Hucho, T. B., Dina, O. A., & Levine, J. D. (2005). Epac mediates a cAMP-to-PKC signaling in inflammatory pain: An isolectin B4(+) neuron-specific mechanism. *The Journal of Neuroscience, 25*, 6119–6126.

Hui, X., Reither, G., Kaestner, L., & Lipp, P. (2014). Targeted activation of conventional and novel protein kinases C through differential translocation patterns. *Molecular and Cellular Biology, 34*, 2370–2381.

Hurd, P. J., Bannister, A. J., Halls, K., Dawson, M. A., Vermeulen, M., Olsen, J. V., et al. (2009). Phosphorylation of histone H3 Thr-45 is linked to apoptosis. *The Journal of Biological Chemistry, 284*, 16575–16583.

Igarashi, M., Wakasaki, H., Takahara, N., Ishii, H., Jiang, Z. Y., Yamauchi, T., et al. (1999). Glucose or diabetes activates p38 mitogen-activated protein kinase via different pathways. *The Journal of Clinical Investigation, 103*, 185–195.

Ikenoue, T., Inoki, K., Yang, Q., Zhou, X., & Guan, K. L. (2008). Essential function of TORC2 in PKC and Akt turn motif phosphorylation, maturation and signalling. *The EMBO Journal*, *27*, 1919–1931.

Inagaki, K., Chen, L., Ikeno, F., Lee, F. H., Imahashi, K., Bouley, D. M., et al. (2003). Inhibition of delta-protein kinase C protects against reperfusion injury of the ischemic heart in vivo. *Circulation*, *108*, 2304–2307.

Inagaki, K., Churchill, E., & Mochly-Rosen, D. (2006). Epsilon protein kinase C as a potential therapeutic target for the ischemic heart. *Cardiovascular Research*, *70*, 222–230.

Inagaki, K., Koyanagi, T., Berry, N. C., Sun, L., & Mochly-Rosen, D. (2008). Pharmacological inhibition of epsilon-protein kinase C attenuates cardiac fibrosis and dysfunction in hypertension-induced heart failure. *Hypertension*, *51*, 1565–1569.

Inagaki, M., Yokokura, H., Itoh, T., Kanmura, Y., Kuriyama, H., & Hidaka, H. (1987). Purified rabbit brain protein kinase C relaxes skinned vascular smooth muscle and phosphorylates myosin light chain. *Archives of Biochemistry and Biophysics*, *254*, 136–141.

Inoguchi, T., Battan, R., Handler, E., Sportsman, J. R., Heath, W., & King, G. L. (1992). Preferential elevation of protein kinase C isoform beta II and diacylglycerol levels in the aorta and heart of diabetic rats: Differential reversibility to glycemic control by islet cell transplantation. *Proceedings of the National Academy of Sciences of the United States of America*, *89*, 11059–11063.

Inoguchi, T., Li, P., Umeda, F., Yu, H. Y., Kakimoto, M., Imamura, M., et al. (2000). High glucose level and free fatty acid stimulate reactive oxygen species production through protein kinase C—Dependent activation of NAD(P)H oxidase in cultured vascular cells. *Diabetes*, *49*, 1939–1945.

Ishida, K., Matsumoto, T., Taguchi, K., Kamata, K., & Kobayashi, T. (2012). Protein kinase C delta contributes to increase in EP3 agonist-induced contraction in mesenteric arteries from type 2 diabetic Goto–Kakizaki rats. *Pflügers Archiv*, *463*, 593–602.

Ivaska, J., Vuoriluoto, K., Huovinen, T., Izawa, I., Inagaki, M., & Parker, P. J. (2005). PKCepsilon-mediated phosphorylation of vimentin controls integrin recycling and motility. *The EMBO Journal*, *24*, 3834–3845.

Jacinto, E., & Lorberg, A. (2008). TOR regulation of AGC kinases in yeast and mammals. *The Biochemical Journal*, *410*, 19–37.

Jakus, V., & Rietbrock, N. (2004). Advanced glycation end-products and the progress of diabetic vascular complications. *Physiological Research*, *53*, 131–142.

Javaid, K., Rahman, A., Anwar, K. N., Frey, R. S., Minshall, R. D., & Malik, A. B. (2003). Tumor necrosis factor-alpha induces early-onset endothelial adhesivity by protein kinase Czeta-dependent activation of intercellular adhesion molecule-1. *Circulation Research*, *92*, 1089–1097.

Je, H. D., Gangopadhyay, S. S., Ashworth, T. D., & Morgan, K. G. (2001). Calponin is required for agonist-induced signal transduction—Evidence from an antisense approach in ferret smooth muscle. *The Journal of Physiology*, *537*, 567–577.

Jernigan, N. L., & Resta, T. C. (2014). Calcium homeostasis and sensitization in pulmonary arterial smooth muscle. *Microcirculation*, *21*, 259–271.

Jialal, I., Machha, A., & Devaraj, S. (2013). Small interfering-RNA to protein kinase C-delta reduces the proinflammatory effects of human C-reactive protein in biobreeding diabetic rats. *Hormone and Metabolic Research*, *45*, 326–328.

Jiao, Y., & Yang, Q. (2015). Downregulation of natriuretic peptide clearance receptor mRNA in vascular smooth muscle cells by angiotensin II. *Fundamental & Clinical Pharmacology*, *29*, 260–268.

Johnson, J. A., & Barman, S. A. (2004). Protein kinase C modulation of cyclic GMP in rat neonatal pulmonary vascular smooth muscle. *Lung*, *182*, 79–89.

Kamm, K. E., & Stull, J. T. (1989). Regulation of smooth muscle contractile elements by second messengers. *Annual Review of Physiology*, *51*, 299–313.

Kanashiro, C. A., Alexander, B. T., Granger, J. P., & Khalil, R. A. (1999). Ca(2+)-insensitive vascular protein kinase C during pregnancy and NOS inhibition. *Hypertension*, *34*, 924–930.

Kanashiro, C. A., Altirkawi, K. A., & Khalil, R. A. (2000). Preconditioning of coronary artery against vasoconstriction by endothelin-1 and prostaglandin F2alpha during repeated downregulation of epsilon-protein kinase C. *Journal of Cardiovascular Pharmacology*, *35*, 491–501.

Kanashiro, C. A., Cockrell, K. L., Alexander, B. T., Granger, J. P., & Khalil, R. A. (2000). Pregnancy-associated reduction in vascular protein kinase C activity rebounds during inhibition of NO synthesis. *American Journal of Physiology. Regulatory, Integrative and Comparative Physiology*, *278*, R295–R303.

Kanashiro, C. A., & Khalil, R. A. (1998). Signal transduction by protein kinase C in mammalian cells. *Clinical and Experimental Pharmacology & Physiology*, *25*, 974–985.

Kanashiro, C. A., & Khalil, R. A. (2001). Gender-related distinctions in protein kinase C activity in rat vascular smooth muscle. *American Journal of Physiology. Cell Physiology*, *280*, C34–C45.

Kappert, K., Furundzija, V., Fritzsche, J., Margeta, C., Kruger, J., Meyborg, H., et al. (2010). Integrin cleavage regulates bidirectional signalling in vascular smooth muscle cells. *Thrombosis and Haemostasis*, *103*, 556–563.

Katzmann, D. J., Odorizzi, G., & Emr, S. D. (2002). Receptor downregulation and multivesicular-body sorting. *Nature Reviews. Molecular Cell Biology*, *3*, 893–905.

Khalil, R. A. (2013). Protein kinase C inhibitors as modulators of vascular function and their application in vascular disease. *Pharmaceuticals (Basel)*, *6*, 407–439.

Khalil, R. A., Lajoie, C., & Morgan, K. G. (1994). In situ determination of [Ca2+]i threshold for translocation of the alpha-protein kinase C isoform. *The American Journal of Physiology*, *266*, C1544–C1551.

Khalil, R. A., Lajoie, C., Resnick, M. S., & Morgan, K. G. (1992). Ca(2+)-independent isoforms of protein kinase C differentially translocate in smooth muscle. *The American Journal of Physiology*, *263*, C714–C719.

Khalil, R. A., Menice, C. B., Wang, C. L., & Morgan, K. G. (1995). Phosphotyrosine-dependent targeting of mitogen-activated protein kinase in differentiated contractile vascular cells. *Circulation Research*, *76*, 1101–1108.

Khan, W. A., Dobrowsky, R., el Touny, S., & Hannun, Y. A. (1990). Protein kinase C and platelet inhibition by D-erythro-sphingosine: Comparison with N,N-dimethylsphingosine and commercial preparation. *Biochemical and Biophysical Research Communications*, *172*, 683–691.

Kheifets, V., Bright, R., Inagaki, K., Schechtman, D., & Mochly-Rosen, D. (2006). Protein kinase C delta (deltaPKC)-annexin V interaction: A required step in deltaPKC translocation and function. *The Journal of Biological Chemistry*, *281*, 23218–23226.

Kheifets, V., & Mochly-Rosen, D. (2007). Insight into intra- and inter-molecular interactions of PKC: Design of specific modulators of kinase function. *Pharmacological Research*, *55*, 467–476.

Kilpatrick, L. E., Sun, S., Mackie, D., Baik, F., Li, H., & Korchak, H. M. (2006). Regulation of TNF mediated antiapoptotic signaling in human neutrophils: Role of delta-PKC and ERK1/2. *Journal of Leukocyte Biology*, *80*, 1512–1521.

Kim, H. L., & Im, D. S. (2008). N, N-dimethyl-D-erythro-sphingosine increases intracellular Ca2+ concentration via Na+-Ca2+-exchanger in HCT116 human colon cancer cells. *Archives of Pharmacal Research*, *31*, 54–59.

Kim, H. R., Appel, S., Vetterkind, S., Gangopadhyay, S. S., & Morgan, K. G. (2008). Smooth muscle signalling pathways in health and disease. *Journal of Cellular and Molecular Medicine*, *12*, 2165–2180.

Kim, H. R., Gallant, C., & Morgan, K. G. (2013). Regulation of PKC autophosphorylation by calponin in contractile vascular smooth muscle tissue. *BioMed Research International, 2013*, 358643.

Kim, J., Koyanagi, T., & Mochly-Rosen, D. (2011). PKCdelta activation mediates angiogenesis via NADPH oxidase activity in PC-3 prostate cancer cells. *Prostate, 71*, 946–954.

Kim, J., Thorne, S. H., Sun, L., Huang, B., & Mochly-Rosen, D. (2011). Sustained inhibition of PKCalpha reduces intravasation and lung seeding during mammary tumor metastasis in an in vivo mouse model. *Oncogene, 30*, 323–333.

Kishimoto, A., Mikawa, K., Hashimoto, K., Yasuda, I., Tanaka, S., Tominaga, M., et al. (1989). Limited proteolysis of protein kinase C subspecies by calcium-dependent neutral protease (calpain). *The Journal of Biological Chemistry, 264*, 4088–4092.

Kizub, I. V., Klymenko, K. I., & Soloviev, A. I. (2014). Protein kinase C in enhanced vascular tone in diabetes mellitus. *International Journal of Cardiology, 174*, 230–242.

Kizub, I. V., Pavlova, O. O., Ivanova, I. V., & Soloviev, A. I. (2010). Protein kinase C-dependent inhibition of BK(Ca) current in rat aorta smooth muscle cells following gamma-irradiation. *International Journal of Radiation Biology, 86*, 291–299.

Klein, G., Schaefer, A., Hilfiker-Kleiner, D., Oppermann, D., Shukla, P., Quint, A., et al. (2005). Increased collagen deposition and diastolic dysfunction but preserved myocardial hypertrophy after pressure overload in mice lacking PKCepsilon. *Circulation Research, 96*, 748–755.

Klymenko, K., Novokhatska, T., Kizub, I., Parshikov, A., Dosenko, V., & Soloviev, A. (2014). PKC-delta isozyme gene silencing restores vascular function in diabetic rat. *Journal of Basic and Clinical Physiology and Pharmacology*, 1–9.

Kong, L., Shen, X., Lin, L., Leitges, M., Rosario, R., Zou, Y. S., et al. (2013). PKCbeta promotes vascular inflammation and acceleration of atherosclerosis in diabetic ApoE null mice. *Arteriosclerosis, Thrombosis, and Vascular Biology, 33*, 1779–1787.

Konishi, H., Tanaka, M., Takemura, Y., Matsuzaki, H., Ono, Y., Kikkawa, U., et al. (1997). Activation of protein kinase C by tyrosine phosphorylation in response to H2O2. *Proceedings of the National Academy of Sciences of the United States of America, 94*, 11233–11237.

Konishi, H., Yamauchi, E., Taniguchi, H., Yamamoto, T., Matsuzaki, H., Takemura, Y., et al. (2001). Phosphorylation sites of protein kinase C delta in H2O2-treated cells and its activation by tyrosine kinase in vitro. *Proceedings of the National Academy of Sciences of the United States of America, 98*, 6587–6592.

Koponen, S., Kurkinen, K., Akerman, K. E., Mochly-Rosen, D., Chan, P. H., & Koistinaho, J. (2003). Prevention of NMDA-induced death of cortical neurons by inhibition of protein kinase Czeta. *Journal of Neurochemistry, 86*, 442–450.

Kostyak, J. C., Bhavanasi, D., Liverani, E., McKenzie, S. E., & Kunapuli, S. P. (2014). Protein kinase C delta deficiency enhances megakaryopoiesis and recovery from thrombocytopenia. *Arteriosclerosis, Thrombosis, and Vascular Biology, 34*, 2579–2585.

Kouroedov, A., Eto, M., Joch, H., Volpe, M., Luscher, T. F., & Cosentino, F. (2004). Selective inhibition of protein kinase Cbeta2 prevents acute effects of high glucose on vascular cell adhesion molecule-1 expression in human endothelial cells. *Circulation, 110*, 91–96.

Koya, D., Jirousek, M. R., Lin, Y. W., Ishii, H., Kuboki, K., & King, G. L. (1997). Characterization of protein kinase C beta isoform activation on the gene expression of transforming growth factor-beta, extracellular matrix components, and prostanoids in the glomeruli of diabetic rats. *The Journal of Clinical Investigation, 100*, 115–126.

Koya, D., & King, G. L. (1998). Protein kinase C activation and the development of diabetic complications. *Diabetes, 47*, 859–866.

Koyanagi, T., Noguchi, K., Ootani, A., Inagaki, K., Robbins, R. C., & Mochly-Rosen, D. (2007). Pharmacological inhibition of epsilon PKC suppresses chronic inflammation in murine cardiac transplantation model. *Journal of Molecular and Cellular Cardiology, 43*, 517–522.

Kraft, A. S., & Anderson, W. B. (1983). Phorbol esters increase the amount of Ca2+, phospholipid-dependent protein kinase associated with plasma membrane. *Nature*, *301*, 621–623.

Kraft, A. S., Smith, J. B., & Berkow, R. L. (1986). Bryostatin, an activator of the calcium phospholipid-dependent protein kinase, blocks phorbol ester-induced differentiation of human promyelocytic leukemia cells HL-60. *Proceedings of the National Academy of Sciences of the United States of America*, *83*, 1334–1338.

Krupinski, J., Slevin, M. A., Kumar, P., Gaffney, J., & Kaluza, J. (1998). Protein kinase C expression and activity in the human brain after ischaemic stroke. *Acta Neurobiologiae Experimentalis (Wars)*, *58*, 13–21.

Lahn, M., Sundell, K., & Moore, S. (2003). Targeting protein kinase C-alpha (PKC-alpha) in cancer with the phosphorothioate antisense oligonucleotide aprinocarsen. *Annals of the New York Academy of Sciences*, *1002*, 263–270.

Lange, A., Gebremedhin, D., Narayanan, J., & Harder, D. (1997). 20-Hydroxyeicosatetraenoic acid-induced vasoconstriction and inhibition of potassium current in cerebral vascular smooth muscle is dependent on activation of protein kinase C. *The Journal of Biological Chemistry*, *272*, 27345–27352.

Large, W. A., Saleh, S. N., & Albert, A. P. (2009). Role of phosphoinositol 4,5-bisphosphate and diacylglycerol in regulating native TRPC channel proteins in vascular smooth muscle. *Cell Calcium*, *45*, 574–582.

Le Good, J. A., Ziegler, W. H., Parekh, D. B., Alessi, D. R., Cohen, P., & Parker, P. J. (1998). Protein kinase C isotypes controlled by phosphoinositide 3-kinase through the protein kinase PDK1. *Science*, *281*, 2042–2045.

Ledoux, J., Werner, M. E., Brayden, J. E., & Nelson, M. T. (2006). Calcium-activated potassium channels and the regulation of vascular tone. *Physiology (Bethesda)*, *21*, 69–78.

Lee, H. W., Smith, L., Pettit, G. R., & Bingham Smith, J. (1996). Dephosphorylation of activated protein kinase C contributes to downregulation by bryostatin. *The American Journal of Physiology*, *271*, C304–C311.

Lee, K., Kim, H., & Jeong, D. (2014). Protein kinase C regulates vascular calcification via cytoskeleton reorganization and osteogenic signaling. *Biochemical and Biophysical Research Communications*, *453*, 793–797.

Lehel, C., Olah, Z., Jakab, G., & Anderson, W. B. (1995a). Protein kinase C epsilon is localized to the Golgi via its zinc-finger domain and modulates Golgi function. *Proceedings of the National Academy of Sciences of the United States of America*, *92*, 1406–1410.

Lehel, C., Olah, Z., Jakab, G., Szallasi, Z., Petrovics, G., Harta, G., et al. (1995b). Protein kinase C epsilon subcellular localization domains and proteolytic degradation sites. A model for protein kinase C conformational changes. *The Journal of Biological Chemistry*, *270*, 19651–19658.

Lehel, C., Olah, Z., Petrovics, G., Jakab, G., & Anderson, W. B. (1996). Influence of various domains of protein kinase C epsilon on its PMA-induced translocation from the Golgi to the plasma membrane. *Biochemical and Biophysical Research Communications*, *223*, 98–103.

Leitges, M., Mayr, M., Braun, U., Mayr, U., Li, C., Pfister, G., et al. (2001). Exacerbated vein graft arteriosclerosis in protein kinase Cdelta-null mice. *The Journal of Clinical Investigation*, *108*, 1505–1512.

Leitges, M., Plomann, M., Standaert, M. L., Bandyopadhyay, G., Sajan, M. P., Kanoh, Y., et al. (2002). Knockout of PKC alpha enhances insulin signaling through PI3K. *Molecular Endocrinology*, *16*, 847–858.

Lengfeld, J., Wang, Q., Zohlman, A., Salvarezza, S., Morgan, S., Ren, J., et al. (2012). Protein kinase C delta regulates the release of collagen type I from vascular smooth muscle cells via regulation of Cdc42. *Molecular Biology of the Cell*, *23*, 1955–1963.

Levesque, L., Dean, N. M., Sasmor, H., & Crooke, S. T. (1997). Antisense oligonucleotides targeting human protein kinase C-alpha inhibit phorbol ester-induced reduction of bradykinin-evoked calcium mobilization in A549 cells. *Molecular Pharmacology, 51*, 209–216.

Levitan, I. B. (1994). Modulation of ion channels by protein phosphorylation and dephosphorylation. *Annual Review of Physiology, 56*, 193–212.

Lewis, J. M., Cheresh, D. A., & Schwartz, M. A. (1996). Protein kinase C regulates alpha v beta 5-dependent cytoskeletal associations and focal adhesion kinase phosphorylation. *The Journal of Cell Biology, 134*, 1323–1332.

Li, C., Wernig, F., Leitges, M., Hu, Y., & Xu, Q. (2003). Mechanical stress-activated PKCdelta regulates smooth muscle cell migration. *The FASEB Journal, 17*, 2106–2108.

Li, T., Xiao, X., Zhang, J., Zhu, Y., Hu, Y., Zang, J., et al. (2014). Age and sex differences in vascular responsiveness in healthy and trauma patients: Contribution of estrogen receptor-mediated Rho kinase and PKC pathways. *American Journal of Physiology. Heart and Circulatory Physiology, 306*, H1105–H1115.

Li, W., Zhang, J., Bottaro, D. P., & Pierce, J. H. (1997). Identification of serine 643 of protein kinase C-delta as an important autophosphorylation site for its enzymatic activity. *The Journal of Biological Chemistry, 272*, 24550–24555.

Light, P. (1996). Regulation of ATP-sensitive potassium channels by phosphorylation. *Biochimica et Biophysica Acta, 1286*, 65–73.

Limas, C. J. (1980). Phosphorylation of cardiac sarcoplasmic reticulum by a calcium-activated, phospholipid-dependent protein kinase. *Biochemical and Biophysical Research Communications, 96*, 1378–1383.

Linch, M., Riou, P., Claus, J., Cameron, A. J., de Naurois, J., Larijani, B., et al. (2014). Functional implications of assigned, assumed and assembled PKC structures. *Biochemical Society Transactions, 42*, 35–41.

Ling, S., Little, P. J., Williams, M. R., Dai, A., Hashimura, K., Liu, J. P., et al. (2002). High glucose abolishes the antiproliferative effect of 17beta-estradiol in human vascular smooth muscle cells. *American Journal of Physiology. Endocrinology and Metabolism, 282*, E746–E751.

Liou, Y. M., & Morgan, K. G. (1994). Redistribution of protein kinase C isoforms in association with vascular hypertrophy of rat aorta. *The American Journal of Physiology, 267*, C980–C989.

Littler, C. M., Morris, K. G., Jr., Fagan, K. A., McMurtry, I. F., Messing, R. O., & Dempsey, E. C. (2003). Protein kinase C-epsilon-null mice have decreased hypoxic pulmonary vasoconstriction. *American Journal of Physiology. Heart and Circulatory Physiology, 284*, H1321–H1331.

Littler, C. M., Wehling, C. A., Wick, M. J., Fagan, K. A., Cool, C. D., Messing, R. O., et al. (2005). Divergent contractile and structural responses of the murine PKC-epsilon null pulmonary circulation to chronic hypoxia. *American Journal of Physiology. Lung Cellular and Molecular Physiology, 289*, L1083–L1093.

Liu, B., Ryer, E. J., Kundi, R., Kamiya, K., Itoh, H., Faries, P. L., et al. (2007). Protein kinase C-delta regulates migration and proliferation of vascular smooth muscle cells through the extracellular signal-regulated kinase 1/2. *Journal of Vascular Surgery, 45*, 160–168.

Liu, J. Y., Lin, S. J., & Lin, J. K. (1993). Inhibitory effects of curcumin on protein kinase C activity induced by 12-O-tetradecanoyl-phorbol-13-acetate in NIH 3T3 cells. *Carcinogenesis, 14*, 857–861.

Liu, M., Large, W. A., & Albert, A. P. (2005). Stimulation of beta-adrenoceptors inhibits store-operated channel currents via a cAMP-dependent protein kinase mechanism in rabbit portal vein myocytes. *The Journal of Physiology, 562*, 395–406.

Liu, W. S., & Heckman, C. A. (1998). The sevenfold way of PKC regulation. *Cellular Signalling, 10*, 529–542.

Liu, Y., Graham, C., Li, A., Fisher, R. J., & Shaw, S. (2002). Phosphorylation of the protein kinase C-theta activation loop and hydrophobic motif regulates its kinase activity, but only activation loop phosphorylation is critical to in vivo nuclear-factor-kappaB induction. *The Biochemical Journal*, *361*, 255–265.

Liu, Y., Terata, K., Rusch, N. J., & Gutterman, D. D. (2001). High glucose impairs voltage-gated K(+) channel current in rat small coronary arteries. *Circulation Research*, *89*, 146–152.

Lizotte, F., Pare, M., Denhez, B., Leitges, M., Guay, A., & Geraldes, P. (2013). PKCdelta impaired vessel formation and angiogenic factor expression in diabetic ischemic limbs. *Diabetes*, *62*, 2948–2957.

Lu, T., Chai, Q., Yu, L., d'Uscio, L. V., Katusic, Z. S., He, T., et al. (2012). Reactive oxygen species signaling facilitates FOXO-3a/FBXO-dependent vascular BK channel beta1 subunit degradation in diabetic mice. *Diabetes*, *61*, 1860–1868.

Lu, W., Finnis, S., Xiang, C., Lee, H. K., Markowitz, Y., Okhrimenko, H., et al. (2007). Tyrosine 311 is phosphorylated by c-Abl and promotes the apoptotic effect of PKCdelta in glioma cells. *Biochemical and Biophysical Research Communications*, *352*, 431–436.

Lu, Z. G., Liu, H., Yamaguchi, T., Miki, Y., & Yoshida, K. (2009). Protein kinase Cdelta activates RelA/p65 and nuclear factor-kappaB signaling in response to tumor necrosis factor-alpha. *Cancer Research*, *69*, 5927–5935.

Lum, M. A., Balaburski, G. M., Murphy, M. E., Black, A. R., & Black, J. D. (2013). Heat shock proteins regulate activation-induced proteasomal degradation of the mature phosphorylated form of protein kinase C. *The Journal of Biological Chemistry*, *288*, 27112–27127.

Lyle, A. N., & Griendling, K. K. (2006). Modulation of vascular smooth muscle signaling by reactive oxygen species. *Physiology (Bethesda)*, *21*, 269–280.

Ma, W., Baumann, C., & Viveiros, M. M. (2015). Lack of protein kinase C-delta (PKCdelta) disrupts fertilization and embryonic development. *Molecular Reproduction and Development*, *82*, 797–808.

Magid, R., & Davies, P. F. (2005). Endothelial protein kinase C isoform identity and differential activity of PKCzeta in an athero-susceptible region of porcine aorta. *Circulation Research*, *97*, 443–449.

Magness, R. R., Rosenfeld, C. R., & Carr, B. R. (1991). Protein kinase C in uterine and systemic arteries during ovarian cycle and pregnancy. *The American Journal of Physiology*, *260*, E464–E470.

Manetta, A., Emma, D., Gamboa, G., Liao, S., Berman, M., & DiSaia, P. (1993). Failure to enhance the in vivo killing of human ovarian carcinoma by sequential treatment with dequalinium chloride and tumor necrosis factor. *Gynecologic Oncology*, *50*, 38–44.

Manna, P. T., Smith, A. J., Taneja, T. K., Howell, G. J., Lippiat, J. D., & Sivaprasadarao, A. (2010). Constitutive endocytic recycling and protein kinase C-mediated lysosomal degradation control K(ATP) channel surface density. *The Journal of Biological Chemistry*, *285*, 5963–5973.

Mao, Y., Su, J., Lei, L., Meng, L., Qi, Y., Huo, Y., et al. (2013). Adrenomedullin and adrenotensin increase the transcription of regulator of Gprotein signaling 2 in vascular smooth muscle cells via the cAMPdependent and PKC pathways. *Molecular Medicine Reports*, *9*, 323–327.

Martin, P., Villares, R., Rodriguez-Mascarenhas, S., Zaballos, A., Leitges, M., Kovac, J., et al. (2005). Control of T helper 2 cell function and allergic airway inflammation by PKCzeta. *Proceedings of the National Academy of Sciences of the United States of America*, *102*, 9866–9871.

Martiny-Baron, G., Kazanietz, M. G., Mischak, H., Blumberg, P. M., Kochs, G., Hug, H., et al. (1993). Selective inhibition of protein kinase C isozymes by the indolocarbazole Go 6976. *The Journal of Biological Chemistry*, *268*, 9194–9197.

Masoumi, K. C., Cornmark, L., Lonne, G. K., Hellman, U., & Larsson, C. (2012). Identification of a novel protein kinase Cdelta-Smac complex that dissociates during paclitaxel-induced cell death. *FEBS Letters*, *586*, 1166–1172.

Mattila, P., Majuri, M. L., Tiisala, S., & Renkonen, R. (1994). Expression of six protein kinase C isotypes in endothelial cells. *Life Sciences*, *55*, 1253–1260.
Meggio, F., Donella Deana, A., Ruzzene, M., Brunati, A. M., Cesaro, L., Guerra, B., et al. (1995). Different susceptibility of protein kinases to staurosporine inhibition. Kinetic studies and molecular bases for the resistance of protein kinase CK2. *European Journal of Biochemistry*, *234*, 317–322.
Mehta, K. D. (2014). Emerging role of protein kinase C in energy homeostasis: A brief overview. *World Journal of Diabetes*, *5*, 385–392.
Mehta, N. N., Sheetz, M., Price, K., Comiskey, L., Amrutia, S., Iqbal, N., et al. (2009). Selective PKC beta inhibition with ruboxistaurin and endothelial function in type-2 diabetes mellitus. *Cardiovascular Drugs and Therapy*, *23*, 17–24.
Meier, M., & King, G. L. (2000). Protein kinase C activation and its pharmacological inhibition in vascular disease. *Vascular Medicine*, *5*, 173–185.
Meller, N., Liu, Y. C., Collins, T. L., Bonnefoy-Berard, N., Baier, G., Isakov, N., et al. (1996). Direct interaction between protein kinase C theta (PKC theta) and 14-3-3 tau in T cells: 14-3-3 Overexpression results in inhibition of PKC theta translocation and function. *Molecular and Cellular Biology*, *16*, 5782–5791.
Mellor, H., & Parker, P. J. (1998). The extended protein kinase C superfamily. *The Biochemical Journal*, *332*(Pt. 2), 281–292.
Mendonca, M. C., Doi, S. Q., Glerum, S., & Sellitti, D. F. (2006). Increase of C-type natriuretic peptide expression by serum and platelet-derived growth factor-BB in human aortic smooth muscle cells is dependent on protein kinase C activation. *Endocrinology*, *147*, 4169–4178.
Mendonca, M. C., Koles, N., & Sellitti, D. F. (2012). Protein kinase C-delta (PKC-delta) and PKC-alpha mediate Ca(2+)-dependent increases in CNP mRNA in human vascular cells. *Vascular Pharmacology*, *57*, 98–104.
Meotti, F. C., Luiz, A. P., Pizzolatti, M. G., Kassuya, C. A., Calixto, J. B., & Santos, A. R. (2006). Analysis of the antinociceptive effect of the flavonoid myricitrin: Evidence for a role of the L-arginine-nitric oxide and protein kinase C pathways. *The Journal of Pharmacology and Experimental Therapeutics*, *316*, 789–796.
Metzger, E., Imhof, A., Patel, D., Kahl, P., Hoffmeyer, K., Friedrichs, N., et al. (2010). Phosphorylation of histone H3T6 by PKCbeta(I) controls demethylation at histone H3K4. *Nature*, *464*, 792–796.
Mii, S., Khalil, R. A., Morgan, K. G., Ware, J. A., & Kent, K. C. (1996). Mitogen-activated protein kinase and proliferation of human vascular smooth muscle cells. *The American Journal of Physiology*, *270*, H142–H150.
Millward, M. J., House C, Bowtell, D., Webster, L., Olver, I. N., Gore, M., et al. (2006). The multikinase inhibitor midostaurin (PKC412A) lacks activity in metastatic melanoma: A phase IIA clinical and biologic study. *British Journal of Cancer*, *95*, 829–834.
Minami, K., Fukuzawa, K., & Nakaya, Y. (1993). Protein kinase C inhibits the Ca(2+)-activated K+ channel of cultured porcine coronary artery smooth muscle cells. *Biochemical and Biophysical Research Communications*, *190*, 263–269.
Mineo, C., Ying, Y. S., Chapline, C., Jaken, S., & Anderson, R. G. (1998). Targeting of protein kinase Calpha to caveolae. *The Journal of Cell Biology*, *141*, 601–610.
Mochly-Rosen, D., Das, K., & Grimes, K. V. (2012). Protein kinase C, an elusive therapeutic target? *Nature Reviews. Drug Discovery*, *11*, 937–957.
Mochly-Rosen, D., & Kauvar, L. M. (2000). Pharmacological regulation of network kinetics by protein kinase C localization. *Seminars in Immunology*, *12*, 55–61.
Mochly-Rosen, D., Khaner, H., & Lopez, J. (1991). Identification of intracellular receptor proteins for activated protein kinase C. *Proceedings of the National Academy of Sciences of the United States of America*, *88*, 3997–4000.

Monti, M., Donnini, S., Giachetti, A., Mochly-Rosen, D., & Ziche, M. (2010). deltaPKC inhibition or varepsilonPKC activation repairs endothelial vascular dysfunction by regulating eNOS post-translational modification. *Journal of Molecular and Cellular Cardiology, 48*, 746–756.

Monti, M., Donnini, S., Morbidelli, L., Giachetti, A., Mochly-Rosen, D., Mignatti, P., et al. (2013). PKCepsilon activation promotes FGF-2 exocytosis and induces endothelial cell proliferation and sprouting. *Journal of Molecular and Cellular Cardiology, 63*, 107–117.

Moreno-Dominguez, A., El-Yazbi, A. F., Zhu, H. L., Colinas, O., Zhong, X. Z., Walsh, E. J., et al. (2014). Cytoskeletal reorganization evoked by Rho-associated kinase- and protein kinase C-catalyzed phosphorylation of cofilin and heat shock protein 27, respectively, contributes to myogenic constriction of rat cerebral arteries. *The Journal of Biological Chemistry, 289*, 20939–20952.

Morgan, S., Yamanouchi, D., Harberg, C., Wang, Q., Keller, M., Si, Y., et al. (2012). Elevated protein kinase C-delta contributes to aneurysm pathogenesis through stimulation of apoptosis and inflammatory signaling. *Arteriosclerosis, Thrombosis, and Vascular Biology, 32*, 2493–2502.

Mukherjee, A., Roy, S., Saha, B., & Mukherjee, D. (2016). Spatio-temporal regulation of PKC isoforms imparts signaling specificity. *Frontiers in Immunology, 7*, 45.

Naitoh, K., Yano, T., Miura, T., Itoh, T., Miki, T., Tanno, M., et al. (2009). Roles of Cx43-associated protein kinases in suppression of gap junction-mediated chemical coupling by ischemic preconditioning. *American Journal of Physiology. Heart and Circulatory Physiology, 296*, H396–H403.

Nakano, S., Uchida, K., Kigoshi, T., Azukizawa, S., Iwasaki, R., Kaneko, M., et al. (1991). Circadian rhythm of blood pressure in normotensive NIDDM subjects. Its relationship to microvascular complications. *Diabetes Care, 14*, 707–711.

Nakashima, H., Frank, G. D., Shirai, H., Hinoki, A., Higuchi, S., Ohtsu, H., et al. (2008). Novel role of protein kinase C-delta Tyr 311 phosphorylation in vascular smooth muscle cell hypertrophy by angiotensin II. *Hypertension, 51*, 232–238.

Nakayama, K., & Tanaka, Y. (1993). Stretch-induced contraction and Ca2+ mobilization in vascular smooth muscle. *Biological Signals, 2*, 241–252.

Naruse, K., Rask-Madsen, C., Takahara, N., Ha, S. W., Suzuma, K., Way, K. J., et al. (2006). Activation of vascular protein kinase C-beta inhibits Akt-dependent endothelial nitric oxide synthase function in obesity-associated insulin resistance. *Diabetes, 55*, 691–698.

Nauli, S. M., Williams, J. M., Akopov, S. E., Zhang, L., & Pearce, W. J. (2001). Developmental changes in ryanodine- and IP(3)-sensitive Ca(2+) pools in ovine basilar artery. *American Journal of Physiology Cell Physiology, 281*, C1785–C1796.

Navarro-Nunez, L., Lozano, M. L., Martinez, C., Vicente, V., & Rivera, J. (2010). Effect of quercetin on platelet spreading on collagen and fibrinogen and on multiple platelet kinases. *Fitoterapia, 81*, 75–80.

Naylor, T. L., Tang, H., Ratsch, B. A., Enns, A., Loo, A., Chen, L., et al. (2011). Protein kinase C inhibitor sotrastaurin selectively inhibits the growth of CD79 mutant diffuse large B-cell lymphomas. *Cancer Research, 71*, 2643–2653.

Nelson, M. T., & Quayle, J. M. (1995). Physiological roles and properties of potassium channels in arterial smooth muscle. *The American Journal of Physiology, 268*, C799–C822.

Newton, A. C. (1995). Protein kinase C: Structure, function, and regulation. *The Journal of Biological Chemistry, 270*, 28495–28498.

Newton, A. C. (2001). Protein kinase C: Structural and spatial regulation by phosphorylation, cofactors, and macromolecular interactions. *Chemical Reviews, 101*, 2353–2364.

Newton, A. C. (2003). Regulation of the ABC kinases by phosphorylation: Protein kinase C as a paradigm. *The Biochemical Journal, 370*, 361–371.

Newton, A. C. (2010). Protein kinase C: Poised to signal. *American Journal of Physiology. Endocrinology and Metabolism, 298*, E395–E402.
Ng, T., Squire, A., Hansra, G., Bornancin, F., Prevostel, C., Hanby, A., et al. (1999). Imaging protein kinase Calpha activation in cells. *Science, 283*, 2085–2089.
Nishizuka, Y. (1992). Intracellular signaling by hydrolysis of phospholipids and activation of protein kinase C. *Science, 258*, 607–614.
Nishizuka, Y. (1995). Protein kinase C and lipid signaling for sustained cellular responses. *The FASEB Journal, 9*, 484–496.
Noh, K. M., Hwang, J. Y., Shin, H. C., & Koh, J. Y. (2000). A novel neuroprotective mechanism of riluzole: Direct inhibition of protein kinase C. *Neurobiology of Disease, 7*, 375–383.
Novokhatska, T., Tishkin, S., Dosenko, V., Boldyriev, A., Ivanova, I., Strielkov, I., et al. (2013). Correction of vascular hypercontractility in spontaneously hypertensive rats using shRNAs-induced delta protein kinase C gene silencing. *European Journal of Pharmacology, 718*, 401–407.
Ogiwara, T., Negishi, T., Chik, C. L., & Ho, A. K. (1998). Differential effects of two protein kinase C inhibitors, calphostin C and Go6976, on pineal cyclic nucleotide accumulation. *Journal of Neurochemistry, 71*, 1405–1412.
Ohanian, V., Ohanian, J., Shaw, L., Scarth, S., Parker, P. J., & Heagerty, A. M. (1996). Identification of protein kinase C isoforms in rat mesenteric small arteries and their possible role in agonist-induced contraction. *Circulation Research, 78*, 806–812.
Oka, N., Yamamoto, M., Schwencke, C., Kawabe, J., Ebina, T., Ohno, S., et al. (1997). Caveolin interaction with protein kinase C. Isoenzyme-dependent regulation of kinase activity by the caveolin scaffolding domain peptide. *The Journal of Biological Chemistry, 272*, 33416–33421.
Okazaki, J., Mawatari, K., Liu, B., & Kent, K. C. (2000). The effect of protein kinase C and its alpha subtype on human vascular smooth muscle cell proliferation, migration and fibronectin production. *Surgery, 128*, 192–197.
Osada, H., Sonoda, T., Tsunoda, K., & Isono, K. (1989). A new biological role of sangivamycin; inhibition of protein kinases. *The Journal of Antibiotics, 42*, 102–106.
Padanilam, B. J. (2001). Induction and subcellular localization of protein kinase C isozymes following renal ischemia. *Kidney International, 59*, 1789–1797.
Palmas, W., Pickering, T., Teresi, J., Schwartz, J. E., Eguchi, K., Field, L., et al. (2008). Nocturnal blood pressure elevation predicts progression of albuminuria in elderly people with type 2 diabetes. *Journal of Clinical Hypertension (Greenwich, Conn.), 10*, 12–20.
Pande, V., Ramos, M. J., & Gago, F. (2008). The Protein kinase inhibitor balanol: Structure–activity relationships and structure-based computational studies. *Anti-Cancer Agents in Medicinal Chemistry, 8*, 638–645.
Parekh, D. B., Ziegler, W., & Parker, P. J. (2000). Multiple pathways control protein kinase C phosphorylation. *The EMBO Journal, 19*, 496–503.
Parissenti, A. M., Kirwan, A. F., Kim, S. A., Colantonio, C. M., & Schimmer, B. P. (1998). Inhibitory properties of the regulatory domains of human protein kinase Calpha and mouse protein kinase Cepsilon. *The Journal of Biological Chemistry, 273*(15), 8940–8945.
Park, W. S., Han, J., Kim, N., Youm, J. B., Joo, H., Kim, H. K., et al. (2005). Endothelin-1 inhibits inward rectifier K+ channels in rabbit coronary arterial smooth muscle cells through protein kinase C. *Journal of Cardiovascular Pharmacology, 46*, 681–689.
Park, W. S., Kim, N., Youm, J. B., Warda, M., Ko, J. H., Kim, S. J., et al. (2006). Angiotensin II inhibits inward rectifier K+ channels in rabbit coronary arterial smooth muscle cells through protein kinase Calpha. *Biochemical and Biophysical Research Communications, 341*, 728–735.

Parker, C. A., Takahashi, K., Tao, T., & Morgan, K. G. (1994). Agonist-induced redistribution of calponin in contractile vascular smooth muscle cells. *The American Journal of Physiology*, *267*, C1262–C1270.

Pears, C. J., Kour, G., House C, Kemp, B. E., & Parker, P. J. (1990). Mutagenesis of the pseudosubstrate site of protein kinase C leads to activation. *European Journal of Biochemistry*, *194*, 89–94.

Pedersen, D. J., Diakanastasis, B., Stockli, J., & Schmitz-Peiffer, C. (2013). Protein kinase Cepsilon modulates insulin receptor localization and trafficking in mouse embryonic fibroblasts. *PLoS One*, *8*, e58046.

Perez-Moreno, M., Avila, A., Islas, S., Sanchez, S., & Gonzalez-Mariscal, L. (1998). Vinculin but not alpha-actinin is a target of PKC phosphorylation during junctional assembly induced by calcium. *Journal of Cell Science*, *111*(Pt. 23), 3563–3571.

Perez-Vizcaino, F., Cogolludo, A., & Moreno, L. (2010). Reactive oxygen species signaling in pulmonary vascular smooth muscle. *Respiratory Physiology & Neurobiology*, *174*, 212–220.

Persaud, S. D., Hoang, V., Huang, J., & Basu, A. (2005). Involvement of proteolytic activation of PKCdelta in cisplatin-induced apoptosis in human small cell lung cancer H69 cells. *International Journal of Oncology*, *27*, 149–154.

Peterman, E. E., Taormina, P., 2nd., Harvey, M., & Young, L. H. (2004). Go 6983 exerts cardioprotective effects in myocardial ischemia/reperfusion. *Journal of Cardiovascular Pharmacology*, *43*, 645–656.

Pfeifhofer, C., Gruber, T., Letschka, T., Thuille, N., Lutz-Nicoladoni, C., Hermann-Kleiter, N., et al. (2006). Defective IgG2a/2b class switching in PKC alpha$^{-/-}$ mice. *The Journal of Immunology*, *176*, 6004–6011.

Piccoletti, R., Bendinelli, P., Arienti, D., & Bernelli-Zazzera, A. (1992). State and activity of protein kinase C in postischemic reperfused liver. *Experimental and Molecular Pathology*, *56*, 219–228.

Poli, A., Mongiorgi, S., Cocco, L., & Follo, M. Y. (2014). Protein kinase C involvement in cell cycle modulation. *Biochemical Society Transactions*, *42*, 1471–1476.

Pomero, F., Allione, A., Beltramo, E., Buttiglieri, S., D'Alu, F., Ponte, E., et al. (2003). Effects of protein kinase C inhibition and activation on proliferation and apoptosis of bovine retinal pericytes. *Diabetologia*, *46*, 416–419.

Posti, J. P., Salo, P., Ruohonen, S., Valve, L., Muszkat, M., Sofowora, G. G., et al. (2013). A polymorphism in the protein kinase C gene PRKCB is associated with alpha2-adrenoceptor-mediated vasoconstriction. *Pharmacogenetics and Genomics*, *23*, 127–134.

Pourmahram, G. E., Snetkov, V. A., Shaifta, Y., Drndarski, S., Knock, G. A., Aaronson, P. I., et al. (2008). Constriction of pulmonary artery by peroxide: Role of Ca2+ release and PKC. *Free Radical Biology & Medicine*, *45*, 1468–1476.

Prekeris, R., Mayhew, M. W., Cooper, J. B., & Terrian, D. M. (1996). Identification and localization of an actin-binding motif that is unique to the epsilon isoform of protein kinase C and participates in the regulation of synaptic function. *The Journal of Cell Biology*, *132*, 77–90.

Pula, G., Schuh, K., Nakayama, K., Nakayama, K. I., Walter, U., & Poole, A. W. (2006). PKCdelta regulates collagen-induced platelet aggregation through inhibition of VASP-mediated filopodia formation. *Blood*, *108*, 4035–4044.

Qi, W., Twigg, S., Chen, X., Polhill, T. S., Poronnik, P., Gilbert, R. E., et al. (2005). Integrated actions of transforming growth factor-beta1 and connective tissue growth factor in renal fibrosis. *American Journal of Physiology. Renal Physiology*, *288*, F800–F809.

Quayle, J. M., Nelson, M. T., & Standen, N. B. (1997). ATP-sensitive and inwardly rectifying potassium channels in smooth muscle. *Physiological Reviews*, 77, 1165–1232.

Qvit, N., & Mochly-Rosen, D. (2010). Highly specific modulators of protein kinase C localization: Applications to heart failure. *Drug Discovery Today. Disease Mechanisms, 7*, e87–e93.

Radhakrishnan, Y., Maile, L. A., Ling, Y., Graves, L. M., & Clemmons, D. R. (2008). Insulin-like growth factor-I stimulates Shc-dependent phosphatidylinositol 3-kinase activation via Grb2-associated p85 in vascular smooth muscle cells. *The Journal of Biological Chemistry, 283*, 16320–16331.

Rainbow, R., Parker, A., & Davies, N. (2011). Protein kinase C-independent inhibition of arterial smooth muscle K(+) channels by a diacylglycerol analogue. *British Journal of Pharmacology, 163*, 845–856.

Rainbow, R. D., Hardy, M. E., Standen, N. B., & Davies, N. W. (2006). Glucose reduces endothelin inhibition of voltage-gated potassium channels in rat arterial smooth muscle cells. *The Journal of Physiology, 575*, 833–844.

Rainbow, R. D., Norman, R. I., Everitt, D. E., Brignell, J. L., Davies, N. W., & Standen, N. B. (2009). Endothelin-I and angiotensin II inhibit arterial voltage-gated K+ channels through different protein kinase C isoenzymes. *Cardiovascular Research, 83*, 493–500.

Rask-Madsen, C., & King, G. L. (2008). Differential regulation of VEGF signaling by PKC-alpha and PKC-epsilon in endothelial cells. *Arteriosclerosis, Thrombosis, and Vascular Biology, 28*, 919–924.

Rathore, R., Zheng, Y. M., Li, X. Q., Wang, Q. S., Liu, Q. H., Ginnan, R., et al. (2006). Mitochondrial ROS-PKCepsilon signaling axis is uniquely involved in hypoxic increase in [Ca2+]i in pulmonary artery smooth muscle cells. *Biochemical and Biophysical Research Communications, 351*, 784–790.

Rathore, R., Zheng, Y. M., Niu, C. F., Liu, Q. H., Korde, A., Ho, Y. S., et al. (2008). Hypoxia activates NADPH oxidase to increase [ROS]i and [Ca2+]i through the mitochondrial ROS-PKCepsilon signaling axis in pulmonary artery smooth muscle cells. *Free Radical Biology & Medicine, 45*, 1223–1231.

Ratz, P. H., & Miner, A. S. (2009). Role of protein kinase Czeta and calcium entry in KCl-induced vascular smooth muscle calcium sensitization and feedback control of cellular calcium levels. *The Journal of Pharmacology and Experimental Therapeutics, 328*, 399–408.

Reddig, P. J., Dreckschmidt, N. E., Ahrens, H., Simsiman, R., Tseng, C. P., Zou, J., et al. (1999). Transgenic mice overexpressing protein kinase Cdelta in the epidermis are resistant to skin tumor promotion by 12-O-tetradecanoylphorbol-13-acetate. *Cancer Research, 59*, 5710–5718.

Rembold, C. M., & Murphy, R. A. (1988). Myoplasmic [Ca2+] determines myosin phosphorylation in agonist-stimulated swine arterial smooth muscle. *Circulation Research, 63*, 593–603.

Ren, J., Wang, Q., Morgan, S., Si, Y., Ravichander, A., Dou, C., et al. (2014). Protein kinase C-δ (PKCδ) regulates proinflammatory chemokine expression through cytosolic interaction with the NF-κB subunit p65 in vascular smooth muscle cells. *The Journal of Biological Chemistry, 289*(13), 9013–9026.

Reynolds, N. J., McCombie, S. W., Shankar, B. B., Bishop, W. R., & Fisher, G. J. (1997). SCH 47112, a novel staurosporine derivative, inhibits 12-O-tetradecanoylphorbol-13-acetate-induced inflammation and epidermal hyperplasia in hairless mouse skin. *Archives of Dermatological Research, 289*, 540–546.

Ritter, O., Jelenik, T., & Roden, M. (2015). Lipid-mediated muscle insulin resistance: Different fat, different pathways? *Journal of Molecular Medicine, 93*, 831–843.

Rodriguez-Pla, A., Bosch-Gil, J. A., Rossello-Urgell, J., Huguet-Redecilla, P., Stone, J. H., & Vilardell-Tarres, M. (2005). Metalloproteinase-2 and -9 in giant cell arteritis: Involvement in vascular remodeling. *Circulation, 112*, 264–269.

Rodriguez, M. M., Chen, C. H., Smith, B. L., & Mochly-Rosen, D. (1999). Characterization of the binding and phosphorylation of cardiac calsequestrin by epsilon protein kinase C. *FEBS Letters*, *454*, 240–246.

Roffey, J., Rosse, C., Linch, M., Hibbert, A., McDonald, N. Q., & Parker, P. J. (2009). Protein kinase C intervention: The state of play. *Current Opinion in Cell Biology*, *21*, 268–279.

Ron, D., & Kazanietz, M. G. (1999). New insights into the regulation of protein kinase C and novel phorbol ester receptors. *The FASEB Journal*, *13*, 1658–1676.

Ron, D., Luo, J., & Mochly-Rosen, D. (1995). C2 region-derived peptides inhibit translocation and function of beta protein kinase C in vivo. *The Journal of Biological Chemistry*, *270*, 24180–24187.

Ron, D., & Mochly-Rosen, D. (1994). Agonists and antagonists of protein kinase C function, derived from its binding proteins. *The Journal of Biological Chemistry*, *269*, 21395–21398.

Rosoff, P. M., Stein, L. F., & Cantley, L. C. (1984). Phorbol esters induce differentiation in a pre-B-lymphocyte cell line by enhancing Na+/H+ exchange. *The Journal of Biological Chemistry*, *259*, 7056–7060.

Ross, A. M., Gibbons, R. J., Stone, G. W., Kloner, R. A., & Alexander, R. W. (2005). A randomized, double-blinded, placebo-controlled multicenter trial of adenosine as an adjunct to reperfusion in the treatment of acute myocardial infarction (AMISTAD-II). *Journal of the American College of Cardiology*, *45*, 1775–1780.

Ross, R. (1999). Atherosclerosis—An inflammatory disease. *The New England Journal of Medicine*, *340*, 115–126.

Rovedo, M. A., Krett, N. L., & Rosen, S. T. (2011). Inhibition of glycogen synthase kinase-3 increases the cytotoxicity of enzastaurin. *The Journal of Investigative Dermatology*, *131*, 1442–1449.

Rovin, B. H., Yoshiumura, T., & Tan, L. (1992). Cytokine-induced production of monocyte chemoattractant protein-1 by cultured human mesangial cells. *The Journal of Immunology*, *148*, 2148–2153.

Rozengurt, E. (2011). Protein kinase D signaling: Multiple biological functions in health and disease. *Physiology (Bethesda)*, *26*, 23–33.

Rybin, V. O., Guo, J., Sabri, A., Elouardighi, H., Schaefer, E., & Steinberg, S. F. (2004). Stimulus-specific differences in protein kinase C delta localization and activation mechanisms in cardiomyocytes. *The Journal of Biological Chemistry*, *279*, 19350–19361.

Rybin, V. O., Sabri, A., Short, J., Braz, J. C., Molkentin, J. D., & Steinberg, S. F. (2003). Cross-regulation of novel protein kinase C (PKC) isoform function in cardiomyocytes. Role of PKC epsilon in activation loop phosphorylations and PKC delta in hydrophobic motif phosphorylations. *The Journal of Biological Chemistry*, *278*, 14555–14564.

Saito, K., Ito, E., Takakuwa, Y., Tamura, M., & Kinjo, M. (2003). In situ observation of mobility and anchoring of PKCbetaI in plasma membrane. *FEBS Letters*, *541*, 126–131.

Sajan, M. P., Jurzak, M. J., Samuels, V. T., Shulman, G. I., Braun, U., Leitges, M., et al. (2014). Impairment of insulin-stimulated glucose transport and ERK activation by adipocyte-specific knockout of PKC-lambda produces a phenotype characterized by diminished adiposity and enhanced insulin suppression of hepatic gluconeogenesis. *Adipocyte*, *3*, 19–29.

Salamanca, D. A., & Khalil, R. A. (2005). Protein kinase C isoforms as specific targets for modulation of vascular smooth muscle function in hypertension. *Biochemical Pharmacology*, *70*, 1537–1547.

Sampson, S. R., & Cooper, D. R. (2006). Specific protein kinase C isoforms as transducers and modulators of insulin signaling. *Molecular Genetics and Metabolism*, *89*, 32–47.

Sanchez-Bautista, S., Corbalan-Garcia, S., Perez-Lara, A., & Gomez-Fernandez, J. C. (2009). A comparison of the membrane binding properties of C1B domains of PKCgamma, PKCdelta, and PKCepsilon. *Biophysical Journal, 96*, 3638–3647.

Sarbassov, D. D., Ali, S. M., Kim, D. H., Guertin, D. A., Latek, R. R., Erdjument-Bromage, H., et al. (2004). Rictor, a novel binding partner of mTOR, defines a rapamycin-insensitive and raptor-independent pathway that regulates the cytoskeleton. *Current Biology, 14*, 1296–1302.

Saurin, A. T., Durgan, J., Cameron, A. J., Faisal, A., Marber, M. S., & Parker, P. J. (2008). The regulated assembly of a PKCepsilon complex controls the completion of cytokinesis. *Nature Cell Biology, 10*, 891–901.

Schechtman, D., Craske, M. L., Kheifets, V., Meyer, T., Schechtman, J., & Mochly-Rosen, D. (2004). A critical intramolecular interaction for protein kinase Cepsilon translocation. *The Journal of Biological Chemistry, 279*, 15831–15840.

Schmalz, D., Kalkbrenner, F., Hucho, F., & Buchner, K. (1996). Transport of protein kinase C alpha into the nucleus requires intact cytoskeleton while the transport of a protein containing a canonical nuclear localization signal does not. *Journal of Cell Science, 109*(Pt. 9), 2401–2406.

Schubert, R., Lidington, D., & Bolz, S. S. (2008). The emerging role of Ca2+ sensitivity regulation in promoting myogenic vasoconstriction. *Cardiovascular Research, 77*, 8–18.

Shaifta, Y., Snetkov, V. A., Prieto-Lloret, J., Knock, G. A., Smirnov, S. V., Aaronson, P. I., et al. (2015). Sphingosylphosphorylcholine potentiates vasoreactivity and voltage-gated Ca2+ entry via NOX1 and reactive oxygen species. *Cardiovascular Research, 106*, 121–130.

Shen, G. X. (2003). Selective protein kinase C inhibitors and their applications. *Current Drug Targets Cardiovascular & Haematological Disorders, 3*, 301–307.

Shi, J., Mori, E., Mori, Y., Mori, M., Li, J., Ito, Y., et al. (2004). Multiple regulation by calcium of murine homologues of transient receptor potential proteins TRPC6 and TRPC7 expressed in HEK293 cells. *The Journal of Physiology, 561*, 415–432.

Si, Y., Ren, J., Wang, P., Rateri, D. L., Daugherty, A., Shi, X. D., et al. (2012). Protein kinase C-delta mediates adventitial cell migration through regulation of monocyte chemoattractant protein-1 expression in a rat angioplasty model. *Arteriosclerosis, Thrombosis, and Vascular Biology, 32*, 943–954.

Sirikci, O., Ozer, N. K., & Azzi, A. (1996). Dietary cholesterol-induced changes of protein kinase C and the effect of vitamin E in rabbit aortic smooth muscle cells. *Atherosclerosis, 126*, 253–263.

Slish, D. F., Welsh, D. G., & Brayden, J. E. (2002). Diacylglycerol and protein kinase C activate cation channels involved in myogenic tone. *American Journal of Physiology. Heart and Circulatory Physiology, 283*, H2196–H2201.

Slosberg, E. D., Yao, Y., Xing, F., Ikui, A., Jirousek, M. R., & Weinstein, I. B. (2000). The protein kinase C beta-specific inhibitor LY379196 blocks TPA-induced monocytic differentiation of HL60 cells the protein kinase C beta-specific inhibitor LY379196 blocks TPA-induced monocytic differentiation of HL60 cells. *Molecular Carcinogenesis, 27*, 166–176.

Snow, J. B., Gonzalez Bosc, L. V., Kanagy, N. L., Walker, B. R., & Resta, T. C. (2011). Role for PKCbeta in enhanced endothelin-1-induced pulmonary vasoconstrictor reactivity following intermittent hypoxia. *American Journal of Physiology. Lung Cellular and Molecular Physiology, 301*, L745–L754.

Snow, J. B., Kanagy, N. L., Walker, B. R., & Resta, T. C. (2009). Rat strain differences in pulmonary artery smooth muscle Ca(2+) entry following chronic hypoxia. *Microcirculation, 16*, 603–614.

Snow, J. B., Kitzis, V., Norton, C. E., Torres, S. N., Johnson, K. D., Kanagy, N. L., et al. (2008). Differential effects of chronic hypoxia and intermittent hypocapnic and eucapnic hypoxia on pulmonary vasoreactivity. *Journal of Applied Physiology, 104*(1), 110–118.

Sobhia, M. E., Grewal, B. K., Ml, S. P., Patel, J., Kaur, A., Haokip, T., et al. (2013). Protein kinase C inhibitors: A patent review (2008–2009). *Expert Opinion on Therapeutic Patents*, *23*, 1297–1315.

Song, H. F., Tang, Z. M., Yuan, S. J., Zhu, B. Z., & Liu, X. W. (2003). Antisense candidates against protein kinase C-alpha designed based on phylogenesis and simulant structure of mRNA. *Acta Pharmacologica Sinica*, *24*, 269–276.

Sontag, E., Sontag, J. M., & Garcia, A. (1997). Protein phosphatase 2A is a critical regulator of protein kinase C zeta signaling targeted by SV40 small t to promote cell growth and NF-kappaB activation. *The EMBO Journal*, *16*, 5662–5671.

Speechly-Dick, M. E., Mocanu, M. M., & Yellon, D. M. (1994). Protein kinase C. Its role in ischemic preconditioning in the rat. *Circulation Research*, *75*, 586–590.

Stebbins, E. G., & Mochly-Rosen, D. (2001). Binding specificity for RACK1 resides in the V5 region of beta II protein kinase C. *The Journal of Biological Chemistry*, *276*, 29644–29650.

Steinberg, S. F. (2008). Structural basis of protein kinase C isoform function. *Physiological Reviews*, *88*, 1341–1378.

Steinberg, S. F. (2015). Mechanisms for redox-regulation of protein kinase C. *Frontiers in Pharmacology*, *6*, 128.

Straub, S. V., Girouard, H., Doetsch, P. E., Hannah, R. M., Wilkerson, M. K., & Nelson, M. T. (2009). Regulation of intracerebral arteriolar tone by K(v) channels: Effects of glucose and PKC. *American Journal of Physiology. Cell Physiology*, *297*, C788–C796.

Suzuki, T., Elias, B. C., Seth, A., Shen, L., Turner, J. R., Giorgianni, F., et al. (2009). PKC eta regulates occludin phosphorylation and epithelial tight junction integrity. *Proceedings of the National Academy of Sciences of the United States of America*, *106*, 61–66.

Sweitzer, S. M., Wong, S. M., Peters, M. C., Mochly-Rosen, D., Yeomans, D. C., & Kendig, J. J. (2004). Protein kinase C epsilon and gamma: Involvement in formalin-induced nociception in neonatal rats. *The Journal of Pharmacology and Experimental Therapeutics*, *309*, 616–625.

Symonds, J. M., Ohm, A. M., Carter, C. J., Heasley, L. E., Boyle, T. A., Franklin, W. A., et al. (2011). Protein kinase C delta is a downstream effector of oncogenic K-ras in lung tumors. *Cancer Research*, *71*, 2087–2097.

Taguchi, K., Kaneko, K., & Kubo, T. (2000). Protein kinase C modulates Ca2+-activated K + channels in cultured rat mesenteric artery smooth muscle cells. *Biological & Pharmaceutical Bulletin*, *23*, 1450–1454.

Takai, Y., Kishimoto, A., Inoue, M., & Nishizuka, Y. (1977). Studies on a cyclic nucleotide-independent protein kinase and its proenzyme in mammalian tissues. I. Purification and characterization of an active enzyme from bovine cerebellum. *The Journal of Biological Chemistry*, *252*, 7603–7609.

Takai, Y., Kishimoto, A., Kikkawa, U., Mori, T., & Nishizuka, Y. (1979). Unsaturated diacylglycerol as a possible messenger for the activation of calcium-activated, phospholipid-dependent protein kinase system. *Biochemical and Biophysical Research Communications*, *91*, 1218–1224.

Tamaoki, T. (1991). Use and specificity of staurosporine, UCN-01, and calphostin C as protein kinase inhibitors. *Methods in Enzymology*, *201*, 340–347.

Tamaoki, T., Nomoto, H., Takahashi, I., Kato, Y., Morimoto, M., & Tomita, F. (1986). Staurosporine, a potent inhibitor of phospholipid/Ca++dependent protein kinase. *Biochemical and Biophysical Research Communications*, *135*, 397–402.

Tanaka, M., Terry, R. D., Mokhtari, G. K., Inagaki, K., Koyanagi, T., Kofidis, T., et al. (2004). Suppression of graft coronary artery disease by a brief treatment with a selective epsilonPKC activator and a deltaPKC inhibitor in murine cardiac allografts. *Circulation*, *110*, II194–II199.

Teng, B., Duong, M., Tossidou, I., Yu, X., & Schiffer, M. (2014). Role of protein kinase C in podocytes and development of glomerular damage in diabetic nephropathy. *Frontiers in Endocrinology*, *5*, 179.

Thallas-Bonke, V., Thorpe, S. R., Coughlan, M. T., Fukami, K., Yap, F. Y., Sourris, K. C., et al. (2008). Inhibition of NADPH oxidase prevents advanced glycation end product-mediated damage in diabetic nephropathy through a protein kinase C-alpha-dependent pathway. *Diabetes*, *57*, 460–469.

Thelen, M., Rosen, A., Nairn, A. C., & Aderem, A. (1991). Regulation by phosphorylation of reversible association of a myristoylated protein kinase C substrate with the plasma membrane. *Nature*, *351*, 320–322.

Thomas, A. C., & Newby, A. C. (2010). Effect of matrix metalloproteinase-9 knockout on vein graft remodelling in mice. *Journal of Vascular Research*, *47*, 299–308.

Thorneloe, K. S., Maruyama, Y., Malcolm, A. T., Light, P. E., Walsh, M. P., & Cole, W. C. (2002). Protein kinase C modulation of recombinant ATP-sensitive K(+) channels composed of Kir6.1 and/or Kir6.2 expressed with SUR2B. *The Journal of Physiology*, *541*, 65–80.

Toker, A., Ellis, C. A., Sellers, L. A., & Aitken, A. (1990). Protein kinase C inhibitor proteins. Purification from sheep brain and sequence similarity to lipocortins and 14-3-3 protein. *European Journal of Biochemistry*, *191*, 421–429.

Toullec, D., Pianetti, P., Coste, H., Bellevergue, P., Grand-Perret, T., Ajakane, M., et al. (1991). The bisindolylmaleimide GF 109203X is a potent and selective inhibitor of protein kinase C. *The Journal of Biological Chemistry*, *266*, 15771–15781.

Tumey, L. N., Bhagirath, N., Brennan, A., Brooijmans, N., Lee, J., Yang, X., et al. (2009). 5-Vinyl-3-pyridinecarbonitrile inhibitors of PKCtheta: Optimization of enzymatic and functional activity. *Bioorganic & Medicinal Chemistry*, *17*, 7933–7948.

Tuttle, K. R., Bakris, G. L., Toto, R. D., McGill, J. B., Hu, K., & Anderson, P. W. (2005). The effect of ruboxistaurin on nephropathy in type 2 diabetes. *Diabetes Care*, *28*, 2686–2690.

Twigg, S. M., Chen, M. M., Joly, A. H., Chakrapani, S. D., Tsubaki, J., Kim, H. S., et al. (2001). Advanced glycosylation end products up-regulate connective tissue growth factor (insulin-like growth factor-binding protein-related protein 2) in human fibroblasts: A potential mechanism for expansion of extracellular matrix in diabetes mellitus. *Endocrinology*, *142*, 1760–1769.

Tykocki, N. R., Wu, B., Jackson, W. F., & Watts, S. W. (2014). Divergent signaling mechanisms for venous versus arterial contraction as revealed by endothelin-1. *Journal of Vascular Surgery*, *62*, 721–733.

Valovka, T., Verdier, F., Cramer, R., Zhyvoloup, A., Fenton, T., Rebholz, H., et al. (2003). Protein kinase C phosphorylates ribosomal protein S6 kinase betaII and regulates its subcellular localization. *Molecular and Cellular Biology*, *23*, 852–863.

Vinik, A. I., Bril, V., Kempler, P., Litchy, W. J., Tesfaye, S., Price, K. L., et al. (2005). Treatment of symptomatic diabetic peripheral neuropathy with the protein kinase C beta-inhibitor ruboxistaurin mesylate during a 1-year, randomized, placebo-controlled, double-blind clinical trial. *Clinical Therapeutics*, *27*, 1164–1180.

Violin, J. D., Zhang, J., Tsien, R. Y., & Newton, A. C. (2003). A genetically encoded fluorescent reporter reveals oscillatory phosphorylation by protein kinase C. *The Journal of Cell Biology*, *161*, 899–909.

Wallukat, G., Homuth, V., Fischer, T., Lindschau, C., Horstkamp, B., Jupner, A., et al. (1999). Patients with preeclampsia develop agonistic autoantibodies against the angiotensin AT1 receptor. *The Journal of Clinical Investigation*, *103*, 945–952.

Wang, H. Y., Bashore, T. R., & Friedman, E. (1995). Exercise reduces age-dependent decrease in platelet protein kinase C activity and translocation. *The Journals of Gerontology. Series A, Biological Sciences and Medical Sciences*, *50A*, M12–M16.

Wang, H. Y., Bashore, T. R., Tran, Z. V., & Friedman, E. (2000). Age-related decreases in lymphocyte protein kinase C activity and translocation are reduced by aerobic fitness. *The Journals of Gerontology. Series A, Biological Sciences and Medical Sciences*, *55*, B545–B551.

Wang, H. Y., Pisano, M. R., & Friedman, E. (1994). Attenuated protein kinase C activity and translocation in Alzheimer's disease brain. *Neurobiology of Aging*, *15*, 293–298.

Wang, Y., Zhou, H., Wu, B., Zhou, Q., Cui, D., & Wang, L. (2015). Protein kinase C isoforms distinctly regulate propofol-induced endothelium-dependent and endothelium-independent vasodilation. *Journal of Cardiovascular Pharmacology*, *66*, 276–284.

Ward, N. E., Pierce, D. S., Chung, S. E., Gravitt, K. R., & O'Brian, C. A. (1998). Irreversible inactivation of protein kinase C by glutathione. *The Journal of Biological Chemistry*, *273*, 12558–12566.

Watanabe, M., Hachiya, T., Hagiwara, M., & Hidaka, H. (1989). Identification of type III protein kinase C in bovine aortic tissue. *Archives of Biochemistry and Biophysics*, *273*, 165–169.

Weidemann, W., & Hanke, H. (2002). Cardiovascular effects of androgens. *Cardiovascular Drug Reviews*, *20*, 175–198.

Welch, E. J., Jones, B. W., & Scott, J. D. (2010). Networking with AKAPs: Context-dependent regulation of anchored enzymes. *Molecular Interventions*, *10*, 86–97.

Wetsel, W. C., Khan, W. A., Merchenthaler, I., Rivera, H., Halpern, A. E., Phung, H. M., et al. (1992). Tissue and cellular distribution of the extended family of protein kinase C isoenzymes. *The Journal of Cell Biology*, *117*, 121–133.

White, W. O., Seibenhener, M. L., & Wooten, M. W. (2002). Phosphorylation of tyrosine 256 facilitates nuclear import of atypical protein kinase C. *Journal of Cellular Biochemistry*, *85*, 42–53.

Whitson, E. L., Bugni, T. S., Chockalingam, P. S., Concepcion, G. P., Feng, X., Jin, G., et al. (2009). Fibrosterol sulfates from the Philippine sponge Lissodendoryx (Acanthodoryx) fibrosa: Sterol dimers that inhibit PKCzeta. *The Journal of Organic Chemistry*, *74*, 5902–5908.

Wilkinson, S. E., Parker, P. J., & Nixon, J. S. (1993). Isoenzyme specificity of bisindolylmaleimides, selective inhibitors of protein kinase C. *The Biochemical Journal*, *294*(Pt. 2), 335–337.

Williams, B., Gallacher, B., Patel, H., & Orme, C. (1997). Glucose-induced protein kinase C activation regulates vascular permeability factor mRNA expression and peptide production by human vascular smooth muscle cells in vitro. *Diabetes*, *46*, 1497–1503.

Williamson, J. R., Chang, K., Frangos, M., Hasan, K. S., Ido, Y., Kawamura, T., et al. (1993). Hyperglycemic pseudohypoxia and diabetic complications. *Diabetes*, *42*, 801–813.

Woodsome, T. P., Eto, M., Everett, A., Brautigan, D. L., & Kitazawa, T. (2001). Expression of CPI-17 and myosin phosphatase correlates with Ca(2+) sensitivity of protein kinase C-induced contraction in rabbit smooth muscle. *The Journal of Physiology*, *535*, 553–564.

Wray, S., & Smith, R. D. (2004). Mechanisms of action of pH-induced effects on vascular smooth muscle. *Molecular and Cellular Biochemistry*, *263*, 163–172.

Xi, G., Shen, X., Maile, L. A., Wai, C., Gollahon, K., & Clemmons, D. R. (2012). Hyperglycemia enhances IGF-I-stimulated Src activation via increasing Nox4-derived reactive oxygen species in a PKCzeta-dependent manner in vascular smooth muscle cells. *Diabetes*, *61*, 104–113.

Xiao, D., Buchholz, J. N., & Zhang, L. (2006). Pregnancy attenuates uterine artery pressure-dependent vascular tone: Role of PKC/ERK pathway. *American Journal of Physiology. Heart and Circulatory Physiology*, *290*, H2337–H2343.

Xiao, D., Zhu, R., & Zhang, L. (2014). Gestational hypoxia up-regulates protein kinase C and inhibits calcium-activated potassium channels in ovine uterine arteries. *International Journal of Medical Sciences*, *11*, 886–892.

Xu, B., Zhao, H., Wang, S., Sun, X., & Qin, X. (2015). Increased ADRP expression in human atherosclerotic lesions correlates with plaque instability. *International Journal of Clinical & Experiment Medicine, 8*, 5414–5421.

Xu, T. R., & Rumsby, M. G. (2004). Phorbol ester-induced translocation of PKC epsilon to the nucleus in fibroblasts: Identification of nuclear PKC epsilon-associating proteins. *FEBS Letters, 570*, 20–24.

Xu, Z. B., Chaudhary, D., Olland, S., Wolfrom, S., Czerwinski, R., Malakian, K., et al. (2004). Catalytic domain crystal structure of protein kinase C-theta (PKCtheta). *The Journal of Biological Chemistry, 279*, 50401–50409.

Yamada, S., Kimura, R., Harada, Y., & Nakayama, K. (1990). Calcium channel receptor sites for (+)-[3H]PN 200-110 in coronary artery. *The Journal of Pharmacology and Experimental Therapeutics, 252*, 327–332.

Yamaguchi, H., Igarashi, M., Hirata, A., Sugae, N., Tsuchiya, H., Jimbu, Y., et al. (2004). Altered PDGF-BB-induced p38 MAP kinase activation in diabetic vascular smooth muscle cells: Roles of protein kinase C-delta. *Arteriosclerosis, Thrombosis, and Vascular Biology, 24*, 2095–2101.

Yan, S. F., Harja, E., Andrassy, M., Fujita, T., & Schmidt, A. M. (2006). Protein kinase C beta/early growth response-1 pathway: A key player in ischemia, atherosclerosis, and restenosis. *Journal of the American College of Cardiology, 48*, A47–A55.

Yang, X., Teguh, D., Wu, J. P., He, B., Kirk, T. B., Qin, S., et al. (2015). Protein kinase C delta null mice exhibit structural alterations in articular surface, intra-articular and subchondral compartments. *Arthritis Research & Therapy, 17*, 210.

Yang, Y., & Igumenova, T. I. (2013). The C-terminal V5 domain of protein kinase Calpha is intrinsically disordered, with propensity to associate with a membrane mimetic. *PLoS One, 8*, e65699.

Yang, Y. C., Wang, X. D., Huang, K., Wang, L., Jiang, Z. L., & Qi, Y. X. (2014). Temporal phosphoproteomics to investigate the mechanotransduction of vascular smooth muscle cells in response to cyclic stretch. *Journal of Biomechanics, 47*, 3622–3629.

Yevseyenkov, V. V., Das, S., Lin, D., Willard, L., Davidson, H., Sitaramayya, A., et al. (2009). Loss of protein kinase Cgamma in knockout mice and increased retinal sensitivity to hyperbaric oxygen. *Archives of Ophthalmology, 127*, 500–506.

Yokota, T., Ma, R. C., Park, J. Y., Isshiki, K., Sotiropoulos, K. B., Rauniyar, R. K., et al. (2003). Role of protein kinase C on the expression of platelet-derived growth factor and endothelin-1 in the retina of diabetic rats and cultured retinal capillary pericytes. *Diabetes, 52*, 838–845.

Yonezawa, T., Kurata, R., Kimura, M., & Inoko, H. (2009). PKC delta and epsilon in drug targeting and therapeutics. *Recent Patents on DNA & Gene Sequences, 3*, 96–101.

Young, L. H., Balin, B. J., & Weis, M. T. (2005). Go 6983: A fast acting protein kinase C inhibitor that attenuates myocardial ischemia/reperfusion injury. *Cardiovascular Drug Reviews, 23*, 255–272.

Yu, C., Chen, Y., Cline, G. W., Zhang, D., Zong, H., Wang, Y., et al. (2002). Mechanism by which fatty acids inhibit insulin activation of insulin receptor substrate-1 (IRS-1)-associated phosphatidylinositol 3-kinase activity in muscle. *The Journal of Biological Chemistry, 277*, 50230–50236.

Zarate, C. A., Jr., Singh, J. B., Carlson, P. J., Quiroz, J., Jolkovsky, L., Luckenbaugh, D. A., et al. (2007). Efficacy of a protein kinase C inhibitor (tamoxifen) in the treatment of acute mania: A pilot study. *Bipolar Disorders, 9*, 561–570.

Zeidman, R., Lofgren, B., Pahlman, S., & Larsson, C. (1999). PKCepsilon, via its regulatory domain and independently of its catalytic domain, induces neurite-like processes in neuroblastoma cells. *The Journal of Cell Biology, 145*, 713–726.

Zhang, G., Kazanietz, M. G., Blumberg, P. M., & Hurley, J. H. (1995). Crystal structure of the cys2 activator-binding domain of protein kinase C delta in complex with phorbol ester. *Cell*, *81*, 917–924.

Zhang, Y., Hermanson, M. E., & Eddinger, T. J. (2013). Tonic and phasic smooth muscle contraction is not regulated by the PKCalpha–CPI-17 pathway in swine stomach antrum and fundus. *PLoS One*, *8*, e74608.

Zhao, M., Xia, L., & Chen, G. Q. (2012). Protein kinase cdelta in apoptosis: A brief overview. *Archivum Immunologiae et Therapiae Experimentalis*, *60*, 361–372.

Zhu, R., Xiao, D., & Zhang, L. (2013). Potassium channels and uterine vascular adaptation to pregnancy and chronic hypoxia. *Current Vascular Pharmacology*, *11*, 737–747.

Zhu, S., White, R. E., & Barman, S. A. (2008). Role of phosphodiesterases in modulation of BKCa channels in hypertensive pulmonary arterial smooth muscle. *Therapeutic Advances in Respiratory Disease*, *2*, 119–127.

Ziegler, W. H., Parekh, D. B., Le Good, J. A., Whelan, R. D., Kelly, J. J., Frech, M., et al. (1999). Rapamycin-sensitive phosphorylation of PKC on a carboxy-terminal site by an atypical PKC complex. *Current Biology*, *9*, 522–529.

CHAPTER SEVEN

Rho-Mancing to Sensitize Calcium Signaling for Contraction in the Vasculature: Role of Rho Kinase

T. Szasz[1], R.C. Webb
Augusta University, Augusta, GA, United States
[1]Corresponding author: e-mail address: iszasz@augusta.edu

Contents

Abstract

Vascular smooth muscle contraction is an important physiological process contributing to cardiovascular homeostasis. The principal determinant of smooth muscle contraction is the intracellular free Ca^{2+} concentration, and phosphorylation of myosin light chain (MLC) by activated myosin light chain kinase (MLCK) in response to increased Ca^{2+} is the main pathway by which vasoconstrictor stimuli induce crossbridge cycling of myosin and actin filaments. A secondary pathway for vascular smooth muscle contraction that is not directly dependent on Ca^{2+} concentration, but rather mediating Ca^{2+} sensitization, is the RhoA/Rho kinase pathway. In response to contractile stimuli, the small GTPase RhoA activates its downstream effector Rho kinase which, in turn, promotes contraction via myosin light chain phosphatase (MLCP) inhibition. RhoA/Rho kinase-mediated MLCP inhibition occurs mainly by phosphorylation and inhibition of MYPT1, the regulatory subunit of MLCP, or by CPI-17-mediated inhibition of the catalytic subunit of MLCP. In this review, we describe the molecular mechanisms underlying the pivotal role exerted by Rho kinase on vascular smooth muscle contraction and discuss the main regulatory pathways for its activity.

Advances in Pharmacology, Volume 78
ISSN 1054-3589
http://dx.doi.org/10.1016/bs.apha.2016.09.001

ABBREVIATIONS

CaM calmodulin
CPI-17 C kinase potentiated protein phosphatase inhibitor
GAP GTPase-activating protein
GDI guanine nucleotide dissociation inhibitor
GEF guanine nucleotide exchange factor
GPCR G protein-coupled receptor
MLC myosin light chain
MLCK myosin light chain kinase
MLCP myosin light chain phosphatase
MYPT myosin phosphatase targeting subunit

1. INTRODUCTION

Vascular smooth muscle contraction is caused by the crossbridge cycling of intracellular filaments of myosin and actin. Upon phosphorylation and activation of the 20 kDa myosin light chain (MLC_{20}) subunit, the myosin head attached to an adjacent actin filament undergoes a conformational change that makes it tilt and slide along the actin filament causing a shortening fuelled by ATP hydrolysis. The state of MLC_{20} phosphorylation is therefore equivalent to the contractile state of the smooth muscle cell and its balance is maintained by the activity of MLC kinase (MLCK) and MLC phosphatase (MLCP), involving Ca^{2+} entry signaling for the former and Ca^{2+} sensitization signaling for the latter (Fig. 1, briefly outlined later and extensively reviewed elsewhere, for reference, see Khalil, 2010; Webb, 2003).

The smooth muscle isoform of MLCK is a serine/threonine kinase which phosphorylates the regulatory MLC_{20} subunit of myosin II at Ser-19 to cause contraction. Activation of MLCK occurs when free intracellular Ca^{2+} concentration rises and the Ca^{2+}-binding protein calmodulin (CaM) is activated, the $(Ca^{2+})_4$–CaM complex then interacting with MLCK via its CaM-binding domain and inducing a conformational change in the MLCK structure that allows it to act upon its target, MLC_{20}. Various vasoconstrictor stimuli such as contractile agonists, mechanical stress, and changes in membrane potential induce increases in intracellular Ca^{2+} that result in activation of CaM. Upon ligand binding to G protein-coupled receptors (GPCRs) and activation of heterotrimeric G protein, the membrane-associated phospholipase C (PLC) induces transformation of phosphatidyl inositol biphosphate

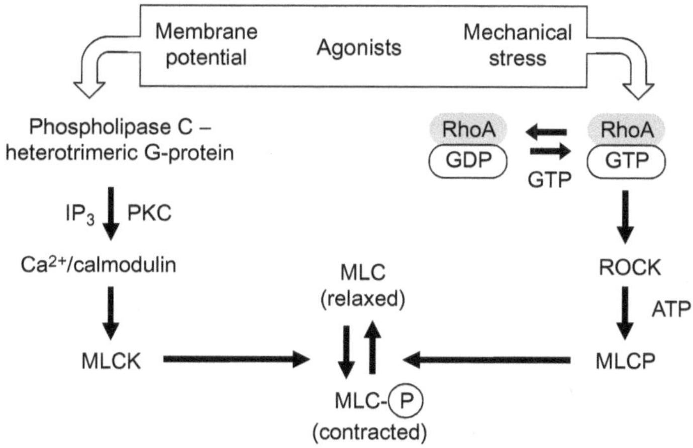

Fig. 1 RhoA/ROCK and contraction of vascular smooth muscle. Various agonists (neurotransmitters, hormones, etc.) bind to specific membrane receptors to activate contraction in smooth muscle. Subsequent to this binding, the prototypical response of the cell is to increase PLC activity via coupling through a heterotrimeric G-protein. PLC produces two potent second messengers: diacylglycerol (DAG) and inositol 1,4,5-trisphosphate (IP_3). IP_3 binds to specific receptors on the sarcoplasmic reticulum causing release of activator Ca^{2+}. DAG along with Ca^{2+} activates protein kinase C (PKC) which phosphorylates specific target proteins. PKC promotes contraction through other effects such as phosphorylation of Ca^{2+} channels or other proteins that regulate crossbridge cycling (i.e., CPI-17, Fig. 4). Activator Ca^{2+} binds to CaM leading to activation of myosin light chain kinase (MLCK). MLCK phosphorylates myosin light chain (MLC) and in conjunction with actin, crossbridge cycling occurs initiating contraction of the smooth muscle cell. However, the elevation in Ca^{2+} concentration is transient and the contractile response is maintained by a Ca^{2+} sensitizing mechanism brought about by the inhibition of myosin light chain phosphatase (MLCP) activity by Rho kinase. This Ca^{2+} sensitizing mechanism is initiated at the same time that PLC is activated and it involves the activation of the small GTP-binding protein RhoA. The precise nature of the activation of RhoA by the G protein-coupled receptor is not entirely clear but involves a guanine nucleotide exchange factor (RhoGEF) and migration of RhoA to the plasma membrane. Following activation, RhoA increases Rho kinase activity leading to inhibition of MLCP promoting the contractile state. In addition to agonist-induced activation, membrane depolarization and mechanical stress have been shown to initiate contraction by these two signaling cascades in a nonagonist-dependent manner.

(PIP_2) into 1,2 diacylglycerol (DAG), and inositol 1,4,5 triphosphate (IP_3). Subsequently, DAG activates protein kinase C (PKC) and IP_3 binds to the IP_3 receptor on the sarcoplasmic reticulum, inducing Ca^{2+} release and a transient increase in intracellular Ca^{2+} concentration which activates CaM and triggers contraction. For a sustained contraction, intracellular Ca^{2+} concentration is maintained high by subsequent entry of Ca^{2+} from the extracellular space via voltage-gated, receptor or ligand-gated, non-specific, and store-operated Ca^{2+} channels.

On the other side of the equation for MLC_{20} phosphorylation stands MLCP, which acts to dephosphorylate MLC_{20} and limit contraction. MLCP is composed of three subunits: a serine/threonine phosphatase catalytic subunit termed PP1c, a regulatory subunit which binds phoshorylated MLC_{20} termed myosin phosphatase targeting subunit (MYPT1), and a smaller subunit the function of which has not been clarified yet.

Since Ca^{2+} sensitivity of vascular smooth muscle is a function of the ratio of the activity of MLCK and that of MLCP, inhibition of MLCP shifts the balance in favor of MLCK-mediated MLC_{20} phosphorylation and contraction. Phosphorylation of MYPT1 and inhibition of PP1c by C kinase potentiated protein phosphatase inhibitor (CPI-17) are two major mechanisms of MLCP inhibition implicated in Rho kinase-mediated Ca^{2+} sensitization which will be discussed in detail later.

2. RhoA/Rho KINASE STRUCTURE AND EXPRESSION

RhoA is a small GTPase and member of the Ras superfamily of GTPases. Small GTPases are homologous to the G_α subunit of heterotrimeric G proteins and function as GTP hydrolases. Because they exist either bound to GTP (active) or to GDP (inactive), small GTPases like RhoA act as molecular switches (Fig. 2). This switch is turned on by guanine nucleotide exchange factors (GEFs), which exchange GDP-Rho with GTP-Rho which is then tethered to the plasma membrane by geranylgeranylation of its C-terminus where it can interact with its targets (Katayama et al., 1991). Conversely, the switch is returned to the inactive state by GTPase-activating proteins (GAPs), which induce GTP hydrolase activity and generation of GDP-Rho that is maintained inactive and located in the cytoplasm by guanine nucleotide dissociation inhibitors (GDIs). Regulation of the molecular switch is thus provided by Rho-specific GAPs and GEFs and additionally by posttranslational modifications and GDIs, which may target RhoA to a specific subcellular localization or modulate its turnover.

RhoA is involved in a wide range of basic cellular processes like cytoskeleton organization, focal adhesions, stress fiber formation, migration, contraction, as well as cell cycle maintenance and transcriptional regulation (Amano, Nakayama, & Kaibuchi, 2010). In vascular smooth muscle, RhoA is activated in response to GPCRs activation (Seasholtz, Majumdar, &

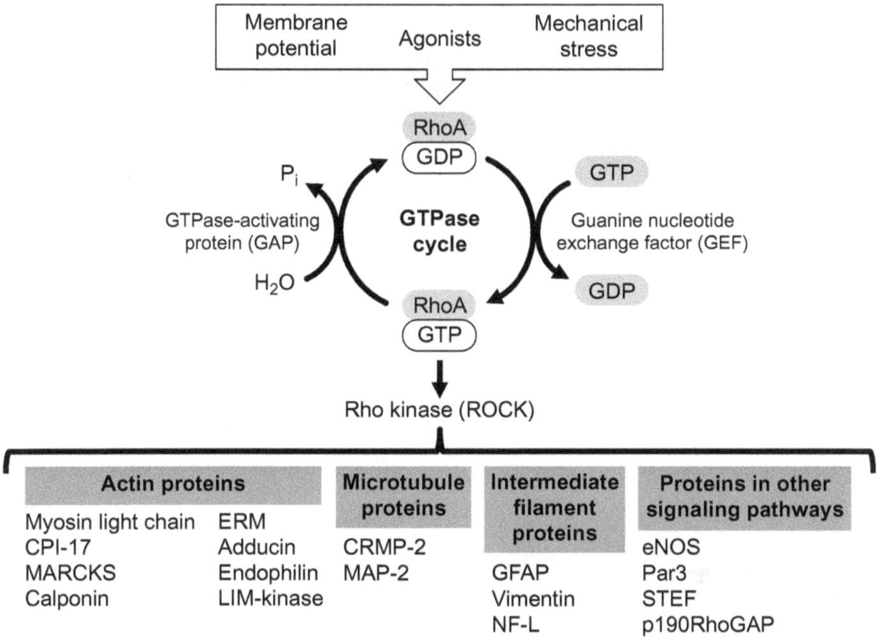

Fig. 2 Substrates of Rho kinase. As described in the text, Rho kinase inhibits MLCP activity through both phosphorylation of MYPT1 and phosphorylation of CPI-17 to cause contraction. Rho kinase and MLCP share several substrates, such as MLC, ERM proteins, adducin, and other proteins. The substrates reported to be phosphorylated by Rho kinase can be classified into four categories: (1) actin-associated proteins (*far left*); (2) microtubule-associated proteins (*second from left*); (3) intermediate filament-associated proteins (*third from left*); and (4) proteins in other signaling pathways (*far right*). Abbreviations: *MARCKS*, myristoylate alanine-rich C kinase; *ERM*, ezrin, radixin, and moesin proteins; *CRMP2*, collapsing response mediator proteins; *MAP-2*, microtubule-associated proteins; *GFAP*, glial fibrillary acidic protein; *NF-L*, neurofilament-light protein; *eNOS*, endothelial nitric oxide synthase; *Par-3*, partitioning defective protein 3; *STEF*, SIF and Tiam-1-like exchange factor; *GEF*, a guanine nucleotide exchange factor; *p190RhoGAP*, a GTPase-activating protein.

Brown, 1999; Woodsome, Polzin, Kitazawa, Eto, & Kitazawa, 2006), depolarization (Liu, Zuo, Pertens, Helli, & Janssen, 2005; Sakurada et al., 2003), or mechanical stress (Smith, Roy, Zhang, & Chauduri, 2003) (Fig. 2).

The main downstream effector of activated RhoA is Rho-associated coiled-coil protein kinase, or Rho kinase. Rho kinase was first identified in 1996 as a 164 kDa GTPγS (nonhydrolyzable form of GTP)-binding protein (Matsui et al., 1996). Rho kinase contains a Rho-binding domain within a central coiled-coil region, flanked by a catalytic kinase domain at the N-terminus responsible for the Ser/Thr kinase activity and a pleckstrin homology domain at the C-terminus which allows for binding to the cell membrane. In the resting state, the Rho-binding domain and the pleckstrin homology domain at the C-terminus each bind the N-terminal catalytic region, exerting an autoinhibitory effect. Consequently, deletion of the C-terminal region renders Rho kinase constitutively active (Amano et al., 1997), caspase 3, or granzyme-induced proteolytic cleavage of the same region activates Rho kinase during apoptosis (Coleman et al., 2001; Sebbagh, Hamelin, Bertoglio, Solary, & Breard, 2005; Sebbagh et al., 2001), and the isolated C-terminal region is capable of inhibiting Rho kinase activity in vitro and in vivo (Amano et al., 1999).This autoinhibition is removed upon Rho-GTP binding to the Rho-binding region, which disrupts the C-terminus binding to the N-terminus and induces activation. Extension segments on the N- and C-termini allow for a head-to-head homodimer formation keeping the enzyme in an active state (Jacobs et al., 2006; Yamaguchi, Kasa, Amano, Kaibuchi, & Hakoshima, 2006).

Rho kinase is evolutionarily conserved from invertebrates to mammals and has two ubiquitously expressed isoforms encoded by different genes and termed ROCK1 (or ROKα) and ROCK2 (or ROKβ). The homology between ROCK1 and ROCK2 is 65% for the overall amino acid sequence identity, 58% for the Rho-binding domain, and 92% for the kinase domain (Amano et al., 2010). Various tissues exhibit preferential expression of one of the two isoforms, however vascular smooth muscle expresses both ROCK1 and ROCK2 and there is little evidence demonstrating isoform-specific effects. Knockout of either of the ROCK isoforms has important deleterious effects during development, however neither is lethal in the heterozygous form, likely demonstrating some compensation or functional redundancy between isoforms (Shimizu et al., 2005; Thumkeo et al., 2003; Thumkeo, Shimizu, Sakamoto, Yamada, & Narumiya, 2005; Thumkeo, Watanabe, & Narumiya, 2013; Zhang et al., 2006).

In addition to RhoA, other small G proteins can modulate Rho kinase, among them RhoE, Gem, and Rad. RhoE inhibits ROCK1 by binding to the N-terminal catalytic domain and preventing RhoA from attaching to the Rho-binding domain, however the mechanism for Gem and Rad inhibition is less clear. Conversely, Rho kinase activity can also be regulated by lipids disrupting the autoinhibitory action of the C-terminal regulatory region and increasing activation, such as arachidonic acid (Araki et al., 2001) and sphingosine phosphorylcholine (Shirao et al., 2015). Evidence indicates that Rho kinase may also be regulated at the transcriptional level. For example, in response to angiotensin II, ROCK is increased in vascular smooth muscle

(Hiroki et al., 2004; Hiroki, Shimokawa, Mukai, Ichiki, & Takeshita, 2005; Jin et al., 2006).

3. Rho KINASE FUNCTION

The consensus sequence of Rho kinase phosphorylation sites is R/KXS/T or R/KXXS/T and to date numerous substrates have been identified (Amano et al., 1996, 2003; Fukata et al., 1999; Goto et al., 1998; Kaneko et al., 2000; Matsui et al., 1998). In vascular tissues, Rho kinase substrates include proteins involved in actomyosin contraction and cytoskeleton organization, as well as microtubule-associated proteins, intermediate filaments, and proteins in other pathways important for vascular function (Fig. 2), thus contributing to regulation of cellular contraction, motility, morphology, polarity, cell division, and gene expression.

Although the intracellular Ca^{2+} concentration in vascular smooth muscle is the main factor determining the contractile state of the cell, Somlyo and coworkers (Kitazawa, Gaylinn, Denney, & Somlyo, 1991; Kitazawa, Masuo, & Somlyo, 1991) observed more than 30 years ago that vasoconstrictor stimuli can induce changes in MLC_{20} phosphorylation and tension in the absence of changes in Ca^{2+} concentration, through what is known as the Ca^{2+} sensitization pathway, the main effector of which is Rho kinase. This is demonstrated by the leftward shift in the Ca^{2+} concentration–response curve in arteries in the presence of the nonhydrolyzable form of GTP, GTPγS (Fig. 3). The effect of GTPγS is RhoA mediated, since bacterial toxins which inactivate Rho proteins inhibit both GTPγS-induced and agonist-induced Ca^{2+}-sensitization in permeabilized smooth muscle, and recombinant RhoA rescues this Ca^{2+} sensitization (Puetz, Lubomirov, & Pfitzer, 2009). Finally, the Rho kinase inhibitor Y27632 reverses GTPγS-induced Ca^{2+} sensitization (Fu, Gong, Jia, Somlyo, & Somlyo, 1998).

Interestingly, some evidence points to a convergence of Ca^{2+} entry and Ca^{2+} sensitization pathways in the case of depolarization-induced RhoA/Rho kinase activation, when L-type Ca^{2+} channels seem to play a role in Ca^{2+}-induced Ca^{2+} release and maintenance of tonic contraction (Fernandez-Tenorio et al., 2011).

RhoA/Rho kinase-mediated Ca^{2+} sensitization is responsible for the maintenance of prolonged contraction in response to vasoactive agonists in several vascular beds including mesenteric and pulmonary arteries, as demonstrated by various studies employing the use of Rho kinase inhibitors. Similar studies implicate Rho kinase in the maintenance of basal tone of

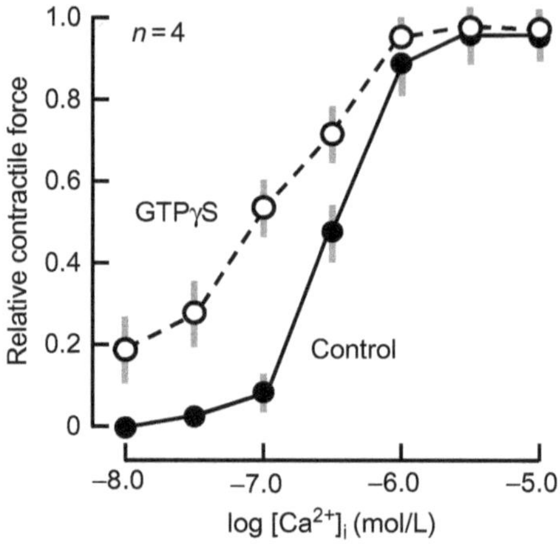

Fig. 3 GTPγS-induced Ca^{2+} sensitization of contraction in α-toxin permeabilized rat tail artery strip. In these experiments, rat tail artery strips were placed in muscle chambers and permeabilized with α-toxin. Molecules of α-toxin form small channels in the membrane (~2.5 nm in diameter) permitting passage of only low molecular-weight substances (less than 1000 molecular weight). Thus, membrane-bound and soluble proteins such as calmodulin, RhoA, and Rho kinase are retained in the cell. Following permeabilization, Ca^{2+} stores in the sarcoplasmic reticulum were depleted by treatment with A23187 (10 μmol/L), a Ca^{2+} ionophore. Following treatment, the permeabilized segments were made to contract by addition of Ca^{2+} (10 nmol/L to 10 μmol/L, buffered with EGTA). Activation of the RhoA/Rho kinase pathway with GTPγS (10 μmol/L) increased sensitivity of the contractile response to Ca^{2+} as evidenced by the shift in the concentration–response curve to the left ($n=4$; asterisks indicate a significant difference from control value, $p<0.05$).

small pressurized vessels (Schubert, Kalentchuk, & Krien, 2002; VanBavel, van der Meulen, & Spaan, 2001), and of myogenic tone of cerebral arteries (Bolz et al., 2003; Dubroca, You, Levy, Loufrani, & Henrion, 2005; Gokina, Park, McElroy-Yaggy, & Osol, 2005). The main mechanisms believed to underlie RhoA/Rho kinase-mediated Ca^{2+} sensitization are phosphorylation and inhibition of MLCP subunits (Kitazawa, Gaylinn, et al., 1991; Kitazawa, Masuo, et al., 1991). The main Rho kinase substrates influencing contraction are *highlighted* later.

Rho kinase phosphorylates the myosin-binding subunit of MLCP, *MYPT1*, at several potential sites, including Ser853, Thr853, and Thr696, inducing inhibition of MLCP. Of these sites, only Ser853 is specific for Rho kinase, while Thr850/855 and Thr695/697 are also targeted by other kinases (Kitazawa, Eto, Woodsome, & Khalequzzaman, 2003; Muranyi et al., 2005). In addition, Rho kinase also phosphorylates zipper-interacting protein kinase (*ZIP kinase*) (Hagerty et al., 2007) which may separately phosphorylate Thr 696 to inhibit MYPT1.

CPI-17 is a small inhibitor of PP1c, the catalytic phosphatase subunit of MLCP (Fig. 4). CPI-17 is activated by phosphorylation at the Thr38 site (Kitazawa, Eto, Woodsome, & Brautigan, 2000; Kitazawa et al., 2003), which is a target for Rho kinase (Koyama et al., 2000) and other kinases such as PKC (Woodsome, Eto, Everett, Brautigan, & Kitazawa, 2001). PKC may also indirectly mediate CLI-17 phosphorylation via activation of integrin-linked kinase (ILK) (Deng, Sutherland, Brautigan, Eto, & Walsh, 2002). Activation of CPI-17 increases its inhibitory activity on MLCP by about 7000-fold. Studies using Rho kinase inhibitors in vivo demonstrated the Rho kinase-specific CPI-17 activation in vascular smooth muscle (Xie et al., 2010). Depending on tissue-specific expression, there is both overlap and crosstalk between Rho/Rho kinase and PKC in mediating Ca^{2+} sensitization via CPI-17 phosphorylation (Fig. 4).

The first reports on the identification of Rho kinase already indicated that it is capable of directly phosphorylating MLC_{20} at Ser19, which would cause smooth muscle contraction (Amano et al., 1996; Kureishi et al., 1997), however the extent and relevance of this mechanism in vivo is still unclear.

A more recent concept is that the actin cytoskeleton remodeling contributes to agonist-induced contraction in vascular smooth muscle cells by increasing the fibrillar (F) to globular (G) actin ratio at the cell cortex, a phenomenon partly mediated by RhoA/Rho kinase. Rho kinase phosphorylation of *LIM kinase* isoforms LIMK1 and LIMK2 at Thr508 and Thr505, respectively, stabilizes actin filaments and contributes to stress fiber

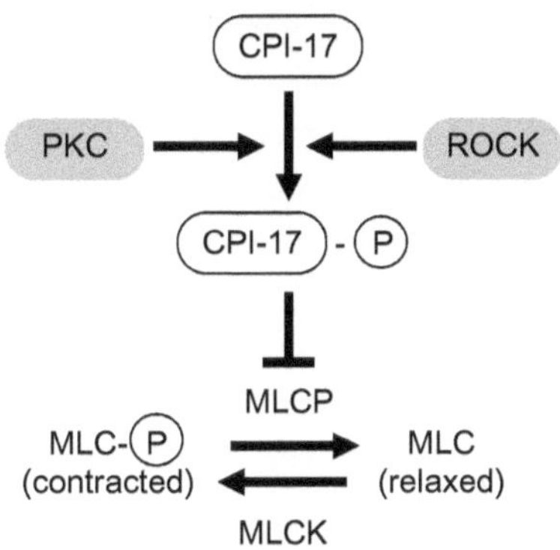

Fig. 4 Protein kinase C (PKC) and Rho kinase (ROCK) phosphorylate CPI-17 to inhibit MLCP. PKC and ROCK have an inhibitory effect on MLCP to maintain the contractile state of vascular smooth muscle.

formation and contractility via LIM kinase-mediated phosphorylation and inhibition of cofilin, an actin-depolymerizing factor (Maekawa et al., 1999; Ohashi et al., 2000).

It has recently been demonstrated that Rho kinase also phosphorylates and inhibits *p190RhoGAP*, thus stabilizing GTP-Rho in a positive feedback loop, explaining sustained activation of Rho kinase in vasospasm (Mori et al., 2009).

Calponin is a Ca^{2+}-binding protein that acts to limit contraction by inhibiting the ATPase activity of the myosin head. Rho kinase-mediated calponin phosphorylation (Kaneko et al., 2000) results in the release of this inhibition, thus further contributing to contraction.

Indirect effects of RhoA/Rho kinase on contraction may be exerted via the phosphorylation of *endothelial NO synthase (eNOS)* at the inhibitory site Ser495 (Sugimoto et al., 2007) as well as via reductions in the expression of eNOS protein (Bivalacqua et al., 2004; Ming et al., 2002), thus decreasing endothelial NO production and augmenting contraction. Conversely, eNOS may inhibit angiotensin II-mediated Rho kinase activation through $G_{12/13}$ (Suzuki et al., 2009) and decreasing contraction, therefore illustrating a mutual inhibitory effect of Rho kinase and eNOS.

4. REGULATION OF RhoA/Rho KINASE ACTIVITY VIA POSTTRANSLATIONAL MODIFICATIONS

A regulatory mechanism of small GTPases like RhoA that has received less attention but is potentially crucial is that of posttranslational modifications (Fig. 5). As mentioned earlier, geranylgeranylation of the C-terminus of GTP-bound Rho makes this region hydrophobic thus targeting it to the membrane where it may interact with activators and effectors. Geranylgeranylation and farnesylation are two different types of prenylation, which is a type of posttranslational modification of proteins by lipids. Prenylation occurs when 20-carbon lipophilic geranylgeranyl or 15-carbon farnesyl isoprene moieties from geranylgeranyl diphosphate (by the action of geranylgeranyltransferase) or farnesyl pyrophosphate (by the action of farnesyl transferase), respectively, are irreversibly attached to cysteine residues on the C-terminus of certain proteins, targeting them to membranes. RhoA is subjected to these two types of prenylation with no specific functional consequence differentiating them (Michaelson et al., 2001; Solski, Helms, Keely, Su, & Der, 2002). Like prenylation, palmitoylation modifies the

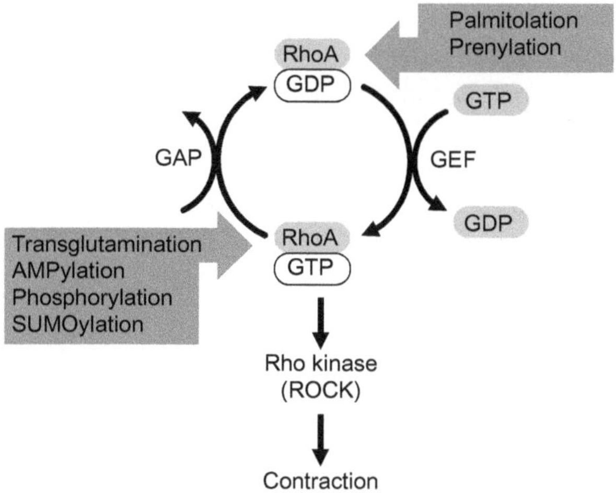

Fig. 5 Posttranslational modification of Rho signaling. GEFs and GAPs are important regulatory components controlling the GTPase cycle and signaling activities of Rho. GEFs activate Rho GTPases by catalyzing the exchange of bound GDP for GTP, whereas GAPs stimulate GTP hydrolysis. Posttranslational modifications are also involved in controlling the activity of Rho GTPase. Palmitoylation and prenylation can target Rho to specific intracellular compartments. Posttranslational covalent modifications, including phosphorylation, transglutamination, AMPylation, and SUMOylation, can induce constitutive activation or inactivation of Rho GTPases. Combined, these regulatory mechanisms contribute to modulation of the signaling activity of Rho, with an effect on actin and microtubule dynamics, cell adhesion, cell cycle progression, cell survival, gene expression, and other cellular processes.

CAAX motif within the hypervariable region at the C-terminus of RhoA and blocks binding of RhoGDI (Michaelson et al., 2001).

Phosphorylation of RhoA at Ser188 by PKA and PKG increases its binding to RhoGDI, thus maintaining it inactive, and on the other hand protecting it from ubiquitylation and subsequent proteasome degradation (Ellerbroek, Wennerberg, & Burridge, 2003; Rolli-Derkinderen et al., 2005; Rolli-Derkinderen, Toumaniantz, Pacaud, & Loirand, 2010).

Other, less well understood posttranslational modifications regulate the activity of small GTPases like RhoA, including sumoylation, ubiquitylation, AMPylation, and transglutamination. These modifications appear to either increase or interfere with RhoA turnover and thus may prolong or limit RhoA/Rho kinase activation. For example, polyubiquitylation of RhoA in response to PKC decreases RhoA/Rho kinase-mediated stress fiber formation by targeting RhoA to the 26S proteasome for degradation, thus decreasing its expression levels (Deng & Huang, 2014). Transamidation of RhoA with serotonin, or serotonylation, induces activation of RhoA and promotes its association with E3 ubiquitin ligase SMURF1 and proteasomal-mediated degradation (Guilluy et al., 2007). Recent evidence indicates that geranylgeranylation may also enhance proteasomal degradation of RhoA (Stubbs & Von Zee, 2012).

Furthermore, posttranslational modifications of RhoGEFs and RhoGAPs introduce a new layer of regulation of RhoA activity (for a recent comprehensive review, please see Hodge & Ridley, 2016).

5. RhoA/Rho KINASE-MEDIATED Ca^{2+} SENSITIZATION IN VASCULAR DISEASE AND Rho KINASE INHIBITORS: FOCUS ON HYPERTENSION

A hallmark of the vascular dysfunction associated with hypertension is increased vascular contraction, which is believed to underlie the increase in total peripheral resistance that defines most cases of essential hypertension. Due to its central role mediating Ca^{2+} sensitization in vascular smooth muscle, the involvement of RhoA/Rho kinase pathway in hypertension-associated vascular dysfunction was extensively studied (reviewed by Loirand, Guerin, & Pacaud, 2006; Mukai et al., 2001).

RhoA/Rho kinase activation is increased in animal models of hypertension (Chrissobolis & Sobey, 2001; Ito et al., 2004; Jin et al., 2006; Moriki et al., 2004; Uehata et al., 1997) and in hypertensive humans (Calo et al., 2014; Masumoto et al., 2001). An early report on the potent effect of Rho kinase inhibition in hypertensive animal models came soon after the identification of Rho kinase as the RhoA effector, in a study where the Rho kinase inhibitor Y27632 was administered in vivo to spontaneously hypertensive rats (SHR), DOCA-salt and two kidney-one clip renal hypertensive rats (Uehata et al., 1997). Y27632 significantly and dose-dependently lowered blood pressure in all these hypertension models, indicating that Rho kinase-mediated Ca^{2+} sensitization is increased in hypertension. Arteries from SHR and mineralocorticoid-induced hypertensive animals also exhibit increased relaxation to Y27643 (Asano & Nomura, 2003; Weber & Webb, 2001). Currently, extensive proof of the increased activation of the RhoA/Rho kinase pathway in various vascular beds in hypertension supports the idea that this pathway contributes to blood pressure regulation (Chrissobolis & Sobey, 2001; Jin et al., 2006; Moriki et al., 2004; Uehata et al., 1997). Consequently, efforts were made to test the efficacy of Rho kinase inhibition as a therapeutical approach for hypertension.

However, the two most widely used Rho kinase inhibitors, Y27632 and fasudil, although invaluable in in vitro and in vivo studies of Rho kinase function, and available clinically in limited situations (fasudil is approved for treatment of cerebral vasospasm following subarachnoidal hemorrhage in Japan), did not succeed as viable antihypertensive therapies. In addition to incomplete safety/tolerability profiles, competitive Rho kinase inhibitors may potentially concomitantly inhibit other kinases like PKA, PKC, and PKC-related kinase 2 (PRK2).

Although not intended to inhibit Rho kinase, therapies administered for other indications may secondarily influence Rho kinase activity. For example, HMG-CoA reductase inhibitors (statins) inhibit prenylation of Rho-GTP and thus decrease Rho kinase activation. Angiotensin receptor blockers like olmesartan also decrease Rho kinase activation (Ravarotto et al., 2015).

New Rho kinase pharmacological inhibitors are being developed with variable results, of them SAR407899 (Grisk et al., 2012; Lohn et al., 2009), WAR-5 (Li et al., 2015), DW1865 (Oh et al., 2013), and K115 or ripasudil (Garnock-Jones, 2014) being the most recent, and novel therapy development approaches underway (Mishra et al., 2014). Noteworthy, DW1865 was shown to selectively and potently inhibit Rho kinase phosphorylation of MYPT1, induce vasorelaxation and decrease blood pressure (Oh et al., 2013). Recently, a ROCK2-specific inhibitor, KD025, was tested in focal cerebral ischemia in mice, with promising results (Lee et al., 2014). However, in the development of potential future Rho kinase inhibitors, consideration should be given to the pleiotropic effects of Rho kinase activation, its ubiquitous expression, and similarity of its active site to that of other kinases.

In addition to calcium sensitization mechanisms in vascular smooth muscle cells, Rho kinase also participates in other important processes within the cardiovascular system, such as in regulation of endothelial function via the aforementioned complex interaction with eNOS. It is thus not surprising that Rho kinase is implicated in other conditions associated with both augmented vasoconstriction and endothelial dysfunction (Yao et al., 2010) besides hypertension, such as diabetes (Mishra, Alokam, Sriram, & Yogeeswari, 2013) and atherosclerosis (Zhou & Liao, 2009).

Therapeutical approaches involving Rho kinase inhibitors have also been tested or used in various other conditions in addition to vascular dysfunction. Reflective of the wide range of phosphorylation targets and cellular effects of Rho kinase depicted in Fig. 2, studies show that its inhibitors ameliorated manifestations of erectile dysfunction (Chitaley, Webb, & Mills, 2001; Sopko, Hannan, & Bivalacqua, 2014), cerebral vasospasm (Naraoka, Munakata, Matsuda, Shimamura, & Ohkuma, 2013), glaucoma (Inoue & Tanihara, 2013), and asthma (Kume, 2008). Additionally, Rho kinase inhibitors revealed the important role of Rho kinase in the central nervous system in growth cone collapse and dendrite formation, and there may be Rho kinase inhibitor therapies on the horizon for neurological disorders like Alzheimer disease, stroke, and spinal cord injury (Mueller, Mack, & Teusch, 2005).

6. CONCLUSION

RhoA and its effector, Rho kinase, play a central role in the highly complex cellular machinery of Ca^{2+} signaling and Ca^{2+} sensitization pathways that converge to effect vascular smooth muscle contraction in response to vasoactive stimuli. The best known effect of Rho kinase activation in vascular smooth muscle is maintenance of contraction through Ca^{2+} sensitization caused by Rho kinase-mediated inhibition of MLCP. However, other Rho kinase targets, including actin proteins, intermediate filament, and microtubule-associated proteins mediate a broad range of cellular effects. Rho kinase is regulated at various levels by multiple mechanisms that overlap or provide positive and negative feedbacks, including RhoGEF, RhoGAP, and RhoGDI regulation of RhoA activity, as well as posttranslational modifications regulation of its subcellular localization and turnover.

CONFLICT OF INTEREST

The authors have no conflict of interest to declare.

REFERENCES

Amano, M., Chihara, K., Kimura, K., Fukata, Y., Nakamura, N., Matsuura, Y., et al. (1997). Formation of actin stress fibers and focal adhesions enhanced by Rho-kinase. *Science*, *275*(5304), 1308–1311.

Amano, M., Chihara, K., Nakamura, N., Kaneko, T., Matsuura, Y., & Kaibuchi, K. (1999). The COOH terminus of Rho-kinase negatively regulates rho-kinase activity. *The Journal of Biological Chemistry*, *274*(45), 32418–32424.

Amano, M., Ito, M., Kimura, K., Fukata, Y., Chihara, K., Nakano, T., et al. (1996). Phosphorylation and activation of myosin by Rho-associated kinase (Rho-kinase). *The Journal of Biological Chemistry*, *271*(34), 20246–20249.

Amano, M., Kaneko, T., Maeda, A., Nakayama, M., Ito, M., Yamauchi, T., et al. (2003). Identification of Tau and MAP2 as novel substrates of Rho-kinase and myosin phosphatase. *Journal of Neurochemistry*, *87*(3), 780–790. http://dx.doi.org/2054 [pii].

Amano, M., Nakayama, M., & Kaibuchi, K. (2010). Rho-kinase/ROCK: A key regulator of the cytoskeleton and cell polarity. *Cytoskeleton (Hoboken, N.J.)*, *67*(9), 545–554. http://dx.doi.org/10.1002/cm.20472.

Araki, S., Ito, M., Kureishi, Y., Feng, J., Machida, H., Isaka, N., et al. (2001). Arachidonic acid-induced Ca2+ sensitization of smooth muscle contraction through activation of Rho-kinase. *Pflügers Archiv*, *441*(5), 596–603.

Asano, M., & Nomura, Y. (2003). Comparison of inhibitory effects of Y-27632, a Rho kinase inhibitor, in strips of small and large mesenteric arteries from spontaneously hypertensive and normotensive Wistar–Kyoto rats. *Hypertension Research*, *26*(1), 97–106.

Bivalacqua, T. J., Champion, H. C., Usta, M. F., Cellek, S., Chitaley, K., Webb, R. C., et al. (2004). RhoA/Rho-kinase suppresses endothelial nitric oxide synthase in the penis: A mechanism for diabetes-associated erectile dysfunction. *Proceedings of the National Academy of Sciences of the United States of America*, *101*(24), 9121–9126. http://dx.doi.org/10.1073/pnas.0400520101.

Bolz, S. S., Vogel, L., Sollinger, D., Derwand, R., Boer, C., Pitson, S. M., et al. (2003). Sphingosine kinase modulates microvascular tone and myogenic responses through activation of RhoA/Rho kinase. *Circulation*, *108*(3), 342–347. http://dx.doi.org/10.1161/01.CIR.0000080324.12530.0D.

Calo, L. A., Davis, P. A., Pagnin, E., Dal Maso, L., Maiolino, G., Seccia, T. M., et al. (2014). Increased level of p63RhoGEF and RhoA/Rho kinase activity in hypertensive patients. *Journal of Hypertension*, *32*(2), 331–338. http://dx.doi.org/10.1097/HJH.0000000000000075.

Chitaley, K., Webb, R. C., & Mills, T. M. (2001). Rho-kinase as a potential target for the treatment of erectile dysfunction. *Drug News & Perspectives*, *14*(10), 601–606. http://dx.doi.org/660508 [pii].

Chrissobolis, S., & Sobey, C. G. (2001). Evidence that Rho-kinase activity contributes to cerebral vascular tone in vivo and is enhanced during chronic hypertension: Comparison with protein kinase C. *Circulation Research*, *88*(8), 774–779.

Coleman, M. L., Sahai, E. A., Yeo, M., Bosch, M., Dewar, A., & Olson, M. F. (2001). Membrane blebbing during apoptosis results from caspase-mediated activation of ROCK I. *Nature Cell Biology*, *3*(4), 339–345. http://dx.doi.org/10.1038/35070009.

Deng, S., & Huang, C. (2014). E3 ubiquitin ligases in regulating stress fiber, lamellipodium, and focal adhesion dynamics. *Cell Adhesion & Migration*, *8*(1), 49–54. http://dx.doi.org/27480 [pii].

Deng, J. T., Sutherland, C., Brautigan, D. L., Eto, M., & Walsh, M. P. (2002). Phosphorylation of the myosin phosphatase inhibitors, CPI-17 and PHI-1, by integrin-linked kinase. *The Biochemical Journal*, *367*(Pt. 2), 517–524. http://dx.doi.org/10.1042/BJ20020522.

Dubroca, C., You, D., Levy, B. I., Loufrani, L., & Henrion, D. (2005). Involvement of RhoA/Rho kinase pathway in myogenic tone in the rabbit facial vein. *Hypertension*, *45*(5), 974–979. http://dx.doi.org/01.HYP.0000164582.63421.2d [pii].

Ellerbroek, S. M., Wennerberg, K., & Burridge, K. (2003). Serine phosphorylation negatively regulates RhoA in vivo. *The Journal of Biological Chemistry*, *278*(21), 19023–19031. http://dx.doi.org/10.1074/jbc.M213066200.

Fernandez-Tenorio, M., Porras-Gonzalez, C., Castellano, A., Del Valle-Rodriguez, A., Lopez-Barneo, J., & Urena, J. (2011). Metabotropic regulation of RhoA/Rho-associated kinase by L-type Ca2+ channels: New mechanism for depolarization-evoked mammalian arterial contraction. *Circulation Research*, *108*(11), 1348–1357. http://dx.doi.org/CIRCRESAHA.111.240127 [pii].

Fu, X., Gong, M. C., Jia, T., Somlyo, A. V., & Somlyo, A. P. (1998). The effects of the Rho-kinase inhibitor Y-27632 on arachidonic acid-, GTPgammaS-, and phorbol ester-induced Ca2+-sensitization of smooth muscle. *FEBS Letters*, *440*(1-2), 183–187. http://dx.doi.org/S0014-5793(98)01455-0 [pii].

Fukata, Y., Oshiro, N., Kinoshita, N., Kawano, Y., Matsuoka, Y., Bennett, V., et al. (1999). Phosphorylation of adducin by Rho-kinase plays a crucial role in cell motility. *The Journal of Cell Biology*, *145*(2), 347–361.

Garnock-Jones, K. P. (2014). Ripasudil: First global approval. *Drugs*, *74*(18), 2211–2215. http://dx.doi.org/10.1007/s40265-014-0333-2.

Gokina, N. I., Park, K. M., McElroy-Yaggy, K., & Osol, G. (2005). Effects of Rho kinase inhibition on cerebral artery myogenic tone and reactivity. *Journal of Applied Physiology (Bethesda, Md.: 1985)*, *98*(5), 1940–1948. http://dx.doi.org/01104.2004 [pii].

Goto, H., Kosako, H., Tanabe, K., Yanagida, M., Sakurai, M., Amano, M., et al. (1998). Phosphorylation of vimentin by Rho-associated kinase at a unique amino-terminal site that is specifically phosphorylated during cytokinesis. *The Journal of Biological Chemistry*, *273*(19), 11728–11736.

Grisk, O., Schluter, T., Reimer, N., Zimmermann, U., Katsari, E., Plettenburg, O., et al. (2012). The Rho kinase inhibitor SAR407899 potently inhibits endothelin-1-induced constriction of renal resistance arteries. *Journal of Hypertension*, *30*(5), 980–989. http://dx.doi.org/10.1097/HJH.0b013e328351d459.

Guilluy, C., Rolli-Derkinderen, M., Tharaux, P. L., Melino, G., Pacaud, P., & Loirand, G. (2007). Transglutaminase-dependent RhoA activation and depletion by serotonin in vascular smooth muscle cells. *The Journal of Biological Chemistry*, *282*(5), 2918–2928. http://dx.doi.org/M604195200 [pii].

Hagerty, L., Weitzel, D. H., Chambers, J., Fortner, C. N., Brush, M. H., Loiselle, D., et al. (2007). ROCK1 phosphorylates and activates zipper-interacting protein kinase. *The Journal of Biological Chemistry*, *282*(7), 4884–4893. http://dx.doi.org/M609990200 [pii].

Hiroki, J., Shimokawa, H., Higashi, M., Morikawa, K., Kandabashi, T., Kawamura, N., et al. (2004). Inflammatory stimuli upregulate Rho-kinase in human coronary vascular smooth muscle cells. *Journal of Molecular and Cellular Cardiology*, *37*(2), 537–546. http://dx.doi.org/10.1016/j.yjmcc.2004.05.008.

Hiroki, J., Shimokawa, H., Mukai, Y., Ichiki, T., & Takeshita, A. (2005). Divergent effects of estrogen and nicotine on Rho-kinase expression in human coronary vascular smooth muscle cells. *Biochemical and Biophysical Research Communications*, *326*(1), 154–159. http://dx.doi.org/S0006-291X(04)02566-5 [pii].

Hodge, R. G., & Ridley, A. J. (2016). Regulating Rho GTPases and their regulators. *Nature Reviews Molecular Cell Biology*, *17*(8), 496–510. http://dx.doi.org/nrm.2016.67 [pii].

Inoue, T., & Tanihara, H. (2013). Rho-associated kinase inhibitors: A novel glaucoma therapy. *Progress in Retinal and Eye Research*, *37*, 1–12. http://dx.doi.org/S1350-9462(13)00039-6 [pii].

Ito, K., Hirooka, Y., Kishi, T., Kimura, Y., Kaibuchi, K., Shimokawa, H., et al. (2004). Rho/Rho-kinase pathway in the brainstem contributes to hypertension caused by chronic nitric oxide synthase inhibition. *Hypertension*, *43*(2), 156–162. http://dx.doi.org/10.1161/01.HYP.0000114602.82140.a4.

Jacobs, M., Hayakawa, K., Swenson, L., Bellon, S., Fleming, M., Taslimi, P., et al. (2006). The structure of dimeric ROCK I reveals the mechanism for ligand selectivity. *The Journal of Biological Chemistry*, *281*(1), 260–268. http://dx.doi.org/M508847200 [pii].

Jin, L., Ying, Z., Hilgers, R. H., Yin, J., Zhao, X., Imig, J. D., et al. (2006). Increased RhoA/Rho-kinase signaling mediates spontaneous tone in aorta from angiotensin II-induced hypertensive rats. *The Journal of Pharmacology and Experimental Therapeutics*, *318*(1), 288–295. http://dx.doi.org/jpet.105.100735 [pii].

Kaneko, T., Amano, M., Maeda, A., Goto, H., Takahashi, K., Ito, M., et al. (2000). Identification of calponin as a novel substrate of Rho-kinase. *Biochemical and Biophysical Research Communications*, *273*(1), 110–116. http://dx.doi.org/10.1006/bbrc.2000.2901.

Katayama, M., Kawata, M., Yoshida, Y., Horiuchi, H., Yamamoto, T., Matsuura, Y., et al. (1991). The posttranslationally modified C-terminal structure of bovine aortic smooth muscle rhoA p21. *The Journal of Biological Chemistry*, *266*(19), 12639–12645.

Khalil, R. A. (2010). *Regulation of vascular smooth muscle function*. Retrieved from *Integrated systems physiology: From molecule to function* . San Rafael, CA: Morgan & Claypool Life Sciences. *http://www.ncbi.nlm.nih.gov/entrez/query.fcgi?cmd=Retrieve&db=PubMed&dopt=Citation&list_uids=21634065. http://dx.doi.org/NBK54586* [bookaccession].

Kitazawa, T., Eto, M., Woodsome, T. P., & Brautigan, D. L. (2000). Agonists trigger G protein-mediated activation of the CPI-17 inhibitor phosphoprotein of myosin light chain phosphatase to enhance vascular smooth muscle contractility. *The Journal of Biological Chemistry, 275*(14), 9897–9900.

Kitazawa, T., Eto, M., Woodsome, T. P., & Khalequzzaman, M. (2003). Phosphorylation of the myosin phosphatase targeting subunit and CPI-17 during Ca2+ sensitization in rabbit smooth muscle. *The Journal of Physiology, 546*(Pt. 3), 879–889. http://dx.doi.org/PHY_029306 [pii].

Kitazawa, T., Gaylinn, B. D., Denney, G. H., & Somlyo, A. P. (1991). G-protein-mediated Ca2+ sensitization of smooth muscle contraction through myosin light chain phosphorylation. *The Journal of Biological Chemistry, 266*(3), 1708–1715.

Kitazawa, T., Masuo, M., & Somlyo, A. P. (1991). G protein-mediated inhibition of myosin light-chain phosphatase in vascular smooth muscle. *Proceedings of the National Academy of Sciences of the United States of America, 88*(20), 9307–9310.

Koyama, M., Ito, M., Feng, J., Seko, T., Shiraki, K., Takase, K., et al. (2000). Phosphorylation of CPI-17, an inhibitory phosphoprotein of smooth muscle myosin phosphatase, by Rho-kinase. *FEBS Letters, 475*(3), 197–200. http://dx.doi.org/S0014-5793(00)01654-9 [pii].

Kume, H. (2008). RhoA/Rho-kinase as a therapeutic target in asthma. *Current Medicinal Chemistry, 15*(27), 2876–2885.

Kureishi, Y., Kobayashi, S., Amano, M., Kimura, K., Kanaide, H., Nakano, T., et al. (1997). Rho-associated kinase directly induces smooth muscle contraction through myosin light chain phosphorylation. *The Journal of Biological Chemistry, 272*(19), 12257–12260.

Lee, J. H., Zheng, Y., von Bornstadt, D., Wei, Y., Balcioglu, A., Daneshmand, A., et al. (2014). Selective ROCK2 inhibition in focal cerebral ischemia. *Annals of Clinical and Translational Neurology, 1*(1), 2–14. http://dx.doi.org/10.1002/acn3.19.

Li, Y. H., Yu, J. Z., Xin, Y. L., Feng, L., Chai, Z., Liu, J. C., et al. (2015). Protective effect of a novel Rho kinase inhibitor WAR-5 in experimental autoimmune encephalomyelitis by modulating inflammatory response and neurotrophic factors. *Experimental and Molecular Pathology, 99*(2), 220–228. http://dx.doi.org/S0014-4800(15)00134-3 [pii].

Liu, C., Zuo, J., Pertens, E., Helli, P. B., & Janssen, L. J. (2005). Regulation of Rho/ROCK signaling in airway smooth muscle by membrane potential and [Ca2+]i. *American Journal of Physiology Lung Cellular and Molecular Physiology, 289*(4), L574–L582. http://dx.doi.org/00134.2005 [pii].

Lohn, M., Plettenburg, O., Ivashchenko, Y., Kannt, A., Hofmeister, A., Kadereit, D., et al. (2009). Pharmacological characterization of SAR407899, a novel rho-kinase inhibitor. *Hypertension, 54*(3), 676–683. http://dx.doi.org/HYPERTENSIONAHA.109.134353 [pii].

Loirand, G., Guerin, P., & Pacaud, P. (2006). Rho kinases in cardiovascular physiology and pathophysiology. *Circulation Research, 98*(3), 322–334. http://dx.doi.org/98/3/322 [pii].

Maekawa, M., Ishizaki, T., Boku, S., Watanabe, N., Fujita, A., Iwamatsu, A., et al. (1999). Signaling from Rho to the actin cytoskeleton through protein kinases ROCK and LIM-kinase. *Science, 285*(5429), 895–898. http://dx.doi.org/7729 [pii].

Masumoto, A., Hirooka, Y., Shimokawa, H., Hironaga, K., Setoguchi, S., & Takeshita, A. (2001). Possible involvement of Rho-kinase in the pathogenesis of hypertension in humans. *Hypertension, 38*(6), 1307–1310.

Matsui, T., Amano, M., Yamamoto, T., Chihara, K., Nakafuku, M., Ito, M., et al. (1996). Rho-associated kinase, a novel serine/threonine kinase, as a putative target for small GTP binding protein Rho. *The EMBO Journal, 15*(9), 2208–2216.

Matsui, T., Maeda, M., Doi, Y., Yonemura, S., Amano, M., Kaibuchi, K., et al. (1998). Rho-kinase phosphorylates COOH-terminal threonines of ezrin/radixin/moesin (ERM) proteins and regulates their head-to-tail association. *The Journal of Cell Biology*, *140*(3), 647–657.

Michaelson, D., Silletti, J., Murphy, G., D'Eustachio, P., Rush, M., & Philips, M. R. (2001). Differential localization of Rho GTPases in live cells: Regulation by hypervariable regions and RhoGDI binding. *The Journal of Cell Biology*, *152*(1), 111–126.

Ming, X. F., Viswambharan, H., Barandier, C., Ruffieux, J., Kaibuchi, K., Rusconi, S., et al. (2002). Rho GTPase/Rho kinase negatively regulates endothelial nitric oxide synthase phosphorylation through the inhibition of protein kinase B/Akt in human endothelial cells. *Molecular and Cellular Biology*, *22*(24), 8467–8477.

Mishra, R. K., Alokam, R., Singhal, S. M., Srivathsav, G., Sriram, D., Kaushik-Basu, N., et al. (2014). Design of novel rho kinase inhibitors using energy based pharmacophore modeling, shape-based screening, in silico virtual screening, and biological evaluation. *Journal of Chemical Information and Modeling*, *54*(10), 2876–2886. http://dx.doi.org/10.1021/ci5004703.

Mishra, R. K., Alokam, R., Sriram, D., & Yogeeswari, P. (2013). Potential role of Rho kinase inhibitors in combating diabetes-related complications including diabetic neuropathy—A review. *Current Diabetes Reviews*, *9*(3), 249–266. http://dx.doi.org/CDR-EPUB-20130314-2 [pii].

Mori, K., Amano, M., Takefuji, M., Kato, K., Morita, Y., Nishioka, T., et al. (2009). Rho-kinase contributes to sustained RhoA activation through phosphorylation of p190A RhoGAP. *The Journal of Biological Chemistry*, *284*(8), 5067–5076. http://dx.doi.org/M806853200 [pii].

Moriki, N., Ito, M., Seko, T., Kureishi, Y., Okamoto, R., Nakakuki, T., et al. (2004). RhoA activation in vascular smooth muscle cells from stroke-prone spontaneously hypertensive rats. *Hypertension Research*, *27*(4), 263–270.

Mueller, B. K., Mack, H., & Teusch, N. (2005). Rho kinase, a promising drug target for neurological disorders. *Nature Reviews Drug Discovery*, *4*(5), 387–398. http://dx.doi.org/nrd1719 [pii].

Mukai, Y., Shimokawa, H., Matoba, T., Kandabashi, T., Satoh, S., Hiroki, J., et al. (2001). Involvement of Rho-kinase in hypertensive vascular disease: A novel therapeutic target in hypertension. *The FASEB Journal*, *15*(6), 1062–1064.

Muranyi, A., Derkach, D., Erdodi, F., Kiss, A., Ito, M., & Hartshorne, D. J. (2005). Phosphorylation of Thr695 and Thr850 on the myosin phosphatase target subunit: Inhibitory effects and occurrence in A7r5 cells. *FEBS Letters*, *579*(29), 6611–6615. http://dx.doi.org/S0014-5793(05)01328-1 [pii].

Naraoka, M., Munakata, A., Matsuda, N., Shimamura, N., & Ohkuma, H. (2013). Suppression of the Rho/Rho-kinase pathway and prevention of cerebral vasospasm by combination treatment with statin and fasudil after subarachnoid hemorrhage in rabbit. *Translational Stroke Research*, *4*(3), 368–374. http://dx.doi.org/10.1007/s12975-012-0247-9.

Oh, K. S., Oh, B. K., Park, C. H., Seo, H. W., Kang, N. S., Lee, J. H., et al. (2013). Cardiovascular effects of a novel selective Rho kinase inhibitor, 2-(1H-indazole-5-yl) amino-4-methoxy-6-piperazino triazine (DW1865). *European Journal of Pharmacology*, *702*(1–3), 218–226. http://dx.doi.org/S0014-2999(13)00042-3 [pii].

Ohashi, K., Nagata, K., Maekawa, M., Ishizaki, T., Narumiya, S., & Mizuno, K. (2000). Rho-associated kinase ROCK activates LIM-kinase 1 by phosphorylation at threonine 508 within the activation loop. *The Journal of Biological Chemistry*, *275*(5), 3577–3582.

Puetz, S., Lubomirov, L. T., & Pfitzer, G. (2009). Regulation of smooth muscle contraction by small GTPases. *Physiology (Bethesda)*, *24*, 342–356. http://dx.doi.org/24/6/342 [pii].

Ravarotto, V., Pagnin, E., Maiolino, G., Fragasso, A., Carraro, G., Rossi, B., et al. (2015). The blocking of angiotensin II type 1 receptor and RhoA/Rho kinase activity in hypertensive patients: Effect of olmesartan medoxomil and implication with cardiovascular-renal remodeling. *Journal of the Renin-Angiotensin-Aldosterone System, 16*(4), 1245–1250. http://dx.doi.org/1470320315594324 [pii].

Rolli-Derkinderen, M., Sauzeau, V., Boyer, L., Lemichez, E., Baron, C., Henrion, D., et al. (2005). Phosphorylation of serine 188 protects RhoA from ubiquitin/proteasome-mediated degradation in vascular smooth muscle cells. *Circulation Research, 96*(11), 1152–1160. http://dx.doi.org/01.RES.0000170084.88780.ea [pii].

Rolli-Derkinderen, M., Toumaniantz, G., Pacaud, P., & Loirand, G. (2010). RhoA phosphorylation induces Rac1 release from guanine dissociation inhibitor alpha and stimulation of vascular smooth muscle cell migration. *Molecular and Cellular Biology, 30*(20), 4786–4796. http://dx.doi.org/MCB.00381-10 [pii].

Sakurada, S., Takuwa, N., Sugimoto, N., Wang, Y., Seto, M., Sasaki, Y., et al. (2003). Ca2+-dependent activation of Rho and Rho kinase in membrane depolarization-induced and receptor stimulation-induced vascular smooth muscle contraction. *Circulation Research, 93*(6), 548–556. http://dx.doi.org/10.1161/01.RES.0000090998.08629.60.

Schubert, R., Kalentchuk, V. U., & Krien, U. (2002). Rho kinase inhibition partly weakens myogenic reactivity in rat small arteries by changing calcium sensitivity. *American Journal of Physiology Heart and Circulatory Physiology, 283*(6), H2288–H2295. http://dx.doi.org/10.1152/ajpheart.00549.2002.

Seasholtz, T. M., Majumdar, M., & Brown, J. H. (1999). Rho as a mediator of G protein-coupled receptor signaling. *Molecular Pharmacology, 55*(6), 949–956.

Sebbagh, M., Hamelin, J., Bertoglio, J., Solary, E., & Breard, J. (2005). Direct cleavage of ROCK II by granzyme B induces target cell membrane blebbing in a caspase-independent manner. *The Journal of Experimental Medicine, 201*(3), 465–471. http://dx.doi.org/jem.20031877 [pii].

Sebbagh, M., Renvoize, C., Hamelin, J., Riche, N., Bertoglio, J., & Breard, J. (2001). Caspase-3-mediated cleavage of ROCK I induces MLC phosphorylation and apoptotic membrane blebbing. *Nature Cell Biology, 3*(4), 346–352. http://dx.doi.org/10.1038/35070019.

Shimizu, Y., Thumkeo, D., Keel, J., Ishizaki, T., Oshima, H., Oshima, M., et al. (2005). ROCK-I regulates closure of the eyelids and ventral body wall by inducing assembly of actomyosin bundles. *The Journal of Cell Biology, 168*(6), 941–953. http://dx.doi.org/jcb.200411179 [pii].

Shirao, S., Yoneda, H., Shinoyama, M., Sugimoto, K., Koizumi, H., Ishihara, H., et al. (2015). A novel trigger for cholesterol-dependent smooth muscle contraction mediated by the sphingosylphosphorylcholine-Rho-kinase pathway in the rat basilar artery: A mechanistic role for lipid rafts. *Journal of Cerebral Blood Flow and Metabolism, 35*(5), 835–842. http://dx.doi.org/jcbfm2014260[pii].

Smith, P. G., Roy, C., Zhang, Y. N., & Chauduri, S. (2003). Mechanical stress increases RhoA activation in airway smooth muscle cells. *American Journal of Respiratory Cell and Molecular Biology, 28*(4), 436–442. http://dx.doi.org/10.1165/rcmb.4754.

Solski, P. A., Helms, W., Keely, P. J., Su, L., & Der, C. J. (2002). RhoA biological activity is dependent on prenylation but independent of specific isoprenoid modification. *Cell Growth & Differentiation, 13*(8), 363–373.

Sopko, N. A., Hannan, J. L., & Bivalacqua, T. J. (2014). Understanding and targeting the Rho kinase pathway in erectile dysfunction. *Nature Reviews Urology, 11*(11), 622–628. http://dx.doi.org/nrurol.2014.278 [pii].

Stubbs, E. B., Jr., & Von Zee, C. L. (2012). Prenylation of Rho G-proteins: A novel mechanism regulating gene expression and protein stability in human trabecular meshwork cells. *Molecular Neurobiology, 46*(1), 28–40. http://dx.doi.org/10.1007/s12035-012-8249-x.

Sugimoto, M., Nakayama, M., Goto, T. M., Amano, M., Komori, K., & Kaibuchi, K. (2007). Rho-kinase phosphorylates eNOS at threonine 495 in endothelial cells.

Biochemical and Biophysical Research Communications, 361(2), 462–467. http://dx.doi.org/S0006-291X(07)01501-X [pii].
Suzuki, H., Kimura, K., Shirai, H., Eguchi, K., Higuchi, S., Hinoki, A., et al. (2009). Endothelial nitric oxide synthase inhibits G12/13 and rho-kinase activated by the angiotensin II type-1 receptor: Implication in vascular migration. *Arteriosclerosis, Thrombosis, and Vascular Biology, 29*(2), 217–224. http://dx.doi.org/ATVBAHA.108.181024 [pii].
Thumkeo, D., Keel, J., Ishizaki, T., Hirose, M., Nonomura, K., Oshima, H., et al. (2003). Targeted disruption of the mouse rho-associated kinase 2 gene results in intrauterine growth retardation and fetal death. *Molecular and Cellular Biology, 23*(14), 5043–5055.
Thumkeo, D., Shimizu, Y., Sakamoto, S., Yamada, S., & Narumiya, S. (2005). ROCK-I and ROCK-II cooperatively regulate closure of eyelid and ventral body wall in mouse embryo. *Genes to Cells, 10*(8), 825–834. http://dx.doi.org/GTC882 [pii].
Thumkeo, D., Watanabe, S., & Narumiya, S. (2013). Physiological roles of Rho and Rho effectors in mammals. *European Journal of Cell Biology, 92*(10-11), 303–315. http://dx.doi.org/S0171-9335(13)00058-7 [pii].
Uehata, M., Ishizaki, T., Satoh, H., Ono, T., Kawahara, T., Morishita, T., et al. (1997). Calcium sensitization of smooth muscle mediated by a Rho-associated protein kinase in hypertension. *Nature, 389*(6654), 990–994. http://dx.doi.org/10.1038/40187.
VanBavel, E., van der Meulen, E. T., & Spaan, J. A. (2001). Role of Rho-associated protein kinase in tone and calcium sensitivity of cannulated rat mesenteric small arteries. *Experimental Physiology, 86*(5), 585–592. http://dx.doi.org/EPH_2217 [pii].
Webb, R. C. (2003). Smooth muscle contraction and relaxation. *Advances in Physiology Education, 27*(1–4), 201–206.
Weber, D. S., & Webb, R. C. (2001). Enhanced relaxation to the rho-kinase inhibitor Y-27632 in mesenteric arteries from mineralocorticoid hypertensive rats. *Pharmacology, 63*(3), 129–133. http://dx.doi.org/56123 [pii].
Woodsome, T. P., Eto, M., Everett, A., Brautigan, D. L., & Kitazawa, T. (2001). Expression of CPI-17 and myosin phosphatase correlates with Ca(2+) sensitivity of protein kinase C-induced contraction in rabbit smooth muscle. *The Journal of Physiology, 535*(Pt. 2), 553–564. http://dx.doi.org/PHY_11963 [pii].
Woodsome, T. P., Polzin, A., Kitazawa, K., Eto, M., & Kitazawa, T. (2006). Agonist- and depolarization-induced signals for myosin light chain phosphorylation and force generation of cultured vascular smooth muscle cells. *Journal of Cell Science, 119*(Pt. 9), 1769–1780. http://dx.doi.org/jcs.02805 [pii].
Xie, Z., Gong, M. C., Su, W., Xie, D., Turk, J., & Guo, Z. (2010). Role of calcium-independent phospholipase A2beta in high glucose-induced activation of RhoA, Rho kinase, and CPI-17 in cultured vascular smooth muscle cells and vascular smooth muscle hypercontractility in diabetic animals. *The Journal of Biological Chemistry, 285*(12), 8628–8638. http://dx.doi.org/M109.057711 [pii].
Yamaguchi, H., Kasa, M., Amano, M., Kaibuchi, K., & Hakoshima, T. (2006). Molecular mechanism for the regulation of rho-kinase by dimerization and its inhibition by fasudil. *Structure, 14*(3), 589–600. http://dx.doi.org/S0969-2126(06)00094-3 [pii].
Yao, L., Romero, M. J., Toque, H. A., Yang, G., Caldwell, R. B., & Caldwell, R. W. (2010). The role of RhoA/Rho kinase pathway in endothelial dysfunction. *Journal of Cardiovascular Disease Research, 1*(4), 165–170. http://dx.doi.org/10.4103/0975-3583.74258.
Zhang, Y. M., Bo, J., Taffet, G. E., Chang, J., Shi, J., Reddy, A. K., et al. (2006). Targeted deletion of ROCK1 protects the heart against pressure overload by inhibiting reactive fibrosis. *FASEB Journal: Official Publication of the Federation of American Societies for Experimental Biology, 20*(7), 916–925. http://dx.doi.org/20/7/916 [pii].
Zhou, Q., & Liao, J. K. (2009). Rho kinase: An important mediator of atherosclerosis and vascular disease. *Current Pharmaceutical Design, 15*(27), 3108–3115.

CHAPTER EIGHT

Vascular Cells in Blood Vessel Wall Development and Disease

R. Mazurek[1], J.M. Dave[1], R.R. Chandran, A. Misra, A.Q. Sheikh, D.M. Greif[2]

Yale Cardiovascular Research Center, Section of Cardiovascular Medicine, Yale University School of Medicine, New Haven, CT, United States

[2]Corresponding author: e-mail address: daniel.greif@yale.edu

Contents

Abstract

The vessel wall is composed of distinct cellular layers, yet communication among individual cells within and between layers results in a dynamic and versatile structure. The morphogenesis of the normal vascular wall involves a highly regulated process of cell proliferation, migration, and differentiation. The use of modern developmental biological and genetic approaches has markedly enriched our understanding of the molecular and cellular mechanisms underlying these developmental events. Additionally, the application of similar approaches to study diverse vascular diseases has resulted in paradigm-shifting insights into pathogenesis. Further investigations into the biology

[1] Co-first authors.

Advances in Pharmacology, Volume 78
ISSN 1054-3589
http://dx.doi.org/10.1016/bs.apha.2016.08.001

of vascular cells in development and disease promise to have major ramifications on therapeutic strategies to combat pathologies of the vasculature.

ABBREVIATIONS

BMPR2 bone morphogenetic protein receptor 2
CNS central nervous system
ECM extracellular matrix
ECs endothelial cells
GM germinal matrix
GMH germinal matrix hemorrhage
IPAH idiopathic pulmonary arterial hypertension
KLF4 Kruppel-like factor 4
MMP matrix metalloproteinase
NG2 neuron–glial antigen 2
Oct4 octamer-binding transcriptional factor 4
PCs pericytes
PDGF platelet-derived growth factor
PDGFR platelet-derived growth factor receptor
PH pulmonary hypertension
Sca stem cell antigen
SMA α-smooth muscle actin
SMCs smooth muscle cells
SMMHC smooth muscle myosin heavy chain
SRF serum response factor
SVAS supravalvular aortic stenosis
TGF transforming growth factor
VEGF vascular endothelial growth factor
VEGFR vascular endothelial growth factor receptor
WBS Williams–Beuren syndrome

1. INTRODUCTION

The cardiovascular system forms early in development, and during embryogenesis and postnatal life, it serves the critical functions of both delivering oxygen and nutrients to support the metabolic activity of tissues and of removing waste products. The vasculature of each organ is comprised of a series of blood vessels that have a specialized structure and form a particular spatial network to facilitate organ-specific physiological functions. The morphogenesis of blood vessel walls is intricately configured to meet the needs of the surrounding tissues. Cardiovascular diseases may result from deficiencies in the initial construction of the vasculature. Alternatively, the integrity and structure of the mature vascular wall may become compromised through diverse mechanisms, including inappropriate recurrence of developmental programs, compensatory responses, and/or independent pathological processes.

The cardiovascular network is comprised of a hierarchy of blood vessels, and each vessel-type has a distinct structure (Fig. 1). Arteries carry blood from the heart into the systemic or pulmonary circulation to arterioles. In turn, small arterioles feed capillaries, which exchange oxygen and carbon dioxide as well as nutrients and metabolic waste products with tissues. Blood is then collected into venules, transported to veins, and finally returned to the heart. Endothelial cells (ECs) are common to all vessels and form the inner tunica intima layer that lines the vascular lumen. The middle layer, or tunica media, consists of circumferentially oriented smooth muscle cells (SMCs), and in large elastic arteries, multiple circular smooth muscle layers alternate with rings of elastic lamellae. Arterioles have fewer smooth muscle layers, and capillaries are covered by a discontinuous coat of pericytes (PCs) instead of SMCs. In comparison to similar-sized arteries, veins have a thinner media and are more

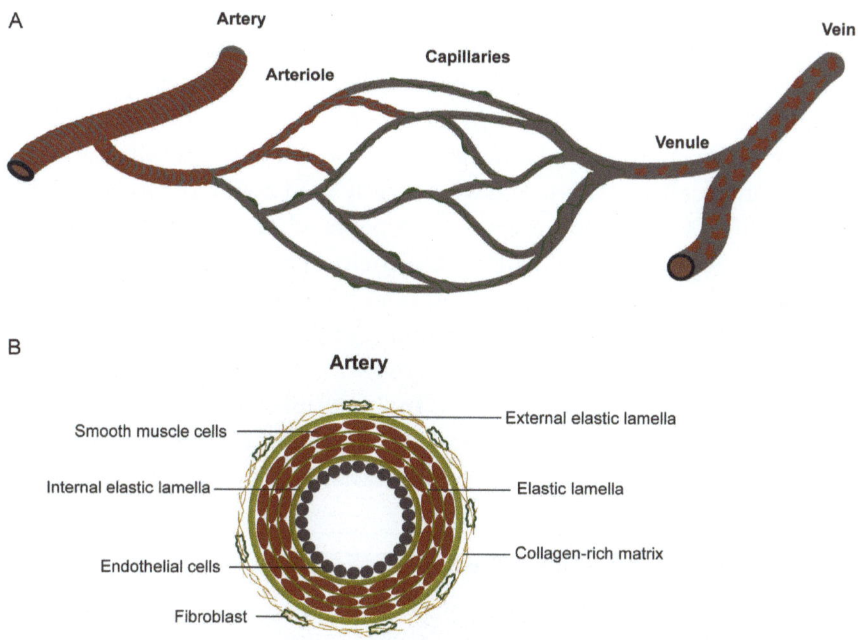

Fig. 1 Structure of the vasculature. (A) The EC tubes of arteries, arterioles, venules, and veins are coated by SMCs (*red*), whereas PCs (*green*) are the mural cells of capillaries. (B) Transverse section of an artery highlighting major constituents of the vascular wall. Although not labeled in the figure, the tunica intima is the region of the vessel that is internal to the internal elastic lamella, the tunica media is sandwiched between the internal and external elastic lamellae, and the tunica adventitia is outside the external elastic lamella.

compliant. Both arteries and veins have an outer tunica adventitia layer, which contains extracellular matrix (ECM), fibroblasts, and progenitor cells.

The morphogenesis and homeostasis of the blood vessel wall requires precise gene expression and cellular signaling to meticulously orchestrate vascular cell migration, proliferation, apoptosis, differentiation, and ECM synthesis. Our ability to study these processes in vivo during vascular development, maintenance, and disease has been markedly enhanced by the use of model systems and fundamental developmental biological and genetic approaches. These approaches include timelines of developmental and pathological processes, mosaic analysis, fate mapping, clonal analysis, and conditional control of gene expression in a temporal and cell type-specific manner. For instance, careful histological and immunohistochemical timelines of multiple stages during development and disease of the murine pulmonary artery have proven essential in delineating underlying processes (Greif et al., 2012; Sheikh, Lighthouse, & Greif, 2014). In addition, many biological processes involve competition between cells for a specific position or a role (e.g., tip vs stalk cells in the morphogenesis of either the trachea in *Drosophila melanogaster* or capillaries in the mouse or zebrafish), and mosaic analyses have helped delineate the cellular and molecular mechanisms underlying this competition (Ghabrial & Krasnow, 2006; Herbert, Cheung, & Stainier, 2012; Jakobsson et al., 2010). Fate mapping facilitates the analysis of cell derivatives and was recently used in mouse models to illustrate that SMCs give rise to diverse cell types in atherosclerotic plaques (Feil et al., 2014; Shankman et al., 2015). Using clonal analysis, we recently identified a novel pool of SMC progenitors in pulmonary arterioles, and with hypoxia-induced pulmonary hypertension (PH), one of these cells migrates distally and clonally expands to give rise to pathological SMCs (Sheikh, Misra, Rosas, Adams, & Greif, 2015). In this chapter, we discuss the cellular components and mechanisms of vascular wall morphogenesis in development as well as pathogenesis in select diseases.

2. BLOOD VESSEL DEVELOPMENT

2.1 Endothelial Cells

The tunica intima consists of a monolayer of ECs that lines the entire vasculature, and the endothelium of a human adult is estimated to consist of approximately 1×10^{13} ECs (Augustin, Kozian, & Johnson, 1994). Several well-characterized markers are employed to identify ECs, including vascular endothelial cadherin, platelet endothelial cell adhesion molecule 1, vascular endothelial growth factor receptors (VEGFRs), and isolectinB4. During development, most ECs derive from the lateral plate mesoderm (Pouget, Gautier, Teillet, & Jaffredo, 2006), and through the process of vasculogenesis, primitive ECs coalesce into the initial blood vessel tubes (Risau & Flamme, 1995). Subsequently, these initial EC tubes give rise to further vessels through angiogenesis, a multistep process consisting of EC proliferation, migration, invasion, lumen formation, and tube stabilization. Newly formed vessels recruit mural cells (SMCs or PCs) inducing stabilization and EC quiescence (Benjamin, Hemo, & Keshet, 1998), whereas some uncoated nascent vessels are refined through pruning and regression. EC tube morphogenesis results in hierarchically branched and functionally perfused vascular beds (Risau & Flamme, 1995).

Angiogenesis is a dynamic process that requires strict coordination of leading "tip" cells with following "stalk" cells (Gerhardt et al., 2003). Tip cells are located at the growing ends of sprouting vessels and display long filopodia facilitating EC migration. Tip cells sense pro- and antiangiogenic directional cues in their environment through cell surface receptors and integrate downstream signaling to migrate in a specific direction. In contrast, stalk cells exhibit fewer filopodia and higher proliferation. These cells establish adherent and tight junctions with neighboring ECs (Dejana, Tournier-Lasserve, & Weinstein, 2009) and form the nascent vascular lumen (Iruela-Arispe & Davis, 2009). Intricate cross talk between VEGF (Gerhardt et al., 2003) and Notch signaling pathways (Phng & Gerhardt, 2009) govern tip vs stalk cell fate. Briefly, ECs of quiescent vessels sense a VEGF gradient in the surrounding environment through VEGFR2. This interaction upregulates expression of the Notch ligand Delta-like 4 in the tip cells. In turn, Notch signaling in the surrounding stalk cells is activated, leading to suppression of both VEGFR2 expression and tip cell phenotype and to induction of another Notch ligand Jagged1. Jagged1 antagonizes Delta-like 4–Notch signaling in tip cells, thereby enhancing differential Notch activity between tip and stalk cells (Blanco & Gerhardt, 2013).

In addition to angiogenesis, ECs play key roles in diverse processes, such as coagulation, inflammation, barrier function, blood flow regulation, and synthesis/degradation of ECM components (Cines et al., 1998). These myriad functions make healthy ECs indispensable for normal vascular development and homeostasis. In mature vessels, ECs are quiescent unless activated by proangiogenic signals: an extensive list of such factors has recently been provided (Dave & Bayless, 2014). Given the critical role of ECs in vascular homeostasis, it is not surprising that perturbed angiogenic balance and EC dysfunction are common findings in several pathological disorders, including systemic, hypertension, PH, atherosclerosis, allograft vasculopathy, stroke, inflammatory syndromes, and cancer (Cines et al., 1998).

2.2 Smooth Muscle Cells

SMCs are the major cell type of the tunica media and through dynamic cell contraction and relaxation regulate vascular tone and hence, blood flow. The contraction–relaxation state of SMCs is dictated by a spectrum of contractile and cytoskeletal proteins. During embryogenesis, α-smooth muscle actin (SMA) is considered the first SMC marker to be expressed and ultimately is the most abundant protein in SMCs. For instance, the developing pulmonary artery forms in a field of cells expressing the undifferentiated mesenchyme marker platelet-derived growth factor receptor (PDGFR)-β (Greif et al., 2012). Shortly thereafter PDGFR-β^+ cells adjacent to the nascent EC tube downregulate PDGFR-β and upregulate SMA (Greif et al., 2012). Early developing SMCs also express SM22α (also known as transgelin), which influences the actin cytoskeleton by stabilizing actin filaments. In addition to SMCs, SMA and SM22α are expressed in other cell types as well, whereas smooth muscle myosin heavy chain (SMMHC) and smoothelin are expressed later in SMC differentiation and are generally considered to be specific to the SMC lineage. Yet, our recent studies suggest that SMMHC is also expressed in alveolar myofibroblasts of adult mice exposed to hypoxia (Sheikh et al., 2014).

Despite the expression of similar markers in SMCs throughout the arterial vasculature, the origins of SMCs in different vessels and even in different regions of the same vessel are diverse. The range of SMC sources is perhaps best exemplified in the aorta. Fate-mapping studies have determined that SMCs in the root, arch, and descending regions of the adult aorta derive from the secondary heart field, neural crest, and presomitic mesoderm, respectively (Majesky, 2007). The axial borders between these regions of the aortic media are clearly demarcated with essentially no mixing of SMCs from different origins, and interestingly, these borders are especially prone to pathological dissection (Cheung, Bernardo, Trotter, Pedersen, & Sinha, 2012; Majesky, 2007). Beyond the aortic arch, the neural crest also gives rise to SMCs of the cranial vasculature. In most organs, the local mesenchyme is considered a key source of vascular smooth muscle, as is the case for pulmonary artery SMCs (Greif et al., 2012), whereas the serosal mesothelium is implicated as an important source of vascular SMCs of the gut

(Wilm, Ipenberg, Hastie, Burch, & Bader, 2005). In addition, much attention has been paid to the proepicardium, a transitory developmental structure that arises as an outgrowth of coelomic mesothelium near the sinoatrial junction of the heart and contributes to coronary artery SMCs (Majesky, 2004). Importantly, the origins of SMCs appear to have functional ramifications as indicated by the differing responses to cytokines of either: (i) SMCs isolated from the arch vs the descending aorta (Topouzis & Majesky, 1996) or (ii) SMCs derived from human embryonic stem cells following differentiation to lineage-specific fates (Cheung et al., 2012).

After cells fated to be SMCs are recruited to a given EC tube, these cells must be assembled into a functional layer and differentiate. In arteries with multiple smooth muscle layers, cells undergo radial patterning sequentially layer by layer from the inside outward with regard to morphology and marker expression (Greif et al., 2012). A number of mechanisms have been implicated to contribute to this radial patterning, such as diffusion of an EC-derived morphogen (e.g., PDGF-B), Jagged1-Notch-mediated lateral induction, and cell migration (Greif et al., 2012; Hoglund & Majesky, 2012; Manderfield et al., 2012). In addition, transforming growth factor (TGF) β plays a key role in SMC differentiation. Upon coculture with ECs, undifferentiated embryonic mesenchymal cells undergo TGFβ-dependent elongation and SMC marker expression (Hirschi, Rohovsky, & D'Amore, 1998).

As differentiated SMCs complete development, their migration, proliferation, and ECM synthesis are downregulated; yet, this relative quiescence is reversible. In contrast to mature skeletal muscle cells or cardiomyocytes, which are believed to have limited plasticity, in response to injury or disease, adult SMCs can undergo phenotypic modulation and markedly change their morphology, gene expression, and rates of proliferation and migration (Owens, Kumar, & Wamhoff, 2004). Thus, depending on specific cues, SMCs apparently can exist within the continuum between a differentiated contractile state and an undifferentiated, highly migratory and proliferative synthetic state. For instance, we recently identified a specific pool of SMC progenitors in the normal adult murine lung that express both SMC markers SMA and SMMHC and the undifferentiated mesenchyme marker PDGFR-β (Sheikh et al., 2015). Upon hypoxia exposure, one of these progenitors downregulates SMMHC and clonally expands giving rise to pathological distal pulmonary arteriole SMCs (see Section 3.3) (Sheikh et al., 2015).

The transcriptional underpinning of smooth muscle contractile and synthetic gene expression has been intensely studied. The ubiquitous transcription factor serum response factor (SRF) plays a key role in modulating SMC gene expression. In the presence of the transcriptional coactivator myocardin, SRF binds a 10-base pair ($CC(A/T)_6GG$) *cis* regulatory element, which is known as the CArG (i.e., C, AT rich, G) box and induces the expression of contractile genes. Myocardin has been termed a "master regulator" of SMC gene expression as ectopic expression of myocardin in some nonmuscle cell types induces contractile gene expression (Wang, Wang, Pipes, & Olson, 2003), and *myocardin* null murine embryos lack SMCs and die by E10.5 (Li, Wang, Wang, Richardson, & Olson, 2003). In addition, Kruppel-like factor 4 (KLF4) is a pluripotency transcription factor that inhibits myocardin-induced SMC contractile gene expression and is critical for PDGF-B-induced SMC dedifferentiation (Deaton, Gan, & Owens, 2009; Liu et al., 2005). Moreover, in vivo mouse studies demonstrate that KLF4 plays an integral role in vascular SMC remodeling in diverse pathologies (Salmon et al., 2013; Shankman et al., 2015; Sheikh et al., 2015).

2.3 Pericytes

Instead of SMCs, the mural cells of capillaries are PCs, which are embedded in the basement membrane at the abluminal surface of ECs. PCs have long cytoplasmic processes, which often interface with multiple adjacent ECs through gaps in the basement membrane and, on occasion, extend to neighboring capillaries (Armulik, Genove, & Betsholtz, 2011; Hill et al., 2015). PC abundance varies in a tissue- and vascular bed-specific manner with the highest PC density generally thought to be in the central nervous system (CNS), where the ratio of ECs:PCs is considered to be $\sim$2:1 (Armulik et al., 2011). Although PDGFR-β, neuron–glial antigen 2 (NG2), and regulator of G-protein signaling 5 are expressed in PCs, the unequivocal identification of PCs is often challenging largely due to a lack of PC-specific markers and dynamic expression. Indeed, SMCs and PCs both have a peri-EC position and depending on location and developmental or pathological state, their molecular markers overlap. This limitation has hindered PC investigations. For instance, fate-mapping studies suggest that vascular mural cells in the CNS (Etchevers, Vincent, Le Douarin, & Couly, 2001; Korn, Christ, & Kurz, 2002) and thymus (Foster et al., 2008) derive from the neuroectoderm and in other organs (e.g., liver, intestine, and lung) derive from mesoderm (Asahina, Zhou, Pu, & Tsukamoto, 2011; Que et al., 2008; Wilm et al., 2005); however, most of these studies do not explicitly distinguish PCs from SMCs.

During development, nascent EC tubes utilize PDGF-B-mediated signaling to recruit PCs, which subsequently stabilize the vessel. PDGF-B is secreted by tip ECs and signals through PDGFR-β on PCs to induce PC proliferation and migration. *Pdgfrb* null mice have markedly reduced PC coverage of capillaries and EC hyperproliferation (Hellstrom et al., 2001; Hellstrom, Kalen, Lindahl, Abramsson, & Betsholtz, 1999; Soriano, 1994) as do EC-specific *Pdgfb* knockouts (Enge et al., 2002). Secreted PDGF-B is ECM bound, and targeted deletion of the C-terminal *Pdgfb* retention motif leads to partial PC detachment in mice, indicating that ECM-bound PDGF-B is required for proper PC recruitment (Lindblom et al., 2003). Following PC recruitment, the molecular mechanisms resulting in EC tube stabilization are incompletely understood; however, Angiopoietin-1/Tie-2 signaling is widely implicated (Armulik et al., 2011; Gaengel, Genove, Armulik, & Betsholtz, 2009). In addition, PC-mediated regulation of EC-derived matrix metalloproteinase (MMP) activity may be critical. Tip ECs produce MMPs that degrade the surrounding ECM and facilitate EC sprouting (Yana et al., 2007), and PC-derived tissue inhibitor of metalloproteinase-3 stabilizes EC tubes (Kamei et al., 2006; Schrimpf et al., 2012).

Given their relative abundance, PCs have been intensely studied in the brain. Embryonic mice null for *Pdgfb* or *Pdgfrb* lack cerebral PCs and have microaneurysms (Hellstrom et al., 1999; Lindahl, Johansson, Leveen, & Betsholtz, 1997). PCs play an integral role in formation and maintenance of the blood–brain barrier and modulate permeability of this barrier largely by regulating EC functions (e.g., tight junction formation and transcytosis) (Armulik et al., 2010; Bell et al., 2010; Daneman, Zhou, Kebede, & Barres, 2010). The direct involvement of PCs in blood flow regulation has been the focus of recent controversy largely, once again, because of ambiguity in identifying this cell type. Atwell and colleagues suggested that in response to neuronal activity, brain PCs relax to induce capillary dilation and increase blood flow (Hall et al., 2014). Recently, these findings have been challenged by a study indicating that brain PCs express NG2 and PDGFR-β but not SMA, and cerebral blood flow is regulated predominately by arteriole SMCs, which are SMA^{+}, but not by PCs (Hill et al., 2015).

In addition to their critical roles in regulating capillary formation, maintenance, and function, PCs have been implicated as a source of diverse cell types in development and disease. PCs and SMCs are generally considered to share a common lineage; however, as noted earlier, distinguishing these cell types can be cumbersome. Recently, based on lineage tracing and clonal analysis, it was suggested that developing coronary artery SMCs are derived, at least partly, from NG2$^+$ PCs via a Notch3-dependent process (Volz et al., 2015). Mural cells isolated from the murine brain can be reprogrammed in culture into cells assuming properties of neuronal cells or stem cells (Karow et al., 2012; Nakagomi et al., 2015). Additionally, there is controversy regarding whether PCs are a substantial source of myofibroblasts during pathological fibrosis of the kidney or lung (Humphreys et al., 2010; Hung et al., 2013; LeBleu et al., 2013; Rock et al., 2011). Thus, PCs play a fundamental role in blood vessel development, but study of their embryonic origins, functions, and fate is hindered by a lack of specific PC markers.

2.4 Adventitial Cells

The tunica adventitia or outermost vascular layer is the least well studied of the blood vessel layers yet is implicated as playing important roles in health and disease of the vasculature (Majesky, Dong, Hoglund, Daum, & Mahoney, 2012). The adventitia consists of a collagen-rich ECM that harbors nerves, lymphatics, and in larger vessels, a microvascular network known as the vasa vasorum. The cells of the adventitia are diverse, including resident macrophages, lymphocytes, mast cells, dendritic cells, stem cell antigen (Sca)-1$^+$ progenitor cells, and fibroblasts, with this latter cell type comprising the largest proportion of adventitial cells (Hu et al., 2004; Majesky et al., 2012). The embryonic origins of fibroblasts are generally not defined; however, the proepicardium has been shown to serve as a source of both fibroblasts and SMCs of the developing avian coronary artery (Dettman, Denetclaw, Ordahl, & Bristow, 1998).

Sca-1$^+$ adventitial cells have been implicated as a major progenitor cell population in the vascular wall (Majesky et al., 2012; Passman et al., 2008). These cells are first detected in the perivascular space between the aorta and pulmonary trunk at E (embryonic day) 15.5–18.5 and increase in numbers postnatally (Passman et al., 2008). The origins of adventitial Sca1$^+$ cells are not defined, but a number of tissues have been excluded as potential sources, including the bone marrow (Hu et al., 2004), cardiac neural crest (Passman et al., 2008), and somites (Wasteson et al., 2008). Sca-1$^+$ adventitial cells are heterogeneous, and it has been suggested that there are at least two main types of such progenitors with regard to cell fate, giving rise to mural cells or alternatively, macrophage-like cells (Majesky, 2015). Substantial further investigation into this intriguing progenitor pool is undoubtedly warranted.

2.5 Vascular ECM

The ECM is a key constituent of the vascular wall, providing the vessel with elasticity (via elastic lamellae) and tensile strength (via collagen fibers) (Wagenseil & Mecham, 2009) and also influencing cellular signaling and behavior. In addition to elastin and collagens which comprise ~50% of the dry weight of larger arteries (Harkness, Harkness, & McDonald, 1957), the vascular ECM also includes microfibrils, fibronectin, proteoglycans, and glycoproteins. Expression array analysis of the murine aorta indicates that most ECM components are initially detected at E14, and subsequently, their mRNA levels increase until they peak at postnatal days 7–14 (McLean, Mecham, Kelleher, Mariani, & Mecham, 2005). Over the following few months, ECM transcript levels decrease and persist at low levels in adults (McLean et al., 2005). Under homeostatic conditions, ECM proteins are generally quite stable: remarkably, the half-life of elastic fibers in human arteries is believed to be ~50–70 years (Arribas, Hinek, & Gonzalez, 2006). However, many vascular pathologies exhibit aberrant ECM levels due to altered gene expression and/or imbalance between proteases that degrade ECM components and protease inhibitors (Jacob, 2003).

The composition and organization of the vascular ECM are specialized based on the radial position within the vessel. Starting at the innermost portion of the vessel, ECs of the tunica intima rest on a thin basement membrane, comprised of laminin, type IV collagen, nidogen, fibronectin, perlecan, heparin sulfate proteoglycans, and other proteins. The interaction of PCs and ECs in microvessels has been implicated in inducing synthesis of ECM components and basement membrane assembly (Stratman, Malotte, Mahan, Davis, & Davis, 2009). In turn, the basement membrane plays a pivotal role in regulating EC migration, proliferation, and tube formation (Davis & Senger, 2005). Moving radially outward from the basement membrane, the internal elastic lamella separates the intima and media. Within the media, circumferential layers of SMCs alternate with elastic lamellae, and collagen bundles are located between lamellae (Wagenseil & Mecham, 2009). These collagen bundles lack a discernible pattern at physiological pressure but with increasing pressure become circumferentially aligned (Wagenseil & Mecham, 2009). Elastin and collagens in the media are thought to be primarily secreted by SMCs (Xu & Shi, 2014). The adventitia is located outside the external elastic lamella and is rich in type I and III collagens, which provide vascular wall rigidity and prevent rupture at high pressure.

3. CARDIOVASCULAR DISEASES

3.1 Supravalvular Aortic Stenosis

Diverse arterial disorders, including atherosclerosis, restenosis, and supravalvular aortic stenosis (SVAS), are plagued by defective elastic lamellae and excess and aberrant SMCs (Brooke, Bayes-Genis, & Li, 2003; Curran et al., 1993; Karnik et al., 2003; Owens et al., 2004; Sandberg, Soskel, & Leslie, 1981). Elastin is the major component of elastic lamellae, and SVAS, a devastating human disease with occlusions and hypermuscularization of large arteries, results from loss-of-function mutations in one allele of the elastin gene *ELN* (Curran et al., 1993). SVAS occurs as an isolated entity or as part of Williams–Beuren syndrome (WBS), a multiorgan system disorder caused by heterozygous deletion of ~27 genes (including *ELN*) on chromosome 7 (Pober, 2010). Arterial obstruction is the major cause of morbidity in WBS.

Elastin mutant mice phenocopy many aspects of the arterial pathology of SVAS (Li, Brooke, et al., 1998; Li, Faury, et al., 1998; Li et al., 1997) and thus are a useful model to study pathogenesis and potential therapies. Similar to SVAS patients, late-stage embryonic or early neonatal $Eln^{(-/-)}$ mice have a relatively disorganized, hyperproliferative, and hypercellular vascular media resulting in luminal obstruction (Li, Brooke, et al., 1998). In comparison to controls, the descending aorta of SVAS patients or $Eln^{(+/-)}$ mice have thinner elastic lamellae but more lamellar units (Li, Faury, et al., 1998). We recently demonstrated that integrin β3 expression and activation are increased in the elastin mutant aortic media in humans and mice and in induced pluripotent stem cell-derived SMCs from SVAS patients (Misra et al., 2016). Furthermore, genetic or pharmacological inhibition of integrin β3 in elastin mutant mice attenuates aortic SMC misalignment and hyperproliferation and, hence, hypermuscularization and luminal stenosis (Misra et al., 2016). Inhibiting integrin β3-mediated signaling is an attractive potential therapeutic strategy for SVAS patients (Fig. 2).

3.2 Aortic Aneurysms

Aneurysms are defined as permanent, focal dilations of greater than 50% of the normal arterial diameter. Most aortic aneurysms occur caudal to the renal arteries, and abdominal aortic aneurysm (AAA) rupture carries a mortality rate of 80–90%. Thoracic aortic aneurysms (TAAs) are less common than

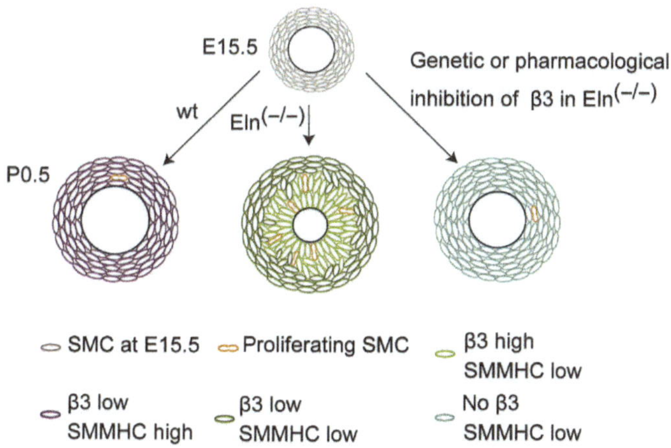

Fig. 2 Schematic of integrin β3 inhibition in elastin null mice. The elastin null aortic pathology develops after ∼E15.5 and is characterized by subendothelial SMCs that have increased integrin β3 levels and are misaligned (*radially oriented*). In addition, SMMHC expression is reduced whereas proliferation and radial migration is increased resulting in hypermuscularization. Genetic or pharmacological inhibition of β3 prevents most of this pathology.

AAAs but are also life-threatening. Risk factors for AAAs include smoking, age greater than 60 years, male gender, atherosclerosis, hypertension, chronic obstructive pulmonary disease, and family history. Descending, but not ascending, TAAs share many of these risk factors. AAAs are influenced by multiple environmental and genetic factors, but no single causative gene has been identified. In contrast, ascending TAAs are primarily due to cystic medial necrosis and often result from mutations of single genes, which encode structural ECM and SMC cytoskeletal proteins or proteins that regulate signaling in the tunica media (Lindsay & Dietz, 2011).

Pathological changes in aneurysms include arterial wall thinning due to SMC loss and ECM remodeling. A study of human infrarenal aortic samples demonstrated that aneurysmal tissue exhibits an increase in SMC apoptosis and a decrease in SMC density (Rowe et al., 2000). In addition, MMP-induced degradation of elastin contributes to AAA pathogenesis (Kadoglou & Liapis, 2004), and the diseased aorta has altered collagen levels with an increase in type I and a decrease in type III (Rodella et al., 2016). Mutations in collagen and fibrillin-1, a key protein in elastic fiber-associated microfibrils, as in Ehlers–Danlos and Marfan syndromes, respectively, predispose to aortic aneurysm, dissection, and rupture. Perturbations in TGFβ-mediated signaling are widely implicated in TAA syndromes in humans and in mouse models (Andelfinger, Loeys, & Dietz, 2016). Furthermore, in an elastase-mediated AAA mouse model, SMC-specific deletion of *TGFβ receptor 2* protects against aneurysm formation and attenuates medial SMC loss, MMP expression, and elastin degradation (Gao et al., 2014). Interestingly, a potential role of vascular wall progenitor cells in aneurysm pathogenesis or treatment has recently been raised (Amato et al., 2015), and thus, further investigation into this cell population in the context of aortic aneurysm is warranted.

3.3 Pulmonary Hypertension

PH is defined by a mean pulmonary arterial pressure greater than 25 mmHg. It is a devastating disease leaving almost half of all patients dead within 3 years of initial diagnosis (Humbert et al., 2010). PH has multiple causes including cardiac, parenchymal lung, thromboembolic, infectious and autoimmune diseases, hypoxia, genetic mutations, drugs, and idiopathic pulmonary arterial hypertension (IPAH) (Simonneau et al., 2013). The histologic changes of the pulmonary vasculature in PH include pruning of small vessels, muscularization of normally nonmuscular distal arterioles, increased smooth muscle in proximal pulmonary vessels, and obliterative intimal lesions composed of ECs, SMCs, and ECM. Moreover, reduced compliance of the pulmonary arterial vasculature is a strong independent predictor of mortality in IPAH (Mahapatra, Nishimura, Sorajja, Cha, & McGoon, 2006), and the increased smooth muscle burden contributes to this reduced compliance. Most existing therapies for PH lower pulmonary artery pressure through vasodilation but have modest clinical efficacy and do not directly target SMC recruitment, dedifferentiation, and migration.

Excess SMCs in PH have been proposed to derive from diverse cell types. With varying levels of evidence, mesenchymal progenitor cells, interstitial fibroblasts, PCs, SMCs, and ECs have been implicated (Qiao et al., 2014; Ricard et al., 2014; Sheikh et al., 2014, 2015; Stenmark, Fagan, & Frid, 2006). Our group recently identified a pool of novel SMC progenitors in the pulmonary vasculature that are located at the muscular–unmuscular arteriole border and express SMA, SMMHC, and PDGFR-β (Sheikh et al., 2015). Based on their location and PDGFR-β expression, we hypothesized that these SMCs are poised to migrate distally into the unmuscular arteriole and proliferate and thus termed them "primed" cells (Sheikh et al., 2015). Indeed, in mice exposed to hypoxia, primed cells express

KLF4, and one of them migrates into the normally unmuscularized distal arteriole, dedifferentiates, and clonally expands (Fig. 3) (Sheikh et al., 2015). In addition to SMC progenitors, there has recently been substantial interest in endothelial-to-mesenchymal transition in vascular diseases in general, and specifically in PH (Stenmark, Frid, & Perros, 2016). *Bone morphogenetic protein receptor 2* (*BMPR2*) mutations are prevalent in human

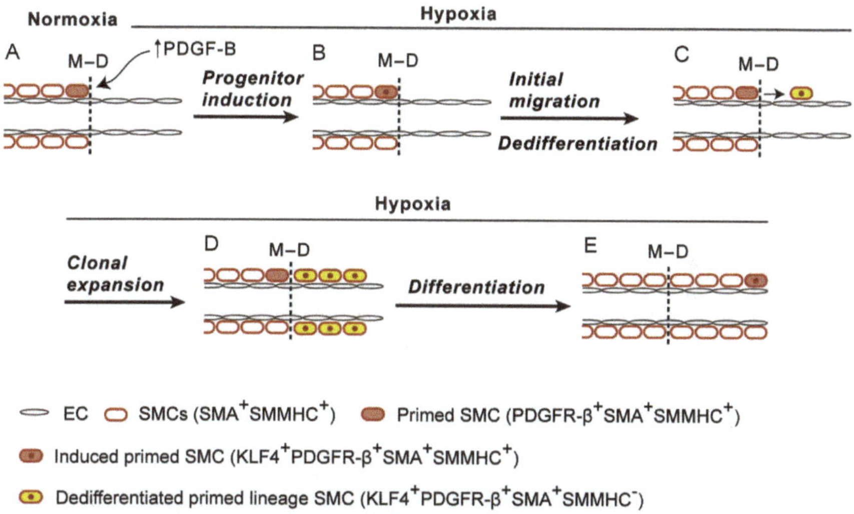

Fig. 3 Summary of molecular and cellular events in hypoxia-induced distal pulmonary arteriole muscularization (Sheikh et al., 2015). (A) In normoxia, SMCs (*red outline*) coat proximal and middle (*M*), but not distal (*D*), pulmonary arteriole EC tubes and express SMA and SMMHC (*no fill*). In addition, rare PDGFR-β^+SMA$^+$SMMHC$^+$ primed SMCs (*pink fill*) are located at each M–D border, which coincides with the transition from muscularized to unmuscularized arteriole. (B) Upon initial hypoxic exposure, lung PDGF-B expression is markedly increased, which is required for KLF4 induction in primed cells (*pink fill with red dot*). (C) Within a day after KLF4 expression, an induced primed SMC (a KLF4$^+$PDGFR-β^+SMA$^+$SMMHC$^+$ cell) migrates distally across the M–D border and dedifferentiates as indicated by downregulating SMMHC expression (*yellow fill with red dot*). (D) Subsequently, the dedifferentiated cell clonally expands, giving rise to the vast majority of distal arteriole SMCs. (E) These cells then reexpress SMMHC, and again rare primed SMCs are localized at the now distally located muscular–unmuscular vascular border. *Reprinted with permission from AAAS: Sheikh, A. Q., Misra, A., Rosas, I. O., Adams, R. H., & Greif, D. M. (2015). Smooth muscle cell progenitors are primed to muscularize in pulmonary hypertension.* Science Translational Medicine, *7(308), 308ra159. http://dx.doi.org/10.1126/scitranslmed.aaa9712.*

pulmonary arterial hypertension, and ECs isolated from EC-specific *Bmpr2* null mice express SM22α protein (Hopper et al., 2016). Moreover, in mice subjected to pneumonectomy and monocrotaline injection to induce PH, fate-mapped ECs contribute to SMC marker$^+$ neointimal cells (Qiao et al., 2014).

Beyond the role of ECs as a source of excess SMCs in PH, EC–SMC interactions are undoubtedly a critical, yet understudied, area in PH. A number of EC-derived factors, including serotonin, PDGF-B, endothelin-1, fibroblast growth factor-2, and interleukin-6, have receptors on SMCs and have been implicated in pathological PA remodeling (Chen & Oparil, 2000; Eddahibi et al., 2001; Izikki et al., 2009; Savale et al., 2009; Schermuly et al., 2005). For example, in cells isolated from the lungs of IPAH patients as comparison to controls, ECs generate more serotonin, and arterial SMCs express more of the serotonin transporter (Eddahibi et al., 2006, 2001). These IPAH SMCs have an enhanced proliferative response to EC-conditioned media, which is abrogated by blocking the serotonin transporter or inhibiting the synthesis of serotonin (Eddahibi et al., 2006). In addition, mice exposed to hypoxia have increased PDGF-B expression in lung ECs, and in *Pdgfb*$^{(+/-)}$ mice, hypoxia does not induce primed SMC KLF4 expression, distal arteriole muscularization, and PH (Sheikh et al., 2015).

3.4 Atherosclerosis

The initiating events in atherogenesis involve the retention of lipoproteins in the subendothelial space of arteries and EC activation (Libby, Ridker, & Hansson, 2011; Tabas, Garcia-Cardena, & Owens, 2015). Circulating monocytes adhere to the activated ECs, enter the vascular wall, and differentiate into tissue macrophages. These macrophages phagocytose lipoproteins and become foam cells. In addition, synthetic SMCs accumulate in atheromas and secrete ECM proteins, and SMCs and collagen are important components of the fibrous cap that covers the atherosclerotic plaque. Plaques with a reduced ratio of SMCs to foam cells are thought to be vulnerable to rupture, which is the inciting event for thrombosis and, thus, myocardial infarction.

SMCs and macrophages are intricately linked in atherogenesis. Preexisting SMCs give rise to SMC marker$^+$ cells in atherosclerotic plaques (Bentzon et al., 2006; Feil, Hofmann, & Feil, 2004; Shankman et al., 2015), and plaque macrophages largely accumulate via local proliferation

(Robbins et al., 2013); however, the crossover between SMCs and macrophages is extensive. Fate-mapping studies demonstrate that in advanced atherosclerotic lesions of *ApoE*$^{(-/-)}$ mice, the majority of SMC-derived cells do not express SMA (Shankman et al., 2015), and many of these SMA^{-} cells express macrophage markers LGALS3 and CD68 (Fig. 4) (Feil et al., 2014; Shankman et al., 2015). In advanced human atherosclerotic plaques, 40% of cells expressing the macrophage marker CD68 are also labeled by SMA (Allahverdian, Chehroudi, McManus, Abraham, & Francis, 2014), and based on results of in situ hybridization proximity ligation assay, Owens and colleagues suggest that SMC-derived macrophage-like cells are present in human coronary artery lesions (Gomez, Shankman, Nguyen, & Owens, 2013; Shankman et al., 2015). Conversely, a bone marrow transplantation study in high-fat fed *ApoE*$^{(-/-)}$ mice indicated that within the atherosclerotic plaque, bone marrow-derived cells give rise to 5% of SMA^{+} cells, but the *Myosin heavy chain 11* transcriptional program (encoding SMMHC) is not active in these cells (Iwata et al., 2010). Furthermore, cross gender bone marrow transplantation in humans revealed that ~10% of SMA^{+} cells in advanced coronary artery plaques derive from hematopoietic cells (Caplice et al., 2003).

The molecular mechanisms underlying SMC-macrophage interconversion are critically important to understand, and studies are beginning to shed

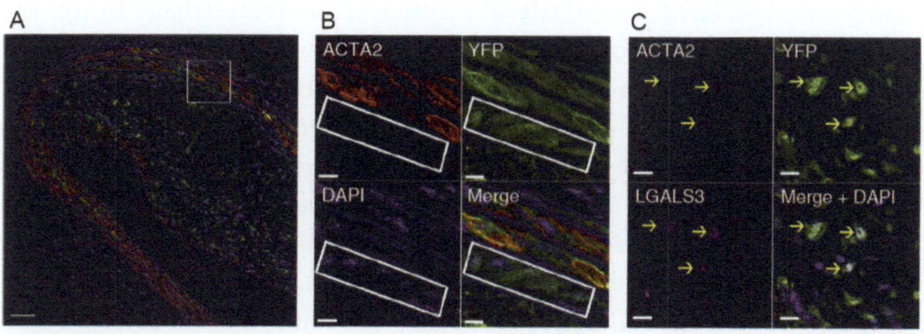

Fig. 4 SMCs transdifferentiate into SMA^{-} macrophage-like cells in atherosclerosis (Shankman et al., 2015). Immunohistochemistry of atherosclerotic brachiocephalic arteries of *ApoE*$^{(-/-)}$, *SMMHC-CreER*^{T2}, and *ROSA26R*$^{(YFP/YFP)}$ mice that were induced with tamoxifen and then fed a Western diet for 18 weeks. (A, B) The *boxed region* in (A) is shown as a close-up in (B). The *rectangle* in (B) contains SMC-derived cells (i.e., YFP^{+} cells) that do not express SMA (ACTA2). (C) *Arrows* indicate SMC-derived cells that express the macrophage marker LGALS3 but not SMA. *Reprinted by permission from Macmillan Publishers Ltd: Shankman, L. S., Gomez, D., Cherepanova, O. A., Salmon, M., Alencar, G. F., Haskins, R. M., . . . Owens, G. K. (2015). KLF4-dependent phenotypic modulation of smooth muscle cells has a key role in atherosclerotic plaque pathogenesis.* Nature Medicine, 21*(6), 628–637. http://dx.doi.org/10.1038/nm.3866, copyright 2015.*

light. For instance, cholesterol loading converts cultured aortic SMCs to a macrophage-like phenotype by upregulating KLF4 (Shankman et al., 2015) and downregulating the microRNA-143/145-myocardin axis (Vengrenyuk et al., 2015). SMC-specific deletion of *Klf4* in *ApoE*$^{(-/-)}$ mice prior to high-fat feeding results in atherosclerotic plaques of reduced size that have an increased percentage of SMA$^+$ cells and reduced SMC-derived macrophage-like cells (Shankman et al., 2015). In contrast, deletion of another pluripotency factor *Octamer-binding transcriptional factor 4* (*Oct4*) in SMCs of *ApoE*$^{(-/-)}$ mice increases plaque burden and reduces SMC-derived smooth muscle marker$^+$ cells in the fibrous cap area (Cherepanova et al., 2016). Interestingly, in comparison to wild-type SMCs, a higher percentage of *Oct4*-deleted SMCs transition to macrophage marker$^+$ cells in the tunica media but not in the neointima (Cherepanova et al., 2016). Less is known regarding the mechanisms involved in SMC marker induction in macrophages; however, treatment of isolated monocytes with thrombin has been shown to induce expression of myocardin and SMMHC (Martin et al., 2009).

3.5 Germinal Matrix Hemorrhage

The germinal matrix (GM) is a highly vascularized region of the developing brain located underneath the lateral ventricles, and hemorrhage in this area (i.e., germinal matrix hemorrhage, GMH) is a devastating neurological disease in premature infants that results in substantial mortality and morbidity. In the United States, one-fifth of neonates born before 35 weeks of gestation and weighing less than 1 kg develop GMH (Guyer, Martin, MacDorman, Anderson, & Strobino, 1997). With substantial GMH, the ventricular ependyma is compromised, and bleeding extends into the ventricles, leading to intraventricular hemorrhage (Ballabh, 2010). Unfortunately, there is no treatment for neonatal GMH, and the only preventive intervention is perinatal glucocorticoids, which have deleterious effects on neuronal development in animal models (Uno et al., 1994). Thus, effective and safe therapies that combat GMH are desperately needed.

Within the GM and associated vasculature, the ECs, PCs, basement membrane, and astrocytes have characteristics that are thought to convey susceptibility to neonatal GMH. First, ECs in the GM are highly proliferative (Ballabh et al., 2007), and EC hyperproliferation has also been reported in embryonic mouse models with reduced PCs and/or cerebral hemorrhage (Arnold et al., 2014; Hellstrom et al., 2001). Second, studies in humans and rabbits suggest that during mid-gestation, PC density and coverage of EC vessels are reduced in the GM relative to the white matter and cortex (Braun et al., 2007). Third, the basement membrane of GM vessels has reduced levels of fibronectin, an ECM protein that is critical for vessel stability (Xu et al., 2008). Finally, astrocytes are a key blood–brain barrier component, and in the human GM, there is reduced perivascular coverage by astrocytes expressing glial fibrillary acidic protein (El-Khoury et al., 2006).

Beyond these cellular characteristics of the GM, molecular pathways responsible for GMH pathogenesis remain poorly understood; however, VEGF and TGFβ are broadly implicated. Studies in premature human neonates and rabbit pups indicate that VEGF is selectively induced in the GM (Ballabh et al., 2007), and Yang and colleagues demonstrated that VEGF overexpression in the murine GM results in upregulated neurovascular proteases and GMH (Yang et al., 2013). Conversely, in a rabbit model of hyperosmolality-induced intracranial hemorrhage, treatment with a VEGFR2 inhibitor attenuated GMH (Ballabh et al., 2007). In addition, premature human neonates have low TGFβ1 levels in the GM relative to other regions of the brain (Braun et al., 2007). In murine embryos, EC-specific deletion of *TGFβ receptor 1* or *2* (Nguyen et al., 2011) or the downstream effector gene *SMAD4* results in intracranial hemorrhage (Li et al., 2011).

4. CONCLUSION

Despite substantial advances in therapeutic and preventive approaches, cardiovascular disease remains the most common cause of death globally. We suggest that a major reason for this lethality is that the vasculature is deceptively complex and is integral to development and homeostasis of the organism. Blood vessels are not simple conduits but instead dynamic tubes of diverse sizes and structure that are joined together to form a functional network delivering nutrients to metabolically active tissues and removing waste products. The formation and maintenance of this vascular network requires the integration of diverse molecular signals that regulate multiple cell types.

Our knowledge of blood vessel morphogenesis and pathogenesis has dramatically improved over the last decade, and further insights are needed to advance therapeutic strategies. The recent increased use of fundamental developmental biological and genetic approaches (e.g., lineage, mosaic and clonal analyses, and conditional gene deletion and misexpression) to study vessels during morphogenesis and disease has enhanced our understanding of underlying cellular and molecular events. In building upon these studies, we suggest two key areas for future investigations of vascular pathogenesis that warrant major research efforts. First, many vascular diseases are characterized by perturbed interactions between different cell types, including ECs and SMCs in PH, SMCs and macrophages in atherosclerosis, and ECs and PCs in GMH. Such cell–cell interactions are poorly understood and likely to be integral to disease mechanisms. Second, many vascular diseases such as atherosclerosis, PH, SVAS, and intracranial hemorrhage are characterized by hyperproliferation of vascular cells. A fundamental question follows: do these excessive cells arise from rare prespecified progenitor cells or are preexisting cells equipotent? Our laboratory recently identified a specialized pool of SMC progenitors in normal pulmonary arterioles and demonstrated their central role in hypoxia-induced PH (Sheikh et al., 2015). However, existence of similar specialized progenitor cells and their fate in other vascular beds during development and other diseases remain unexplored. Only through continued rigorous and diverse investigations of vascular development, maintenance, and disease will we be able to reduce the substantial impact of cardiovascular pathologies on human health.

CONFLICT OF INTEREST

The authors have no conflicts of interest to declare.

REFERENCES

Allahverdian, S., Chehroudi, A. C., McManus, B. M., Abraham, T., & Francis, G. A. (2014). Contribution of intimal smooth muscle cells to cholesterol accumulation and macrophage-like cells in human atherosclerosis. *Circulation*, *129*(15), 1551–1559. http://dx.doi.org/10.1161/CIRCULATIONAHA.113.005015.

Amato, B., Compagna, R., Amato, M., Grande, R., Butrico, L., Rossi, A., … Serra, R. (2015). Adult vascular wall resident multipotent vascular stem cells, matrix metalloproteinases, and arterial aneurysms. *Stem Cells International*, *2015*, 434962. http://dx.doi.org/10.1155/2015/434962.

Andelfinger, G., Loeys, B., & Dietz, H. (2016). A decade of discovery in the genetic understanding of thoracic aortic disease. *The Canadian Journal of Cardiology*, *32*(1), 13–25. http://dx.doi.org/10.1016/j.cjca.2015.10.017.

Armulik, A., Genove, G., & Betsholtz, C. (2011). Pericytes: Developmental, physiological, and pathological perspectives, problems, and promises. *Developmental Cell*, *21*(2), 193–215. http://dx.doi.org/10.1016/j.devcel.2011.07.001. S1534-5807(11)00269-3 [pii].

Armulik, A., Genove, G., Mae, M., Nisancioglu, M. H., Wallgard, E., Niaudet, C., … Betsholtz, C. (2010). Pericytes regulate the blood-brain barrier. *Nature*, *468*(7323), 557–561. http://dx.doi.org/10.1038/nature09522. nature09522 [pii].

Arnold, T. D., Niaudet, C., Pang, M. F., Siegenthaler, J., Gaengel, K., Jung, B., … Reichardt, L. F. (2014). Excessive vascular sprouting underlies cerebral hemorrhage in mice lacking alphaVbeta8-TGFbeta signaling in the brain. *Development*, *141*(23), 4489–4499. http://dx.doi.org/10.1242/dev.107193.

Arribas, S. M., Hinek, A., & Gonzalez, M. C. (2006). Elastic fibres and vascular structure in hypertension. *Pharmacology and Therapeutics*, *111*(3), 771–791. http://dx.doi.org/10.1016/j.pharmthera.2005.12.003.

Asahina, K., Zhou, B., Pu, W. T., & Tsukamoto, H. (2011). Septum transversum-derived mesothelium gives rise to hepatic stellate cells and perivascular mesenchymal cells in developing mouse liver. *Hepatology*, *53*(3), 983–995. http://dx.doi.org/10.1002/hep.24119.

Augustin, H. G., Kozian, D. H., & Johnson, R. C. (1994). Differentiation of endothelial cells: Analysis of the constitutive and activated endothelial cell phenotypes. *Bioessays*, *16*(12), 901–906. http://dx.doi.org/10.1002/bies.950161208.

Ballabh, P. (2010). Intraventricular hemorrhage in premature infants: Mechanism of disease. *Pediatric Research*, *67*(1), 1–8. http://dx.doi.org/10.1203/PDR.0b013e3181c1b176.

Ballabh, P., Xu, H., Hu, F., Braun, A., Smith, K., Rivera, A., … Nedergaard, M. (2007). Angiogenic inhibition reduces germinal matrix hemorrhage. *Nature Medicine*, *13*(4), 477–485. http://dx.doi.org/10.1038/nm1558.

Bell, R. D., Winkler, E. A., Sagare, A. P., Singh, I., LaRue, B., Deane, R., & Zlokovic, B. V. (2010). Pericytes control key neurovascular functions and neuronal phenotype in the adult brain and during brain aging. *Neuron*, *68*(3), 409–427. http://dx.doi.org/10.1016/j.neuron.2010.09.043. S0896-6273(10)00824-X [pii].

Benjamin, L. E., Hemo, I., & Keshet, E. (1998). A plasticity window for blood vessel remodelling is defined by pericyte coverage of the preformed endothelial network and is regulated by PDGF-B and VEGF. *Development*, *125*(9), 1591–1598.

Bentzon, J. F., Weile, C., Sondergaard, C. S., Hindkjaer, J., Kassem, M., & Falk, E. (2006). Smooth muscle cells in atherosclerosis originate from the local vessel wall and not circulating progenitor cells in ApoE knockout mice. *Arteriosclerosis, Thrombosis, and Vascular Biology*, *26*(12), 2696–2702. http://dx.doi.org/10.1161/01.ATV.0000247243.48542.9d.

Blanco, R., & Gerhardt, H. (2013). VEGF and Notch in tip and stalk cell selection. *Cold Spring Harbor Perspectives in Medicine*, *3*(1), a006569. http://dx.doi.org/10.1101/cshperspect.a006569.

Braun, A., Xu, H., Hu, F., Kocherlakota, P., Siegel, D., Chander, P., … Ballabh, P. (2007). Paucity of pericytes in germinal matrix vasculature of premature infants. *The Journal of Neuroscience*, *27*(44), 12012–12024. http://dx.doi.org/10.1523/JNEUROSCI.3281-07.2007. 27/44/12012 [pii].

Brooke, B. S., Bayes-Genis, A., & Li, D. Y. (2003). New insights into elastin and vascular disease. *Trends in Cardiovascular Medicine*, *13*(5), 176–181. S1050173803000653 [pii].

Caplice, N. M., Bunch, T. J., Stalboerger, P. G., Wang, S., Simper, D., Miller, D. V., … Edwards, W. D. (2003). Smooth muscle cells in human coronary atherosclerosis can originate from cells administered at marrow transplantation. *Proceedings of the National Academy of Sciences of the United States of America*, *100*(8), 4754–4759. http://dx.doi.org/10.1073/pnas.0730743100.

Chen, Y. F., & Oparil, S. (2000). Endothelin and pulmonary hypertension. *Journal of Cardiovascular Pharmacology*, *35*(4 Suppl. 2), S49–S53.

Cherepanova, O. A., Gomez, D., Shankman, L. S., Swiatlowska, P., Williams, J., Sarmento, O. F., … Owens, G. K. (2016). Activation of the pluripotency factor OCT4 in smooth muscle cells is atheroprotective. *Nature Medicine*, *22*(6), 657–665. http://dx.doi.org/10.1038/nm.4109.

Cheung, C., Bernardo, A. S., Trotter, M. W., Pedersen, R. A., & Sinha, S. (2012). Generation of human vascular smooth muscle subtypes provides insight into embryological origin-dependent disease susceptibility. *Nature Biotechnology*, *30*(2), 165–173. http://dx.doi.org/10.1038/nbt.2107. nbt.2107 [pii].

Cines, D. B., Pollak, E. S., Buck, C. A., Loscalzo, J., Zimmerman, G. A., McEver, R. P., … Stern, D. M. (1998). Endothelial cells in physiology and in the pathophysiology of vascular disorders. *Blood*, *91*(10), 3527–3561.

Curran, M. E., Atkinson, D. L., Ewart, A. K., Morris, C. A., Leppert, M. F., & Keating, M. T. (1993). The elastin gene is disrupted by a translocation associated with supravalvular aortic stenosis. *Cell, 73*(1), 159–168. http://dx.doi.org/10.1016/0092-8674(93)90168-P.

Daneman, R., Zhou, L., Kebede, A. A., & Barres, B. A. (2010). Pericytes are required for blood-brain barrier integrity during embryogenesis. *Nature, 468*(7323), 562–566. http://dx.doi.org/10.1038/nature09513. nature09513 [pii].

Dave, J. M., & Bayless, K. J. (2014). Vimentin as an integral regulator of cell adhesion and endothelial sprouting. *Microcirculation, 21*(4), 333–344. http://dx.doi.org/10.1111/micc.12111.

Davis, G. E., & Senger, D. R. (2005). Endothelial extracellular matrix: Biosynthesis, remodeling, and functions during vascular morphogenesis and neovessel stabilization. *Circulation Research, 97*(11), 1093–1107. http://dx.doi.org/10.1161/01.RES.0000191547.64391.e3.

Deaton, R. A., Gan, Q., & Owens, G. K. (2009). Sp1-dependent activation of KLF4 is required for PDGF-BB-induced phenotypic modulation of smooth muscle. *American Journal of Physiology. Heart and Circulatory Physiology, 296*(4), H1027–H1037. http://dx.doi.org/10.1152/ajpheart.01230.2008.

Dejana, E., Tournier-Lasserve, E., & Weinstein, B. M. (2009). The control of vascular integrity by endothelial cell junctions: Molecular basis and pathological implications. *Developmental Cell, 16*(2), 209–221. http://dx.doi.org/10.1016/j.devcel.2009.01.004.

Dettman, R. W., Denetclaw, W., Jr., Ordahl, C. P., & Bristow, J. (1998). Common epicardial origin of coronary vascular smooth muscle, perivascular fibroblasts, and intermyocardial fibroblasts in the avian heart. *Developmental Biology, 193*(2), 169–181. http://dx.doi.org/10.1006/dbio.1997.8801. S0012-1606(97)98801-1 [pii].

Eddahibi, S., Guignabert, C., Barlier-Mur, A. M., Dewachter, L., Fadel, E., Dartevelle, P., … Adnot, S. (2006). Cross talk between endothelial and smooth muscle cells in pulmonary hypertension: Critical role for serotonin-induced smooth muscle hyperplasia. *Circulation, 113*(15), 1857–1864. http://dx.doi.org/10.1161/CIRCULATIONAHA.105.591321.

Eddahibi, S., Humbert, M., Fadel, E., Raffestin, B., Darmon, M., Capron, F., … Adnot, S. (2001). Serotonin transporter overexpression is responsible for pulmonary artery smooth muscle hyperplasia in primary pulmonary hypertension. *Journal of Clinical Investigation, 108*(8), 1141–1150. http://dx.doi.org/10.1172/JCI12805.

El-Khoury, N., Braun, A., Hu, F., Pandey, M., Nedergaard, M., Lagamma, E. F., & Ballabh, P. (2006). Astrocyte end-feet in germinal matrix, cerebral cortex, and white matter in developing infants. *Pediatric Research, 59*(5), 673–679. http://dx.doi.org/10.1203/01.pdr.0000214975.85311.9c.

Enge, M., Bjarnegard, M., Gerhardt, H., Gustafsson, E., Kalen, M., Asker, N., … Betsholtz, C. (2002). Endothelium-specific platelet-derived growth factor-B ablation mimics diabetic retinopathy. *EMBO Journal, 21*(16), 4307–4316.

Etchevers, H. C., Vincent, C., Le Douarin, N. M., & Couly, G. F. (2001). The cephalic neural crest provides pericytes and smooth muscle cells to all blood vessels of the face and forebrain. *Development, 128*(7), 1059–1068.

Feil, S., Fehrenbacher, B., Lukowski, R., Essmann, F., Schulze-Osthoff, K., Schaller, M., & Feil, R. (2014). Transdifferentiation of vascular smooth muscle cells to macrophage-like cells during atherogenesis. *Circulation Research, 115*(7), 662–667. http://dx.doi.org/10.1161/CIRCRESAHA.115.304634.

Feil, S., Hofmann, F., & Feil, R. (2004). SM22alpha modulates vascular smooth muscle cell phenotype during atherogenesis. *Circulation Research, 94*(7), 863–865. http://dx.doi.org/10.1161/01.RES.0000126417.38728.F6. 01.RES.0000126417.38728.F6 [pii].

Foster, K., Sheridan, J., Veiga-Fernandes, H., Roderick, K., Pachnis, V., Adams, R., … Coles, M. (2008). Contribution of neural crest-derived cells in the embryonic and adult thymus. *Journal of Immunology*, *180*(5), 3183–3189.

Gaengel, K., Genove, G., Armulik, A., & Betsholtz, C. (2009). Endothelial-mural cell signaling in vascular development and angiogenesis. *Arteriosclerosis, Thrombosis, and Vascular Biology*, *29*(5), 630–638. http://dx.doi.org/10.1161/ATVBAHA.107.161521. ATVBAHA.107.161521 [pii].

Gao, F., Chambon, P., Offermanns, S., Tellides, G., Kong, W., Zhang, X., & Li, W. (2014). Disruption of TGF-beta signaling in smooth muscle cell prevents elastase-induced abdominal aortic aneurysm. *Biochemical and Biophysical Research Communications*, *454*(1), 137–143. http://dx.doi.org/10.1016/j.bbrc.2014.10.053.

Gerhardt, H., Golding, M., Fruttiger, M., Ruhrberg, C., Lundkvist, A., Abramsson, A., … Betsholtz, C. (2003). VEGF guides angiogenic sprouting utilizing endothelial tip cell filopodia. *Journal of Cell Biology*, *161*(6), 1163–1177. http://dx.doi.org/10.1083/jcb.200302047. jcb.200302047 [pii].

Ghabrial, A. S., & Krasnow, M. A. (2006). Social interactions among epithelial cells during tracheal branching morphogenesis. *Nature*, *441*(7094), 746–749. http://dx.doi.org/10.1038/nature04829. nature04829 [pii].

Gomez, D., Shankman, L. S., Nguyen, A. T., & Owens, G. K. (2013). Detection of histone modifications at specific gene loci in single cells in histological sections. *Nature Methods*, *10*(2), 171–177. http://dx.doi.org/10.1038/nmeth.2332.

Greif, D. M., Kumar, M., Lighthouse, J. K., Hum, J., An, A., Ding, L., … Krasnow, M. A. (2012). Radial construction of an arterial wall. *Developmental Cell*, *23*(3), 482–493. http://dx.doi.org/10.1016/j.devcel.2012.07.009. S1534-5807(12)00328-0 [pii].

Guyer, B., Martin, J. A., MacDorman, M. F., Anderson, R. N., & Strobino, D. M. (1997). Annual summary of vital statistics—1996. *Pediatrics*, *100*(6), 905–918.

Hall, C., Reynell, C., Gesslein, B., Hamilton, N. B., Mishra, A., Sutherland, B. A., … Attwell, D. (2014). Capillary pericytes regulate cerebral blood flow in health and disease. *Nature*, *508*(7494), 55–60.

Harkness, M. L., Harkness, R. D., & McDonald, D. A. (1957). The collagen and elastin content of the arterial wall in the dog. *Proceedings of the Royal Society of London—Series B: Biological Sciences*, *146*(925), 541–551.

Hellstrom, M., Gerhardt, H., Kalen, M., Li, X., Eriksson, U., Wolburg, H., & Betsholtz, C. (2001). Lack of pericytes leads to endothelial hyperplasia and abnormal vascular morphogenesis. *Journal of Cell Biology*, *153*(3), 543–553.

Hellstrom, M., Kalen, M., Lindahl, P., Abramsson, A., & Betsholtz, C. (1999). Role of PDGF-B and PDGFR-beta in recruitment of vascular smooth muscle cells and pericytes during embryonic blood vessel formation in the mouse. *Development*, *126*(14), 3047–3055.

Herbert, S. P., Cheung, J. Y., & Stainier, D. Y. (2012). Determination of endothelial stalk versus tip cell potential during angiogenesis by H2.0-like homeobox-1. *Current Biology*, *22*(19), 1789–1794. http://dx.doi.org/10.1016/j.cub.2012.07.037.

Hill, R. A., Tong, L., Yuan, P., Murikinati, S., Gupta, S., & Grutzendler, J. (2015). Regional blood flow in the normal and ischemic brain is controlled by arteriolar smooth muscle cell contractility and not by capillary pericytes. *Neuron*, *87*(1), 95–110. http://dx.doi.org/10.1016/j.neuron.2015.06.001.

Hirschi, K. K., Rohovsky, S. A., & D'Amore, P. A. (1998). PDGF, TGF-beta, and heterotypic cell-cell interactions mediate endothelial cell-induced recruitment of 10 T1/2 cells and their differentiation to a smooth muscle fate. *Journal of Cell Biology*, *141*(3), 805–814.

Hoglund, V. J., & Majesky, M. W. (2012). Patterning the artery wall by lateral induction of notch signaling. *Circulation*, *125*(2), 212–215. http://dx.doi.org/10.1161/CIRCULATIONAHA.111.075937. CIRCULATIONAHA.111.075937 [pii].

Hopper, R. K., Moonen, J. R., Diebold, I., Cao, A., Rhodes, C. J., Tojais, N. F., … Rabinovitch, M. (2016). In pulmonary arterial hypertension, reduced BMPR2 promotes endothelial-to-mesenchymal transition via HMGA1 and its target slug. *Circulation, 133*(18), 1783–1794. http://dx.doi.org/10.1161/CIRCULATIONAHA.115.020617.

Hu, Y., Zhang, Z., Torsney, E., Afzal, A. R., Davison, F., Metzler, B., & Xu, Q. (2004). Abundant progenitor cells in the adventitia contribute to atherosclerosis of vein grafts in ApoE-deficient mice. *Journal of Clinical Investigation, 113*(9), 1258–1265. http://dx.doi.org/10.1172/JCI19628.

Humbert, M., Sitbon, O., Chaouat, A., Bertocchi, M., Habib, G., Gressin, V., … Simonneau, G. (2010). Survival in patients with idiopathic, familial, and anorexigen-associated pulmonary arterial hypertension in the modern management era. *Circulation, 122*(2), 156–163. http://dx.doi.org/10.1161/CIRCULATIONAHA.109.911818. CIRCULATIONAHA.109.911818 [pii].

Humphreys, B. D., Lin, S. L., Kobayashi, A., Hudson, T. E., Nowlin, B. T., Bonventre, J. V., … Duffield, J. S. (2010). Fate tracing reveals the pericyte and not epithelial origin of myofibroblasts in kidney fibrosis. *The American Journal of Pathology, 176*(1), 85–97. http://dx.doi.org/10.2353/ajpath.2010.090517.

Hung, C., Linn, G., Chow, Y. H., Kobayashi, A., Mittelsteadt, K., Altemeier, W. A., … Duffield, J. S. (2013). Role of lung pericytes and resident fibroblasts in the pathogenesis of pulmonary fibrosis. *American Journal of Respiratory and Critical Care Medicine, 188*(7), 820–830. http://dx.doi.org/10.1164/rccm.201212-2297OC.

Iruela-Arispe, M. L., & Davis, G. E. (2009). Cellular and molecular mechanisms of vascular lumen formation. *Developmental Cell, 16*(2), 222–231. http://dx.doi.org/10.1016/j.devcel.2009.01.013.

Iwata, H., Manabe, I., Fujiu, K., Yamamoto, T., Takeda, N., Eguchi, K., … Nagai, R. (2010). Bone marrow-derived cells contribute to vascular inflammation but do not differentiate into smooth muscle cell lineages. *Circulation, 122*(20), 2048–2057. http://dx.doi.org/10.1161/CIRCULATIONAHA.110.965202.

Izikki, M., Guignabert, C., Fadel, E., Humbert, M., Tu, L., Zadigue, P., … Eddahibi, S. (2009). Endothelial-derived FGF2 contributes to the progression of pulmonary hypertension in humans and rodents. *Journal of Clinical Investigation, 119*(3), 512–523. http://dx.doi.org/10.1172/JCI35070.

Jacob, M. P. (2003). Extracellular matrix remodeling and matrix metalloproteinases in the vascular wall during aging and in pathological conditions. *Biomedicine and Pharmacotherapy, 57*(5–6), 195–202.

Jakobsson, L., Franco, C. A., Bentley, K., Collins, R. T., Ponsioen, B., Aspalter, I. M., … Gerhardt, H. (2010). Endothelial cells dynamically compete for the tip cell position during angiogenic sprouting. *Nature Cell Biology, 12*(10), 943–953. http://dx.doi.org/10.1038/ncb2103.

Kadoglou, N. P., & Liapis, C. D. (2004). Matrix metalloproteinases: Contribution to pathogenesis, diagnosis, surveillance and treatment of abdominal aortic aneurysms. *Current Medical Research and Opinion, 20*(4), 419–432. http://dx.doi.org/10.1185/030079904125003143.

Kamei, M., Saunders, W. B., Bayless, K. J., Dye, L., Davis, G. E., & Weinstein, B. M. (2006). Endothelial tubes assemble from intracellular vacuoles in vivo. *Nature, 442*(7101), 453–456. http://dx.doi.org/10.1038/nature04923. nature04923 [pii].

Karnik, S. K., Brooke, B. S., Bayes-Genis, A., Sorensen, L., Wythe, J. D., Schwartz, R. S., … Li, D. Y. (2003). A critical role for elastin signaling in vascular morphogenesis and disease. *Development, 130*(2), 411–423.

Karow, M., Sanchez, R., Schichor, C., Masserdotti, G., Ortega, F., Heinrich, C., … Berninger, B. (2012). Reprogramming of pericyte-derived cells of the adult human brain into induced neuronal cells. *Cell Stem Cell, 11*(4), 471–476. http://dx.doi.org/10.1016/j.stem.2012.07.007.

Korn, J., Christ, B., & Kurz, H. (2002). Neuroectodermal origin of brain pericytes and vascular smooth muscle cells. *Journal of Comparative Neurology*, *442*(1), 78–88.

LeBleu, V. S., Taduri, G., O'Connell, J., Teng, Y., Cooke, V. G., Woda, C., ... Kalluri, R. (2013). Origin and function of myofibroblasts in kidney fibrosis. *Nature Medicine*, *19*(8), 1047–1053. http://dx.doi.org/10.1038/nm.3218.

Li, D. Y., Brooke, B., Davis, E. C., Mecham, R. P., Sorensen, L. K., Boak, B. B., ... Keating, M. T. (1998). Elastin is an essential determinant of arterial morphogenesis. *Nature*, *393*(6682), 276–280. http://dx.doi.org/10.1038/30522.

Li, D. Y., Faury, G., Taylor, D. G., Davis, E. C., Boyle, W. A., Mecham, R. P., ... Keating, M. T. (1998). Novel arterial pathology in mice and humans hemizygous for elastin. *Journal of Clinical Investigation*, *102*(10), 1783–1787. http://dx.doi.org/10.1172/JCI4487.

Li, F., Lan, Y., Wang, Y., Wang, J., Yang, G., Meng, F., ... Yang, X. (2011). Endothelial Smad4 maintains cerebrovascular integrity by activating N-cadherin through cooperation with Notch. *Developmental Cell*, *20*(3), 291–302. http://dx.doi.org/10.1016/j.devcel.2011.01.011. S1534-5807(11)00039-6 [pii].

Li, D. Y., Toland, A. E., Boak, B. B., Atkinson, D. L., Ensing, G. J., Morris, C. A., & Keating, M. T. (1997). Elastin point mutations cause an obstructive vascular disease, supravalvular aortic stenosis. *Human Molecular Genetics*, *6*(7), 1021–1028. http://dx.doi.org/10.1093/hmg/6.7.1021.

Li, S., Wang, D. Z., Wang, Z., Richardson, J. A., & Olson, E. N. (2003). The serum response factor coactivator myocardin is required for vascular smooth muscle development. *Proceedings of the National Academy of Sciences of the United States of America*, *100*(16), 9366–9370. http://dx.doi.org/10.1073/pnas.1233635100. 1233635100 [pii].

Libby, P., Ridker, P. M., & Hansson, G. K. (2011). Progress and challenges in translating the biology of atherosclerosis. *Nature*, *473*(7347), 317–325. http://dx.doi.org/10.1038/nature10146.

Lindahl, P., Johansson, B. R., Leveen, P., & Betsholtz, C. (1997). Pericyte loss and microaneurysm formation in PDGF-B-deficient mice. *Science*, *277*(5323), 242–245.

Lindblom, P., Gerhardt, H., Liebner, S., Abramsson, A., Enge, M., Hellstrom, M., ... Betsholtz, C. (2003). Endothelial PDGF-B retention is required for proper investment of pericytes in the microvessel wall. *Genes and Development*, *17*(15), 1835–1840. http://dx.doi.org/10.1101/gad.266803. 17/15/1835 [pii].

Lindsay, M. E., & Dietz, H. C. (2011). Lessons on the pathogenesis of aneurysm from heritable conditions. *Nature*, *473*(7347), 308–316. http://dx.doi.org/10.1038/nature10145. nature10145 [pii].

Liu, Y., Sinha, S., McDonald, O. G., Shang, Y., Hoofnagle, M. H., & Owens, G. K. (2005). Kruppel-like factor 4 abrogates myocardin-induced activation of smooth muscle gene expression. *Journal of Biological Chemistry*, *280*(10), 9719–9727. http://dx.doi.org/10.1074/jbc.M412862200. S0021925819620084 [pii].

Mahapatra, S., Nishimura, R. A., Sorajja, P., Cha, S., & McGoon, M. D. (2006). Relationship of pulmonary arterial capacitance and mortality in idiopathic pulmonary arterial hypertension. *Journal of the American College of Cardiology*, *47*(4), 799–803. http://dx.doi.org/10.1016/j.jacc.2005.09.054.

Majesky, M. W. (2004). Development of coronary vessels. *Current Topics in Developmental Biology*, *62*, 225–259. http://dx.doi.org/10.1016/S0070-2153(04)62008-4.

Majesky, M. W. (2007). Developmental basis of vascular smooth muscle diversity. *Arteriosclerosis, Thrombosis, and Vascular Biology*, *27*(6), 1248–1258. http://dx.doi.org/10.1161/ATVBAHA.107.141069. ATVBAHA.107.141069 [pii].

Majesky, M. W. (2015). Adventitia and perivascular cells. *Arteriosclerosis, Thrombosis, and Vascular Biology*, *35*(8), e31–e35. http://dx.doi.org/10.1161/ATVBAHA.115.306088.

Majesky, M. W., Dong, X. R., Hoglund, V., Daum, G., & Mahoney, W. M., Jr. (2012). The adventitia: A progenitor cell niche for the vessel wall. *Cells, Tissues, Organs*, *195*(1–2), 73–81. http://dx.doi.org/10.1159/000331413. 000331413 [pii].

Manderfield, L. J., High, F. A., Engleka, K. A., Liu, F., Li, L., Rentschler, S., & Epstein, J. A. (2012). Notch activation of Jagged1 contributes to the assembly of the arterial wall. *Circulation, 125*(2), 314–323. http://dx.doi.org/10.1161/CIRCULATIONAHA.111.047159. Epub 2011 Dec 6.

Martin, K., Weiss, S., Metharom, P., Schmeckpeper, J., Hynes, B., O'Sullivan, J., & Caplice, N. (2009). Thrombin stimulates smooth muscle cell differentiation from peripheral blood mononuclear cells via protease-activated receptor-1, RhoA, and myocardin. *Circulation Research, 105*(3), 214–218. http://dx.doi.org/10.1161/CIRCRESAHA.109.199984.

McLean, S. E., Mecham, B. H., Kelleher, C. M., Mariani, T. J., & Mecham, R. P. (2005). Extracellular matrix gene expression in the developing mouse aorta. *Advances in Developmental Biology, 15*, 81–128.

Misra, A., Sheikh, A. Q., Kumar, A., Luo, J., Zhang, J., Hinton, R. B., … Greif, D. M. (2016). Integrin beta3 inhibition is a therapeutic strategy for supravalvular aortic stenosis. *Journal of Experimental Medicine, 213*(3), 451–463. http://dx.doi.org/10.1084/jem.20150688.

Nakagomi, T., Kubo, S., Nakano-Doi, A., Sakuma, R., Lu, S., Narita, A., … Matsuyama, T. (2015). Brain vascular pericytes following ischemia have multipotential stem cell activity to differentiate into neural and vascular lineage cells. *Stem Cells, 33*(6), 1962–1974. http://dx.doi.org/10.1002/stem.1977.

Nguyen, H. L., Lee, Y. J., Shin, J., Lee, E., Park, S. O., McCarty, J. H., & Oh, S. P. (2011). TGF-beta signaling in endothelial cells, but not neuroepithelial cells, is essential for cerebral vascular development. *Laboratory Investigation, 91*(11), 1554–1563. http://dx.doi.org/10.1038/labinvest.2011.124.

Owens, G. K., Kumar, M. S., & Wamhoff, B. R. (2004). Molecular regulation of vascular smooth muscle cell differentiation in development and disease. *Physiological Reviews, 84*(3), 767–801.

Passman, J. N., Dong, X. R., Wu, S. P., Maguire, C. T., Hogan, K. A., Bautch, V. L., & Majesky, M. W. (2008). A sonic hedgehog signaling domain in the arterial adventitia supports resident Sca1 + smooth muscle progenitor cells. *Proceedings of the National Academy of Sciences of the United States of America, 105*(27), 9349–9354. http://dx.doi.org/10.1073/pnas.0711382105. 0711382105 [pii].

Phng, L. K., & Gerhardt, H. (2009). Angiogenesis: A team effort coordinated by notch. *Developmental Cell, 16*(2), 196–208. http://dx.doi.org/10.1016/j.devcel.2009.01.015. S1534-5807(09)00043-4 [pii].

Pober, B. R. (2010). Williams-Beuren syndrome. *New England Journal of Medicine, 362*(3), 239–252. http://dx.doi.org/10.1056/NEJMra0903074. 362/3/239 [pii].

Pouget, C., Gautier, R., Teillet, M. A., & Jaffredo, T. (2006). Somite-derived cells replace ventral aortic hemangioblasts and provide aortic smooth muscle cells of the trunk. *Development, 133*(6), 1013–1022. http://dx.doi.org/10.1242/dev.02269. dev.02269 [pii].

Qiao, L., Nishimura, T., Shi, L., Sessions, D., Thrasher, A., Trudell, J. R., … Kao, P. N. (2014). Endothelial fate mapping in mice with pulmonary hypertension. *Circulation, 129*(6), 692–703. http://dx.doi.org/10.1161/CIRCULATIONAHA.113.003734.

Que, J., Wilm, B., Hasegawa, H., Wang, F., Bader, D., & Hogan, B. L. (2008). Mesothelium contributes to vascular smooth muscle and mesenchyme during lung development. *Proceedings of the National Academy of Sciences of the United States of America, 105*(43), 16626–16630.

Ricard, N., Tu, L., Le Hiress, M., Huertas, A., Phan, C., Thuillet, R., … Guignabert, C. (2014). Increased pericyte coverage mediated by endothelial-derived fibroblast growth factor-2 and interleukin-6 is a source of smooth muscle-like cells in pulmonary hypertension. *Circulation, 129*(15), 1586–1597. http://dx.doi.org/10.1161/CIRCULATIONAHA.113.007469.

Risau, W., & Flamme, I. (1995). Vasculogenesis. *Annual Review of Cell and Developmental Biology, 11*, 73–91. http://dx.doi.org/10.1146/annurev.cb.11.110195.000445.

Robbins, C. S., Hilgendorf, I., Weber, G. F., Theurl, I., Iwamoto, Y., Figueiredo, J. L., … Swirski, F. K. (2013). Local proliferation dominates lesional macrophage accumulation in atherosclerosis. *Nature Medicine*, *19*(9), 1166–1172. http://dx.doi.org/10.1038/nm.3258.

Rock, J. R., Barkauskas, C. E., Cronce, M. J., Xue, Y., Harris, J. R., Liang, J., … Hogan, B. L. (2011). Multiple stromal populations contribute to pulmonary fibrosis without evidence for epithelial to mesenchymal transition. *Proceedings of the National Academy of Sciences of the United States of America*, *108*(52), E1475–E1483. http://dx.doi.org/10.1073/pnas.1117988108.

Rodella, L. F., Rezzani, R., Bonomini, F., Peroni, M., Cocchi, M. A., Hirtler, L., & Bonardelli, S. (2016). Abdominal aortic aneurysm and histological, clinical, radiological correlation. *Acta Histochemica*, *118*(3), 256–262. http://dx.doi.org/10.1016/j.acthis.2016.01.007.

Rowe, V. L., Stevens, S. L., Reddick, T. T., Freeman, M. B., Donnell, R., Carroll, R. C., & Goldman, M. H. (2000). Vascular smooth muscle cell apoptosis in aneurysmal, occlusive, and normal human aortas. *Journal of Vascular Surgery*, *31*(3), 567–576.

Salmon, M., Johnston, W. F., Woo, A., Pope, N. H., Su, G., Upchurch, G. R., Jr., … Ailawadi, G. (2013). KLF4 regulates abdominal aortic aneurysm morphology and deletion attenuates aneurysm formation. *Circulation*, *128*(11 Suppl. 1), S163–S174. http://dx.doi.org/10.1161/CIRCULATIONAHA.112.000238.

Sandberg, L. B., Soskel, N. T., & Leslie, J. G. (1981). Elastin structure, biosynthesis, and relation to disease states. *New England Journal of Medicine*, *304*(10), 566–579. http://dx.doi.org/10.1056/NEJM198103053041004.

Savale, L., Tu, L., Rideau, D., Izziki, M., Maitre, B., Adnot, S., & Eddahibi, S. (2009). Impact of interleukin-6 on hypoxia-induced pulmonary hypertension and lung inflammation in mice. *Respiratory Research*, *10*, 6. http://dx.doi.org/10.1186/1465-9921-10-6. 1465-9921-10-6 [pii].

Schermuly, R. T., Dony, E., Ghofrani, H. A., Pullamsetti, S., Savai, R., Roth, M., … Grimminger, F. (2005). Reversal of experimental pulmonary hypertension by PDGF inhibition. *Journal of Clinical Investigation*, *115*(10), 2811–2821.

Schrimpf, C., Xin, C., Campanholle, G., Gill, S. E., Stallcup, W., Lin, S. L., … Duffield, J. S. (2012). Pericyte TIMP3 and ADAMTS1 modulate vascular stability after kidney injury. *Journal of the American Society of Nephrology*, *23*(5), 868–883. http://dx.doi.org/10.1681/ASN.2011080851. ASN.2011080851 [pii].

Shankman, L. S., Gomez, D., Cherepanova, O. A., Salmon, M., Alencar, G. F., Haskins, R. M., … Owens, G. K. (2015). KLF4-dependent phenotypic modulation of smooth muscle cells has a key role in atherosclerotic plaque pathogenesis. *Nature Medicine*, *21*(6), 628–637. http://dx.doi.org/10.1038/nm.3866.

Sheikh, A. Q., Lighthouse, J. K., & Greif, D. M. (2014). Recapitulation of developing artery muscularization in pulmonary hypertension. *Cell Reports*, *6*(5), 809–817. http://dx.doi.org/10.1016/j.celrep.2014.01.042. S2211-1247(14)00076-X [pii].

Sheikh, A. Q., Misra, A., Rosas, I. O., Adams, R. H., & Greif, D. M. (2015). Smooth muscle cell progenitors are primed to muscularize in pulmonary hypertension. *Science Translational Medicine*, 7, 308ra159. http://dx.doi.org/10.1126/scitranslmed.aaa9712.

Simonneau, G., Gatzoulis, M. A., Adatia, I., Celermajer, D., Denton, C., Ghofrani, A., … Souza, R. (2013). Updated clinical classification of pulmonary hypertension. *Journal of the American College of Cardiology*, *62*(25 Suppl.), D34–D41. http://dx.doi.org/10.1016/j.jacc.2013.10.029.

Soriano, P. (1994). Abnormal kidney development and hematological disorders in PDGF beta-receptor mutant mice. *Genes and Development*, *8*(16), 1888–1896.

Stenmark, K. R., Fagan, K. A., & Frid, M. G. (2006). Hypoxia-induced pulmonary vascular remodeling: Cellular and molecular mechanisms. *Circulation Research*, *99*(7), 675–691. http://dx.doi.org/10.1161/01.RES.0000243584.45145.3f. 99/7/675 [pii].

Stenmark, K. R., Frid, M., & Perros, F. (2016). Endothelial-to-mesenchymal transition: An evolving paradigm and a promising therapeutic target in PAH. *Circulation*, *133*(18), 1734–1737. http://dx.doi.org/10.1161/CIRCULATIONAHA.116.022479.

Stratman, A. N., Malotte, K. M., Mahan, R. D., Davis, M. J., & Davis, G. E. (2009). Pericyte recruitment during vasculogenic tube assembly stimulates endothelial basement membrane matrix formation. *Blood*, *114*(24), 5091–5101. http://dx.doi.org/10.1182/blood-2009-05-222364.

Tabas, I., Garcia-Cardena, G., & Owens, G. K. (2015). Recent insights into the cellular biology of atherosclerosis. *Journal of Cell Biology*, *209*(1), 13–22. http://dx.doi.org/10.1083/jcb.201412052.

Topouzis, S., & Majesky, M. W. (1996). Smooth muscle lineage diversity in the chick embryo. Two types of aortic smooth muscle cell differ in growth and receptor-mediated transcriptional responses to transforming growth factor-beta. *Developmental Biology*, *178*(2), 430–445. http://dx.doi.org/10.1006/dbio.1996.0229. S0012-1606(96)90229-8 [pii].

Uno, H., Eisele, S., Sakai, A., Shelton, S., Baker, E., DeJesus, O., & Holden, J. (1994). Neurotoxicity of glucocorticoids in the primate brain. *Hormones and Behavior*, *28*(4), 336–348. http://dx.doi.org/10.1006/hbeh.1994.1030. S0018-506X(84)71030-0 [pii].

Vengrenyuk, Y., Nishi, H., Long, X., Ouimet, M., Savji, N., Martinez, F. O., … Fisher, E. A. (2015). Cholesterol loading reprograms the microRNA-143/145-myocardin axis to convert aortic smooth muscle cells to a dysfunctional macrophage-like phenotype. *Arteriosclerosis, Thrombosis, and Vascular Biology*, *35*(3), 535–546. http://dx.doi.org/10.1161/ATVBAHA.114.304029.

Volz, K. S., Jacobs, A. H., Chen, H. I., Poduri, A., McKay, A. S., Riordan, D. P., … Red-Horse, K. (2015). Pericytes are progenitors for coronary artery smooth muscle. *eLife*. *4*. http://dx.doi.org/10.7554/eLife.10036.

Wagenseil, J. E., & Mecham, R. P. (2009). Vascular extracellular matrix and arterial mechanics. *Physiological Reviews*, *89*(3), 957–989. http://dx.doi.org/10.1152/physrev.00041.2008. 89/3/957 [pii].

Wang, Z., Wang, D. Z., Pipes, G. C., & Olson, E. N. (2003). Myocardin is a master regulator of smooth muscle gene expression. *Proceedings of the National Academy of Sciences of the United States of America*, *100*(12), 7129–7134. http://dx.doi.org/10.1073/pnas.1232341100. 1232341100 [pii].

Wasteson, P., Johansson, B. R., Jukkola, T., Breuer, S., Akyurek, L. M., Partanen, J., & Lindahl, P. (2008). Developmental origin of smooth muscle cells in the descending aorta in mice. *Development*, *135*(10), 1823–1832.

Wilm, B., Ipenberg, A., Hastie, N. D., Burch, J. B., & Bader, D. M. (2005). The serosal mesothelium is a major source of smooth muscle cells of the gut vasculature. *Development*, *132*(23), 5317–5328. http://dx.doi.org/10.1242/dev.02141. 132/23/5317 [pii].

Xu, H., Hu, F., Sado, Y., Ninomiya, Y., Borza, D. B., Ungvari, Z., … Ballabh, P. (2008). Maturational changes in laminin, fibronectin, collagen IV, and perlecan in germinal matrix, cortex, and white matter and effect of betamethasone. *Journal of Neuroscience Research*, *86*(7), 1482–1500. http://dx.doi.org/10.1002/jnr.21618.

Xu, J., & Shi, G. P. (2014). Vascular wall extracellular matrix proteins and vascular diseases. *Biochimica et Biophysica Acta*, *1842*(11), 2106–2119. http://dx.doi.org/10.1016/j.bbadis.2014.07.008.

Yana, I., Sagara, H., Takaki, S., Takatsu, K., Nakamura, K., Nakao, K., … Seiki, M. (2007). Crosstalk between neovessels and mural cells directs the site-specific expression of MT1-MMP to endothelial tip cells. *Journal of Cell Science*, *120*(Pt. 9), 1607–1614. http://dx.doi.org/10.1242/jcs.000679.

Yang, D., Baumann, J. M., Sun, Y. Y., Tang, M., Dunn, R. S., Akeson, A. L., … Kuan, C. Y. (2013). Overexpression of vascular endothelial growth factor in the germinal matrix induces neurovascular proteases and intraventricular hemorrhage. *Science Translational Medicine*, *5*(193), 193ra190. http://dx.doi.org/10.1126/scitranslmed.3005794. 5/193/193ra90 [pii].

CHAPTER NINE

Notch Signaling in Vascular Smooth Muscle Cells

J.T. Baeten, B. Lilly[1]
The Center for Cardiovascular Research and The Heart Center at Nationwide Children's Hospital, The Ohio State University, Columbus, OH, United States
[1]Corresponding author: e-mail address: brenda.lilly@nationwidechildrens.org

Contents

Abstract

The Notch signaling pathway is a highly conserved pathway involved in cell fate determination in embryonic development and also functions in the regulation of physiological processes in several systems. It plays an especially important role in vascular development and physiology by influencing angiogenesis, vessel patterning,

Advances in Pharmacology, Volume 78
ISSN 1054-3589
http://dx.doi.org/10.1016/bs.apha.2016.07.002

arterial/venous specification, and vascular smooth muscle biology. Aberrant or dysregulated Notch signaling is the cause of or a contributing factor to many vascular disorders, including inherited vascular diseases, such as cerebral autosomal dominant arteriopathy with subcortical infarcts and leukoencephalopathy, associated with degeneration of the smooth muscle layer in cerebral arteries. Like most signaling pathways, the Notch signaling axis is influenced by complex interactions with mediators of other signaling pathways. This complexity is also compounded by different members of the Notch family having both overlapping and unique functions. Thus, it is vital to fully understand the roles and interactions of each Notch family member in order to effectively and specifically target their exact contributions to vascular disease. In this chapter, we will review the Notch signaling pathway in vascular smooth muscle cells as it relates to vascular development and human disease.

ABBREVIATIONS

ACTA2 alpha smooth muscle actin
CADASIL cerebral autosomal dominant arteriopathy with subcortical infarcts and leukoencephalopathy
ECD extracellular domain
ECM extracellular matrix
EGF epidermal growth factor-like
GOM granular osmiophilic material
HDACs histone deacetylases
HES hairy/enhancer of split
HRT hairy-related transcription factors
MAML mastermind-like
MAPK mitogen-activated protein kinase
NICD Notch intracellular domain
NRR negative regulatory region
PAH pulmonary arterial hypertension
PDA patent ductus arteriosus
PDGF-B platelet-derived growth factor B
PEST domain rich in proline [P], glutamic acid [E], serine [S], and threonine [T]
RBPJ recombination signal-binding protein for immunoglobulin kappa J region
TGFβ transforming growth factor β
VSMCs vascular smooth muscle cells
Wnt wingless-related integration site

1. INTRODUCTION

The Notch signaling pathway is an evolutionarily conserved pathway that plays a critical role in cell fate decisions for numerous systems in vertebrates (Artavanis-Tsakonas, Rand, & Lake, 1999; Ehebauer, Hayward, & Arias, 2006). Over the past ~15 years, many studies have demonstrated the importance of Notch signaling in regulating vascular development and physiology, through control of endothelial cells and smooth muscle cells and their interactions with each other. The overall contribution of Notch signaling to the vasculature and especially the role of endothelial cell-derived Notch signaling has been well studied and previously covered in several exceptional reviews (Fouillade, Monet-Lepretre, Baron-Menguy, & Joutel, 2012; Gridley, 2007, 2010; Iso, Hamamori, & Kedes, 2003; Phng & Gerhardt, 2009; Roca & Adams, 2007; Rostama, Peterson, Vary, & Liaw, 2014). In this review, we will focus on the role of Notch signaling in vascular smooth muscle cells (VSMCs) and how it regulates the development and homeostasis of the vasculature, and how when deficient or perturbed can contribute to disease states.

Originally discovered in *Drosophila* (and named for the "notched" wings that resulted from its mutation) (Morgan, 1917), the canonical Notch signaling pathway is relatively straightforward, consisting of a membrane-bound ligand and receptor engagement on neighboring cells leading to cleavage of the Notch intracellular domain (NICD) of the receptor, which shuttles to the nucleus and acts a transcriptional cofactor. This series of events allows for a signal to be passed directly from receptor activation to transcriptional modulation while bypassing the signaling intermediates and kinase cascades necessary for many other signaling pathways. This linear system coupled with its inherent feedback mechanisms allows for the creation of signal-sending: signal-receiving binary cell fate determinations in early developmental processes, such as those seen in the lateral specification of neural and epidermal cells in *Drosophila* (Heitzler & Simpson, 1991). However, beyond its role initially described in lower species, this basic signaling pathway has been adapted and utilized by nearly every cell type and organ system to enact a myriad of functions.

The cardiovascular system is the first organ system to develop in embryogenesis and is made up of the heart and an expansive network of vessels of different types and sizes (Patel-Hett & D'Amore, 2011). Early in development, angioblasts differentiate into endothelial cells and aggregate into a nascent vascular network in a process called vasculogenesis. This network is extended, pruned, and remodeled over the course of development to meet the needs of the developing embryo. To support the growing network of endothelial cell tubes, mural cells are recruited and differentiate into VSMCs or pericytes. These mural cells are vital to the maturation of vessels into functional arteries, veins, and capillaries. VSMCs and pericytes play countless roles in these vessels, from controlling blood flow to regulating permeability, and those roles shift depending upon the state of the vessel, embryological origin, and the environmental context. VSMCs are not "terminally differentiated" cells and are capable of adapting their activities in response to molecular and physiological cues. This ability of VSMCs to radically alter their functions in different contexts is known as VSMC phenotypic switching or plasticity. This concept has been well studied and reviewed (Alexander & Owens, 2012; Gomez & Owens, 2012; Owens, Kumar, & Wamhoff, 2004) and attributed to many of the activities of VSMCs in vascular diseases.

In VSMCs, the Notch receptors Notch2 and Notch3 and the ligand Jagged1 predominate (High et al., 2007; Tang et al., 2010). Notch1 and Notch4 are more highly expressed in endothelial cells, although Notch1 has been shown to play a role in VSMCs in the context of vascular injury (Li, Takeshita, et al., 2009) and in pericytes (Kofler, Cuervo, Uh, Murtomaki, & Kitajewski, 2015). Over the past few decades, extensive work has been performed to ascertain the role of these receptors and Notch signaling as a whole in VSMCs. While significant progress has been made and strong connections have linked the Notch pathway to vascular development, VSMC differentiation and phenotype, and vascular disease, there is still much that is not understood about the precise mechanisms involved (Fig. 1).

2. THE NOTCH SIGNALING PATHWAY

2.1 The Canonical Notch Signaling Pathway

In mammals, there are four Notch receptors (Notch1–4) and five ligands (Jagged1 and 2, Delta-like 1, 3, and 4) which are all Type I transmembrane proteins. Both the ligands and receptors have long extracellular domains (ECDs) made up of epidermal growth factor-like (EGF) repeats which serve as the binding platform for the receptor–ligand interaction. The receptor arrives at the cell membrane as a heterodimer after it is cleaved by furin at the site 1 (S1). After a receptor is bound by ligand, the negative regulatory region (NRR) is displaced, exposing the site 2 (S2) for cleavage by one of the ADAM metalloproteases (a disintegrin and metalloproteinase domain-containing protein), and the site 3 (S3) for

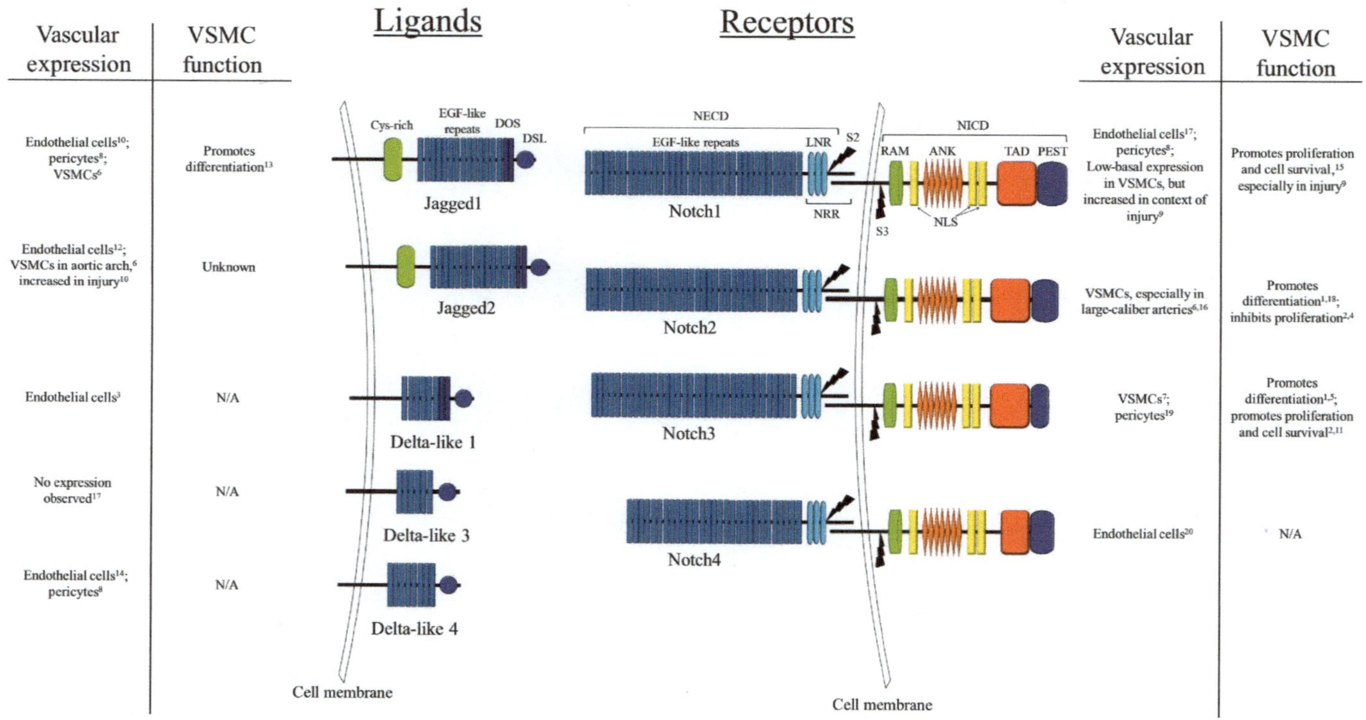

Fig. 1 Notch family of ligands and receptors. Abbreviations: *ANK*, ankyrin repeats; *Cys-rich*, cysteine-rich region; *DOS*, Delta and OSM11-like proteins motif; *DSL*, Delta/Serrate/Lag2 motif; *LNR*, Lin12-Notch repeats; *NECD*, Notch extracellular domain; *NICD*, Notch intracellular domain; *NLS*, nuclear localization signal; *NRR*, negative regulatory region; *PEST*, proline/glutamic acid/serine/threonine-rich motifs; *RAM*, RBPJ-association module; *S2*, cleavage site 2; *S3*, cleavage site 3; *TAD*, transactivation domain. References: 1, Baeten, Jackson, McHugh, and Lilly (2015); 2, Baeten and Lilly (2015); 3, Beckers, Clark, Wunsch, Hrabe De Angelis, and Gossler (1999); 4, Boucher, Harrington, Rostama, Lindner, and Liaw (2013); 5, Domenga et al. (2004); 6, High et al. (2007); 7, Joutel et al. (2000); 8, Kofler et al. (2015); 9, Li, Takeshita, et al. (2009); 10, Lindner et al. (2001); 11, Liu, Zhang, Kennard, Caldwell, and Lilly (2010); 12, Luo, Aster, Hasserjian, Kuo, and Sklar (1997); 13, Manderfield et al. (2012); 14, Shutter et al. (2000); 15, Sweeney et al. (2004); 16, Varadkar et al. (2008); 17, Villa et al. (2001); 18, Wang, Zhao, Kennard, and Lilly (2012); 19, Wang, Pan, Moens, and Appel (2014); 20, Wu et al. (2005).

cleavage by the γ-secretase complex, releasing the NICD fragment. After cleavage at the S3 site, the NICD's nuclear localization signal triggers translocation to the nucleus. The NICD acts as a transcriptional cofactor by forming a complex with the CSL (CBF1/SuH/Lag1) transcription factor named for its homologs in mammals, *Drosophila*, and *Caenorhabditis elegans*, though it is now designated RBPJ (recombination signal-binding protein for immunoglobulin kappa J region) in mammals. This complex binds to its target DNA sequence and to one of the three mastermind-like (MAML) proteins, which act as transcriptional activators. The formation of this complex displaces transcriptional repressors including SKIP (Ski-interacting protein), CIR (CBF1-interacting corepressor), KyoT2, and SHARP (SMRT- and HDAC-associated repressor) and histone deacetylases (HDACs) from RBPJ, leading to expression of Notch target genes (Borggrefe & Oswald, 2009). Among the most highly activated genes are a family of basic helix–loop–helix (bHLH) proteins hairy/enhancer of split (HES) and HES-related proteins (HEY/HRT/HERP) which act as transcriptional repressors.

Though the transduction of the Notch signal via NICD cleavage and nuclear translocation is well established, there are many opportunities along its path for regulation to occur that introduces signaling diversity. One such mechanism is the Fringe family of glycosyltransferases, Lunatic Fringe, Manic Fringe, and Radical Fringe (Moloney et al., 2000). Both Notch ligands and receptors have attached sugar moieties on the EGF repeats of the ECDs, and Fringe glycosyltransferases are able to elongate N- and O-linked glycans at certain positions during posttranslational processing in the Golgi. The length and position of these glycans can determine whether ligands and receptors interact between neighboring cells (*trans* activation) or within the same cell (*cis* inhibition) (LeBon, Lee, Sprinzak, Jafar-Nejad, & Elowitz, 2014). The glycosylation state can also influence the binding affinities for the different receptor–ligand pairs (Hicks et al., 2000; Stanley & Okajima, 2010). This influence on which receptor and ligand interact can then control ultimate effect of Notch signaling, as the different Notch ligands and Notch receptors are capable of producing different signaling outcomes. For instance, the ligands JAG1 and DLL4 have opposing effects on angiogenesis and their activity is regulated by their glycosylation (Benedito et al., 2009), and DLL4, but not DLL1, can induce pericyte differentiation in myoblasts (Cappellari et al., 2013). Also, the NICDs of the different Notch receptors can activate transcription of their target genes with varying efficacy depending on the orientation and number of RBPJ-binding sites in the promoter (Ong et al., 2006). Another potential mechanism of Notch signal diversity is the control of the intensity and duration of the Notch signal. It has been demonstrated that some Notch target genes can be activated by low levels of active NICD, while others require high levels of activity (Ong et al., 2006). The number of NICD molecules present in the nucleus is dependent on both the influx of NICD from receptor activation and the turnover or degradation of NICD. The degradation of NICDs is regulated by the E3 ubiquitin ligases (Lai, 2002) which ubiquitinate NICDs in their PEST domain (rich in proline [P], glutamic acid [E], serine [S], and threonine [T]) and target them for degradation by the proteasome (Fig. 2).

2.2 Interaction with Other Signaling Pathways and Noncanonical Signaling

In addition to the long understood canonical Notch pathway, there is increasing evidence that the Notch pathway can exert effects outside of its prominent role as a transcriptional cofactor (Andersen, Uosaki, Shenje, & Kwon, 2012; Ayaz & Osborne, 2014; Martinez Arias, Zecchini, & Brennan, 2002; Sanalkumar, Dhanesh, & James, 2010). An example of this is the ability of the NICD to bind active β-catenin, leading to its degradation by the lysosome (Kwon et al., 2011). Though non-canonical activities of Notch have been shown in many different cell types and systems, evidence for these activities in VSMCs is relatively sparse and the mechanisms not well defined (Wang, Prince, Mou, & Pollman, 2002). However, based on the ubiquity of noncanonical signaling in other contexts, it seems likely that some of the effects of Notch signaling in VSMCs are due to undiscovered noncanonical interactions.

Most of the observed noncanonical signaling involves interactions of Notch components with members of other signaling pathways, but the overlap of Notch signaling and other pathways is not limited to the non-canonical pathway. Signals from outside of the Notch pathway can regulate Notch expression and activity as well as alter the transcription factors recruited to Notch target genes. Notch signaling can also in turn regulate the expression and activity of other signaling pathways. Much of the diversity in Notch signaling outcomes can be attributed to its interaction with other signaling pathways. Here we will discuss some of the known interactions with other signaling pathways in VSMCs.

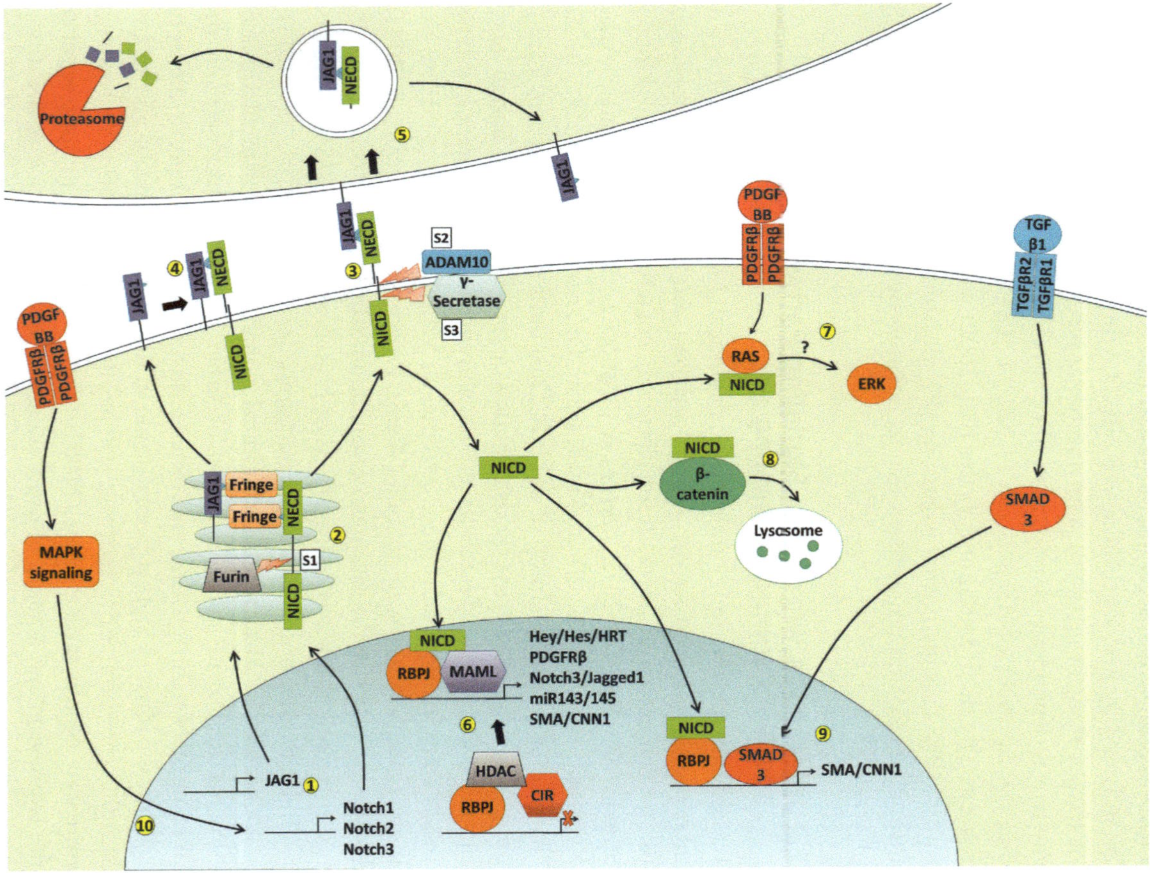

Fig. 2 See legend on opposite page.

2.2.1 Platelet-Derived Growth Factor B

Platelet-derived growth factor B (PDGF-B) is a major mitogen and chemotractant in VSMCs that acts through its receptor, PDGFRβ (Donovan, Abraham, & Norman, 2013). PDGF signaling has been shown to interact with Notch signaling in VSMCs in two major capacities. First, Notch signaling has been shown to directly promote the transcription of PDGFRβ and enhance the signaling effects of PDGF-B (Donovan et al., 2013; Jin et al., 2008). This relationship is especially apparent in pericytes, as both PDGFRβ and Notch3 are considered markers of pericytes and their dysfunction in animal models and humans can produce similar phenotypes (Lee, 2013; Martignetti et al., 2013; Wang et al., 2014). Second, it has been shown that PDGF-B signaling can regulate the expression of Notch components (Baeten & Lilly, 2015; Campos, Wang, Pollman, & Gibbons, 2002). Overall, it is apparent that the PDGF and Notch signaling pathways are strongly intertwined, with significant functional overlap in pericytes and coregulation in VSMCs.

2.2.2 Transforming Growth Factor β

The transforming growth factor β (TGFβ) pathway is a vital contributor to the differentiation of VSMCs (Guo & Chen, 2012). The Notch and TGFβ signaling pathways cooperatively regulate the differentiation of VSMCs through direct interaction of NICD and SMAD2/3 (a transcription factor within the TGFβ pathway) and transcription of smooth muscle contractile

Fig. 2 Notch signaling pathway. (1) Notch receptor and ligand genes are transcribed and translated into protein. (2) During posttranslational processing in the Golgi, Notch receptors are cleaved by furin to create heterodimers; both ligands and receptors are glycosylated on their EGF-like repeats by Fringe family enzymes. (3) *trans* activation of Notch receptor by Notch ligand from a neighboring cell, triggering cleavage by ADAM10 at S2 and γ-secretase at S3, releasing the NICD. (4) *cis* inhibition of Notch receptor by Notch ligand on the same cell, preventing activation. (5) Endocytosis of Notch ligand with associated NECD can be degraded by the proteasome or recycled back to the cell surface. (6) NICD is shuttled to the nucleus where it binds to RBPJ, displacing HDACs and corepressors such as CIR and recruiting MAML to activate transcription of Notch target genes. (7) NICD can both directly interact with Ras and promote ERK phosphorylation, though the mechanism is not yet known. (8) NICD can bind to the Wnt signaling mediator β-catenin and target it for degradation in the lysosome. (9) Transcription of some genes, including smooth muscle contractile genes, is coregulated by NICD/RBPJ and the TGFβ transcription factor SMAD3. (10) Many signaling pathways can regulate expression of Notch components, including PDGFRβ signaling via the MAPK pathway.

genes (Tang et al., 2010). This cooperative induction of VSMC differentiation has also been shown in progenitor cell types (Grieskamp, Rudat, Ludtke, Norden, & Kispert, 2011; Kurpinski et al., 2010), and the direct interaction of NICD and SMAD3 as transcriptional activators has been shown in other systems (Blokzijl et al., 2003). Interestingly, it has also been shown that these two pathways can negatively regulate each other's expression. This is seen in fibroblasts where TGFβ signaling decreases expression of Notch3 (Kennard, Liu, & Lilly, 2008) and in VSMCs where Notch signaling promotes the transcription of miR145 which inhibits the TGFβ pathway by targeting the TGFβ receptor 2 (Boucher, Peterson, Urs, Zhang, & Liaw, 2011; Zhao et al., 2015). Taken together, the Notch and TGFβ pathways are able to both cooperatively promote transcription of their downstream genes and inhibit each other's expression, suggesting a very complex relationship and potentially a mechanism for negative feedback within each pathway.

2.2.3 Mitogen-Activated Protein Kinase

The mitogen-activated protein kinase (MAPK) pathway is one of the major signaling pathways controlling cell growth, survival, and many other processes and is a main transducer of growth factor signals (Pearson et al., 2001). In cultured VSMCs, Notch3 is capable of activating this pathway by phosphorylation of the MAPK component ERK (Baeten & Lilly, 2015; Wang, Campos, Prince, Mou, & Pollman, 2002), through a mechanism that is not yet known. Physical interaction of Notch and MAPK components has previously been shown in other models (Hodkinson et al., 2007). However, it may simply be a result of upregulation of growth factor ligands or receptors by Notch signaling, such as PDGFRβ. Regardless of how it is enacting this effect, it is clear that at least in some contexts Notch signaling is capable of promoting MAPK pathway activity.

2.2.4 Wingless-Related Integration Site

The Wnt (wingless-related integration site) signaling pathway is an important developmental pathway that regulates smooth muscle proliferation, migration, and survival (Mill & George, 2012). The Wnt and Notch pathways have significant cross talk in many systems, by both direct effect on each other's pathway components and shared regulation of developmental processes (Andersen et al., 2012; Duncan et al., 2005; Espinosa, Ingles-Esteve, Aguilera, & Bigas, 2003; Fre et al., 2009; Hayward et al., 2005). These interactions have been shown to extend to VSMCs as well, as early mesoderm cells in chick embryos are pushed toward a VSMC lineage through cooperation of the Notch and Wnt pathway (Shin, Nagai, & Sheng, 2009). While evidence for Wnt/Notch pathway interaction in VSMCs is limited, the results in mesodermal smooth muscle progenitors coupled with the known cross talk between Notch and Wnt signaling in other systems suggest that these pathways cooperate to drive early VSMC differentiation.

3. NOTCH SIGNALING IN VSMC DEVELOPMENT

Notch signaling plays an essential role in the development of the cardiovascular system, from hematopoiesis to cardiac development to endothelial cell sprouting and vasculogenesis. In the development and differentiation of VSMCs, the functions of Notch signaling can be categorized into two main roles: construction of the vessel wall and arterial–venous specification.

3.1 Constructing a Vessel Wall

Following the initial stages of vasculogenesis, nascent vessels are composed of an endothelial tube surrounded by perivascular cells. For a vessel to mature into a conduit capable of withstanding increasing pressure and volume demands, it must recruit mural cells to encompass it and differentiate into one or more layers of vascular smooth muscle. This recruitment is largely driven by a PDGF-B gradient secreted by the endothelial cells (Hellstrom, Kalen, Lindahl, Abramsson, & Betsholtz, 1999). Once recruited to the nascent vessel, endothelial cell-derived Notch ligands can activate Notch signaling in these mural cells, which induces integrin adhesion to the endothelial basement membrane and initiates maturation and differentiation (Scheppke et al., 2012). We have learned a great deal about how endothelial-derived Notch signals affect VSMCs and mural cells from mutant mouse models and in vitro systems. In the great vessels of the outflow tract, cardiac neural crest progenitors give rise to the VSMCs within the vessel walls. With the complete abrogation of Notch signaling by "knocking in" of a dominant-negative MAML1 in cardiac neural crest cells, mice display various defects of the outflow tract and aortic arch associated with decreased differentiation and expression of smooth muscle markers (High et al., 2007). Mice with an endothelial deletion of Jagged1 also display decreased VSMC differentiation and expression of smooth muscle markers (High et al., 2008). Through a combination of in vitro coculture experiments (Liu, Kennard, & Lilly, 2009) and the cardiac neural crest-specific Jagged1 knockout mice (Manderfield et al., 2012), it was demonstrated that induction of Notch signaling in VSMCs by endothelial-expressed Jagged1 leads to the increased expression of both Notch3 and Jagged1. This allows for lateral induction of Notch signaling by homotypic VSMC–VSMC interactions through multiple layers of smooth muscle to promote VSMC differentiation after the initial signal is received from endothelial-derived Jagged1 (Hoglund & Majesky, 2012). This model has been further supported by smooth muscle deletion of Jagged1, RBPJ, and a smooth muscle deletion of Notch2 coupled with heterozygous deletion of Notch3 (Baeten et al., 2015; Feng, Krebs, & Gridley, 2010; Krebs, Norton, & Gridley, 2016). All three of these models develop patent ductus arteriosus (PDA) due to contractile deficits of the VSMCs and present decreased expression of smooth muscle markers. An interesting parallel exists between this model and one proposed from experiments performed by the Krasnow lab (Greif et al., 2012), where they demonstrate that the expression of PDGFRβ and ACTA2 (alpha smooth muscle actin) moves through layers of smooth muscle like a wave. They describe a major role for endothelial-derived PDGF-B in the migration, orientation, and proliferation of these smooth muscle cell layers and suggest that there may be another signaling pathway contributing to the radial pattern of construction. Indeed, Notch signaling would elegantly fit into this proposed model, as it is known to induce PDGFRβ and ACTA2, and could be passed through the layers of smooth muscle via cell–cell contact (Jin et al., 2008; Noseda et al., 2006).

An interesting finding from comparing the knockout phenotypes of mice deficient in Notch2 and Notch3 is their unique roles in different vascular beds. Notch2 is mainly expressed in the outflow tract and other large-caliber vessels, while Notch3 is expressed in all mural cells (High et al., 2007; Joutel et al., 2000). Likewise, in large-caliber vessels, Notch2 signaling is indispensable, with loss of Notch3 only acting to amplify defects in Notch2-null mice (McCright et al., 2001; Wang et al., 2012). This demonstrates that while there may be some functional redundancy of Notch2/3 in large-caliber vessels, Notch3 cannot completely replace Notch2 function. However, in pericytes and VSMCs in smaller caliber vessels where Notch2 expression is significantly diminished, Notch3 plays a major role, especially in the cerebral vasculature (Henshall et al., 2015; Liu et al., 2010; Volz et al., 2015; Wang et al., 2014).

3.2 Arterial–Venous Specification

Another important role for Notch signaling in VSMCs in the developing vasculature is the specialization of vessels into arteries and veins. Arteries and veins have distinct structural properties and express distinct protein markers that contribute to their unique functions in the cardiovascular system (Wang, Chen, & Anderson, 1998). In endothelial cells, Notch signaling regulates arterial specification in cooperation with VEGF (Hirashima, 2009; Kim et al., 2008). Similarly, Notch3 regulates arterial specification in VSMCs. In a mouse model with global knockout of Notch3, arteries acquired a more venous phenotype, with enlarged vessels, thinner and disorganized smooth muscle layers, and less festooned elastic lamina (Domenga et al., 2004). They did not see these changes in the outflow tract or aorta where Notch2 is highly expressed, likely suggesting a functional redundancy for Notch2 and Notch3 in this context. The expression of the arterial marker smoothelin and a LacZ reporter driven by an arterial-specific promoter element was also reduced in these Notch3-null mice, showing both structural and morphological defects in arterial specification. In zebrafish, Notch3 expression was found to be restricted to arteries, and inhibition of Notch signaling in zebrafish leads to decreased arterial specification and similar phenotypes compared to the mouse model (Lawson et al., 2001).

4. NOTCH SIGNALING AND VSMC PHENOTYPE

The ability of VSMCs to display various phenotypes and enact disparate functions depending on their cellular context, known as phenotypic switching or plasticity, is central to their changing roles from early developmental vasculogenesis to vascular homeostasis and remodeling in adult animals. Notch signaling has been closely tied to the many possible phenotypic endpoints of VSMCs. While there are still disparities in the literature about some of the phenotypic effects of Notch signaling in VSMCs, we are starting to obtain a clearer picture of how Notch activation drives VSMCs to different endpoints and the contributions from the different Notch receptors.

4.1 Differentiation

A "differentiated" VSMC is generally described as a quiescent cell expressing contractile proteins (such as ACTA2, smooth muscle myosin heavy chain [MYH11], transgelin [SM22α], and calponin [CNN1]) and primed to provide contractile force to constrict (or dilate) the vessel in which it resides in response to external cues. This is the state necessary for mature vessels to maintain physiological homeostasis. Notch signaling has been tied to regulation of VSMC differentiation and expression of contractile proteins for some time. While initial studies in vitro and in injury models showed that Notch-activated genes of the HRT (hairy-related transcription factors) family could repress myocardin-induced contractile protein expression

(Doi et al., 2005; Proweller, Pear, & Parmacek, 2005; Yoshida et al., 2003), there were also several studies, showing that Notch activity promoted VSMC differentiation and contractile marker expression (Doi et al., 2006; Domenga et al., 2004; High et al., 2007). Although the discrepancy has not been conclusively explained, it is likely that the observed repression of contractile genes by the HRT family is negative feedback, as this family of repressors interferes with the transcription of many Notch target genes, including the HRT genes themselves (King et al., 2006; Nakagawa et al., 2000; Takebayashi et al., 1994). Notch activity has since been shown to directly promote the transcription of ACTA2 (Noseda et al., 2006), and Notch-induced transcription of ACTA2 was repressed by HRT transcription factors (Tang, Urs, & Liaw, 2008). Overall, the preponderance of data supports the hypothesis that Notch signaling in VSMCs promotes contractile differentiation (Baeten et al., 2015; Boscolo et al., 2011; Boucher, Gridley, & Liaw, 2012; Doi et al., 2006; Domenga et al., 2004; High et al., 2008, 2007; Lin & Lilly, 2014b; Liu et al., 2010; Meng, Zhang, Lee, & Wang, 2013; Noseda et al., 2006; Tang et al., 2010; Wang et al., 2012), which fits with our understanding of its role in development of the vessel wall and response to contact-induced signaling with endothelial cells (Baeten et al., 2015; Gaengel, Genove, Armulik, & Betsholtz, 2009; High et al., 2008; Hoglund & Majesky, 2012; Lilly & Kennard, 2009; Lin & Lilly, 2014a, 2014b; Liu et al., 2009; Manderfield et al., 2012; Regan & Majesky, 2009; Wang et al., 2012).

4.2 Proliferation

The proliferation of VSMCs is negatively tied to their differentiation status, so much so that they are often discussed as polar opposites, where mitotically active cells are considered "dedifferentiated" or "synthetic" and differentiated VSMCs are assumed to be quiescent, though there is likely a range of intermediate phenotypes between these extremes (Owens et al., 2004). Since Notch signaling in VSMCs promotes contractile differentiation, one would assume cells with active Notch signaling would not be actively proliferating. However, the initial data suggested the opposite, with several papers showing that Notch signaling promoted proliferation in vitro (Campos et al., 2002; Havrda, Johnson, O'Neill, & Liaw, 2006; Sweeney et al., 2004) and in the context of vascular injury in vivo (Li, Takeshita, et al., 2009; Wang, Campos, et al., 2002). In 2013, findings from Lucy Liaw's lab demonstrated that the Notch2 receptor specifically inhibited proliferation via cell-cycle arrest and Notch2 was localized in non-proliferating regions of injured vessels, suggesting a receptor-specific regulation of VSMC proliferation (Boucher et al., 2013). Indeed, the previous data demonstrating promotion of proliferation were largely focused on the activity of Notch1 and/or Notch3 (Havrda et al., 2006; Li, Takeshita, et al., 2009; Sweeney et al., 2004; Wang, Campos, et al., 2002). Delving further into receptor-specific roles, our own lab demonstrated that manipulation of Notch2 and Notch3 had opposite effects on PDGF-B-dependent proliferation, and that PDGF-B treatment decreased expression of Notch2, but increased expression of Notch3 (Baeten & Lilly, 2015). Interestingly, while Notch2 and Notch3 have disparate roles in regulating VSMC proliferation, they both promote contractile differentiation (Baeten et al., 2015). Together these results indicate that regulation of the expression of the different Notch receptors in VSMCs could have a large impact on phenotypic outcome.

4.3 Survival

The role of Notch signaling in VSMC survival and apoptosis has drawn significant attention due to the human disease CADASIL (cerebral autosomal dominant arteriopathy with subcortical infarcts and leukoencephalopathy), which is involves the loss of VSMCs and pericytes from cerebral vessels and is caused by mutation of the Notch3 receptor (which will be discussed further in the next section). Notch signaling has been implicated in suppression of apoptosis in a number of cell types (Dror et al., 2007; Jehn, Bielke, Pear, & Osborne, 1999; Meurette et al., 2009; Rosati et al., 2009), especially in many forms of cancer (Dang, 2012). This role as an inhibitor or apoptosis and promoter of cell survival appears to be expressed in VSMCs and pericytes as well (Arboleda-Velasquez et al., 2014; Baeten & Lilly, 2015; Henshall et al., 2015; Li, Takeshita, et al., 2009; Liu et al., 2010; Sweeney et al., 2004; Wang, Campos, et al., 2002; Wang et al., 2014; Wang, Prince, et al., 2002). Interestingly, the Notch receptors responsible for promotion of cell survival in mural cells are also those implicated in the promotion of proliferation, with Notch1 and Notch3 promoting survival, but with no known role for Notch2 (Baeten & Lilly, 2015; Henshall et al., 2015; Li, Takeshita, et al., 2009; Liu et al., 2010; Sweeney et al., 2004; Wang, Campos, et al., 2002; Wang et al., 2014; Wang, Prince, et al., 2002). Direct comparison of the effects of Notch2 and Notch3 manipulation in vitro on VSMC survival revealed that Notch3 promoted survival, but Notch2 had no effect (Baeten & Lilly, 2015). Notch3 activity correlates with MAPK/ERK pathway activation and with survival gene expression, which could be abrogated by inhibition of the MAPK/ERK pathway. This suggests that Notch3 may be promoting VSMC survival through a noncanonical association with the MAPK pathway, which had previously been suggested (Wang, Prince, et al., 2002). Overall, Notch activity promotes survival in VSMCs, but this effect is receptor specific and involves a noncanonical interaction with the MAPK pathway.

4.4 Extracellular Matrix Synthesis

VSMCs and their precursors are predominant contributors to the production of the extracellular matrix (ECM) that comprises the basement membrane and supports tensile strength and elasticity in the vessel (Armulik, Genove, & Betsholtz, 2011; Underwood, Bean, & Whitelock, 1998; Wagenseil & Mecham, 2009). The ability of Notch signaling to promote ECM synthesis has been demonstrated in fibroblasts, where it induces collagen production (Dees et al., 2011; Lilly & Kennard, 2009), and in renal podocytes, where inhibition of Notch signaling reduced expression of collagen IV and laminin, but increased expression of MMP-2 and MMP-9 (Yao, Wang, Zhang, Chi, & Gao, 2015). This relationship is maintained in VSMCs, where coculture studies identified that endothelial-derived Notch signaling in VSMCs promotes collagen synthesis and other markers of a synthetic phenotype (Lilly & Kennard, 2009; Lin & Lilly, 2014b). This may be a recapitulation of the developmental program in early vessel development, wherein Notch and other synthetic signaling pathways (such as TGFβ) induce ECM synthesis in perivascular cells to produce the framework of the vessel wall.

4.5 Migration

The migration of VSMCs is primarily important in two situations: recruitment of mural cells to the developing vessel and in response to vessel injury (Hellstrom et al., 1999). The main chemotractant for VSMCs is PDGF-B secreted by endothelial cells as well as platelets and macrophages (Hellstrom et al., 1999; Shimokado et al., 1985). Because Notch signaling has been shown to increase expression of the PDGF receptor β (Jin et al., 2008), it would follow that active Notch signaling in VSMCs would increase migration in response to PDGF-B. Indeed, much of the work done on

Notch and VSMC migration has shown an induction of migratory ability. Similar to their roles in proliferation, Notch1 and Notch3 were shown to promote VSMC migration in vitro (Sweeney et al., 2004) and in a mouse injury model Notch1 was shown to be important for migration, proliferation, and formation of the neointimal layer (Li, Takeshita, et al., 2009). The Notch-responsive gene HEY2 was also shown to promote proliferation, migration, and neointimal formation (Sakata et al., 2004), and Notch signaling has been shown to regulate expression of MMP-2 and MMP-9, which are important in the matrix remodeling necessary for migration through the vessel wall rich in ECM (Delbosc et al., 2008). Notch signaling has also been shown to regulate cellular adhesions between cells via N-cadherin (Li et al., 2011; Lindner et al., 2001) and to the surrounding matrix via integrins (Scheppke et al., 2012). These results support the idea that Notch signaling is important for recruitment of mural cells to the vessel wall in response to a PDGF-B gradient, and maintaining their position via cellular adhesions.

5. NOTCH SIGNALING IN VASCULAR DISEASE

The various roles of Notch signaling in regulating smooth muscle development and phenotypic modulation make it an obvious player in many vascular diseases associated with VSMCs. These include both inherited genetic diseases and ailments brought on by defective physiological homeostasis (Table 1).

5.1 Cerebral Autosomal Dominant Arteriopathy with Subcortical Infarcts and Leukoencephalopathy

Perhaps the most studied Notch-related disease in VSMCs, CADASIL, is caused by mutations in the Notch3 gene through a mechanism we still do not fully understand (Joutel et al., 1996). This disorder presents in the patient as migraines, mood disorders, ischemic attacks, cognitive impairment, and stroke and is marked by lesions within the cerebral and systemic vasculature characterized by accumulation of granular osmiophilic material (GOM) at smooth muscle basement membrane and progressive loss of VSMCs (Chabriat et al., 2009; Joutel et al., 2000, 1997). These GOM deposits appear to be comprised of oligomerized Notch3 ECD, and the mutations found in human patients lead to a gain or loss of a cysteine residue within the EGF-like repeats of the ECD (Joutel et al., 1997). It is not clear what exactly causes the disease progression of CADASIL, whether it is impaired function of Notch3, aberrant effects of GOM deposition, or a

Table 1 Vascular Diseases Associated with Mutations of Notch Pathway Components

Notch Component	Mutation Site	Notch Signaling Effect	Disease Outcome
Jagged1	Identified in every exon and several splice sites	Loss of function	Alagille syndrome[1]
	EGF-like repeat 2	Not known	Tetralogy of Fallot[2]
DLL4	DSL domain, N-terminal domain, and EGF-like repeats	Predicted loss of function	Adams–Oliver syndrome[3]
Notch1	Haploinsufficiency, EGF-like repeat 11, LNR 2, ANK domain 3	Predicted loss of function	Adams–Oliver syndrome[4]
	Haploinsufficiency, EGF-like repeat 29	Predicted loss of function	Bicuspid aortic valve[5]
Notch2	Truncated NICD	Predicted loss of function	Alagille syndrome[6]
	PEST domain	Predicted gain of function	Hajdu–Cheney[7]
Notch3	EGF-like repeats 1–32	Likely loss of function	CADASIL[8-11]
	PEST domain	Predicted gain of function	Lehman[12]
	Heterodimerization domain	Predicted gain of function	Infantile myofibromatosis[13]
	EGF-like repeats 21 and 23	Not known	Childhood PAH[14]
RBPJ	DNA-binding domain	Loss of Function	Adams–Oliver syndrome[15]
EOGT	Domains not defined	Predicted loss of function	Adams–Oliver syndrome[16]

References: 1, Spinner et al. (2001); 2, Eldadah et al. (2001); 3, Meester et al. (2015); 4, Stittrich et al. (2014); 5, Garg et al. (2005); 6, McDaniell et al. (2006); 7, Simpson et al. (2011); 8, Chabriat, Joutel, Dichgans, Tournier-Lasserve, and Bousser (2009); 9, Joutel et al. (2000); 10, Joutel et al. (1996); 11, Joutel et al. (1997); 12, Gripp et al. (2015); 13, Lee (2013); 14, Chida et al. (2014); 15, Hassed et al. (2012); 16, Shaheen et al. (2013).

combination of both. Mouse models expressing Notch3 receptors containing disease causing mutations have had mixed results in recreating CADASIL disease states. A knockin model of a CADASIL Notch3 mutation did not develop disease phenotypes or GOM deposits (Lundkvist et al., 2005), but another mouse with ectopically overexpressed mutant Notch3 protein did produce GOM deposits and mirrored many of the CADASIL disease symptoms (Joutel et al., 2010). Notch3-null mutant mice are viable and fertile and do not show any overt symptoms associated with CADASIL (Krebs et al., 2003), though they do exhibit some disruption of the VSMCs in cerebral arteries (Henshall et al., 2015). It has also been shown that hypomorphic mutations of Notch3 in humans do not cause CADASIL (Rutten et al., 2013). These results would seem to suggest that it is the gained aberrant function of GOM deposits that causes CADASIL phenotypes. However, a homozygous-null Notch3 mutation has been shown to produce a recessive early-onset arteriopathy and cavitating leukoencephalopathy very similar to CADASIL without the deposition of GOM (Pippucci et al., 2015), and a mouse model with combined deficiency of Notch1 and Notch3 causes pericyte dysfunction and models some phenotypes of CADASIL (Kofler et al., 2015). While it is still difficult at this point to definitively rule out any of the possible explanations, it seems likely that impaired Notch3 signaling plays at least some role in the CADASIL phenotypes, based on the known function of Notch3 in regulating cell survival in VSMCs and pericytes (Baeten & Lilly, 2015; Liu et al., 2010).

5.2 Patent Ductus Arteriosus

PDA results from the failure of the ductus arteriosus (DA), which diverts blood around the fetal lungs, to close shortly after the first breath and inflation of the lungs (Schneider & Moore, 2006). There are several signaling pathways involved in regulating the timing of this closure, but ultimately, it relies on the VSMCs of the DA to contract and functionally close the vessel. Because of this, defects in contractile proteins or VSMC differentiation often result in PDA (Guo et al., 2007; Huang et al., 2008; Zhu et al., 2006). Since Notch signaling promotes VSMC differentiation, it is not surprising that certain mouse models of Notch deficiency also present with the PDA phenotype (Baeten et al., 2015; Feng et al., 2010; Krebs et al., 2016). This connection is also seen in human disease, where a subset of patients afflicted with Alagille, Adams–Oliver, or Hajdu–Cheney syndromes, all caused by mutation of Notch family members, present with

PDA (Battelino, Writzl, Bratanic, Irving, & Novljan, 2016; Li et al., 1997; Stittrich et al., 2014).

5.3 Alagille Syndrome

Alagille syndrome is an autosomal dominant disorder caused by loss of Notch2 or Jagged1 and is characterized by many developmental defects, including those of the liver, heart, skeleton, eye, face, and kidney (Li et al., 1997; McDaniell et al., 2006). While many of these defects are caused by the roles of Notch signaling in other cell types, there are some interesting vascular phenotypes that could be related to dysfunctional Notch signaling in VSMCs. In addition to the PDA previously discussed, Alagille syndrome patients often present with defects of the cardiac outflow tract and great vessels (McElhinney et al., 2002). These defects are similarly modeled in mice with abrogated Notch signaling in the cardiac neural crest lineage that gives rise to the VSMCs of the outflow tract and great vessels (High et al., 2007).

5.4 Pulmonary Arterial Hypertension

Pulmonary arterial hypertension (PAH) is characterized by a persistent elevation of pulmonary arterial pressure, pulmonary arterial remodeling, and right ventricular hypertrophy (Crosswhite & Sun, 2014). This pulmonary arterial remodeling presents as increased VSMC proliferation, survival, and expression of contractile proteins such as ACTA2. Notch signaling has been proposed to regulate the pathogenesis of PAH, as elevated Notch3 expression is one of the hallmarks of the disease (Crosswhite & Sun, 2014; Geraci et al., 2001), and this would fit with the proliferation, survival, and differentiation roles that have been shown for Notch3. Two Notch3 mutations were identified in cases of childhood PAH, and in vitro experiments using the mutant proteins demonstrated increased proliferation (Chida et al., 2014). This role for Notch3 in PAH is further confirmed by the discovery that Notch3-null mice do not develop PAH in a hypoxia-induced PAH mouse model and that treatment with the Notch inhibitor DAPT (Li, Zhang, et al., 2009), or with a soluble Jagged1 ligand, which functions as a Notch antagonist, could prevent disease progression (Xiao, Gong, & Wang, 2013).

5.5 Infantile Myofibromatosis

Infantile myofibromatosis is an unusual mesenchymal malignancy that is seen at or near birth and typically spontaneously regresses in early childhood

(Hatzidaki et al., 2001). Although the tumors rarely present any significant harm to the patient, they are of interest to us because of their causative mutations: PDGFRβ and Notch3 (Lee, 2013; Martignetti et al., 2013). These mutations are predicted to create constitutively active forms of the proteins (Lee, 2013), which fits our current understanding of the relationship between Notch3 and PDGFRβ in promotion of VSMC proliferation (Baeten & Lilly, 2015; Jin et al., 2008).

5.6 Vascular Injury

In situations of vascular injury, many signaling pathways are activated to repair and remodel the vessel. However, these changes can become pathological if not properly regulated and can lead to neointimal hyperplasia, an aggressive migration, and proliferation of VSMCs into the lumen of the vessel (Wu & Zhang, 2009). Expression of members of the Notch signaling pathway, including Notch1, Notch3, HEY1, HEY2, and JAG1, is regulated in response to vascular injury, with an initial downregulation, followed by upregulation after several days (Gridley, 2010; Li, Takeshita, et al., 2009; Lindner et al., 2001). Several of these members have been specifically shown to regulate migration and proliferation of the neointima in injury models, including Notch1, Notch3, and HEY2 (Li, Takeshita, et al., 2009; Sakata et al., 2004; Wang, Campos, et al., 2002). Overall, the general consensus is that the Notch signaling pathway drives neointimal formation after vascular injury and is therefore a potential pharmacological target in situations of restenosis or other vascular injuries.

6. CONCLUSION

Our understanding of the molecular mechanisms of Notch signaling in VSMCs is constantly expanding, and in the last few years we have made great strides in uncovering some of the unique functions of the different Notch family members. There is still much we do not know, however, both in what roles each Notch component has in VSMC development and disease, and the precise mechanisms that allow for functional divergence of the Notch signal. How do the different receptors enact different functional outcomes while utilizing common cofactors? Work from the Gridley lab recently showed that the NICDs of Notch1 and Notch2 are functionally equivalent in development by creating Notch1/2 chimeric proteins (Liu et al., 2015), despite the known differences in Notch1 and Notch2 discussed in this review, especially in regard to VSMC proliferation. Can these differences be explained by different expression profiles and ligand affinities alone? Or are there noncanonical interactions yet to be discovered? The Kitajewski lab has developed decoys that specifically target DLL/Notch or JAG/Notch ligand–receptor interactions, and demonstrated their different functions in tumor angiogenesis (Kangsamaksin et al., 2015). Further use of these tools in VSMCs will go a long way toward understanding the differing functionalities of Notch ligand–receptor pairs. Additionally, the advent of CRISPR technology will greatly expedite our ability to make compound and chimeric Notch receptor mouse models to pinpoint the precise source of the unique functions of the different Notch receptors. These are some of the most pressing questions for the Notch field to answer in the coming years as we strive to uncover the intricacies of Notch signaling in vascular development and disease.

CONFLICT OF INTERESTS

We have no conflicts of interest to report.

ACKNOWLEDGMENTS

We would like to acknowledge the Center for Cardiovascular Research and the Heart Center at Nationwide Children's hospital for their support.

REFERENCES

Alexander, M. R., & Owens, G. K. (2012). Epigenetic control of smooth muscle cell differentiation and phenotypic switching in vascular development and disease. *Annual Review of Physiology*, *74*, 13–40. http://dx.doi.org/10.1146/annurev-physiol-012110-142315.

Andersen, P., Uosaki, H., Shenje, L. T., & Kwon, C. (2012). Non-canonical Notch signaling: Emerging role and mechanism. *Trends in Cell Biology*, *22*(5), 257–265. http://dx.doi.org/10.1016/j.tcb.2012.02.003.

Arboleda-Velasquez, J. F., Primo, V., Graham, M., James, A., Manent, J., & D'Amore, P. A. (2014). Notch signaling functions in retinal pericyte survival. *Investigative Ophthalmology & Visual Science*, *55*(8), 5191–5199. http://dx.doi.org/10.1167/iovs.14-14046.

Armulik, A., Genove, G., & Betsholtz, C. (2011). Pericytes: Developmental, physiological, and pathological perspectives, problems, and promises. *Developmental Cell*, *21*(2), 193–215. http://dx.doi.org/10.1016/j.devcel.2011.07.001.

Artavanis-Tsakonas, S., Rand, M. D., & Lake, R. J. (1999). Notch signaling: Cell fate control and signal integration in development. *Science*, *284*(5415), 770–776.

Ayaz, F., & Osborne, B. A. (2014). Non-canonical notch signaling in cancer and immunity. *Frontiers in Oncology*, *4*, 345. http://dx.doi.org/10.3389/fonc.2014.00345.

Baeten, J. T., Jackson, A. R., McHugh, K. M., & Lilly, B. (2015). Loss of Notch2 and Notch3 in vascular smooth muscle causes patent ductus arteriosus. *Genesis*, *53*(12), 738–748. http://dx.doi.org/10.1002/dvg.22904.

Baeten, J. T., & Lilly, B. (2015). Differential regulation of NOTCH2 and NOTCH3 contribute to their unique functions in vascular smooth muscle cells. *The Journal of Biological Chemistry*, *290*, 16226–16237. http://dx.doi.org/10.1074/jbc.M115.655548.

Battelino, N., Writzl, K., Bratanic, N., Irving, M. D., & Novljan, G. (2016). End-stage renal disease in an infant with Hajdu-Cheney syndrome. *Therapeutic Apheresis and Dialysis, 20*(3), 318–321. http://dx.doi.org/10.1111/1744-9987.12444.

Beckers, J., Clark, A., Wunsch, K., Hrabe De Angelis, M., & Gossler, A. (1999). Expression of the mouse Delta1 gene during organogenesis and fetal development. *Mechanisms of Development, 84*(1–2), 165–168.

Benedito, R., Roca, C., Sorensen, I., Adams, S., Gossler, A., Fruttiger, M., & Adams, R. H. (2009). The notch ligands Dll4 and Jagged1 have opposing effects on angiogenesis. *Cell, 137*(6), 1124–1135. http://dx.doi.org/10.1016/j.cell.2009.03.025.

Blokzijl, A., Dahlqvist, C., Reissmann, E., Falk, A., Moliner, A., Lendahl, U., & Ibanez, C. F. (2003). Cross-talk between the Notch and TGF-beta signaling pathways mediated by interaction of the Notch intracellular domain with Smad3. *The Journal of Cell Biology, 163*(4), 723–728. http://dx.doi.org/10.1083/jcb.200305112.

Borggrefe, T., & Oswald, F. (2009). The Notch signaling pathway: Transcriptional regulation at Notch target genes. *Cellular and Molecular Life Sciences, 66*(10), 1631–1646. http://dx.doi.org/10.1007/s00018-009-8668-7.

Boscolo, E., Stewart, C. L., Greenberger, S., Wu, J. K., Durham, J. T., Herman, I. M., … Bischoff, J. (2011). JAGGED1 signaling regulates hemangioma stem cell-to-pericyte/vascular smooth muscle cell differentiation. *Arteriosclerosis, Thrombosis, and Vascular Biology, 31*(10), 2181–2192. http://dx.doi.org/10.1161/ATVBAHA.111.232934.

Boucher, J., Gridley, T., & Liaw, L. (2012). Molecular pathways of notch signaling in vascular smooth muscle cells. *Frontiers in Physiology, 3*, 81. http://dx.doi.org/10.3389/fphys.2012.00081.

Boucher, J. M., Harrington, A., Rostama, B., Lindner, V., & Liaw, L. (2013). A receptor-specific function for Notch2 in mediating vascular smooth muscle cell growth arrest through cyclin-dependent kinase inhibitor 1B. *Circulation Research, 113*(8), 975–985. http://dx.doi.org/10.1161/CIRCRESAHA.113.301272.

Boucher, J. M., Peterson, S. M., Urs, S., Zhang, C., & Liaw, L. (2011). The miR-143/145 cluster is a novel transcriptional target of Jagged-1/Notch signaling in vascular smooth muscle cells. *The Journal of Biological Chemistry, 286*(32), 28312–28321. http://dx.doi.org/10.1074/jbc.M111.221945.

Campos, A. H., Wang, W., Pollman, M. J., & Gibbons, G. H. (2002). Determinants of Notch-3 receptor expression and signaling in vascular smooth muscle cells: Implications in cell-cycle regulation. *Circulation Research, 91*(11), 999–1006.

Cappellari, O., Benedetti, S., Innocenzi, A., Tedesco, F. S., Moreno-Fortuny, A., Ugarte, G., … Cossu, G. (2013). Dll4 and PDGF-BB convert committed skeletal myoblasts to pericytes without erasing their myogenic memory. *Developmental Cell, 24*(6), 586–599. http://dx.doi.org/10.1016/j.devcel.2013.01.022.

Chabriat, H., Joutel, A., Dichgans, M., Tournier-Lasserve, E., & Bousser, M. G. (2009). Cadasil. *Lancet Neurology, 8*(7), 643–653. http://dx.doi.org/10.1016/S1474-4422(09)70127-9.

Chida, A., Shintani, M., Matsushita, Y., Sato, H., Eitoku, T., Nakayama, T., … Nakanishi, T. (2014). Mutations of NOTCH3 in childhood pulmonary arterial hypertension. *Molecular Genetics & Genomic Medicine, 2*(3), 229–239. http://dx.doi.org/10.1002/mgg3.58.

Crosswhite, P., & Sun, Z. (2014). Molecular mechanisms of pulmonary arterial remodeling. *Molecular Medicine, 20*, 191–201. http://dx.doi.org/10.2119/molmed.2013.00165.

Dang, T. P. (2012). Notch, apoptosis and cancer. *Advances in Experimental Medicine and Biology, 727*, 199–209. http://dx.doi.org/10.1007/978-1-4614-0899-4_15.

Dees, C., Tomcik, M., Zerr, P., Akhmetshina, A., Horn, A., Palumbo, K., … Distler, J. H. (2011). Notch signalling regulates fibroblast activation and collagen release in systemic sclerosis. *Annals of the Rheumatic Diseases, 70*(7), 1304–1310. http://dx.doi.org/10.1136/ard.2010.134742.

Delbosc, S., Glorian, M., Le Port, A. S., Bereziat, G., Andreani, M., & Limon, I. (2008). The benefit of docosahexaenoic acid on the migration of vascular smooth muscle cells is partially dependent on Notch regulation of MMP-2/-9. *The American Journal of Pathology*, *172*(5), 1430–1440. http://dx.doi.org/10.2353/ajpath.2008.070951.

Doi, H., Iso, T., Sato, H., Yamazaki, M., Matsui, H., Tanaka, T., … Kurabayashi, M. (2006). Jagged1-selective notch signaling induces smooth muscle differentiation via a RBP-Jkappa-dependent pathway. *The Journal of Biological Chemistry*, *281*(39), 28555–28564. http://dx.doi.org/10.1074/jbc.M602749200.

Doi, H., Iso, T., Yamazaki, M., Akiyama, H., Kanai, H., Sato, H., … Kurabayashi, M. (2005). HERP1 inhibits myocardin-induced vascular smooth muscle cell differentiation by interfering with SRF binding to CArG box. *Arteriosclerosis, Thrombosis, and Vascular Biology*, *25*(11), 2328–2334. http://dx.doi.org/10.1161/01.ATV.0000185829.47163.32.

Domenga, V., Fardoux, P., Lacombe, P., Monet, M., Maciazek, J., Krebs, L. T., … Joutel, A. (2004). Notch3 is required for arterial identity and maturation of vascular smooth muscle cells. *Genes and Development*, *18*(22), 2730–2735. http://dx.doi.org/10.1101/gad.308904.

Donovan, J., Abraham, D., & Norman, J. (2013). Platelet-derived growth factor signaling in mesenchymal cells. *Frontiers in Bioscience (Landmark Edition)*, *18*, 106–119.

Dror, V., Nguyen, V., Walia, P., Kalynyak, T. B., Hill, J. A., & Johnson, J. D. (2007). Notch signalling suppresses apoptosis in adult human and mouse pancreatic islet cells. *Diabetologia*, *50*(12), 2504–2515. http://dx.doi.org/10.1007/s00125-007-0835-5.

Duncan, A. W., Rattis, F. M., DiMascio, L. N., Congdon, K. L., Pazianos, G., Zhao, C., … Reya, T. (2005). Integration of Notch and Wnt signaling in hematopoietic stem cell maintenance. *Nature Immunology*, *6*(3), 314–322. http://dx.doi.org/10.1038/ni1164.

Ehebauer, M., Hayward, P., & Arias, A. M. (2006). Notch, a universal arbiter of cell fate decisions. *Science*, *314*(5804), 1414–1415. http://dx.doi.org/10.1126/science.1134042.

Eldadah, Z. A., Hamosh, A., Biery, N. J., Montgomery, R. A., Duke, M., Elkins, R., & Dietz, H. C. (2001). Familial tetralogy of Fallot caused by mutation in the jagged1 gene. *Human Molecular Genetics*, *10*(2), 163–169.

Espinosa, L., Ingles-Esteve, J., Aguilera, C., & Bigas, A. (2003). Phosphorylation by glycogen synthase kinase-3 beta down-regulates Notch activity, a link for Notch and Wnt pathways. *The Journal of Biological Chemistry*, *278*(34), 32227–32235. http://dx.doi.org/10.1074/jbc.M304001200.

Feng, X., Krebs, L. T., & Gridley, T. (2010). Patent ductus arteriosus in mice with smooth muscle-specific Jag1 deletion. *Development*, *137*(24), 4191–4199. http://dx.doi.org/10.1242/dev.052043.

Fouillade, C., Monet-Lepretre, M., Baron-Menguy, C., & Joutel, A. (2012). Notch signalling in smooth muscle cells during development and disease. *Cardiovascular Research*, *95*(2), 138–146. http://dx.doi.org/10.1093/cvr/cvs019.

Fre, S., Pallavi, S. K., Huyghe, M., Lae, M., Janssen, K. P., Robine, S., … Louvard, D. (2009). Notch and Wnt signals cooperatively control cell proliferation and tumorigenesis in the intestine. *Proceedings of the National Academy of Sciences of the United States of America*, *106*(15), 6309–6314. http://dx.doi.org/10.1073/pnas.0900427106.

Gaengel, K., Genove, G., Armulik, A., & Betsholtz, C. (2009). Endothelial-mural cell signaling in vascular development and angiogenesis. *Arteriosclerosis, Thrombosis, and Vascular Biology*, *29*(5), 630–638. http://dx.doi.org/10.1161/ATVBAHA.107.161521.

Garg, V., Muth, A. N., Ransom, J. F., Schluterman, M. K., Barnes, R., King, I. N., … Srivastava, D. (2005). Mutations in NOTCH1 cause aortic valve disease. *Nature*, *437*(7056), 270–274. http://dx.doi.org/10.1038/nature03940.

Geraci, M. W., Moore, M., Gesell, T., Yeager, M. E., Alger, L., Golpon, H., … Voelkel, N. F. (2001). Gene expression patterns in the lungs of patients with primary pulmonary hypertension: A gene microarray analysis. *Circulation Research*, *88*(6), 555–562.

Gomez, D., & Owens, G. K. (2012). Smooth muscle cell phenotypic switching in atherosclerosis. *Cardiovascular Research*, *95*(2), 156–164. http://dx.doi.org/10.1093/cvr/cvs115.

Greif, D. M., Kumar, M., Lighthouse, J. K., Hum, J., An, A., Ding, L., … Krasnow, M. A. (2012). Radial construction of an arterial wall. *Developmental Cell*, *23*(3), 482–493. http://dx.doi.org/10.1016/j.devcel.2012.07.009.

Gridley, T. (2007). Notch signaling in vascular development and physiology. *Development*, *134*(15), 2709–2718. http://dx.doi.org/10.1242/dev.004184.

Gridley, T. (2010). Notch signaling in the vasculature. *Current Topics in Developmental Biology*, *92*, 277–309. http://dx.doi.org/10.1016/S0070-2153(10)92009-7.

Grieskamp, T., Rudat, C., Ludtke, T. H., Norden, J., & Kispert, A. (2011). Notch signaling regulates smooth muscle differentiation of epicardium-derived cells. *Circulation Research*, *108*(7), 813–823. http://dx.doi.org/10.1161/CIRCRESAHA.110.228809.

Gripp, K. W., Robbins, K. M., Sobreira, N. L., Witmer, P. D., Bird, L. M., Avela, K., … Sol-Church, K. (2015). Truncating mutations in the last exon of NOTCH3 cause lateral meningocele syndrome. *American Journal of Medical Genetics. Part A*, *167A*(2), 271–281. http://dx.doi.org/10.1002/ajmg.a.36863.

Guo, X., & Chen, S. Y. (2012). Transforming growth factor-beta and smooth muscle differentiation. *World Journal of Biological Chemistry*, *3*(3), 41–52. http://dx.doi.org/10.4331/wjbc.v3.i3.41.

Guo, D. C., Pannu, H., Tran-Fadulu, V., Papke, C. L., Yu, R. K., Avidan, N., … Milewicz, D. M. (2007). Mutations in smooth muscle alpha-actin (ACTA2) lead to thoracic aortic aneurysms and dissections. *Nature Genetics*, *39*(12), 1488–1493. http://dx.doi.org/10.1038/ng.2007.6.

Hassed, S. J., Wiley, G. B., Wang, S., Lee, J. Y., Li, S., Xu, W., … Gaffney, P. M. (2012). RBPJ mutations identified in two families affected by Adams-Oliver syndrome. *The American Journal of Human Genetics*, *91*(2), 391–395. http://dx.doi.org/10.1016/j.ajhg.2012.07.005.

Hatzidaki, E., Korakaki, E., Voloudaki, A., Daskaloyannaki, M., Manoura, A., & Giannakopoulou, C. (2001). Infantile myofibromatosis with visceral involvement and complete spontaneous regression. *The Journal of Dermatology*, *28*(7), 379–382.

Havrda, M. C., Johnson, M. J., O'Neill, C. F., & Liaw, L. (2006). A novel mechanism of transcriptional repression of p27kip1 through Notch/HRT2 signaling in vascular smooth muscle cells. *Thrombosis and Haemostasis*, *96*(3), 361–370. http://dx.doi.org/10.1160/TH06-04-0224.

Hayward, P., Brennan, K., Sanders, P., Balayo, T., DasGupta, R., Perrimon, N., & Martinez Arias, A. (2005). Notch modulates Wnt signalling by associating with Armadillo/beta-catenin and regulating its transcriptional activity. *Development*, *132*(8), 1819–1830. http://dx.doi.org/10.1242/dev.01724.

Heitzler, P., & Simpson, P. (1991). The choice of cell fate in the epidermis of Drosophila. *Cell*, *64*(6), 1083–1092.

Hellstrom, M., Kalen, M., Lindahl, P., Abramsson, A., & Betsholtz, C. (1999). Role of PDGF-B and PDGFR-beta in recruitment of vascular smooth muscle cells and pericytes during embryonic blood vessel formation in the mouse. *Development*, *126*(14), 3047–3055.

Henshall, T. L., Keller, A., He, L., Johansson, B. R., Wallgard, E., Raschperger, E., … Lendahl, U. (2015). Notch3 is necessary for blood vessel integrity in the central nervous system. *Arteriosclerosis, Thrombosis, and Vascular Biology*, *35*(2), 409–420. http://dx.doi.org/10.1161/ATVBAHA.114.304849.

Hicks, C., Johnston, S. H., diSibio, G., Collazo, A., Vogt, T. F., & Weinmaster, G. (2000). Fringe differentially modulates Jagged1 and Delta1 signalling through Notch1 and Notch2. *Nature Cell Biology*, *2*(8), 515–520. http://dx.doi.org/10.1038/35019553.

High, F. A., Lu, M. M., Pear, W. S., Loomes, K. M., Kaestner, K. H., & Epstein, J. A. (2008). Endothelial expression of the Notch ligand Jagged1 is required for vascular smooth muscle development. *Proceedings of the National Academy of Sciences of the United States of America, 105*(6), 1955–1959. http://dx.doi.org/10.1073/pnas.0709663105.

High, F. A., Zhang, M., Proweller, A., Tu, L., Parmacek, M. S., Pear, W. S., & Epstein, J. A. (2007). An essential role for Notch in neural crest during cardiovascular development and smooth muscle differentiation. *The Journal of Clinical Investigation, 117*(2), 353–363. http://dx.doi.org/10.1172/JCI30070.

Hirashima, M. (2009). Regulation of endothelial cell differentiation and arterial specification by VEGF and Notch signaling. *Anatomical Science International, 84*(3), 95–101. http://dx.doi.org/10.1007/s12565-009-0026-1.

Hodkinson, P. S., Elliott, P. A., Lad, Y., McHugh, B. J., MacKinnon, A. C., Haslett, C., & Sethi, T. (2007). Mammalian NOTCH-1 activates beta1 integrins via the small GTPase R-Ras. *The Journal of Biological Chemistry, 282*(39), 28991–29001. http://dx.doi.org/10.1074/jbc.M703601200.

Hoglund, V. J., & Majesky, M. W. (2012). Patterning the artery wall by lateral induction of Notch signaling. *Circulation, 125*(2), 212–215. http://dx.doi.org/10.1161/CIRCULATIONAHA.111.075937.

Huang, J., Cheng, L., Li, J., Chen, M., Zhou, D., Lu, M. M., … Parmacek, M. S. (2008). Myocardin regulates expression of contractile genes in smooth muscle cells and is required for closure of the ductus arteriosus in mice. *The Journal of Clinical Investigation, 118*(2), 515–525. http://dx.doi.org/10.1172/JCI33304.

Iso, T., Hamamori, Y., & Kedes, L. (2003). Notch signaling in vascular development. *Arteriosclerosis, Thrombosis, and Vascular Biology, 23*(4), 543–553. http://dx.doi.org/10.1161/01.ATV.0000060892.81529.8F.

Jehn, B. M., Bielke, W., Pear, W. S., & Osborne, B. A. (1999). Cutting edge: Protective effects of notch-1 on TCR-induced apoptosis. *The Journal of Immunology, 162*(2), 635–638.

Jin, S., Hansson, E. M., Tikka, S., Lanner, F., Sahlgren, C., Farnebo, F., … Lendahl, U. (2008). Notch signaling regulates platelet-derived growth factor receptor-beta expression in vascular smooth muscle cells. *Circulation Research, 102*(12), 1483–1491. http://dx.doi.org/10.1161/CIRCRESAHA.107.167965.

Joutel, A., Andreux, F., Gaulis, S., Domenga, V., Cecillon, M., Battail, N., … Tournier-Lasserve, E. (2000). The ectodomain of the Notch3 receptor accumulates within the cerebrovasculature of CADASIL patients. *The Journal of Clinical Investigation, 105*(5), 597–605. http://dx.doi.org/10.1172/JCI8047.

Joutel, A., Corpechot, C., Ducros, A., Vahedi, K., Chabriat, H., Mouton, P., … Tournier-Lasserve, E. (1996). Notch3 mutations in CADASIL, a hereditary adult-onset condition causing stroke and dementia. *Nature, 383*(6602), 707–710. http://dx.doi.org/10.1038/383707a0.

Joutel, A., Corpechot, C., Ducros, A., Vahedi, K., Chabriat, H., Mouton, P., … Tournier-Lasserve, E. (1997). Notch3 mutations in cerebral autosomal dominant arteriopathy with subcortical infarcts and leukoencephalopathy (CADASIL), a mendelian condition causing stroke and vascular dementia. *The Annals of the New York Academy of Sciences, 826*, 213–217.

Joutel, A., Monet-Lepretre, M., Gosele, C., Baron-Menguy, C., Hammes, A., Schmidt, S., … Hubner, N. (2010). Cerebrovascular dysfunction and microcirculation rarefaction precede white matter lesions in a mouse genetic model of cerebral ischemic small vessel disease. *The Journal of Clinical Investigation, 120*(2), 433–445. http://dx.doi.org/10.1172/JCI39733.

Kangsamaksin, T., Murtomaki, A., Kofler, N. M., Cuervo, H., Chaudhri, R. A., Tattersall, I. W., … Kitajewski, J. (2015). NOTCH decoys that selectively block

DLL/NOTCH or JAG/NOTCH disrupt angiogenesis by unique mechanisms to inhibit tumor growth. *Cancer Discovery*, *5*(2), 182–197. http://dx.doi.org/10.1158/2159-8290.CD-14-0650.

Kennard, S., Liu, H., & Lilly, B. (2008). Transforming growth factor-beta (TGF-1) down-regulates Notch3 in fibroblasts to promote smooth muscle gene expression. *The Journal of Biological Chemistry*, *283*(3), 1324–1333. http://dx.doi.org/10.1074/jbc.M706651200.

Kim, Y. H., Hu, H., Guevara-Gallardo, S., Lam, M. T., Fong, S. Y., & Wang, R. A. (2008). Artery and vein size is balanced by Notch and ephrin B2/EphB4 during angiogenesis. *Development*, *135*(22), 3755–3764. http://dx.doi.org/10.1242/dev.022475.

King, I. N., Kathiriya, I. S., Murakami, M., Nakagawa, M., Gardner, K. A., Srivastava, D., & Nakagawa, O. (2006). Hrt and Hes negatively regulate Notch signaling through interactions with RBP-Jkappa. *Biochemical and Biophysical Research Communications*, *345*(1), 446–452. http://dx.doi.org/10.1016/j.bbrc.2006.04.097.

Kofler, N. M., Cuervo, H., Uh, M. K., Murtomaki, A., & Kitajewski, J. (2015). Combined deficiency of Notch1 and Notch3 causes pericyte dysfunction, models CADASIL, and results in arteriovenous malformations. *Scientific Reports*, *5*, 16449. http://dx.doi.org/10.1038/srep16449.

Krebs, L. T., Norton, C. R., & Gridley, T. (2016). Notch signal reception is required in vascular smooth muscle cells for ductus arteriosus closure. *Genesis*, *54*, 86–90. http://dx.doi.org/10.1002/dvg.22916.

Krebs, L. T., Xue, Y., Norton, C. R., Sundberg, J. P., Beatus, P., Lendahl, U., … Gridley, T. (2003). Characterization of Notch3-deficient mice: Normal embryonic development and absence of genetic interactions with a Notch1 mutation. *Genesis*, *37*(3), 139–143. http://dx.doi.org/10.1002/gene.10241.

Kurpinski, K., Lam, H., Chu, J., Wang, A., Kim, A., Tsay, E., … Li, S. (2010). Transforming growth factor-beta and notch signaling mediate stem cell differentiation into smooth muscle cells. *Stem Cells*, *28*(4), 734–742. http://dx.doi.org/10.1002/stem.319.

Kwon, C., Cheng, P., King, I. N., Andersen, P., Shenje, L., Nigam, V., & Srivastava, D. (2011). Notch post-translationally regulates beta-catenin protein in stem and progenitor cells. *Nature Cell Biology*, *13*(10), 1244–1251. http://dx.doi.org/10.1038/ncb2313.

Lai, E. C. (2002). Protein degradation: Four E3s for the notch pathway. *Current Biology*, *12*(2), R74–R78.

Lawson, N. D., Scheer, N., Pham, V. N., Kim, C. H., Chitnis, A. B., Campos-Ortega, J. A., & Weinstein, B. M. (2001). Notch signaling is required for arterial-venous differentiation during embryonic vascular development. *Development*, *128*(19), 3675–3683.

LeBon, L., Lee, T. V., Sprinzak, D., Jafar-Nejad, H., & Elowitz, M. B. (2014). Fringe proteins modulate Notch-ligand cis and trans interactions to specify signaling states. *eLife*, *3*, e02950. http://dx.doi.org/10.7554/eLife.02950.

Lee, J. W. (2013). Mutations in PDGFRB and NOTCH3 are the first genetic causes identified for autosomal dominant infantile myofibromatosis. *Clinical Genetics*, *84*(4), 340–341. http://dx.doi.org/10.1111/cge.12238.

Li, L., Krantz, I. D., Deng, Y., Genin, A., Banta, A. B., Collins, C. C., … Spinner, N. B. (1997). Alagille syndrome is caused by mutations in human Jagged1, which encodes a ligand for Notch1. *Nature Genetics*, *16*(3), 243–251. http://dx.doi.org/10.1038/ng0797-243.

Li, F., Lan, Y., Wang, Y., Wang, J., Yang, G., Meng, F., … Yang, X. (2011). Endothelial Smad4 maintains cerebrovascular integrity by activating N-cadherin through cooperation with Notch. *Developmental Cell*, *20*(3), 291–302. http://dx.doi.org/10.1016/j.devcel.2011.01.011.

Li, Y., Takeshita, K., Liu, P. Y., Satoh, M., Oyama, N., Mukai, Y., … Liao, J. K. (2009). Smooth muscle Notch1 mediates neointimal formation after vascular injury. *Circulation*, *119*(20), 2686–2692. http://dx.doi.org/10.1161/CIRCULATIONAHA.108.790485.

Li, X., Zhang, X., Leathers, R., Makino, A., Huang, C., Parsa, P., … Thistlethwaite, P. A. (2009). Notch3 signaling promotes the development of pulmonary arterial hypertension. *Nature Medicine (New York), 15*(11), 1289–1297. http://dx.doi.org/10.1038/nm.2021.

Lilly, B., & Kennard, S. (2009). Differential gene expression in a coculture model of angiogenesis reveals modulation of select pathways and a role for Notch signaling. *Physiological Genomics, 36*(2), 69–78. http://dx.doi.org/10.1152/physiolgenomics.90318.2008.

Lin, C. H., & Lilly, B. (2014a). Endothelial cells direct mesenchymal stem cells toward a smooth muscle cell fate. *Stem Cells and Development, 23*(21), 2581–2590. http://dx.doi.org/10.1089/scd.2014.0163.

Lin, C. H., & Lilly, B. (2014b). Notch signaling governs phenotypic modulation of smooth muscle cells. *Vascular Pharmacology, 63*(2), 88–96. http://dx.doi.org/10.1016/j.vph.2014.09.004.

Lindner, V., Booth, C., Prudovsky, I., Small, D., Maciag, T., & Liaw, L. (2001). Members of the Jagged/Notch gene families are expressed in injured arteries and regulate cell phenotype via alterations in cell matrix and cell-cell interaction. *The American Journal of Pathology, 159*(3), 875–883. http://dx.doi.org/10.1016/S0002-9440(10)61763-4.

Liu, Z., Brunskill, E., Varnum-Finney, B., Zhang, C., Zhang, A., Jay, P. Y., … Kopan, R. (2015). The intracellular domains of Notch1 and Notch2 are functionally equivalent during development and carcinogenesis. *Development, 142*(14), 2452–2463. http://dx.doi.org/10.1242/dev.125492.

Liu, H., Kennard, S., & Lilly, B. (2009). NOTCH3 expression is induced in mural cells through an autoregulatory loop that requires endothelial-expressed JAGGED1. *Circulation Research, 104*(4), 466–475. http://dx.doi.org/10.1161/CIRCRESAHA.108.184846.

Liu, H., Zhang, W., Kennard, S., Caldwell, R. B., & Lilly, B. (2010). Notch3 is critical for proper angiogenesis and mural cell investment. *Circulation Research, 107*(7), 860–870. http://dx.doi.org/10.1161/CIRCRESAHA.110.218271.

Lundkvist, J., Zhu, S., Hansson, E. M., Schweinhardt, P., Miao, Q., Beatus, P., … Lendahl, U. (2005). Mice carrying a R142C Notch 3 knock-in mutation do not develop a CADASIL-like phenotype. *Genesis, 41*(1), 13–22. http://dx.doi.org/10.1002/gene.20091.

Luo, B., Aster, J. C., Hasserjian, R. P., Kuo, F., & Sklar, J. (1997). Isolation and functional analysis of a cDNA for human Jagged2, a gene encoding a ligand for the Notch1 receptor. *Molecular and Cellular Biology, 17*(10), 6057–6067.

Manderfield, L. J., High, F. A., Engleka, K. A., Liu, F., Li, L., Rentschler, S., & Epstein, J. A. (2012). Notch activation of Jagged1 contributes to the assembly of the arterial wall. *Circulation, 125*(2), 314–323. http://dx.doi.org/10.1161/CIRCULATIONAHA.111.047159.

Martignetti, J. A., Tian, L., Li, D., Ramirez, M. C., Camacho-Vanegas, O., Camacho, S. C., … Hakonarson, H. (2013). Mutations in PDGFRB cause autosomal-dominant infantile myofibromatosis. *The American Journal of Human Genetics, 92*(6), 1001–1007. http://dx.doi.org/10.1016/j.ajhg.2013.04.024.

Martinez Arias, A., Zecchini, V., & Brennan, K. (2002). CSL-independent Notch signalling: A checkpoint in cell fate decisions during development? *Current Opinion in Genetics and Development, 12*(5), 524–533.

McCright, B., Gao, X., Shen, L., Lozier, J., Lan, Y., Maguire, M., … Gridley, T. (2001). Defects in development of the kidney, heart and eye vasculature in mice homozygous for a hypomorphic Notch2 mutation. *Development, 128*(4), 491–502.

McDaniell, R., Warthen, D. M., Sanchez-Lara, P. A., Pai, A., Krantz, I. D., Piccoli, D. A., & Spinner, N. B. (2006). NOTCH2 mutations cause Alagille syndrome, a heterogeneous disorder of the notch signaling pathway. *The American Journal of Human Genetics, 79*(1), 169–173. http://dx.doi.org/10.1086/505332.

McElhinney, D. B., Krantz, I. D., Bason, L., Piccoli, D. A., Emerick, K. M., Spinner, N. B., & Goldmuntz, E. (2002). Analysis of cardiovascular phenotype and genotype-phenotype correlation in individuals with a JAG1 mutation and/or Alagille syndrome. *Circulation, 106*(20), 2567–2574.

Meester, J. A., Southgate, L., Stittrich, A. B., Venselaar, H., Beekmans, S. J., den Hollander, N., ... Wuyts, W. (2015). Heterozygous loss-of-function mutations in DLL4 cause Adams-Oliver syndrome. *The American Journal of Human Genetics*, *97*(3), 475–482. http://dx.doi.org/10.1016/j.ajhg.2015.07.015.

Meng, H., Zhang, X., Lee, S. J., & Wang, M. M. (2013). Von Willebrand factor inhibits mature smooth muscle gene expression through impairment of Notch signaling. *PLoS One*, *8*(9), e75808. http://dx.doi.org/10.1371/journal.pone.0075808.

Meurette, O., Stylianou, S., Rock, R., Collu, G. M., Gilmore, A. P., & Brennan, K. (2009). Notch activation induces Akt signaling via an autocrine loop to prevent apoptosis in breast epithelial cells. *Cancer Research*, *69*(12), 5015–5022. http://dx.doi.org/10.1158/0008-5472.CAN-08-3478.

Mill, C., & George, S. J. (2012). Wnt signalling in smooth muscle cells and its role in cardiovascular disorders. *Cardiovascular Research*, *95*(2), 233–240. http://dx.doi.org/10.1093/cvr/cvs141.

Moloney, D. J., Panin, V. M., Johnston, S. H., Chen, J., Shao, L., Wilson, R., ... Vogt, T. F. (2000). Fringe is a glycosyltransferase that modifies Notch. *Nature*, *406*(6794), 369–375. http://dx.doi.org/10.1038/35019000.

Morgan, T. H. (1917). The theory of the gene. *The American Naturalist*, *51*(609), 513–544. http://dx.doi.org/10.1086/279629.

Nakagawa, O., McFadden, D. G., Nakagawa, M., Yanagisawa, H., Hu, T., Srivastava, D., & Olson, E. N. (2000). Members of the HRT family of basic helix-loop-helix proteins act as transcriptional repressors downstream of Notch signaling. *Proceedings of the National Academy of Sciences of the United States of America*, *97*(25), 13655–13660. http://dx.doi.org/10.1073/pnas.250485597.

Noseda, M., Fu, Y., Niessen, K., Wong, F., Chang, L., McLean, G., & Karsan, A. (2006). Smooth muscle alpha-actin is a direct target of Notch/CSL. *Circulation Research*, *98*(12), 1468–1470. http://dx.doi.org/10.1161/01.RES.0000229683.81357.26.

Ong, C. T., Cheng, H. T., Chang, L. W., Ohtsuka, T., Kageyama, R., Stormo, G. D., & Kopan, R. (2006). Target selectivity of vertebrate notch proteins. Collaboration between discrete domains and CSL-binding site architecture determines activation probability. *The Journal of Biological Chemistry*, *281*(8), 5106–5119. http://dx.doi.org/10.1074/jbc.M506108200.

Owens, G. K., Kumar, M. S., & Wamhoff, B. R. (2004). Molecular regulation of vascular smooth muscle cell differentiation in development and disease. *Physiological Reviews*, *84*(3), 767–801. http://dx.doi.org/10.1152/physrev.00041.2003.

Patel-Hett, S., & D'Amore, P. A. (2011). Signal transduction in vasculogenesis and developmental angiogenesis. *The International Journal of Developmental Biology*, *55*(4–5), 353–363. http://dx.doi.org/10.1387/ijdb.103213sp.

Pearson, G., Robinson, F., Beers Gibson, T., Xu, B. E., Karandikar, M., Berman, K., & Cobb, M. H. (2001). Mitogen-activated protein (MAP) kinase pathways: Regulation and physiological functions. *Endocrine Reviews*, *22*(2), 153–183. http://dx.doi.org/10.1210/edrv.22.2.0428.

Phng, L. K., & Gerhardt, H. (2009). Angiogenesis: A team effort coordinated by notch. *Developmental Cell*, *16*(2), 196–208. http://dx.doi.org/10.1016/j.devcel.2009.01.015.

Pippucci, T., Maresca, A., Magini, P., Cenacchi, G., Donadio, V., Palombo, F.,Seri, M. ... (2015). Homozygous NOTCH3 null mutation and impaired NOTCH3 signaling in recessive early-onset arteriopathy and cavitating leukoencephalopathy. *EMBO Molecular Medicine*, 7(6), 848–858. http://dx.doi.org/10.15252/emmm.201404399.

Proweller, A., Pear, W. S., & Parmacek, M. S. (2005). Notch signaling represses myocardin-induced smooth muscle cell differentiation. *The Journal of Biological Chemistry*, *280*(10), 8994–9004. http://dx.doi.org/10.1074/jbc.M413316200.

Regan, J. N., & Majesky, M. W. (2009). Building a vessel wall with notch signaling. *Circulation Research*, *104*(4), 419–421. http://dx.doi.org/10.1161/CIRCRESAHA.109.194233.

Roca, C., & Adams, R. H. (2007). Regulation of vascular morphogenesis by Notch signaling. *Genes and Development*, *21*(20), 2511–2524. http://dx.doi.org/10.1101/gad.1589207.

Rosati, E., Sabatini, R., Rampino, G., Tabilio, A., Di Ianni, M., Fettucciari, K., … Marconi, P. (2009). Constitutively activated Notch signaling is involved in survival and apoptosis resistance of B-CLL cells. *Blood*, *113*(4), 856–865. http://dx.doi.org/10.1182/blood-2008-02-139725.

Rostama, B., Peterson, S. M., Vary, C. P., & Liaw, L. (2014). Notch signal integration in the vasculature during remodeling. *Vascular Pharmacology*, *63*(2), 97–104. http://dx.doi.org/10.1016/j.vph.2014.10.003.

Rutten, J. W., Boon, E. M., Liem, M. K., Dauwerse, J. G., Pont, M. J., Vollebregt, E., … Lesnik Oberstein, S. A. (2013). Hypomorphic NOTCH3 alleles do not cause CADASIL in humans. *Human Mutation*, *34*(11), 1486–1489. http://dx.doi.org/10.1002/humu.22432.

Sakata, Y., Xiang, F., Chen, Z., Kiriyama, Y., Kamei, C. N., Simon, D. I., & Chin, M. T. (2004). Transcription factor CHF1/Hey2 regulates neointimal formation in vivo and vascular smooth muscle proliferation and migration in vitro. *Arteriosclerosis, Thrombosis, and Vascular Biology*, *24*(11), 2069–2074. http://dx.doi.org/10.1161/01.ATV.0000143936.77094.a4.

Sanalkumar, R., Dhanesh, S. B., & James, J. (2010). Non-canonical activation of Notch signaling/target genes in vertebrates. *Cellular and Molecular Life Sciences*, *67*(17), 2957–2968 http://dx.doi.org/10.1007/s00018-010-0391-x.

Scheppke, L., Murphy, E. A., Zarpellon, A., Hofmann, J. J., Merkulova, A., Shields, D. J., … Cheresh, D. A. (2012). Notch promotes vascular maturation by inducing integrin-mediated smooth muscle cell adhesion to the endothelial basement membrane. *Blood*, *119*(9), 2149–2158. http://dx.doi.org/10.1182/blood-2011-04-348706.

Schneider, D. J., & Moore, J. W. (2006). Patent ductus arteriosus. *Circulation*, *114*(17), 1873–1882. http://dx.doi.org/10.1161/CIRCULATIONAHA.105.592063.

Shaheen, R., Aglan, M., Keppler-Noreuil, K., Faqeih, E., Ansari, S., Horton, K., … Alkuraya, F. S. (2013). Mutations in EOGT confirm the genetic heterogeneity of autosomal-recessive Adams-Oliver syndrome. *The American Journal of Human Genetics*, *92*(4), 598–604. http://dx.doi.org/10.1016/j.ajhg.2013.02.012.

Shimokado, K., Raines, E. W., Madtes, D. K., Barrett, T. B., Benditt, E. P., & Ross, R. (1985). A significant part of macrophage-derived growth factor consists of at least two forms of PDGF. *Cell*, *43*(1), 277–286.

Shin, M., Nagai, H., & Sheng, G. (2009). Notch mediates Wnt and BMP signals in the early separation of smooth muscle progenitors and blood/endothelial common progenitors. *Development*, *136*(4), 595–603. http://dx.doi.org/10.1242/dev.026906.

Shutter, J. R., Scully, S., Fan, W., Richards, W. G., Kitajewski, J., Deblandre, G. A., … Stark, K. L. (2000). Dll4, a novel Notch ligand expressed in arterial endothelium. *Genes and Development*, *14*(11), 1313–1318.

Simpson, M. A., Irving, M. D., Asilmaz, E., Gray, M. J., Dafou, D., Elmslie, F. V., … Trembath, R. C. (2011). Mutations in NOTCH2 cause Hajdu-Cheney syndrome, a disorder of severe and progressive bone loss. *Nature Genetics*, *43*(4), 303–305. http://dx.doi.org/10.1038/ng.779.

Spinner, N. B., Colliton, R. P., Crosnier, C., Krantz, I. D., Hadchouel, M., & Meunier-Rotival, M. (2001). Jagged1 mutations in Alagille syndrome. *Human Mutation*, *17*(1), 18–33. http://dx.doi.org/10.1002/1098-1004(2001)17:1<18::AID-HUMU3>3.0.CO;2-T.

Stanley, P., & Okajima, T. (2010). Roles of glycosylation in Notch signaling. *Current Topics in Developmental Biology*, *92*, 131–164. http://dx.doi.org/10.1016/S0070-2153(10)92004-8.

Stittrich, A. B., Lehman, A., Bodian, D. L., Ashworth, J., Zong, Z., Li, H., ... Patel, M. S. (2014). Mutations in NOTCH1 cause Adams-Oliver syndrome. *The American Journal of Human Genetics, 95*(3), 275–284. http://dx.doi.org/10.1016/j.ajhg.2014.07.011.

Sweeney, C., Morrow, D., Birney, Y. A., Coyle, S., Hennessy, C., Scheller, A., ... Cahill, P. A. (2004). Notch 1 and 3 receptor signaling modulates vascular smooth muscle cell growth, apoptosis, and migration via a CBF-1/RBP-Jk dependent pathway. *The FASEB Journal, 18*(12), 1421–1423. http://dx.doi.org/10.1096/fj.04-1700fje.

Takebayashi, K., Sasai, Y., Sakai, Y., Watanabe, T., Nakanishi, S., & Kageyama, R. (1994). Structure, chromosomal locus, and promoter analysis of the gene encoding the mouse helix-loop-helix factor HES-1. Negative autoregulation through the multiple N box elements. *The Journal of Biological Chemistry, 269*(7), 5150–5156.

Tang, Y., Urs, S., Boucher, J., Bernaiche, T., Venkatesh, D., Spicer, D. B., ... Liaw, L. (2010). Notch and transforming growth factor-beta (TGFbeta) signaling pathways cooperatively regulate vascular smooth muscle cell differentiation. *The Journal of Biological Chemistry, 285*(23), 17556–17563. http://dx.doi.org/10.1074/jbc.M109.076414.

Tang, Y., Urs, S., & Liaw, L. (2008). Hairy-related transcription factors inhibit Notch-induced smooth muscle alpha-actin expression by interfering with Notch intracellular domain/CBF-1 complex interaction with the CBF-1-binding site. *Circulation Research, 102*(6), 661–668. http://dx.doi.org/10.1161/CIRCRESAHA.107.165134.

Underwood, P. A., Bean, P. A., & Whitelock, J. M. (1998). Inhibition of endothelial cell adhesion and proliferation by extracellular matrix from vascular smooth muscle cells: Role of type V collagen. *Atherosclerosis, 141*(1), 141–152.

Varadkar, P., Kraman, M., Despres, D., Ma, G., Lozier, J., & McCright, B. (2008). Notch2 is required for the proliferation of cardiac neural crest-derived smooth muscle cells. *Developmental Dynamics, 237*(4), 1144–1152. http://dx.doi.org/10.1002/dvdy.21502.

Villa, N., Walker, L., Lindsell, C. E., Gasson, J., Iruela-Arispe, M. L., & Weinmaster, G. (2001). Vascular expression of Notch pathway receptors and ligands is restricted to arterial vessels. *Mechanisms of Development, 108*(1–2), 161–164.

Volz, K. S., Jacobs, A. H., Chen, H. I., Poduri, A., McKay, A. S., Riordan, D. P., ... Red-Horse, K. (2015). Pericytes are progenitors for coronary artery smooth muscle. *eLife, 4*, e10036. http://dx.doi.org/10.7554/eLife.10036.

Wagenseil, J. E., & Mecham, R. P. (2009). Vascular extracellular matrix and arterial mechanics. *Physiological Reviews, 89*(3), 957–989. http://dx.doi.org/10.1152/physrev.00041.2008.

Wang, W., Campos, A. H., Prince, C. Z., Mou, Y., & Pollman, M. J. (2002). Coordinate Notch3-hairy-related transcription factor pathway regulation in response to arterial injury. Mediator role of platelet-derived growth factor and ERK. *The Journal of Biological Chemistry, 277*(26), 23165–23171. http://dx.doi.org/10.1074/jbc.M201409200.

Wang, H. U., Chen, Z. F., & Anderson, D. J. (1998). Molecular distinction and angiogenic interaction between embryonic arteries and veins revealed by ephrin-B2 and its receptor Eph-B4. *Cell, 93*(5), 741–753.

Wang, Y., Pan, L., Moens, C. B., & Appel, B. (2014). Notch3 establishes brain vascular integrity by regulating pericyte number. *Development, 141*(2), 307–317. http://dx.doi.org/10.1242/dev.096107.

Wang, W., Prince, C. Z., Mou, Y., & Pollman, M. J. (2002). Notch3 signaling in vascular smooth muscle cells induces c-FLIP expression via ERK/MAPK activation. Resistance to Fas ligand-induced apoptosis. *The Journal of Biological Chemistry, 277*(24), 21723–21729. http://dx.doi.org/10.1074/jbc.M202224200.

Wang, Q., Zhao, N., Kennard, S., & Lilly, B. (2012). Notch2 and Notch3 function together to regulate vascular smooth muscle development. *PLoS One*, 7(5), e37365. http://dx.doi.org/10.1371/journal.pone.0037365.

Wu, J., Iwata, F., Grass, J. A., Osborne, C. S., Elnitski, L., Fraser, P., ... Bresnick, E. H. (2005). Molecular determinants of NOTCH4 transcription in vascular endothelium. *Molecular and Cellular Biology, 25*(4), 1458–1474. http://dx.doi.org/10.1128/MCB.25.4.1458-1474.2005.

Wu, J., & Zhang, C. (2009). Neointimal hyperplasia, vein graft remodeling, and long-term patency. *American Journal of Physiology. Heart and Circulatory Physiology, 297*(4), H1194–H1195. http://dx.doi.org/10.1152/ajpheart.00703.2009.

Xiao, Y., Gong, D., & Wang, W. (2013). Soluble JAGGED1 inhibits pulmonary hypertension by attenuating notch signaling. *Arteriosclerosis, Thrombosis, and Vascular Biology, 33*(12), 2733–2739. http://dx.doi.org/10.1161/ATVBAHA.113.302062.

Yao, M., Wang, X., Zhang, T., Chi, Y., & Gao, F. (2015). The Notch pathway mediates the angiotensin II-induced synthesis of extracellular matrix components in podocytes. *International Journal of Molecular Medicine, 36*(1), 294–300. http://dx.doi.org/10.3892/ijmm.2015.2193.

Yoshida, T., Sinha, S., Dandre, F., Wamhoff, B. R., Hoofnagle, M. H., Kremer, B. E., ... Owens, G. K. (2003). Myocardin is a key regulator of CArG-dependent transcription of multiple smooth muscle marker genes. *Circulation Research, 92*(8), 856–864. http://dx.doi.org/10.1161/01.RES.0000068405.49081.09.

Zhao, N., Koenig, S. N., Trask, A. J., Lin, C. H., Hans, C. P., Garg, V., & Lilly, B. (2015). mir145 regulates TGFBR2 expression and matrix synthesis in vascular smooth muscle cells. *Circulation Research, 116,* 23–34. http://dx.doi.org/10.1161/CIRCRESAHA.115.303970.

Zhu, L., Vranckx, R., Khau Van Kien, P., Lalande, A., Boisset, N., Mathieu, F., ... Jeunemaitre, X. (2006). Mutations in myosin heavy chain 11 cause a syndrome associating thoracic aortic aneurysm/aortic dissection and patent ductus arteriosus. *Nature Genetics, 38*(3), 343–349. http://dx.doi.org/10.1038/ng1721.

CHAPTER TEN

Smooth Muscle Phenotypic Diversity: Effect on Vascular Function and Drug Responses

S.A. Fisher[1]
University of Maryland School of Medicine, Baltimore, MD, United States
[1]Corresponding author: e-mail address: sfisher1@medicine.umaryland.edu

Contents

Abstract

At its simplest resistance to blood flow is regulated by changes in the state of contraction of the vascular smooth muscle (VSM), a function of the competing activities of the myosin kinase and phosphatase determining the phosphorylation and activity of the myosin ATPase motor protein. In contrast, the vascular system of humans and other mammals is incredibly complex and highly regulated. Much of this complexity derives

Advances in Pharmacology, Volume 78
ISSN 1054-3589
http://dx.doi.org/10.1016/bs.apha.2016.07.003

from phenotypic diversity within the smooth muscle, reflected in very differing power outputs and responses to signaling pathways that regulate vessel tone, presumably having evolved over the millennia to optimize vascular function and its control. The highly regulated nature of VSM tone, described as pharmacomechanical coupling, likely underlies the many classes of drugs in clinical use to alter vascular tone through activation or inhibition of these signaling pathways. This review will first describe the phenotypic diversity within VSM, followed by presentation of specific examples of how molecular diversity in signaling, myofilament, and calcium cycling proteins impacts arterial smooth muscle function and drug responses.

ABBREVIATIONS

AO aorta
BP blood pressure
CBD calcium-binding domain
CCB calcium channel blockers
CDI calcium-dependent inhibition
DHP dihydropyridine
EDHF endothelium-derived hyperpolarizing factor
EDRF endothelium-derived relaxing factor
GPCR G protein-coupled receptor
GWAS genome-wide association studies
IRAG IP3 receptor-associated gating protein
iPSC induced pluripotent stem cell
LTCC L-type calcium channels
LZ leucine zipper
MA mesenteric arteries
MLCK myosin light chain kinase
MP myosin phosphatase
NCX sodium–calcium exchanger
NO nitric oxide
nt nucleotide
P2R purinergic receptors
PKC protein kinase C
RAA renin–angiotensin–aldosterone
RASM resistance artery smooth muscle
ROCK Rho-associated protein kinase
ROS reactive oxygen species
SMcKO smooth muscle-specific conditional knock out
SMMHC smooth muscle myosin heavy chain
SNP single nucleotide polymorphism
SRA small resistance artery
VDI voltage-dependent inhibition
VSM vascular smooth muscle

1. INTRODUCTION

Relaxation of vascular smooth muscle (VSM) is a common goal of drug therapies whose ultimate goal is to lower vascular resistance, reduce blood pressure (BP), and increase blood flow. In acutely ill patients in states of shock due to infection or inadequate cardiac output a goal of drug therapies is to increase contraction of the VSM so as to raise BP and stabilize perfusion of critical organs including the brain and heart. A number of different classes of drugs targeting different pathways and gene products are available for these purposes. It is a reasonable assumption, and the focus of this chapter, that specificity and precision in pharmacological targeting of VSM gene products and pathways, in specific disease states in specific people, would provide the greatest benefit ("precision and personalized" cardiovascular medicine, see Blaus et al., 2015). However, there is currently no evidence from clinical studies to support this premise in the treatment of hypertension nor most other conditions where VSM is pharmacologically targeted. In the treatment of hypertension, the lowering of BP appears to be more critical than the specific classes of drugs that are used to achieve this (James, Oparil, Carter, et al., 2014; The ALLHAT Officers, 2002). Perhaps this is due to insufficient knowledge to target the right gene product in the right person under the right disease condition, as hopefully will become clear as this chapter develops.

The vascular system is incredibly diverse, anatomically, functionally, and in its cell and molecular composition. This diversity is apparent when comparing the vascular beds of the different organs (lungs, brain, kidney, heart, splanchnic, etc.) and even within a single vascular bed (see Majesky, 2007; Reho, Zheng, & Fisher, 2014) for thorough reviews. The focus here is on how VSM phenotypic diversity influences vascular function and determines the response to drugs that increase or decrease muscle force and thereby raise or lower vascular resistance. Due to the steep inverse relationship between vessel radius and resistance to flow (r^{-4}), resistance to blood flow occurs primarily at the smaller arteries and arterioles, generally of radius of 500 μm and smaller. This review will focus on phenotypic differences between the smooth muscle of large (conduit and capacitance) blood vessels and small resistance arteries (SRAs) and the significance with respect to vascular function and drug responses. In most all VSM force is activated and deactivated by receptor signaling pathways, described as pharmacomechanical coupling. This likely underlies the plethora of drugs and drug classes available for modulating VSM function. However, it should also be noted that there is significant crosstalk between the electro- and pharmacomechanical pathways in activation and deactivation of force in VSM, which thus may now be thought of as an oversimplified yet still useful classification scheme. While all VSM may be controlled by pharmaco- and electromechanical coupling, the nature of the response and drug effects vary significantly among VSMs and is the topic of this review.

There are a number of related topics that may impact VSM pharmacomechanical coupling and drug responses that are not feasibly presented here. First, the focus here is on the VSM response. This of course depends on the strength of the signal (agonist) activating the pathway, as for example the increase in corpus cavernosal blood flow and tumescence due to release of nitric oxide (NO) during sexual arousal that is potentiated by PDE5 inhibition (Sildenafil class). Second, receptor pathways may be acutely or chronically sensitized or desensitized depending on the disease state and history of activation, but this is not a phenotypic trait in the classic sense. Third, pharmacokinetics are important considerations for drug responses, as for example the activation of prodrugs such as NO donors (nitroglycerin) and Plavix, drug catabolism, and effects of polymorphisms in cytochrome P450 (CYPs) and other genes on drug responses, but beyond the scope of this review (see Roden et al., 2011, for review). Lastly, lymphatic smooth muscle will not be reviewed. It has its own unique properties and may be affected by drug therapies, e.g., calcium channel blockers (CCB) causing peripheral edema (reviewed in Chakraborty, Davis, & Muthuchamy, 2015).

2. BRIEF REVIEW OF THE PLAYERS IN VSM CONTRACTILE FUNCTION

This is extensively reviewed elsewhere in this monograph. At its simplest VSM force is activated by calcium flux activating myosin light chain kinase (MLCK), which by phosphorylating myosin activates its ATPase activity resulting in force generation and shortening. Force is deactivated by the dephosphorylation of myosin by myosin phosphatase (MP). In pharmacomechanical coupling, the signals that activate or deactivate force do so by changing the calcium flux, and thus MLCK activity, and by directly signaling to change the activity of the MP, thereby altering the relationship between calcium concentration and force, i.e., calcium sensitivity of force production (reviewed in Dippold & Fisher, 2014b; Gao et al., 2013; Ito, Nakano, Erdodi, & Hartshorne, 2004). Similarly, drugs that are used to alter VSM force may do so by an effect on calcium flux and/or calcium sensitivity of the myofilaments. Complicating these simple relationships are the complexity of the pathways that control calcium flux and myofilament performance, and differences in muscle performance and power output dependent upon VSM phenotype, i.e., its molecular composition. One simple classification scheme is to dichotomize smooth muscle based on force outputs as tonic, in which resting tone is maintained and modulated by signals, as in the large arteries and veins, vs phasic, in which there is periodic contraction and relaxation of the smooth muscle (Somlyo & Somlyo, 1968, 1994). The portal vein contains prototypical phasic VSM, while much of the resistance artery smooth muscle (RASM) has intermediate properties, with some resting tone and superimposed phasic contractions termed vasomotion (Aalkjaer, Boedtkjer, & Matchkov, 2011; Haddock & Hill, 2005). A full description of this topic is not feasible here (see Fisher, 2010, for review), but will be considered within the framework of drug responses.

3. DIVERSITY WITHIN THE VASCULAR SYSTEM

Phylogenetic comparisons of humans with favored model organisms such as lower vertebrates (e.g., fishes) and invertebrates (the worm *Caenorhabditis elegans*; the fly *Drosophila melanogaster*) reveal the incredible diversity and increasing complexity of the human vascular and other organ systems. Admittedly this is an oversimplification of the tremendous phylogenetic diversity and complexity, a subject beyond the scope of this review. In the pregenomic era, it was assumed that this diversity was generated by an increasing number of genes; subsequent genome sequencing suggested this is not the case, as the number of identified genes in species up and down the phylogenetic tree varies by less than twofold (International Human Genome Sequencing Consortium, 2001). Thus current estimates are that there are ~25,000 individual genes in the mammalian genome coding for greater than 1 million unique protein products. Thus some of the VSM diversity is generated by expansion of gene families, as for example the numerous α- and β-subunits of the calcium channels. Much of the complexity and diversity of higher organisms, however, derives from the variable processing of gene transcripts generating unique gene products through alternative transcriptional start sites and alternative splicing of exons, and the highly regulated expression of these diverse gene products (Johnson et al., 2003; Merkin, Russell, Chen, & Burge, 2012; Pal et al., 2011; Wang et al., 2008). For example, the combination of alternative splicing and multiple transcriptional start sites in human tropomyosin alpha gives rise to 18 distinct protein coding transcripts (reviewed in Gooding & Smith, 2008). Similarly the combinations of alternative transcriptional initiation and exon splicing may give rise to 5–10 distinct protein coding transcripts in each of the MP targeting (regulatory) subunits Mypt1 and Mypt2 (Dippold & Fisher, 2014a). Most genes are part of a multigene family which additively increase the number of transcripts and enables their independent expressional regulation, while alternative exon usage through a pre-mRNA transcript produces multiplicative diversity, with Down syndrome cell adhesion molecule perhaps being the most robust example with the possibility of 37,000 unique transcripts (Schmucker & Chen, 2009). The diversity in gene products controlling VSM function will be considered later in the context of pharmacomechanical coupling and drug responses. For a thorough review of diversity of gene products in relation to smooth muscle contractile function see Fisher (2010) and references therein. Interindividual differences in gene sequences may also determine vascular function and drug responses and are briefly considered at the end of this chapter.

4. AGONIST-MEDIATED VASOCONSTRICTION AND ITS ANTAGONISM

Signaling through G protein-coupled receptors (GPCRs) plays a major role in the control of VSM tone and thus vascular resistance. GPCRs are a large family of proteins and a frequent drug target due to their accessibility and success of drug targeting to improve patient outcomes (reviewed in Maguire & Davenport, 2005). Various screens and other high-throughput methods have recently identified novel GPCRs (so-called orphan receptors) that control VSM tone and possibly in a vessel-specific manner (Pluznick et al., 2013), reviewed in Pluznick (2013). These may provide novel- and vessel-specific targets for future drug development. The classic GPCRs mediating VSM constriction are receptors for neural (α-adrenergic), paracrine (endothelin), and endocrine (angiotensin) signaling as well as others (Fig. 1). The diversity in the expression of receptor family members and complexity of their G protein-initiated downstream signaling provides an opportunity for great specificity in the control and targeting of VSM tone (Fig. 1).

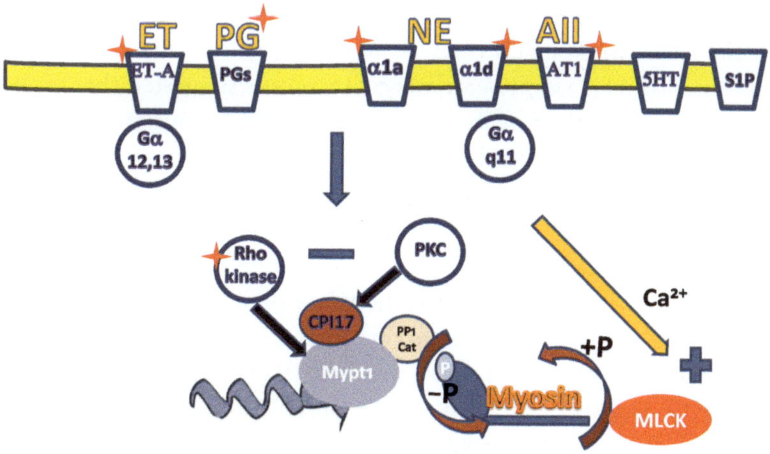

Fig. 1 Diversity in signaling pathways to activate force in vascular smooth muscle (VSM). Shown are plasmalemmal G protein-coupled receptors that when bound by ligand will cause contraction of VSM. The GPCRs may couple to different G proteins as shown. These signaling pathways activate force by (1) calcium flux from intracellular and extracellular sources activating MLCK and (2) inhibition of myosin phosphatase, thereby increasing the sensitivity to calcium. Inhibition of MP may occur through either PKC-dependent phosphorylation of the inhibitory subunit (CPI-17) or Rho kinase-dependent phosphorylation of the regulatory subunit Mypt1. It is the balance of the activities of MLCK and MP that determine the level of phosphorylation of myosin and thus force. Abbreviations: *ET*, endothelin; *PG*, prostaglandin; *NE*, norepinephrine; *AT, AII*, angiotensin (II); *PKC*, protein kinase C; *PP1cat*, Protein phosphatase type 1 catalytic subunit; *MLCK*, myosin light chain kinase; *Mypt1*, myosin phosphatase targeting subunit 1. ✦, drug targets.

4.1 Vasoconstriction

Sympathetic neural activation of α-adrenergic receptors present on the surface of RASM plays a major role in maintaining basal tone and vasoconstriction during exercise and in disease states. In the smaller rat mesenteric resistance arteries studied ex vivo α1a receptors coupled to protein kinase C (PKC) which via the phosphorylation of the MP inhibitory subunit CPI-17, inhibited MP activity thereby increasing VSM force at any given concentration of calcium, i.e., calcium sensitization of force production (Kitazawa & Kitazawa, 2012). In contrast, in the large conduit aortic smooth muscle the α1d receptor-mediated α-adrenergic vasoconstriction and required the activity of Rho kinase and the constitutive phosphorylation of its target the MP regulatory subunit Mypt1 (Kitazawa & Kitazawa, 2012). In intermediate sized arteries both pathways were active, consistent with a study of the intermediate sized mouse femoral artery in vivo (Zacharia, Mauban, Raina, Fisher, & Wier, 2013). Pharmacological inhibition specific for α1a- or α1d-adrenergic receptors suggested that each contributed approximately equally to the resting tone of this artery.

Evidence for specificity in pharmacomechanical coupling of the GPCR to the contractile apparatus also comes from smooth muscle-specific conditional knockout (SMcKO) of the G proteins to which the GPCRs couple (see Loirand, Sauzeau, & Pacaud, 2013) (Fig. 1). SMcKO of Gαq11 completely abrogated constrictor responses to phenylephrine (PE) and angiotensin II (AII) but only partially suppressed constrictor responses to endothelin-1 (ET-1) and the thromboxane analogue U46619 (Wirth et al., 2008). In contrast SMcKO of Gα12,13 had no effect on PE-induced constriction and a partial effect on the other agonists. Perhaps most interestingly, SMcKO of Gα12/13 had no effect on basal BP but suppressed the development of salt-induced hypertension. In a more recent study, these same investigators examined specificity of pharmacomechanical coupling in a mouse model of spontaneous aging-related vascular dysfunction and hypertension (Wirth et al., 2015). At one year of age, the mice had increased vascular resistance as the cause of the increased BP that was specifically suppressed with the ET-A receptor blocker Danusentan or SMcKO of the endothelin-A (ET-A) receptor. α-Adrenergic or AII receptor blockers prazosin and losartan, respectively, caused similar degrees of BP lowering in young and old mice. The hypertension of aging was also suppressed by SMcKO of Gαq11 or Gα12/13 suggesting that GPCRs that couple to both of these pathways are involved. These studies suggest that targeting of specific components of these signaling pathways may be able to suppress the development of hypertension while leaving intact the normal physiological regulation of BP. These studies provide theoretical support for precise targeting of these, and perhaps other, signaling pathways in the treatment of vascular dysfunction and hypertension.

4.2 Unresolved Questions

The mechanisms for the differential activation of vasoconstrictor signaling pathways in diverse VSM and sensitivities to pharmacological inhibition are not well understood. Differential activation could be due to differential expression of gene products in the pathway or differential coupling between receptors and targets. Some studies have supported differential expression of α1-adrenergic receptor subtypes in the vascular system while others have not (reviewed in Docherty, 2010). It has also been suggested that differential expression of the MP inhibitory subunit CPI-17 determines the extent to which the PKC pathway is used in GPCR inhibition of MP (Kitazawa, Polzin, & Eto, 2004; Woodsome, Eto, Everett, Brautigan, & Kitazawa, 2001). However, there has been limited systematic and comprehensive analysis of differences in gene programs between arteries. We used deep sequencing (RNASeq) of RNAs from rat aorta (AO) and mesenteric arteries (MA) for a comprehensive and unbiased comparison of gene expression in these very different arteries. We did not find significant differences in expression of α-adrenergic receptor subtypes nor CPI-17/Mypt1 ratios (Reho, Shetty, Dippold, Mahurkar, & Fisher, 2015). There were significant differences in the expression of purinergic receptors (P2R) that mediate vasoconstriction. In MA P2Rx1 was ~50-fold higher than P2Rx5, and P2Rx1 was 10-fold higher in MA vs AO. In AO P2rx1 and P2Rx5 were expressed at similar levels, and P2Rx5 was approximately fourfold higher in AO vs MA. Comparisons between these and other studies may be confounded by differences between species, including humans, and dependent upon which blood vessels in which vascular beds are examined.

Many of the studies have depended upon chemical inhibitors to probe these pathways. These studies must be interpreted cautiously given that inhibitors have only relative specificity for their intended targets and differences in their metabolism may confound cross-species or even multitissue comparisons. Many of the experiments are performed ex vivo with limitations in methods that may confound extrapolation to in vivo physiology, in particular in terms of how initial tone is developed, presence of pressure and flow, and regulatory and counter-regulatory influences. Gene targeting, usually in mice, provides great specificity for testing specific components of these signaling pathways, and the development of a smooth muscle-specific and Tamoxifen-inducible Cre transgene (Wirth et al., 2008) provides a powerful tool for testing gene function specifically in smooth muscle at any stage from early development through maturity. Genetic manipulation has been used to only a limited extent to test these vasoactive pathways. As one example, a recent study (Chen et al., 2015) examined the effect of mutation of what have been identified as Rho kinase phosphorylation sites in the MP regulatory subunit Mypt1 (discussed later). Surprisingly T852A mutation had no effect on force production or myosin phosphorylation, yet force was inhibited by a Rho-associated protein kinase (ROCK) inhibitor. This suggests that RhoKinase phosphorylation of Mypt1 at T852 is not involved in force activation with the caveat that (1) this study examined bladder smooth muscle and (2) signaling was not tested in models of disease such as hypertension. Along similar lines knockout of Rho kinase has not so far been shown to effect arterial function, though this may be confounded by redundancy in function between ROCK1 + 2 (reviewed in Loirand et al., 2013; Shimokawa, Sunamura, & Satoh, 2016). This is covered in more detail in chapter "Rho-mancing to sensitize calcium signaling for contraction in the vasculature: Role of Rho kinase" by T. Szasz and R. Clinton Webb. The issue of muscle diversity determining tissue-specific responses to signaling pathways has not yet been examined in these genetically modified mice.

4.3 Clinical Significance

A number of different classes of drugs are available to suppress receptor-mediated vasoconstriction and thereby lower vascular resistance and BP. These include α-adrenergic subtype selective blockers, renin–angiotensin–aldosterone (RAA) blockers (ACE inhibitors or AII receptor blockers), ET receptor blockers, and the Rho kinase inhibitor Fasudil. While each may lower BP, they do not appear to be equivalent in their effects. Some disparity may be due to effects beyond the VSM, for example kidney, heart, and brain, while some disparity may be due to vessel-specific effects as described earlier, though this remains conjecture at this point. Fasudil (Y-27632) was identified as a Rho kinase inhibitor that suppressed PE/α-adrenergic inhibition of MP, i.e., inhibited calcium sensitization, and lowered BP in hypertensive rats with minimal effect in control rats (Uehata et al., 1997). It is approved in Japan and used for the treatment of cerebral vasospasm in the setting of subarachnoid hemorrhage and ischemic stroke and is in clinical trials for a number of vascular diseases (reviewed in Loirand et al., 2013; Shimokawa et al., 2016), but is not approved for use in the United States. This and studies reviewed earlier and later support the premise that coupling of signaling pathways, and thus effects of pharmacological blockade on vasoreactivity, may vary dependent upon the model of disease and be disease specific, whereas many studies to define signaling pathways are performed on animals in the absence of established disease. α-Adrenergic receptors antagonists effectively lower vascular resistance and BP in humans with hypertension and heart failure. Unfortunately these drugs provide less improvement in clinical outcomes as compared to other classes of drugs, including vasodilators (Cohn et al., 1986; The ALLHAT Officers, 2002), supporting the premise that lowering of systemic vascular resistance and BP does not bear a simple relationship to patient outcomes. Similarly the ET antagonists are approved for treatment of pulmonary hypertension but not systemic hypertension nor heart failure

(Nasser & El-Mas, 2014). The specific mechanisms explaining the lack of equivalence of different classes of vasodilators are not defined, and further research is indicated to determine if it is a function of which vessels respond and/or how the vasodilation is achieved. Drugs that more selectively block the different α-receptor subtypes are commercially available but have not been tested for their ability to improve cardiovascular outcomes (see Andersson & Gratzke, 2007; Chen & Minneman, 2005).

5. SIGNALING-MEDIATED VASODILATION AND ITS AGONISM

The endothelium-derived relaxing factor (EDRF, now known as NO) and subsequently endothelium-derived hyperpolarizing factor (EDHF) were discovered a century and more after NO-generating chemicals, nitrates (nitroglycerin) and nitrites (amyl nitrite), were synthesized and used medicinally for their vasodilatory properties manifest by intense headaches, flushing, and relief of angina (reviewed in Marsh & Marsh, 2000). Since the discovery of the critical role of endothelium-derived factors in the control of blood flow, tremendous progress has been made in (1) signaling pathways dependent upon the second messenger cGMP (reviewed in Hofmann, Feil, Kleppisch, & Schlossmann, 2006) (Fig. 2), or via direct reactions of NO with substrates (nitrosylation) (reviewed in Lima, Forrester, Hess, & Stamler, 2010), (2) desensitization and pharmacological targeting of this pathway (discussed later), and (3) the role of reduced bioavailability of EDRF/NO in suppressed vasodilator responses in disease, i.e., "endothelial (vascular) dysfunction" (Gimbrone & Garcia-Cardena, 2016; Ludmer et al., 1986).

That smooth muscle tissue may vary in their responses to NO and cGMP signaling was suggested by experiments conducted around the time of their initial discovery. Experiments on isolated vascular and non-VSM tissues suggested resistance or relative resistance to EDRF (NO) and second messenger cGMP using various smooth muscle containing preparations including dog portal vein (Feletou, Hoeffner, & Vanhoutte, 1989), chicken gizzard (Pfitzer, Merkel, Ruegg, & Hofmann, 1986), and non-VSM tissues (Diamond, 1983). Subsequent studies supported the concept that classic EDRF/NO plays a lesser and EDHF(s) a greater role as artery size decreases in sundry regional circulations (Chu et al., 1990; Garland & McPherson, 1992; Kuo, Davis, & Chilian, 1995; Nagao, Illiano, & Vanhoutte, 1992; Sellke, Myers, Bates, & Harrison, 1990; Shimokawa et al., 1996), including in humans (Fok, Jiang, Clapp, & Chowienczyk, 2012). However, a possible role for smooth

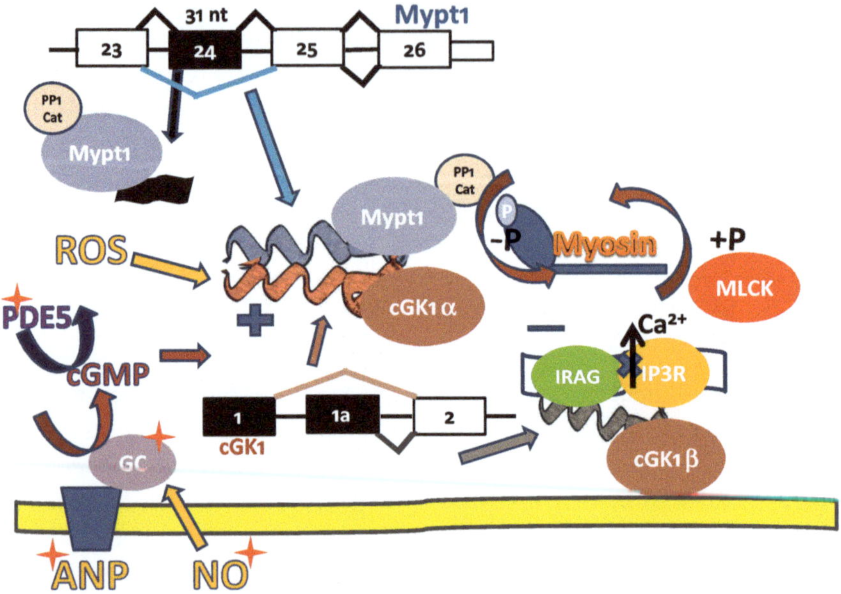

Fig. 2 Diversity in signaling for relaxation of VSM. ANP and NO signal through guanylate cyclase and the second messenger cGMP to activate VSM relaxation. ANP activates a membrane-bound receptor that has endogenous guanylate cyclase activity, while NO diffuses into the cytosol to bind and activate guanylate cyclase. cGMP is degraded by PDEs, particularly PDE5, or bind and activate its target cGK1α. cGK1α is targeted to its substrates by its N-terminus leucine zipper (LZ) motif, represented in the cartoon by a ribbon. One of its targets is the MP. cGK1α binds to Mypt1, the regulatory subunit of MP, by LZ-mediated heterodimerization between the C-terminus of Mypt1 and N-terminus of cGK1α. Activation of MP by cGK1α causes dephosphorylation of myosin and thus relaxation even in the presence of activating calcium, i.e., calcium desensitization of force production. Isoforms of Mypt1 are generated by the alternative splicing of the 31 nt exon 24 (24). Skipping of E24 generates an mRNA that codes for the C-terminal LZ motif (shown as a *blue ribbon*). Inclusion of E24 shifts the reading frame and codes for a distinct C-terminus lacking the LZ motif as well as generating a premature termination codon (shown as a *truncated black ribbon*). This LZ- variant of Mypt1 is expressed in phasic smooth muscle and the smaller arteries and is neither bound nor activated by cGK1α. cGMP also causes relaxation by suppressing calcium flux. In this case cGK1β binds to IRAG and suppresses calcium release from the SR IP3R calcium release channel. As discussed in the text splice variants of IRAG are proposed to determine ability of cGK1β to inhibit IP3R. Variants of cGK1 are generated via alternative first exons coding for distinct N-terminal LZ sequences. ROS also activate cGK1 but only the cGK1α isoform due to the presence of Cys42 within the unique N-terminus of cGK1α. *ANP*, atrial naturetic protein; *NO*, nitric oxide; *GC*, guanylate cyclase; *cGMP*, cyclic guanosine monophosphate; *PDE*, phosphodiesterase; *ROS*, reactive oxygen species; *cGK*, cyclic GMP kinase; *IRAG*, IP3R-associated gating protein; *IP3R*, inositol triphosphate receptor; *nt*, nucleotide. ✦, Drug targets.

muscle phenotypic diversity in determining responses to endogenous or pharmacological activators of this pathway was mostly overlooked as the field focused on bioavailability of NO and acute or chronic desensitization of this pathway (see, for example, Steinhorn, Loscalzo, & Michel, 2015).

5.1 NO-Mediated Vasodilation and Specificity

5.1.1 Calcium Sensitivity and MP

The identification of the intermediate signaling molecules and end targets (effectors) of NO/cGMP-mediated vasodilation provided the opportunity to examine how their regulated expression may determine the response. Classic experiments by Kitazawa (Lee, Li, & Kitazawa, 1997) and Somlyo (Wu, Somlyo, & Somlyo, 1996) and coworkers demonstrated that cGMP relaxes VSM through activation of MP and thus a reduction in the calcium sensitivity of the myofilaments for force production. In these experiments, MLCK and thus force were activated in permeabilized VSM under submaximal calcium clamp. Increasing concentrations of the cGMP analogue 8-Br-cGMP caused dephosphorylation of myosin and relaxation of VSM indicative of activation of MP. A subsequent study by Surks, Mendelsohn, and coworkers (Surks et al., 1999) identified a mechanism for cGMP activation of MP. They showed that the cGMP-dependent protein kinase (cGK1α) is targeted to MP through interactions mediated by the leucine zipper (LZ) dimerization motifs in the C-terminus of Mypt1 and N-terminus of cGK1α (Fig. 2). The specific mechanism by which cGK1α activates MP is still debated (Grassie et al., 2012; Nakamura, Koga, Sakai, Homma, & Ikebe, 2007; Wooldridge et al., 2004; Yuen, Ogut, & Brozovich, 2014), but presumably involves phosphorylation of Ser–Thr residues of Mypt1/MP.

In that same time period (1990s), the era of gene discovery, a 31 nt exon 24 (E24) 3′ splice variant of Mypt1 had been described that, by changing the reading frame, codes for a Mypt1 variant that lacks the C-terminal LZ motif (LZ-) (Chen et al., 1994; Johnson, Cohen, Chen, Chen, & Cohen, 1997) (Fig. 2). We characterized the pattern of expression of the Mypt1 isoforms and correlated them with the functional test, the ability of NO/cGMP to activate MP, and thereby relax smooth muscle, i.e., calcium desensitization of force production (reviewed in Dippold & Fisher, 2014b; Reho et al., 2014). Mypt1 mRNAs containing the E24 alternative exon coding for the LZ-isoform (E24+/LZ−) are exclusively or highly expressed in phasic smooth muscle of the rat and mouse including portal vein, bladder, and intestine (Fu et al., 2012; Khatri, Joyce, Brozovich, & Fisher, 2001; Payne et al., 2006) and are predominant in small arteries of the rat (80:20

ratio E24+/−) (Karim et al., 2004; Lu et al., 2008; Payne et al., 2004; Zhang & Fisher, 2007) and to a lesser extent the mouse (~60:40 ratio) (Reho, Kenchegowda, Asico, & Fisher, 2016; Reho, Zheng, Asico, & Fisher, 2015; Zheng, Reho, Wirth, & Fisher, 2015). These small arteries have a mixture of phasic force production, termed vasomotion, and maintained (tonic) force (reviewed in Aalkjaer et al., 2011; Haddock & Hill, 2005). In contrast in the larger arteries and veins such as the AO, vena cava, and carotid arteries, the Mypt1 E24− variant coding for LZ+ isoform predominates (~80:20 ratio of E24−/+). Similar patterns of expression are observed in humans (unpublished data). We and others have also demonstrated that the expression of these isoforms is highly dynamic (reviewed in Dippold & Fisher, 2014b; Reho et al., 2014). During postnatal maturation of the rodent vascular, gastrointestinal (GI), and genitourinary (GU) systems, smooth muscle tissues destined for a mature phasic or mixed phenotype, including the small arteries, switch from Mypt1 E24−/LZ+ to E24+/LZ− as part of a more generalized shift from a slow (tonic) to fast (phasic) gene programs (Reho & Fisher, 2015; Zheng et al., 2015). In disease models including sepsis (Reho, Zheng, et al., 2015), portal hypertension (Payne et al., 2004), and flow-induced small artery remodeling (Zhang & Fisher, 2007; Zhang, Pakeerappa, Lee, & Fisher, 2009), a switch to Mypt1 E24−/LZ+ isoform correlates with increased sensitivity to NO/cGMP-mediated activation of MP and vasorelaxation (calcium desensitization). The switch to E24−/LZ+ isoform is proposed to allow for maximum sensitivity and response to NO, as for example in sepsis, when full vasodilation is required to maximize blood flow. In other models, such as behavioral stress of neonatal rats induced by separation from their mothers, thereby accelerating maturational programming of these arteries (Reho & Fisher, 2015), and in animal models of heart failure (Karim et al., 2004) and pulmonary hypertension (Konik, Han, & Brozovich, 2013), a switch to the E24+/LZ− isoform is associated with reduced sensitivity to NO/cGMP-mediated vasorelaxation. The shift to the E24+/LZ− isoform during normal arterial maturation is proposed to allow for the increased vascular resistance of the systemic circulation, a necessary condition for the highly regulated high pressure/high resistance state. The skipping of Mypt1 E24 coding for the Mypt1 LZ+ isoform is the evolutionarily primordial and default pattern (Dippold & Fisher, 2014a), suggesting that splicing of Mypt1 E24 arose in parallel with the evolution of high pressure/high resistance circulations.

More recently, we have genetically engineered the mouse to directly test the role of Mypt1 E24 splice variants in determining VSM sensitivity to

NO/cGMP-mediated vasorelaxation and vascular control in vivo. LoxP sites were inserted into the introns flanking Mypt1 E24–300 nt from the splice sites so as to avoid identified and conserved splicing regulatory sequences (Dippold & Fisher, 2014a; Fu et al., 2012; Shukla & Fisher, 2008). Crossing of these floxed mice with the SMMHCCreER^{T2} mice enabled Tamoxifen-inducible deletion of E24 specifically in smooth muscle (SMcKO E24) with high efficiency, thereby converting the muscle to the E24−/LZ+ isoform of Mypt1 (Reho et al., 2016; Reho, Zheng, et al., 2015). The small MAs of these mice were studied ex vivo in a wire myograph under isometric conditions to define the contractile properties of the VSM. SMcKO E24 markedly increased sensitivity of MA1s (1) to the NO donor DEA/NO (EC50: CON: ~18 μm vs SMcKO: ~3 nM) in arteries preconstricted with the α-adrenergic agonist PE and (2) to 8-Br-cGMP (EC50: CON: ~95 nM vs SMcKO: ~3 nM) in MA1s permeabilized with α-toxin and force activated by submaximal calcium clamp (pCa 6) (Reho et al., 2016; Reho, Zheng, et al., 2015). This assay indicates enhanced activation of MP in situ. The SMcKO E24 mice also had reductions in ambulatory BP of ~20 mm Hg measured by telemetry with no change in the diurnal variation. Notably, deletion of 1 or both E24 alleles had the same effect on NO/cGMP responses and BP, indicating high sensitivity to E24 splicing. These experiments establish the role of the Mypt1 E24 splice variants in vasodilator responsiveness and regulation of BP and support a model in which toggling of Mypt1 splice variants determines sensitivity to NO/cGMP-mediated relaxation. Stated differently, the expression of the Mypt1 E24+/LZ− isoform (1) by reducing sensitivity to NO/cGMP, provides a space for myriad other signaling pathways to control VSM tone in a finely tuned and combinatorial fashion and (2) provides reserve within this signaling pathway which can be called upon under conditions when maximal vasodilation to NO/ROS is required as for example in sepsis (Reho, Zheng, et al., 2015).

5.1.2 Calcium Flux and IRAG

Vasodilator signals also reduce calcium flux. The IP3 receptor-associated gating protein (IRAG) was identified as a critical target of cGK1β in mediating this effect (Schlossmann et al., 2000) (Fig. 2). A number of variants of IRAG generated by alternative splicing and alternative transcriptional start sites were more recently identified (Werder et al., 2011). It was suggested based on patterns of expression and in vitro biochemical studies that splice-variant truncations of IRAG may suppress cGK1β binding and activation, and thus lead to smooth muscle tissues-specific responses to NO/cGMP-mediated suppression of calcium release, analogous to the effect of Mypt1 splice variants on calcium sensitivity discussed earlier. In a prior study (Frei et al., 2009), these investigators had shown that mouse jejunal and colon smooth muscle use different pathways for cGMP-dependent relaxation, with jejunum dependent upon reductions in calcium sensitivity (MP activation), and colon dependent upon inhibition of calcium flux. These investigators suggested in their more recent study that the splice variants of IRAG may confer tissue-specific resistance to cGMP inhibition of calcium flux.

5.2 Other Components of Vasorelaxant Signaling

More recently, it has been shown that this vasorelaxant pathway is also activated by reactive oxygen species (ROS) as the signal and cGK1 as the signaling effector (Burgoyne et al., 2007; Burgoyne, Oka, Ale-Agha, & Eaton, 2013; Prysyazhna, Rudyk, & Eaton, 2012; Reho, Zheng, et al., 2015; Rudyk, Prysyazhna, Burgoyne, & Eaton, 2012) (Fig. 2). Molecular diversity again plays a role, as the reactive cysteine that transduces the ROS signal is present in the cGK1α but not the cGK1β isoform, with crystal structure showing that oxidation of the C42 disulfide bond stabilizes cGK1α (Qin et al., 2015). cGK1α and cGK1β differ only in their first exons and thus N-terminal protein sequence (Fig. 2) presumably due to alternative transcriptional start sites (sometimes mistakenly referred to as alternative splicing). However there is limited information regarding the functional significance of the cGK1 isoforms. Isoform specific in vivo substitution experiments showed minimal effect (Weber et al., 2007). Alternatively, it is possible that the two isoforms allow for independent control of their expression through transcriptional control or some other mechanism, but again there is limited information on this topic. There is additional diversity all through the pathway, including the guanylate cyclases that generate cGMP, PDEs that catabolize it, cGK effectors and downstream targets, as well as noncanonical pathways, which cannot feasibly be reviewed here.

5.3 Unresolved Questions

Experiments over the past few decades have identified the targets of NO/cGMP signaling that mediate vasorelaxation through effects on calcium sensitivity and/or calcium flux. There is substantial data to support unique responses of phenotypically distinct smooth muscle to this signaling pathway. Some of the mechanisms have been defined but many questions remain. It remains to be determined how switching to the Mypt1 E24−/LZ+ isoform (cGMP responsive) affects the force production of prototypical phasic smooth muscle in portal vein, bladder, and intestine. Smooth muscles throughout the vascular, GI and GU systems have very different power outputs with phasic, tonic, or mixed patterns, and there is supportive evidence that cGMP (and other signals) may differentially affect the frequency, amplitude, or maintenance of force (tone) dependent upon the muscle type and stimulus history (see Haddock & Hill, 2005). Similarly it is not yet clear to what extent muscle activation history determines the magnitude of the response to this (cGMP activation of MP) or other signaling pathways, as for example in functional sympatholysis when local vasodilator signals must overcome sympathetic vasoconstrictor signaling to increase blood flow to the exercising muscles (Mizuno, Iwamoto, Vongpatanasin, Mitchell, & Smith, 2014; Nyberg et al., 2015; Saltin & Mortensen, 2012). In our studies, we noted that the shift in sensitivity to the NO donor in muscle preactivated with PE was much greater than that to cGMP in permeabilized muscle activated with calcium. While this is an apples-to-oranges comparison with numerous confounding variables, it does support a model in which NO/cGMP causes disinhibition of MP in the context of contractile agonist signaling (Grassie et al., 2012; Nakamura et al., 2007; Wooldridge et al., 2004). Similarly the relative contributions of effect on calcium flux vs calcium sensitivity in mediating vasorelaxation dependent upon muscle type, activation history, and disease state is not well defined. In the studies of Mypt1 variants determining calcium sensitivity, it remains to be determined how much the response to endogenous NO is affected in vivo under conditions where NO signaling is activated, for example during flow-mediated dilation of exercise. Similarly it has yet to be determined how responses to NO donor drugs are affected in vivo (discussed later). Lastly, how the switching of Mypt1 E24 splice variants, and its forced switch in the SMcKO mouse model, affects vascular function and drug responses in models with preexisting disease, particularly those characterized by reduced bioavailability of NO and oxidative/nitrosative stress, has not been studied much. Our preliminary study suggests that forced switch to the Mypt1 E24−/LZ+ isoform may have beneficial effects on the vascular function of mice on a high salt diet (Reho et al., 2016), but further study is required in this and other disease models. The proposed model for IRAG splice variants determining response to cGMP inhibition of calcium flux needs to be tested in in vivo in mice genetically engineered to express specific isoforms, followed by a similar sequence as described earlier for the study of effects on calcium sensitivity.

5.4 Clinical Significance

Current textbooks of pharmacology relate the antianginal effect of nitroglycerin (NTG) placed under the tongue to systemic venodilation and thus unloading of the heart, with coronary artery and systemic arterial vasodilation only occurring at higher concentrations of the drug usually delivered intravenously. However, and surprisingly, after the many years of study, the action of NTG when placed under the tongue remains contentious vis-a-vis unloading of the heart through venodilation vs increasing coronary blood flow via coronary artery dilation. The use of organic nitrate vasodilator drugs have assumed a smaller share of cardiovascular medicine in recent years most likely due to a general absence of studies showing that they improve long-term clinical outcomes, in contrast to the well-established improvement in outcomes with drugs blocking RAA and beta-adrenergic receptor signaling (see Florea & Cohn, 2014). One exception to this is the demonstration that long-acting nitrates when added to the direct-acting vasodilator hydralazine improved outcomes when added to standard therapy in white (Cohn et al., 1986) and black patients (Taylor et al., 2004) with advanced systolic heart failure, though not as effectively as blockers of RAA activation in the white population (Cohn et al., 1991). In contrast to the diminishing role of medicinal nitrates in the cardiovascular armamentarium, the other components of this pathway are being increasingly targeted. Over the past few decades, there have been the development of inhibitors of PDE5 (Sildenafil class) and activators of guanylate cyclase (Ghofrani, Osterloh, & Grimminger, 2006), and most recently endopeptidase inhibitors that increase bioavailability of ANP (Entresto) (McMurray et al., 2014), also signaling through cGMP. Interestingly the PDE5 inhibitors failed for the purpose for which they were developed, treatment of angina, but were rescued by their efficacy in treatment of erectile dysfunction. Since their initial introduction these and the other drugs targeting this pathway have found many uses, in many instances by serendipity, and in particular in the treatment of pulmonary hypertension. Whether this efficacy is a function of smooth muscle phenotype determining responses to this pathway, e.g., in pulmonary vs systemic circulations, remains to be determined (see Francis, Busch, & Corbin, 2010; Konik et al., 2013). There also is interest in new NO-derived compounds that may have unique properties as to their effects on cardiac and smooth muscle and unique clinical efficacies, being developed by Cardioxyl Pharmaceuticals, now a part of Bristol-Myers Squibb.

6. CALCIUM FLUX AND ITS INHIBITION

Cytosolic calcium activates force within all muscle types and is universally used by cells for signaling. Yet how calcium activates force and how calcium flux is controlled varies considerably by muscle type across muscle lineages (smooth, skeletal, and cardiac) and sublineages (fast vs slow muscle; Berridge, 2008; Karaki et al., 1997; Wray & Burdyga, 2010). It is this diversity in muscle with respect to calcium that underlies the differing hemodynamic effects of the different classes of drugs that block the L-type calcium channels (LTCC) as will be discussed later. A functional study compared the conduit rat AO with the muscular bovine tail artery (Ashida, Schaeffer, Goldman, Wade, & Blaustein, 1988). Agonist- and depolarization-induced contractions were substantially more suppressed by Ryanodine in AO and by CCB in tail artery, consistent with electron microscopy studies in concluding that the muscular tail artery has ~60% less sarcoplasmic reticulum than AO and is substantially more dependent on LTCC for calcium entry. This is consistent with the paradigm of phasic vs tonic muscle, with the latter having approximately double the amount of sarcoplasmic reticulum as the former (Devine, Somlyo, & Somlyo, 1972), analogous to the situation in fast vs slow striated muscle (Berridge, 2008).

6.1 LTCC

Within the LTCC itself there is great diversity generated by the expression of different subunits, subunit isoforms generated by multigene families, and alternative transcription and splicing of exons (reviewed in Franz Hofmann, Flockerzi, Kahl, & Wegener, 2014). The α1 subunit (Cav1.2α1) is the pore forming unit that conducts calcium in muscle (Fig. 3). The mammalian gene contains 49–50 exons of which at least 20 are reported to undergo alternative usage with the potential to generate thousands of different isoforms (Hofmann et al., 2014; Tang et al., 2004). The tissue-specific expression of these isoforms may determine differences in LTCC function and sensitivity to dihydropyridine (DHP) class CCBs. Tissue-specific expression of Cav1.2α1 splice variants involves alternative exons 1/1a, 8/8a, 9/9*, and 31/32. One study suggested that the Cav 1.2 "smooth muscle" splice variant is more sensitive to inhibition by DHP, as compared to cardiac isoforms, because it contains exon 8, coding for sequence within transmembrane segment IS6 where DHPs bind (Welling et al., 1997) (Fig. 3). A subsequent

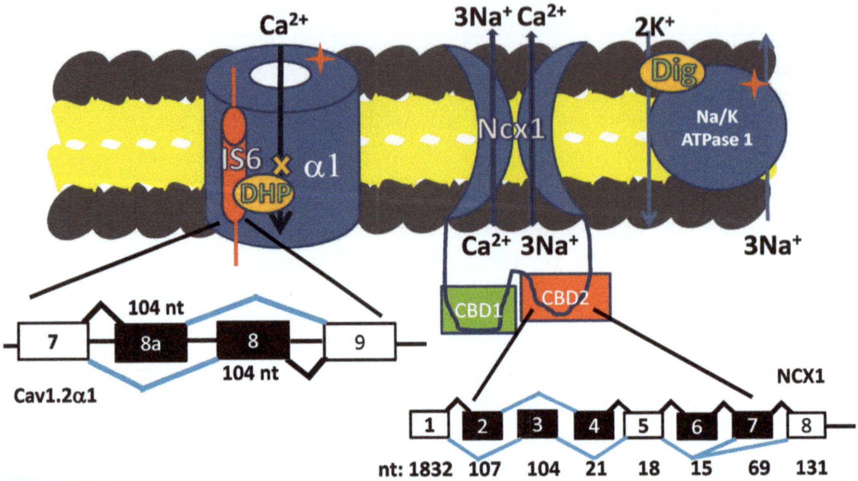

Fig. 3 Diversity within calcium handling proteins in determining calcium flux and drug responses. Calcium entry through the α1-subunit of the LTCC is a major source of calcium for activation of MLCK and force in VSM. A number of splice variants of the ~50 exon α1 subunit have been described. Only mutually exclusive splicing of exons 8/8a coding for sequence within the transmembrane IS6 segment is shown. These code for the sequence and region where DHP class of CCBs bind and inhibit calcium flux through the channel. The alternative splicing of exon 8/8A is proposed to determine increased sensitivity of VSM to DHP CCBs as compared to striated muscle, though other splice variants have also been proposed as causal. NCX1 exchanges calcium for sodium and can function in forward or reverse mode depending upon electrochemical gradients. A very long first coding exon is followed by two mutually exclusive exons that are traditionally designated A and B but here are labeled numerically for consistency. The following exons also undergo cassette-type alternative splicing. These alternative exons code for sequence in the second calcium-binding domain and are thought to determine calcium binding to this low affinity site. The Na-KATPase generates an electrochemical gradient and by setting intracellular sodium concentrations indirectly affects the activity of NCX and thus calcium flux. The sodium–potassium pump is the target of digitalis class of drugs and endogenous oubain, which by blocking Na-KATPase increases NCX activity in forward mode. *IS6*, intramembrane segment 6; *DHP*, dihydropyridine; *NCX*, sodium–calcium exchanger; *Dig*, digoxin; *Na-KATPase*, sodium–potassium ATPase; *CBD*, calcium-binding domain; *nt*, nucleotide. ✦, drug targets.

study suggested that alternative exon 33 coding for part of the extracellular IVS3–4 linker confers this property (Liao et al., 2007). A third study of rat cerebral resistance arteries suggested that alternative exons 1c and 9 encodes the unique hyperpolarizing shift in voltage sensitivity of the LTCC and DHP sensitivity (Cheng et al., 2009). Limitations of these studies include: (1) isoforms were studied in heterologous expression systems (HEK293 cells). Their function has yet to be tested in SMCs and more importantly in vivo through genetic manipulation to force isoform switching and (2) there has been no systematic study of the many different combinations of splice-variant isoforms. In addition to the Cav1.2α1 splice variants, the relatively depolarized state of resting VSM with basal calcium conductance, and channels in an inactivated state, has also been proposed to explain the relative selectivity of DHPs for VSM. In contrast, the non-DHP CCBs (verapamil, diltiazem) suppress cardiac pacemaker and contractile functions with less potent vasodilatory properties.

The expression of these variants may change during development or in disease or vary between humans (Wang, Papp, Binkley, Johnson, & Sadee, 2006) and potentially could affect drug responses (discussed later). Mutations in mutually exclusive alternative exons 8 (G406R or G402S) or 8a (G406R) cause Timothy syndrome, a rare genetic disease characterized by long QT intervals on EKG and ventricular arrhythmias, autism, and immune deficiencies (Splawski et al., 2004). These mutations may inhibit voltage- or calcium-dependent inhibition (VDI and CDI) of the channel (Dick, Joshi-Mukherjee, Yang, & Yue, 2016) reviewed in Dixon, Cheng, Mercado, and Santana (2012), and thus raise the possibility that these splice variants may determine VDI, CDI, or calcium conductance, in addition to CCB responses, in a vessel/tissue-specific manner.

6.2 NCX

The sodium–calcium exchanger (NCX) is another protein in which splice-variant isoforms generate great diversity that could impact vascular function and drug responses. NCX1–3 are members of the calcium/cation antiporter superfamily (predominately NCX and NKX in vertebrates; reviewed in Lytton, 2007). NCX1 is most widely expressed, including in VSM. Numerous isoforms are generated by variation in usage of the six exons following the large, coding first exon (Fig. 3). The first two alternative exons, labeled 2 and 3 in Fig. 3 but historically referred to as A and B, are mutually exclusive and followed by four coding exons each of which may undergo cassette-type alternative splicing (reviewed in Lytton, 2007). NCX cotransports Na^+ and Ca^{2+} in forward or reverse modes dependent upon the electrochemical gradients. The splice variants substantially alter the affinity of the second calcium-binding domain (CBD2) and thus are thought to determine the affinity of NCX1 for calcium (reviewed in Hilge, 2012), though this has yet to be tested in vivo. NCX1 through calcium entry and enhanced myogenic tone of VSM was shown to play a critical role in the development of salt-dependent hypertension in a mouse model (Iwamoto et al., 2004; Zhang et al., 2010), and a number of inhibitors of NCX1 have been developed and tested in animal models but have yet to make it to the clinic. NCX function may be indirectly impacted through the digitalis family of glycosides and their endogenous counterparts, which by inhibiting the Na-KATPase increase the amount of intracellular sodium available to drive Na^{+}–Ca^{2+} exchange (Blaustein et al., 2012; Linde, Antos, Golovina, & Blaustein, 2012). To the best of my knowledge, how splice-variant isoforms of NCX1 may affect vascular function and drug responses have not been investigated.

7. DIVERSITY WITHIN HUMAN POPULATIONS

It is also reasonable to hypothesize that there may be differences among humans in how their VSM responds to these signals and drug therapies. While this is an active area of investigation, these studies are mostly in discovery phase, and there remains little hard evidence that this is the case.

7.1 Human Diversity and Vascular Function

The vast majority of dominant genetic mutations that have been found to cause familial forms of hypertension do so through defective handling of salt by the kidneys (reviewed in Lifton, Gharavi, & Geller, 2001; Padmanabhan, Newton-Cheh, & Dominiczak, 2012). This has led to considerable support for the hypothesis that hypertension is primarily a disorder of sodium handling by the kidneys. However, others have argued that increase in systemic vascular resistance is a primary mechanism for the initiation and maintenance of increased BP (reviewed in Blaustein et al., 2012; Morris, Schmidlin, Sebastian, Tanaka, & Kurtz, 2016). In support of the vascular hypothesis is a family of Turkish ethnicity in which an autosomal dominant form of hypertension causes severe hypertension (and brachydactyly) through increase in vascular resistance (Toka et al., 2015). The affected individuals have a gain-of-function T445N mutation in PDE3A, the cGMP-inhibited cAMP phosphodiesterase 3A. These patients have increased vascular resistance, abnormal pressor responses, and increased sensitivity to nitroprusside infusion, thought to be due to defects in baroreflex buffering (Jordan et al., 2000), though the precise mechanism has not been determined. Genome-wide association studies (GWAS) also suggest that variation in genes in the vasoconstrictor and -dilator pathways could be causal in the pathogenesis of hypertension, though this remains speculative at the present time.

In the largest GWAS study of hypertension of 200,000 individuals of European descent, single nucleotide polymorphisms (SNPs) were identified in a number of genes that could exert an effect through increased vascular tone (Ehret et al., 2011). These include SNPs in CACNB2, GUCY1A3/B3, NPR3, and others. These and other association studies are limited by (1) uncertainty as to whether the SNP is causal or only a marker, in linkage with another SNP, (2) lack of mechanism, and (3) small effect size (generally less than 1 mm Hg). Smaller association studies have implicated SNPs in other genes in the vasoactive pathways, including cGK1 and NCX1 (Citterio et al., 2011). Since these genes are widely expressed their effect could be in other organs including the kidney. A major limitation of many of the studies is in the phenotyping of hypertension given its complexity. Studies of a subset of hypertensives, for example those with increased vasoreactivity in the cold pressor test, may lead to better definition of genetic variants that associate with this trait.

7.2 Diversity and VSM Drug Responses

This is also an active area of investigation that is mostly still in discovery phase. Some studies implicate variation in genes that control VSM tone in determining BP and drug response (reviewed in Cooper-DeHoff & Johnson, 2016; Roden et al., 2011). The INVEST-GENES substudy found that patients with the genotype Rs1051375 A/A in Cav1.2α1 randomized to the non-DHP CCB verapamil SR had a 46% reduction in primary outcomes (death, MI, and stroke) as compared to those randomized to treatment with the β-blocker atenolol (Beitelshees et al., 2009). Patients of the genotype G/G had a greater than fourfold increase in the primary outcome when randomized to verapamil as compared to atenolol. Whether this SNP is causal in the observed effect has not been determined. This SNP is synonymous (Thr1835Thr), but is found in a region with substantial alternative splicing (exons 40–46). The SNP is predicted to alter an exon splicing enhancer. However the splice variants expressed in the heart were not altered by the SNP (Beitelshees et al., 2009). The pharmacogenomics field continues to investigate this question, as for example in the NIH sponsored Cardiovascular Health Study (CHS). The NHLBI GO-ESP Heart Cohorts Exome Sequencing Project has DHP CCB response as a variable in BP response to treatment (available at dbGaP).

Other studies have used race as a surrogate for genotype in examining drug responses in human populations. It has been recognized and recommended in the latest guidelines (Joint National Committee, JNC8) that CCB or thiazide diuretic class of drugs should be initial therapy in black hypertensives, as these patients tend to have low renin hypertension with reduced response to RAA blockade (James et al., 2014) reviewed in Vanichakarn, Hwa, and Stitham (2014). In contrast in nonblack populations any of the five classes of antihypertensives may be selected as initial therapy in what remains essentially a random process. There also could be a genetic basis for the difference in the response of patients with heart failure to vasodilator drugs. The AHEFT trial demonstrated a significant clinical benefit in the addition of hydralazine plus nitrates to standard therapy (ACEI, β-blocker) in the treatment of blacks with classes 3 and 4 heart failure (Taylor et al., 2004). Whether or not these beneficial effects would also accrue to nonblacks has not been tested. A final note is the differences between men and women in vasoactive pathways and drug responses. Prior to menopause women have low vascular resistance and BP and reduced hypotensive responses to drugs that block β-adrenergic receptors, thought to be due to enhanced basal β-adrenergic receptor-mediated and NO-dependent vasodilation (reviewed in Charkoudian & Wallin, 2014).

7.3 Newer Methodology: iPS (Induced Pluripotent Stem) Cells

One limitation to the study of human VSM function and drug responses is the difficulty in obtaining resistance size arteries for functional and molecular studies. Investigators have studied SRAs obtained from fat and muscle biopsies (Heagerty, Heerkens, & Izzard, 2010) or from the heart (atria) at the time of surgery (Gutterman et al., 2016) and observed a number of changes in vasoreactivity that generally have not involved altered VSM contractility, though this remains a contentious issue (see Blaustein et al., 2012; Folkow, 1995; Morris et al., 2016). A newer methodology is the derivation of differentiated VSMs from iPSCs from patients with hypertension (Biel et al., 2015), see also Cheung, Bernardo, Pedersen, and Sinha (2014) and Cheung, Bernardo, Trotter, Pedersen, and Sinha (2012). This may provide a tool to study hypertension pharmacogenomics in a simpler system, though it remains to be determined how faithfully these cells recapitulate RASM function from the patients from which they were derived. The smooth muscle phenotype is quite diverse throughout the vascular system; a prior study indicated that diversity in smooth muscle phenotypes can be generated from human iPSCs using differing combinations of growth factor treatments (Cheung et al., 2014, 2012).

8. CONCLUSION

There is great diversity in the gene products that control VSM function and are the targets of drugs that relax smooth muscle and lower vascular resistance and BP. Here I have presented only a few of perhaps the best studied examples of the thousands of protein isoforms that are generated by differential exon usage in multigene families. Constrictor and dilator signaling pathways and regulation of calcium flux in relation to phenotypic (functional) diversity of VSM and drug responses were discussed. A particular emphasis was on the MP, the effector of smooth muscle relaxation and a key target of the signaling pathways that regulate vascular tone, and how splice variants of the regulatory subunit and upstream mediators may tune VSM sensitivity to endogenous and pharmacological activation of this pathway through NO and other reactive moieties. The role of splice variants and mutations in the α1 subunit of the LTCC in determining unique properties of VSM, sensitivity to DHP CCBs, and their role in Timothy syndrome were reviewed. The review finished with a discussion of interindividual (genotypic) variability in the genes in these pathways in human populations, their potential role in the increased vascular resistance of hypertension and patient-specific responses to the many drug therapies that target VSM contraction. In conclusion, it is my hope and expectation that continued investigation of VSM phenotypic diversity and its modulation in diseases of vascular dysfunction, such as hypertension and heart failure, will enable precision targeting of these therapies in a highly personalized manner.

CONFLICT OF INTEREST

The National Institute of Health supported this work (R01-HL066171).

ACKNOWLEDGMENT

I would like to thank Kimberly Oslin for her help in the preparation of this manuscript.

REFERENCES

Aalkjaer, C., Boedtkjer, D., & Matchkov, V. (2011). Vasomotion—What is currently thought? *Acta Physiologica (Oxford, England)*, *202*(3), 253–269.

Andersson, K. E., & Gratzke, C. (2007). Pharmacology of alpha1-adrenoceptor antagonists in the lower urinary tract and central nervous system. *Nature Clinical Practice. Urology*, *4*(7), 368–378.

Ashida, T., Schaeffer, J., Goldman, W. F., Wade, J. B., & Blaustein, M. P. (1988). Role of sarcoplasmic reticulum in arterial contraction: Comparison of ryanodines's effect in a conduit and a muscular artery. *Circulation Research*, *62*(4), 854–863.

Beitelshees, A. L., Navare, H., Wang, D., Gong, Y., Wessel, J., Moss, J. I., ... Johnson, J. A. (2009). CACNA1C gene polymorphisms, cardiovascular disease outcomes, and treatment response. *Circulation. Cardiovascular Genetics*, *2*(4), 362–370.

Berridge, M. J. (2008). Smooth muscle cell calcium activation mechanisms. *The Journal of Physiology*, *586*(21), 5047–5061.

Biel, N. M., Santostefano, K. E., DiVita, B. B., El Rouby, N., Carrasquilla, S. D., Simmons, C., ... Terada, N. (2015). Vascular smooth muscle cells from hypertensive patient-derived induced pluripotent stem cells to advance hypertension pharmacogenomics. *Stem Cells Translational Medicine*, *4*(12), 1380–1390.

Blaus, A., Madabushi, R., Pacanowski, M., Rose, M., Schuck, R. N., Stockbridge, N., ... Unger, E. F. (2015). Personalized cardiovascular medicine today: A food and drug administration/center for drug evaluation and research perspective. *Circulation*, *132*(15), 1425–1432.

Blaustein, M. P., Leenen, F. H., Chen, L., Golovina, V. A., Hamlyn, J. M., Pallone, T. L., ... Wier, W. G. (2012). How NaCl raises blood pressure: A new paradigm for the pathogenesis of salt-dependent hypertension. *American Journal of Physiology. Heart and Circulatory Physiology*, *302*(5), 4.

Burgoyne, J. R., Madhani, M., Cuello, F., Charles, R. L., Brennan, J. P., Schröder, E., ... Eaton, P. (2007). Cysteine redox sensor in PKGIa enables oxidant-induced activation. *Science*, *317*(5843), 1393–1397.

Burgoyne, J. R., Oka, S., Ale-Agha, N., & Eaton, P. (2013). Hydrogen peroxide sensing and signaling by protein kinases in the cardiovascular system. *Antioxidants and Redox Signaling*, *18*(9), 1042–1052. http://dx.doi.org/10.1089/ars.2012.4817.

Chakraborty, S., Davis, M. J., & Muthuchamy, M. (2015). Emerging trends in the pathophysiology of lymphatic contractile function. *Seminars in Cell and Developmental Biology*, *38*, 55–66.

Charkoudian, N., & Wallin, B. G. (2014). Sympathetic neural activity to the cardiovascular system: Integrator of systemic physiology and interindividual characteristics. *Comprehensive Physiology*, *4*(2), 825–850.

Chen, C.-P., Chen, X., Qiao, Y.-N., Wang, P., He, W.-Q., Zhang, C.-H., ... Zhu, M.-S. (2015). In vivo roles for myosin phosphatase targeting subunit-1 phosphorylation sites T694 and T852 in bladder smooth muscle contraction. *The Journal of Physiology*, *593*(3), 681–700.

Chen, Y. H., Chen, M. X., Alessi, D., Campbell, D. G., Shanahan, C., Cohen, P., & Cohen, P. T. W. (1994). Molecular cloning of cDNA encoding the 110 kDa and 21 kDa regulatory subunits of smooth muscle protein phosphatase 1. *FEBS Letters*, *356*, 51–55.

Chen, Z. J., & Minneman, K. P. (2005). Recent progress in alpha1-adrenergic receptor research. *Acta Pharmacologica Sinica*, *26*(11), 1281–1287.

Cheng, X., Pachuau, J., Blaskova, E., Asuncion-Chin, M., Liu, J., Dopico, A. M., & Jaggar, J. H. (2009). Alternative splicing of Cav1.2 channel exons in smooth muscle cells of resistance-size arteries generates currents with unique electrophysiological properties. *American Journal of Physiology. Heart and Circulatory Physiology*, *297*(2), H680–H688.

Cheung, C., Bernardo, A. S., Pedersen, R. A., & Sinha, S. (2014). Directed differentiation of embryonic origin-specific vascular smooth muscle subtypes from human pluripotent stem cells. *Nature Protocols*, *9*(4), 929–938.

Cheung, C., Bernardo, A. S., Trotter, M. W. B., Pedersen, R. A., & Sinha, S. (2012). Generation of human vascular smooth muscle subtypes provides insight into embryological origin-dependent disease susceptibility. *Nature Biotechnology*, *30*(2), 165–173.

Chu, A., Murray, J. J., Lin, C. C., Russell, M., Hagen, P. O., & Cobb, F. R. (1990). Preferential proximal coronary dilation by activators of guanylate cyclase in awake dogs. *The American Journal of Physiology*, *259*(2 Pt. 2), H340–H345.

Citterio, L., Simonini, M., Zagato, L., Salvi, E., Delli Carpini, S., Lanzani, C., ... Manunta, P. (2011). Genes involved in vasoconstriction and vasodilation system affect salt-sensitive hypertension. *PLoS One*, *6*(5), e19620.

Cohn, J. N., Archibald, D. G., Ziesche, S., Franciosa, J. A., Harston, W. E., Tristani, F. E., et al. (1986). Effect of vasodilator therapy on mortality in chronic congestive heart failure. Results of a veterans administration cooperative study. *The New England Journal of Medicine*, *314*(24), 1547–1552.

Cohn, J. N., Johnson, G., Ziesche, S., Cobb, F., Francis, G., Tristani, F., et al. (1991). A comparison of enalapril with hydralazine-isosorbide dinitrate in the treatment of chronic congestive heart failure. *The New England Journal of Medicine*, *325*(5), 303–310.

Cooper-DeHoff, R. M., & Johnson, J. A. (2016). Hypertension pharmacogenomics: In search of personalized treatment approaches. *Nature Reviews Nephrology*, *12*(2), 110–122.

Devine, C. E., Somlyo, A. V., & Somlyo, A. P. (1972). Sarcoplasmic reticulum and excitation–contraction coupling in mammalian smooth muscles. *The Journal of Cell Biology*, *52*(3), 690–718.

Diamond, J. (1983). Lack of correlation between cyclic GMP elevation and relaxation of nonvascular smooth muscle by nitroglycerin, nitroprusside, hydroxylamine and sodium azide. *The Journal of Pharmacology and Experimental Therapeutics*, *225*(2), 422–426.

Dick, I. E., Joshi-Mukherjee, R., Yang, W., & Yue, D. T. (2016). Arrhythmogenesis in Timothy syndrome is associated with defects in Ca^{2+}-dependent inactivation. *Nature Communications*, 7(10370), 1–12 (Article).

Dippold, R. P., & Fisher, S. A. (2014a). A bioinformatic and computational study of myosin phosphatase subunit diversity. *American Journal of Physiology. Regulatory, Integrative and Comparative Physiology*, *307*(3), R256–R270.

Dippold, R. P., & Fisher, S. A. (2014b). Myosin phosphatase isoforms as determinants of smooth muscle contractile function and calcium sensitivity of force production. *Microcirculation*, *21*(3), 239–248.

Dixon, R. E., Cheng, E. P., Mercado, J. L., & Santana, L. F. (2012). L-type Ca^{2+} channel function during Timothy syndrome. *Trends in Cardiovascular Medicine*, *22*(3), 72–76.

Docherty, J. R. (2010). Subtypes of functional alpha1-adrenoceptor. *Cellular and Molecular Life Sciences*, *67*(3), 405–417.

Ehret, G. B., Munroe, P. B., Rice, K. M., Bochud, M., Johnson, A. D., Chasman, D. I., ... Johnson, T. (2011). Genetic variants in novel pathways influence blood pressure and cardiovascular disease risk. *Nature*, *478*(7367), 103–109.

Feletou, M., Hoeffner, U., & Vanhoutte, P. M. (1989). Endothelium-dependent relaxing factors do not affect the smooth muscle of portal vein. *Blood Vessels*, *26*, 21–32.

Fisher, S. A. (2010). Vascular smooth muscle phenotypic diversity and function. *Physiological Genomics*, *42A*(3), 169–187.

Florea, V. G., & Cohn, J. N. (2014). The autonomic nervous system and heart failure. *Circulation Research*, *114*(11), 1815–1826.

Fok, H., Jiang, B., Clapp, B., & Chowienczyk, P. (2012). Regulation of vascular tone and pulse wave velocity in human muscular conduit arteries: Selective effects of nitric oxide donors to dilate muscular arteries relative to resistance vessels. *Hypertension*, *60*(5), 1220–1225.

Folkow, B. (1995). Hypertensive structural changes in systemic precapillary resistance vessels: How important are they for in vivo haemodynamics? *Journal of Hypertension*, *13*(12 Pt. 2), 1546–1559.

Francis, S. H., Busch, J. L., & Corbin, J. D. (2010). cGMP-dependent protein kinases and cGMP phosphodiesterases in nitric oxide and cGMP action. *Pharmacological Reviews*, *62*(3), 525–563.

Frei, E., Huster, M., Smital, P., Schlossmann, J., Hofmann, F., & Wegener, J. Ý. W. (2009). Calcium-dependent and calcium-independent inhibition of contraction by cGMP/

cGKI in intestinal smooth muscle. *American Journal of Physiology. Gastrointestinal and Liver Physiology, 297*(4), G834–G839.

Fu, K., Mende, Y., Bhetwal, B. P., Baker, S., Perrino, B. A., Wirth, B., & Fisher, S. A. (2012). Tra2beta protein is required for tissue-specific splicing of a smooth muscle myosin phosphatase targeting subunit alternative exon. *The Journal of Biological Chemistry, 287*(20), 16575–16585.

Gao, N., Huang, J., He, W., Zhu, M., Kamm, K. E., & Stull, J. T. (2013). Signaling through myosin light chain kinase in smooth muscles. *The Journal of Biological Chemistry, 288*(11), 7596–7605.

Garland, J. G., & McPherson, G. A. (1992). Evidence that nitric oxide does not mediate the hyperpolarization and relaxation to acetylcholine in the rat small mesenteric artery. *British Journal of Pharmacology, 105*(2), 429–435.

Ghofrani, H. A., Osterloh, I. H., & Grimminger, F. (2006). Sildenafil: From angina to erectile dysfunction to pulmonary hypertension and beyond. *Nature Reviews. Drug Discovery, 5*(8), 689–702.

Gimbrone, M. A., Jr., & Garcia-Cardena, G. (2016). Endothelial cell dysfunction and the pathobiology of atherosclerosis. *Circulation Research, 118*(4), 620–636.

Gooding, C., & Smith, C. W. (2008). Tropomyosin exons as models for alternative splicing. *Advances in Experimental Medicine & Biology, 644*, 27–42.

Grassie, M. E., Sutherland, C., Ulke-Lemée, A., Chappellaz, M., Kiss, E., Walsh, M. P., & MacDonald, J. A. (2012). Cross-talk between Rho-associated kinase and cyclic nucleotide-dependent kinase signaling pathways in the regulation of smooth muscle myosin light chain phosphatase. *Journal of Biological Chemistry, 287*(43), 36356–36369.

Gutterman, D. D., Chabowski, D. S., Kadlec, A. O., Durand, M. J., Freed, J. K., Ait-Aissa, K., & Beyer, A. M. (2016). The human microcirculation: Regulation of flow and beyond. *Circulation Research, 118*(1), 157–172.

Haddock, R. E., & Hill, C. E. (2005). Rhythmicity in arterial smooth muscle. *The Journal of Physiology, 566*(3), 645–656.

Heagerty, A. M., Heerkens, E. H., & Izzard, A. S. (2010). Small artery structure and function in hypertension. *Journal of Cellular and Molecular Medicine, 14*(5), 1037–1043.

Hilge, M. (2012). Ca^{2+} regulation of ion transport in the Na^{+}/Ca^{2+} exchanger. *The Journal of Biological Chemistry, 287*(38), 31641–31649.

Hofmann, F., Feil, R., Kleppisch, T., & Schlossmann, J. (2006). Function of cGMP-dependent protein kinases as revealed by gene deletion. *Physiological Reviews, 86*(1), 1–23.

Hofmann, F., Flockerzi, V., Kahl, S., & Wegener, J. W. (2014). L-Type CaV1.2 calcium channels: From in vitro findings to in vivo function. *Physiological Reviews, 94*(1), 303–326.

International Human Genome Sequencing Consortium. (2001). Initial sequencing and analysis of the human genome. *Nature, 409*(6822), 860–921.

Ito, M., Nakano, T., Erdodi, F., & Hartshorne, D. J. (2004). Myosin phosphatase: Structure, regulation and function. *Molecular and Cellular Biochemistry, 259*(1-2), 197–209.

Iwamoto, T., Kita, S., Zhang, J., Blaustein, M. P., Arai, Y., Yoshida, S., … Katsuragi, T. (2004). Salt-sensitive hypertension is triggered by Ca^{2+} entry via Na^{+}/Ca^{2+} exchanger type-1 in vascular smooth muscle. *Nature Medicine, 10*(11), 1193–1199.

James, P. A., Oparil, S., Carter, B. L., et al. (2014). 2014 Evidence-based guideline for the management of high blood pressure in adults: Report from the panel members appointed to the eighth joint national committee (jnc 8). *JAMA, 311*(5), 507–520.

Johnson, D., Cohen, P., Chen, Y. H., Chen, M. X., & Cohen, P. T. W. (1997). Identification of the regions on the M110 subunit of protein phosphatase 1 M that interacts with the M21 subunit and with myosin. *European Journal of Biochemistry, 244*, 931–939.

Johnson, J. M., Castle, J., Garrett-Engele, P., Kan, Z., Loerch, P. M., Armour, C. D., … Shoemaker, D. D. (2003). Genome-wide survey of human alternative pre-mRNA splicing with exon junction microarrays. *Science, 302*(5653), 2141–2144.

Jordan, J., Toka, H. R., Heusser, K., Toka, O., Shannon, J. R., Tank, J., ... Luft, F. C. (2000). Severely impaired baroreflex-buffering in patients with monogenic hypertension and neurovascular contact. *Circulation*, *102*(21), 2611–2618.

Karaki, H., Ozaki, H., Hori, M., Mitsui-Saito, M., Amano, K.-I., Harada, K.-I., ... Sato, K. (1997). Calcium movements, distribution, and functions in smooth muscle. *Pharmacological Reviews*, *49*(2), 157–230.

Karim, S. M., Rhee, A. Y., Given, A. M., Faulx, M. D., Hoit, B. D., & Brozovich, F. V. (2004). Vascular reactivity in heart failure: Role of myosin light chain phosphatase. *Circulation Research*, *95*(6), 612–618.

Khatri, J. J., Joyce, K. M., Brozovich, F. V., & Fisher, S. A. (2001). Role of myosin phosphatase isoforms in cGMP-mediated smooth muscle relaxation. *The Journal of Biological Chemistry*, *276*(40), 37250–37257.

Kitazawa, T., & Kitazawa, K. (2012). Size-dependent heterogeneity of contractile Ca^{2+} sensitization in rat arterial smooth muscle. *The Journal of Physiology*, *590*(Pt. 21), 5401–5423.

Kitazawa, T., Polzin, A. N., & Eto, M. (2004). CPI-17-deficient smooth muscle of chicken. *The Journal of Physiology*, *557*(2), 515–528.

Konik, E. A., Han, Y. S., & Brozovich, F. V. (2013). The role of pulmonary vascular contractile protein expression in pulmonary arterial hypertension. *Journal of Molecular and Cellular Cardiology*, *65*, 147–155.

Kuo, L., Davis, M. J., & Chilian, W. M. (1995). Longitudinal gradients for endothelium-dependent and -independent vascular responses in the coronary microcirculation. *Circulation*, *92*(3), 518–525.

Lee, M. R., Li, L., & Kitazawa, T. (1997). Cyclic GMP causes Ca^{2+} desensitization in vascular smooth muscle by activating the myosin light chain phosphatase. *Journal of Biological Chemistry*, *272*, 5063–5068.

Liao, P., Yu, D., Li, G., Yong, T. F., Soon, J. L., Chua, Y. L., & Soong, T. W. (2007). A smooth muscle Cav1.2 calcium channel splice variant underlies hyperpolarized window current and enhanced state-dependent inhibition by nifedipine. *Journal of Biological Chemistry*, *282*(48), 35133–35142.

Lifton, R. P., Gharavi, A. G., & Geller, D. S. (2001). Molecular mechanisms of human hypertension. *Cell*, *104*(4), 545–556.

Lima, B., Forrester, M. T., Hess, D. T., & Stamler, J. S. (2010). S-nitrosylation in cardiovascular signaling. *Circulation Research*, *106*(4), 633–646.

Linde, C. I., Antos, L. K., Golovina, V. A., & Blaustein, M. P. (2012). Nanomolar ouabain increases NCX1 expression and enhances Ca^{2+} signaling in human arterial myocytes: A mechanism that links salt to increased vascular resistance? *American Journal of Physiology. Heart and Circulatory Physiology*, *303*(7), H784–H794.

Loirand, G., Sauzeau, V., & Pacaud, P. (2013). Small G proteins in the cardiovascular system: Physiological and pathological aspects. *Physiological Reviews*, *93*(4), 1659–1720.

Lu, Y., Zhang, H., Gokina, N., Mandala, M., Sato, O., Ikebe, M., ... Fisher, S. A. (2008). Uterine artery myosin phosphatase isoform switching and increased sensitivity to SNP in a rat L-NAME model of hypertension of pregnancy. *American Journal of Physiology. Cell Physiology*, *294*(2), C564–C571.

Ludmer, P. L., Selwyn, A. P., Shook, T. L., Wayne, R. R., Mudge, G. H., Alexander, R. W., & Ganz, P. (1986). Paradoxical vasoconstriction induced by acetylcholine in atherosclerotic coronary arteries. *The New England Journal of Medicine*, *315*(17), 1046–1051.

Lytton, J. (2007). Na^{+}/Ca^{2+} exchangers: Three mammalian gene families control Ca^{2+} transport. *The Biochemical Journal*, *406*(3), 365–382.

Maguire, J. J., & Davenport, A. P. (2005). Regulation of vascular reactivity by established and emerging GPCRs. *Trends in Pharmacological Sciences*, *26*(9), 448–454.

Majesky, M. W. (2007). Developmental basis of vascular smooth muscle diversity. *Arteriosclerosis, Thrombosis, and Vascular Biology, 27*(6), 1248–1258.
Marsh, N., & Marsh, A. (2000). A short history of nitroglycerine and nitric oxide in pharmacology and physiology. *Clinical and Experimental Pharmacology and Physiology, 27*(4), 313–319.
McMurray, J. J., Packer, M., Desai, A. S., Gong, J., Lefkowitz, M. P., Rizkala, A. R., … Zile, M. R. (2014). Angiotensin-neprilysin inhibition versus enalapril in heart failure. *The New England Journal of Medicine, 371*(11), 993–1004.
Merkin, J., Russell, C., Chen, P., & Burge, C. B. (2012). Evolutionary dynamics of gene and isoform regulation in Mammalian tissues. *Science, 338*(6114), 1593–1599.
Mizuno, M., Iwamoto, G. A., Vongpatanasin, W., Mitchell, J. H., & Smith, S. A. (2014). Exercise training improves functional sympatholysis in spontaneously hypertensive rats through a nitric oxide-dependent mechanism. *American Journal of Physiology. Heart and Circulatory Physiology., 307*(2), 9–16.
Morris, R. C., Schmidlin, O., Sebastian, A., Tanaka, M., & Kurtz, T. W. (2016). Vasodysfunction that involves renal vasodysfunction, not abnormally increased renal retention of sodium, accounts for the initiation of salt-induced hypertension. *Circulation, 133*(9), 881–893.
Nagao, T., Illiano, S., & Vanhoutte, P. M. (1992). Heterogeneous distribution of endothelium-dependent relaxations resistant to NG-nitro L-arginine in rats. *The American Journal of Physiology, 263*(4 Pt. 2), H1090–H1094.
Nakamura, K., Koga, Y., Sakai, H., Homma, K., & Ikebe, M. (2007). cGMP-dependent relaxation of smooth muscle is coupled with the change in the phosphorylation of myosin phosphatase. *Circulation Research, 101*(7), 712–722.
Nasser, S. A., & El-Mas, M. M. (2014). Endothelin ETA receptor antagonism in cardiovascular disease. *European Journal of Pharmacology, 737*, 210–213.
Nyberg, M., Piil, P., Egelund, J., Sprague, R. S., Mortensen, S. P., & Hellsten, Y. (2015). Effect of PDE5 inhibition on the modulation of sympathetic α-adrenergic vasoconstriction in contracting skeletal muscle of young and older recreationally active humans. *American Journal of Physiology. Heart and Circulatory Physiology, 309*(11), H1867–H1875.
Padmanabhan, S., Newton-Cheh, C., & Dominiczak, A. F. (2012). Genetic basis of blood pressure and hypertension. *Trends in Genetics, 28*(8), 397–408.
Pal, S., Gupta, R., Kim, H., Wickramasinghe, P., Baubet, V., Showe, L. C., … Davuluri, R. V. (2011). Alternative transcription exceeds alternative splicing in generating the transcriptome diversity of cerebellar development. *Genome Research, 21*(8), 1260–1272.
Payne, M. C., Zhang, H. Y., Prosdocimo, T., Joyce, K. M., Koga, Y., Ikebe, M., & Fisher, S. A. (2006). Myosin phosphatase isoform switching in vascular smooth muscle development. *Journal of Molecular and Cellular Cardiology, 40*(2), 274–282.
Payne, M. C., Zhang, H. Y., Shirasawa, Y., Koga, Y., Ikebe, M., Benoit, J. N., & Fisher, S. A. (2004). Dynamic changes in expression of myosin phosphatase in a model of portal hypertension. *American Journal of Physiology. Heart and Circulatory Physiology, 286*(5), H1801–H1810.
Pfitzer, G., Merkel, L., Ruegg, J. C., & Hofmann, F. (1986). Cyclic GMP-dependent protein kinase relaxes skinned fibers from guinea pig taenia coli but not from chicken gizzard. *Pflügers Archiv—European Journal of Physiology, 407*(1), 87–91.
Pluznick, J. L. (2013). Renal and cardiovascular sensory receptors and blood pressure regulation. *American Journal of Physiology. Renal Physiology, 305*(4), 12–24.
Pluznick, J. L., Protzko, R. J., Gevorgyan, H., Peterlin, Z., Sipos, A., Han, J., … Caplan, M. J. (2013). Olfactory receptor responding to gut microbiota-derived signals plays a role in renin secretion and blood pressure regulation. *Proceedings of the National Academy of Sciences, 110*(11), 4410–4415.

Prysyazhna, O., Rudyk, O., & Eaton, P. (2012). Single atom substitution in mouse protein kinase G eliminates oxidant sensing to cause hypertension. *Nature Medicine*, *18*(2), 286–290.

Qin, L., Reger, A. S., Guo, E., Yang, M. P., Zwart, P., Casteel, D. E., & Kim, C. (2015). Structures of cGMP-dependent protein kinase (PKG) Iα leucine zippers reveal an interchain disulfide bond important for dimer stability. *Biochemistry*, *54*(29), 4419–4422.

Reho, J. J., & Fisher, S. A. (2015). The stress of maternal separation causes mis-programming in the post-natal maturation of rat resistance arteries. *American Journal of Physiology. Heart and Circulatory Physiology*, *309*(9), H1468–H1478.

Reho, J. J., Kenchegowda, D., Asico, L., & Fisher, S. A. (2016). A splice variant of the myosin phosphatase regulatory subunit tunes arterial reactivity and suppresses response to salt loading. *American Journal of Physiology. Heart and Circulatory Physiology*, *310*(11), H1715–H1724.

Reho, J. J., Shetty, A., Dippold, R. P., Mahurkar, A., & Fisher, S. A. (2015). Unique gene program of rat small resistance mesenteric arteries as revealed by deep RNA sequencing. *Physiological Reports*, *3*(7).

Reho, J. J., Zheng, X., Asico, L. D., & Fisher, S. A. (2015). Redox signaling and splicing dependent change in myosin phosphatase underlie early versus late changes in NO vasodilator reserve in a mouse LPS model of sepsis. *American Journal of Physiology. Heart and Circulatory Physiology*, *308*(9), H1039–H1050.

Reho, J. J., Zheng, X., & Fisher, S. A. (2014). Smooth muscle contractile diversity in the control of regional circulations. *American Journal of Physiology. Heart and Circulatory Physiology*, *306*(2), H163–H172.

Roden, D. M., Johnson, J. A., Kimmel, S. E., Krauss, R. M., Medina, M. W., Shuldiner, A., & Wilke, R. A. (2011). Cardiovascular pharmacogenomics. *Circulation Research*, *109*(7), 807–820.

Rudyk, O., Prysyazhna, O., Burgoyne, J. R., & Eaton, P. (2012). Nitroglycerin fails to lower blood pressure in redox-dead Cys42Ser PKG1alpha knock-in mouse. *Circulation*, *126*(3), 287–295.

Saltin, B., & Mortensen, S. P. (2012). Inefficient functional sympatholysis is an overlooked cause of malperfusion in contracting skeletal muscle. *The Journal of Physiology*, *590*(24), 6269–6275.

Schlossmann, J., Ammendola, A., Ashman, K., Zong, X., Huber, A., Neubauer, G., … Ruth, P. (2000). Regulation of intracellular calcium by a signalling complex of IRAG, IP3 receptor and cGMP kinase Ibeta. *Nature*, *404*(6774), 197–201.

Schmucker, D., & Chen, B. (2009). Dscam and DSCAM: Complex genes in simple animals, complex animals yet simple genes. *Genes & Development*, *23*(2), 147–156.

Sellke, F. W., Myers, P. R., Bates, J. N., & Harrison, D. G. (1990). Influence of vessel size on the sensitivity of porcine coronary microvessels to nitroglycerin. *The American Journal of Physiology*, *258*(2 Pt. 2), H515–H520.

Shimokawa, H., Sunamura, S., & Satoh, K. (2016). RhoA/Rho-Kinase in the cardiovascular system. *Circulation Research*, *118*(2), 352–366.

Shimokawa, H., Yasutake, H., Fujii, K., Owada, M. K., Nakaike, R., Fukumoto, Y., … Takeshita, A. (1996). The importance of the hyperpolarizing mechanism increases as the vessel size decreases in endothelium-dependent relaxations in rat mesenteric circulation. *Journal of Cardiovascular Pharmacology*, *28*(5), 703–711.

Shukla, S., & Fisher, S. A. (2008). Tra2beta as a novel mediator of vascular smooth muscle diversification. *Circulation Research*, *103*(5), 485–492.

Somlyo, A. P., & Somlyo, A. V. (1994). Signal transduction and regulation in smooth muscle. *Nature*, *372*, 231–236.

Somlyo, A. V., & Somlyo, A. P. (1968). Electromechanical and pharmacomechanical coupling in vascular smooth muscle. *The Journal of Pharmacology and Experimental Therapeutics*, *159*, 129–145.

Splawski, I., Timothy, K. W., Sharpe, L. M., Decher, N., Kumar, P., Bloise, R., … Keating, M. T. (2004). Ca(V)1.2 calcium channel dysfunction causes a multisystem disorder including arrhythmia and autism. *Cell*, *119*(1), 19–31.

Steinhorn, B. S., Loscalzo, J., & Michel, T. (2015). Nitroglycerin and nitric oxide—A rondo of themes in cardiovascular therapeutics. *The New England Journal of Medicine*, *373*(3), 277–280.

Surks, H. K., Mochizuki, N., Kasai, Y., Georgescu, S. P., Tang, K. M., Ito, M., … Mendelsohn, M. E. (1999). Regulation of myosin phosphatase by a specific interaction with cGMP-dependent protein kinase Ialpha. *Science*, *286*(5444), 1583–1587.

Tang, Z. Z., Liang, M. C., Lu, S., Yu, D., Yu, C. Y., Yue, D. T., & Soong, T. W. (2004). Transcript scanning reveals novel and extensive splice variations in human L-type voltage-gated calcium channel, Cav1.2 α1 subunit. *Journal of Biological Chemistry*, *279*(43), 44335–44343.

Taylor, A. L., Ziesche, S., Yancy, C., Carson, P., D'Agostino, R. J., Ferdinand, K., … Cohn, J. N. (2004). Combination of isosorbide dinitrate and hydralazine in blacks with heart failure. *New England Journal of Medicine*, *351*(20), 2049–2057.

The ALLHAT Officers. (2002). Major outcomes in high-risk hypertensive patients randomized to angiotensin-converting enzyme inhibitor or calcium channel blocker vs diuretic. *JAMA: The Journal of the American Medical Association*, *288*(23), 2981–2997.

Toka, O., Tank, J., Schächterle, C., Aydin, A., Maass, P. G., Elitok, S., … Luft, F. C. (2015). Clinical effects of phosphodiesterase 3A mutations in inherited hypertension with brachydactyly. *Hypertension*, *66*(4), 800–808.

Uehata, M., Ishizaki, T., Satoh, H., Ono, T., Kawahara, T., Morishita, T., … Narumiya, S. (1997). Calcium sensitization of smooth muscle mediated by a Rho-associated protein kinase in hypertension. *Nature*, *389*(6654), 990–994.

Vanichakarn, P., Hwa, J., & Stitham, J. (2014). Cardiovascular pharmacogenetics of antihypertensive and lipid- lowering therapies. *Current Molecular Medicine*, *14*(7), 849–879.

Wang, D., Papp, A. C., Binkley, P. F., Johnson, J. A., & Sadee, W. (2006). Highly variable mRNA expression and splicing of L-type voltage-dependent calcium channel alpha subunit 1C in human heart tissues. *Pharmacogenetics and Genomics*, *16*(10), 735–745.

Wang, E. T., Sandberg, R., Luo, S., Khrebtukova, I., Zhang, L., Mayr, C., … Burge, C. B. (2008). Alternative isoform regulation in human tissue transcriptomes. *Nature*, *456*(7221), 470–476.

Weber, S., Bernhard, D., Lukowski, R., Weinmeister, P., Worner, R., Wegener, J. W., … Feil, R. (2007). Rescue of cGMP kinase I knockout mice by smooth muscle specific expression of either isozyme. *Circulation Research*, *101*(11), 1096–1103.

Welling, A., Ludwig, A., Zimmer, S., Klugbauer, N., Flockerzi, V., & Hofmann, F. (1997). Alternatively spliced IS6 segments of the α1C gene determine the tissue-specific dihydropyridine sensitivity of cardiac and vascular smooth muscle L-type Ca^{2+} channels. *Circulation Research*, *81*(4), 526–532.

Werder, A. v., Mayr, M., Schneider, G., Oesterle, D., Fritsch, R. M., Seidler, B., … Saur, D. (2011). Truncated IRAG variants modulate cGMP-mediated inhibition of human colonic smooth muscle cell contraction. *American Journal of Physiology. Cell Physiology*, *301*(6), C1445–C1457.

Wirth, A., Benyo, Z., Lukasova, M., Leutgeb, B., Wettschureck, N., Gorbey, S., … Offermanns, S. (2008). G12-G13-LARG-mediated signaling in vascular smooth muscle is required for salt-induced hypertension. *Nature Medicine*, *14*(1), 64–68.

Wirth, A., Wang, S., Takefuji, M., Tang, C., Althoff, T. F., Schweda, F., … Offermanns, S. (2015). Age-dependent blood pressure elevation is due to increased vascular smooth muscle tone mediated by G-protein signalling. *Cardiovascular Research*, *109*(1), 131–140.

Woodsome, T. P., Eto, M., Everett, A., Brautigan, D. L., & Kitazawa, T. (2001). Expression of CPI-17 and myosin phosphatase correlates with Ca(2+) sensitivity of protein kinase C-induced contraction in rabbit smooth muscle. *The Journal of Physiology*, *535*(Pt. 2), 553–564.

Wooldridge, A. A., MacDonald, J. A., Erdodi, F., Ma, C., Borman, M. A., Hartshorne, D. J., & Haystead, T. A. J. (2004). Smooth muscle phosphatase is regulated in vivo by exclusion of phosphorylation of threonine 696 of MYPT1 by phosphorylation of serine 695 in response to cyclic nucleotides. *Journal of Biological Chemistry*, *279*(33), 34496–34504.

Wray, S., & Burdyga, T. (2010). Sarcoplasmic reticulum function in smooth muscle. *Physiological Reviews*, *90*(1), 113–178.

Wu, X., Somlyo, A. V., & Somlyo, A. P. (1996). c-GMP-dependent stimulation reverses G-protein-coupled inhibition of smooth muscle myosin light chain phosphatase. *Biochemical and Biophysical Research Communications*, *220*, 658–663.

Yuen, S. L., Ogut, O., & Brozovich, F. V. (2014). Differential phosphorylation of LZ^+/LZ^- MYPT1 isoforms regulates MLC phosphatase activity. *Archives of Biochemistry and Biophysics*, *562*, 37–42.

Zacharia, J., Mauban, J. R., Raina, H., Fisher, S. A., & Wier, W. G. (2013). High vascular tone of mouse femoral arteries in vivo is determined by sympathetic nerve activity via alpha1A- and alpha1D-adrenoceptor subtypes. *PLoS One*, *8*(6), e65969.

Zhang, H., & Fisher, S. A. (2007). Conditioning effect of blood flow on resistance artery smooth muscle myosin phosphatase. *Circulation Research*, *100*(5), 730–737.

Zhang, H., Pakeerappa, P., Lee, H. J., & Fisher, S. A. (2009). Induction of PDE5 and de-sensitization to endogenous NO signaling in a systemic resistance artery under altered blood flow. *Journal of Molecular and Cellular Cardiology*, *47*(1), 57–65.

Zhang, J., Ren, C., Chen, L., Navedo, M. F., Antos, L. K., Kinsey, S. P., … Blaustein, M. P. (2010). Knockout of Na^+/Ca^{2+} exchanger in smooth muscle attenuates vasoconstriction and L-type Ca^{2+} channel current and lowers blood pressure. *American Journal of Physiology Heart and Circulatory Physiology*, *298*(5), H1472–H1483.

Zheng, X., Reho, J. J., Wirth, B., & Fisher, S. A. (2015). TRA2beta controls Mypt1 exon 24 splicing in the developmental maturation of mouse mesenteric artery smooth muscle. *American Journal of Physiology Cell Physiology*, *308*(4), C289–C296.

FSC
www.fsc.org
MIX
Papier | Fördert
gute Waldnutzung
FSC® C083411

Kolibri 360 GmbH
Ferdinand-Jühlke-Straße 7
99095 Erfurt, Deutschland
produktsicherheit@kolibri360.de